MODELING FOR INSIGHT

MODELING FOR INSIGHT

A Master Class for Business Analysts

Stephen G. Powell
Robert J. Batt
Tuck School of Business
Dartmouth College
Hanover, NH

A JOHN WILEY & SONS, INC. PUBLICATION

Published by John Wiley & Sons, Inc., Hoboken, New Jersey
Published simultaneously in Canada

Library of Congress Cataloging-in-Publication Data

Powell, Stephen G.
Modeling for insight : a master class for business analysts /
Stephen G. Powell, Robert J. Batt.
p. cm.
Includes bibliographical references and index.
ISBN 978-0-470-17555-2
1. Decision making. 2. Business planning. 3. Business—Computer simulation. 4. Electronic spreadsheets. I. Batt, Robert J. II. Title.
HD30.23.P69 2008
658.4′0352—dc22

2008009579

10 9 8 7 6 5 4 3 2

For Nick and Annika

CONTENTS

PREFACE

STEPHEN G. POWELL

Origins of the Book

This book has its origins in a class I took in the mid-1970s from Dick Smallwood and Pete Morris in Stanford University's Department of Engineering-Economic Systems. Smallwood and Morris taught a course they called "The Art of Modeling," in which they taught modeling as a craft. It was an eye-opening experience for me. All my other courses were devoted to formal mathematical techniques, and although there was a certain abstract beauty to the quantitative methods I was learning, it was hard to see how I could apply them to practical problems. Modeling as Smallwood and Morris taught it could be applied anywhere. Moreover, it was a creative, sometimes playful, enterprise. It was the most challenging course I had ever taken, and it struck a deep chord.

Some years later, as a professor at the Tuck School teaching management science to MBAs, I created my own version of the course. I borrowed the "Art of Modeling" title, although it was unpopular with some of my colleagues, who thought it sounded more like a New Age seminar than a serious course for a business school. I tried to keep the mathematical level modest and added an emphasis on presentation of quantitative insights. Even so, it was just as challenging for Tuck MBAs as the Stanford course had been for engineering PhD students. Nonetheless, many of the students loved it, which was more than sufficient reason to persist with it.

Smallwood and Morris taught their course in the style of an art studio. That is, there was almost no modeling theory to learn, just repeated practical exercises. We worked for a week, solo or in pairs, on each modeling challenge. At the end of the week, we handed in a paper describing our work and our recommendations to the client. Later we would listen as Smallwood and Morris described their approach to the problem, which was always far more clever and insightful than our own.

Each week in this course we were forced to make the entire modeling journey from developing a problem statement, through modeling and analysis, and concluding with recommendations to the client. The problems Smallwood and Morris posed for us were not the kind of homework problems encountered in other courses, which always had a well-defined solution and only one approach (that was sure to be whatever mathematical topic we were studying that week). These were ill-defined problems, typical of consulting practice, with no right answer. And where homework problems *ended* with the numerical solution, the hard work of interpretation and presentation in these problems *began* with the numerical solution.

In the early years of my Tuck modeling course, I followed the same principles (described in Powell, 1998). I told my students that modeling was an art and a craft, and therefore, I could not *teach* them how to do it, but I could *coach* them to develop their modeling skills by giving them appropriate problems and by helping them during the modeling process. To this end, I required them to come to my office halfway through the week to show me their work in progress. This meeting gave me an invaluable opportunity to observe them at work, to learn about the challenges they faced in modeling the problem of the week, and of course, to assist them over the rough spots. One skill I developed during these appointments was to not answer when they asked if they were "on the right track" but instead to ask what *their goals* were in the problem at hand. Then, instead of showing them how I had approached the problem, I tried to help them achieve their goals. These mid-week interviews were more like tutoring sessions because, with just one or two students to deal with, I could tailor my interaction precisely to what they needed. I could further challenge students who had already made great progress or provide detailed guidance to those who were struggling.

Through this process, I began to see how well trained our students were to believe that every problem had one right answer (known only to the professor) and that their task was to find that answer. I espoused a radically different philosophy in the course—that is, the problems we worked on had many credible approaches and the students were free to pick their own approach as long as it led to sound insights. It took some students weeks to finally accept that they could set their own goals and be creative in achieving those goals. I even went so far as to avoid discussing my own solution in class, so as to further discourage the notion that the professor has the one right answer. This way the class saw many different approaches to a problem, and sometimes different recommendations, all developed by students and each as credible as the next.

Modeling Heuristics

Through this experience of coaching students, I began to develop some rules of thumb for good modeling. These were pithy statements of techniques I found myself using over and over, and I tried to develop a shorthand way of stating them to help students remember. I eventually identified six tricks of the trade or heuristics and wrote a short article in *Interfaces* (Powell, 1995) to describe them. They are as follows:

1. Decomposition: Divide and Conquer
2. Prototyping: Get Something Working

3. Sketch a Graph: Visualize
4. Parameterization: Call It Alpha
5. Separate Idea Generation from Evaluation: Quiet the Critic
6. Model the Data: Be Skeptical

These heuristics were my first attempt to reduce certain aspects of the modeling craft to principles in order to eliminate some of the mystery of modeling. However, at that time, I had no idea how to use these heuristics in teaching, except to repeat them endlessly and hope students somehow learned to translate them into effective modeling behaviors.

Research on Modelers and Modeling

In the mid-1990s, I came under the influence of Chris Jernstedt, a learning theorist at Dartmouth College. Jernstedt opened my eyes to the idea that if we want our students to learn a skill, we first must understand how novices acquire that skill. This understanding requires study of and reflection on how novices, including both strong and weak students, proceed step-by-step from an initial state of incompetence to a state of competence. Although this principle may not seem revolutionary, it shocked me into admitting how little I knew about how novices learn modeling skills. I had been convinced for so many years that modeling was an art acquired only through practice that I had not really given much thought to breaking it down into learnable stages.

A great deal of research had been done by psychologists on the differences between novices and experts in such skills as swinging a golf club. Some research had been done specifically on novice and expert *problem solving*, such as Chi's work on solving physics problems (Chi et al., 1981). No research had been done on understanding how expert modelers approach their task until Tom Willemain took on the challenge in the early 1990s. He interviewed a small number of recognized modeling experts and taped their conversations while they worked on a modeling problem for an hour. Willemain's papers on this research (Willemain, 1994, 1995) revealed how experts shift their attention frequently from one aspect of problem solving to another and how little time they actually spend building a model compared with the time they spend evaluating whether their work meets the needs of the client.

Several years ago, I joined forces with Tom to extend his research on expert modelers to novice modelers. We presented some of the same problems he had used with experts to Tuck student volunteers, some of whom had taken my "Modeling" course. Despite my many years teaching modeling to MBAs, I was somewhat shocked to read transcripts of these sessions. I had no idea how much students floundered when they first encountered an ill-structured problem. This was almost as true of students who had just completed my course as it was of those who had not taken it. We identified the following five behaviors in our subjects that impeded their progress toward a functioning model (Powell and Willemain, 2007; Willemain and Powell, 2007):

- Over-reliance on given numerical data
- Taking shortcuts to an answer
- Insufficient use of abstract variables and relationships
- Ineffective self-regulation
- Overuse of brainstorming relative to structured problem solving

In many ways, this book is my response to our findings in this research.

Modeling for Insight

About the time I began thinking about how novice modelers learn, I was fortunate to have Jay Goldman, then of Strategic Decisions Group, come to Tuck—first to teach the modeling course on his own and then to coteach it with me. Jay is both a master modeler and a natural teacher, so he was enthusiastic when I told him what I was learning from the theorists, and he was eager to work with me to develop a step-wise approach to modeling. Jay coined the phrase "modeling for insight" to stress that the ultimate goal in modeling ill-structured problems is not to find the "solution" but to generate insights that will help a client understand the problem and take action. Jay also convinced me of the power of influence diagrams, the importance of spreadsheet hygiene (which I have renamed spreadsheet engineering), and the usefulness of strategy analysis. He also introduced several excellent cases, some of which appear here. To a large degree, this book is a direct descendant of the modeling course we developed together; Jay's influence on this book is profound.

STEPHEN G. POWELL

Hanover, NH
February 2008

ROBERT J. BATT

Steve and I met when I was a first-year MBA student in his Decision Science class at Tuck. I was struck by how Steve refused to allow the course to be simply about how to use Excel. Most of the assignments in the class were business cases that required more than a simple, numeric answer. They required a recommended course of action and an understanding of the business implications of that recommendation. Students were regularly called on to present their findings to the class. These were not presentations of a model but, rather, of the thinking behind the model and the relevance of the results it produced. Steve was pushing us to model for insight. Later that year, I also took Steve's "Applications of Simulation" course, which similarly was not just about methodology and techniques but, rather, about how to use simulation to gain insight into the problem at hand.

The next year, Steve asked me to help him redesign the simulation course. We decided that in order to allow the students more time to work on bigger, more realistic cases, we would cancel half the lectures. Instead of class time we required office hours where he and I met with the students to review their modeling progress and to help coach them along the way. The days the class did meet, class time was primarily devoted to students presenting their work. While working together on this course, I saw the power of the modeling methods Steve had developed.

After graduating from Tuck, I joined the school as a research fellow. Steve had already drafted the first few chapters of this book, and he enlisted my help in reviewing them. Together we refined and improved the conceptualization and delivery of the core principles of modeling for insight. We went on to share the load of writing and editing the rest of the book to such an extent that Steve was generous enough to allow me to be a full coauthor of the book. Without exception, the roots of the ideas presented in this book are Steve's. It has been my great pleasure to come alongside and help give voice and expression to these ideas.

ROBERT J. BATT

Hanover, NH
February 2008

USING THIS BOOK

ORGANIZATION OF THE BOOK

This book is organized into three parts. Part I consists of Chapters 1–3. Chapter 1 introduces the ideas of models and modeling, describes the class of problem this book addresses, and explains what "modeling for insight" means. Chapter 2 lays out the modeling and presentation methods on which the book rests. Chapter 3 describes the basics of spreadsheet engineering: how to design, build, test, and analyze a spreadsheet model with efficiency and effectiveness. If you are an experienced spreadsheet user, some of these ideas may be familiar to you, but you are sure to find something new here. In particular, the sensitivity analysis tools we discuss in this chapter are vital throughout the book and are likely to be new to many readers. We recommend that you skim familiar sections of this chapter but study new concepts carefully.

In Part II, which consists of Chapters 4–6, we take you through a series of increasingly complex modeling cases and show how the procedures and tools of modeling for insight work. Chapter 4 presents a first example, worked out as we would actually carry out a modeling assignment. This chapter demonstrates the type of problem that modeling for insight was developed to deal with and illustrates the kinds of results you can develop with these methods. If you want to know whether this book is for you, read Chapter 4 now.

Chapters 5 and 6 address different modeling problems, beginning in each case with the *simplest* version of the problem. Expert modelers begin by reducing a problem to its essentials. We believe learning how to do this is critical for novices, so we begin each chapter with a radically simplified version of the problem and we show how to analyze this version effectively. Then we introduce a few complexities so that the problem becomes one step more realistic and we modify our model and analysis and develop additional insights. This process is repeated several times until we find ourselves modeling the problem in all its real-world richness. Part II is the heart of the book and is essential for all readers. We recommend you do not attempt the later chapters before you have worked carefully through Part II of the book.

Part III, Chapters 7–12, consists of a series of modeling cases worked out in detail. Each problem is presented in all its complexity so the modeling challenge is as realistic as possible. We offer our own approach to each problem, but we offer you many chances to sketch out your own approach along the way. Our solutions are intended to show one way to approach each problem, not to present the perfect solution. Thus, we share our thoughts about the right approach to take and occasionally present work that leads down a dead end. Although we do not want to burden you with our mistakes, we do want you to understand how competent modelers iterate their way to a good final result through a process of intelligent trial and error.

REQUIRED BACKGROUND

This book is intended primarily for modelers in the business and nonprofit sectors, although the techniques are applicable to engineering and other technical disciplines. Accordingly, most of our examples originate in actual business situations. Some background in business and economics will be helpful, but none of the examples assumes detailed knowledge of any particular business sector or functional area. Where necessary, we explain enough of the background to a problem to make it accessible for general readers.

We assume you have considerable experience with Microsoft Excel® and basic skills in the following areas:

- Windows operations
- Workbook navigation
- Entering text and data
- Editing cells
- Formatting cells
- Writing formulas
- Using functions (e.g., SUM, IF, NPV)
- Creating graphs

We do not assume you have extensive experience in creating Excel models from scratch, but some prior experience in this area will be highly beneficial. When we use advanced Excel tools, we explain them thoroughly in the text. We make routine use of Solver® for optimization, Crystal Ball® for simulation, and the Sensitivity Toolkit for sensitivity analysis; short appendices are provided that describe these tools. If you have never used them, we recommend you study the appendices independently before tackling this book.

You do not need knowledge of advanced mathematics to understand this book, but you should be comfortable with basic algebra and probability. Familiarity with simple functions, like the straight line and the power function, as well as basic probability concepts, such as expected value, conditional probability, and distributions is required.

If you need to bolster your background, you will find extensive coverage of many of the basic skills in *Management Science: The Art of Modeling with Spreadsheets* (Powell and Baker, 2007). Chapter 3 on basic Excel skills and

Chapters 5 and 6 on spreadsheet engineering and analysis will be helpful, as will Chapters 9 and 10 on optimization (Solver) and Chapter 15 on simulation (Crystal Ball).

SOFTWARE

All models in this book were built in Excel, specifically in Excel 2007 for Windows. We also rely heavily on three Excel add-ins: Premium Solver, Crystal Ball, and the Sensitivity Toolkit. Trial versions of Solver and Crystal Ball can be downloaded from <www.solver.com> and <www.decisioneering.com>, respectively. The Sensitivity Toolkit is available at <http://mba.tuck.dartmouth.edu/toolkit/>. The spreadsheets referred to in the book are available at <www.modelingforinsight.com> along with additional materials related to modeling.

HOW TO LEARN TO MODEL FOR INSIGHT

No one can become an expert modeler merely by reading a book, even this one. However, you can dramatically improve your modeling skills if you work though this book in a dedicated and organized fashion. Furthermore, the skills you learn here will provide a firm foundation for your efforts to refine your modeling skills in school or on the job.

The key to using this book effectively is to *read actively*. Instead of reading large chunks of the text with your feet up, read in small chunks and stop frequently to consider what you have read. When you see "To the reader" and the stopsign icon, stop and take some time to do the exercise we suggest. You will internalize what you have learned better if you try it first yourself. You may feel clumsy initially, but try the problems before you read our suggested approaches. Do not be discouraged if you do not make much progress with these problems at first. As you make your way through the book and learn the tools and approaches we develop here, you will become more comfortable and skilled in developing your own creative approach to the problems.

Our final word of advice is to look for opportunities to apply modeling everywhere you go. Some problems in this book originated from articles in newspapers or public radio news programs. Modeling opportunities are everywhere, and you will increase your capacity to model well if you think about how you would model situations as they arise. Here are some examples:

- The state legislature is considering raising the cigarette tax. Can you model the impacts on retailers and consumers?
- A local museum is considering a major capital campaign. Can you model the impact on the museum's endowment in the future?

- Chrysler has amassed a multibillion-dollar cash fund, and major investors are demanding it be distributed to the shareholders. Can you use modeling to determine whether Chrysler should declare a large dividend, or save its cash to help maintain R&D spending during the next economic downturn?

Now, on to modeling!

REFERENCES

Chi M, Feltovich P, Glaster R. Categorization and representation of physics problems by experts and novices. *Cognitive Science* 1981;**5**:121–152.

Powell S. The teacher's forum: Six key modeling heuristics. *Interfaces* 1995;**25**(4): 114–125.

Powell S. The studio approach to teaching the craft of modeling. *Annals of Operations Research* 1998;**82**:29–47.

Powell S, Baker K. *Management Science: The Art of Modeling with Spreadsheets*. 2nd edition. Hoboken, NJ: Wiley, 2007.

Powell S, Willemain T. How novices formulate models. Part I: qualitative insights and implications for teaching. *Journal of the Operational Research Society* 2007;**58**: 983–995.

Willemain T. Insights on modeling from a dozen experts. *Operations Research* 1994; **42**:213–222.

Willemain T. Model formulation: What experts think about and when. *Operations Research* 1995;**43**:916–932.

Willemain T, Powell S. How novices formulate models. Part II: a quantitative description of behavior. *Journal of the Operational Research Society* 2007;**58**:1271–1283.

ACKNOWLEDGMENTS

This book is not so much our invention as our interpretation of ideas developed by others. We therefore owe more than the usual authorial debts. Some of the most important ones are as follows:

Dick Smallwood and *Peter Morris*: who were among the first to recognize that mathematical technique is not enough for a modeler and to find a way to teach craft skills.

Mike Magazine, *Steve Pollock*, and *Seth Bonder*: all pioneers in teaching modeling.

Mike Rothkopf: who encouraged Steve to write a column for *Interfaces* on teaching and modeling and thereby helped stimulate additional thought on the topic.

Peter Regan and *Jay Goldman*: Peter taught "Art of Modeling" when Steve was on sabbatical and developed several new cases. He has since become a valued teaching colleague. Jay taught the course several times and developed new cases and new concepts. His ideas on modeling and presenting have strongly influenced the book.

Ken Baker and *Rob Shumsky*: Ken has been a colleague at Tuck for many years. Steve and Ken have taught management science together and have developed a *Management Science* textbook that contains some of the ideas presented here. Rob Shumsky joined Tuck in 2005 and has embraced the spirit of the modeling courses enthusiastically.

Tom Willemain: who did the first work on expert modelers and opened the way for joint research on novices.

Julie Lang and Jason Romeo: who shared their expertise on communication and presentation design.

Beth Golub: an editor at John Wiley & Sons, who encouraged Ken Baker and Steve to write the management science textbook. She has supported our efforts for many years and kept reminding us that the rest of the world would eventually see things our way if we just persevered.

Susanne Steitz-Filler: the editor of this book and an enthusiastic supporter throughout.

Anita Warren: who gave us valuable advice on the design of the book and did an excellent job of copyediting our drafts.

Bob Hansen and *Dave Pyke*: Associate Deans at the Tuck School who supported our work and made it possible for Bob to devote time to this project.

ACKNOWLEDGMENTS FOR CASES

Chapter 6, Technology Option.
Developed by Jay Goldman, Senior Engagement Manager, Strategic Decisions Group, and Professor Stephen G. Powell of the Tuck School of Business Administration, Dartmouth College, Hanover, NH.

Chapter 7, MediDevice.
Developed by Jay Goldman, Senior Engagement Manager, Strategic Decisions Group, and Professor Stephen G. Powell of the Tuck School of Business Administration, Dartmouth College, Hanover, NH.

Chapter 8, Draft Commercials.
Developed by Professor Stephen G. Powell of the Tuck School of Business Administration, Dartmouth College, Hanover, NH, and Professor Thomas Willemain of Rensselaer Polytechnic Institute, Troy, NY.

Chapter 9, New England College Skiway.
Developed by Professor Stephen G. Powell of the Tuck School of Business Administration, Dartmouth College, Hanover, NH. Adapted from an earlier case by Dick Smallwood and Peter Morris of Stanford University. The assistance of Don Cutter and Peter Riess of Dartmouth College, and Christy Bieber of the Tuck School, is gratefully acknowledged.

Chapter 10, National Leasing, Inc.
Developed by Adjunct Professor Peter Regan and Professor Stephen G. Powell of the Tuck School of Business Administration, Dartmouth College, Hanover, NH, and Jay Goldman, Senior Engagement Manager, Strategic Decisions Group.

Chapter 11, Pharma X and Pharma Y.
Developed by Adjunct Professor Peter Regan and Professor Stephen G. Powell of the Tuck School of Business Administration, Dartmouth College, Hanover, NH.

Chapter 12, Invivo Diagnostics, Inc.
Developed by Stephanie Bichet of Andersen Consulting and Professor Stephen G. Powell of the Tuck School of Business Administration, Dartmouth College, Hanover, NH.

ABOUT THE AUTHORS

Steve Powell is a professor at the Tuck School of Business at Dartmouth College. His primary research interest lies in modeling production and service processes, but he has also been active in research in energy economics, marketing, and operations. At Tuck he has developed a variety of courses in management science, including the core "Decision Science" course and electives in the "Art of Modeling," "Business Process Redesign," and "Applications of Simulation." He originated the Teacher's Forum column in *Interfaces*, and he has written several articles on teaching modeling to practitioners. He was the Academic Director of the annual INFORMS Teaching of Management Science Workshops. In 2001 he was awarded the INFORMS Prize for the Teaching of Operations Research/Management Science Practice. Prof. Powell holds an AB degree from Oberlin College and MS and PhD degrees from Stanford University.

Bob Batt is a Tuck Fellow and researcher at the Tuck School of Business at Dartmouth College. His work is focused on operations management and finance. He has written case studies on Steinway & Sons and Hurricane Katrina. He also serves as a teaching fellow for the core MBA Corporate Finance course and the Corporate Financial Management course. Prior to coming to Tuck, Mr. Batt worked as a manufacturing engineer designing manufacturing and quality control processes. He holds a BA degree from Wheaton College (IL) and BE, MEM, and MBA degrees from Dartmouth College. He graduated from the Tuck School of Business at Dartmouth in 2006 as an Edward Tuck Scholar with High Distinction. Beginning in the fall of 2008, Mr. Batt will be a doctoral student in Operations Management at the Wharton School of the University of Pennsylvania.

PART I

CHAPTER 1

Introduction

1.0 MODELS AND MODELING

A *model* is a simplified representation of reality. We use models all the time, although often we are not aware of doing so. Any map is a model of some part of the natural or human landscape. A decision to merge two companies is likely to be based on a model of how the combined companies will operate. Predictions that the U.S. Social Security program will go bankrupt are based on models, as are decisions to recommend evacuation in the face of a hurricane. Models, in fact, are a ubiquitous feature of modern life, and everyone is affected by them.

Modeling is the process of building and using models. Some models are built for a single purpose. For example, someone might build a simple spreadsheet model to test the impact of various investment alternatives on their personal tax liabilities. Other models are built by modeling experts to help a group of managers make a one-time decision. An example of this might be a consulting company that builds a model to value a potential acquisition target for a large corporation. Still other models are used repetitively, as part of everyday operations. State legislatures use models routinely, for instance, to forecast the annual budget surplus based on tax receipts collected to date.

Just as there are many different types of models and purposes for modeling, there are also many different types of modelers. *Professional modelers* often specialize in a particular industry or modeling method. Professional modelers work on airline-crew scheduling problems, marketing media-selection problems, and stock-option valuation problems. Other professionals specialize in particular tools, such as econometrics, optimization, or simulation. Also, millions of modelers build and use models as part of their jobs or in their personal lives but do not think of themselves as modelers. Most spreadsheet users probably fall into this category. We refer to these individuals as *end-user modelers*. End users often are professionals in a field other than modeling. They may be consultants, lawyers, accountants, marketing or financial analysts, plant managers, or hospital administrators. This book is designed to help end users

Modeling for Insight: A Master Class for Business Analysts, by Stephen G. Powell & Robert J. Batt

become more like professional modelers, without investing the time to acquire an advanced degree or serve an apprenticeship.

Modeling is used by different people in different circumstances for different purposes. Some modeling is routine, in that the problem is well structured and the procedure for getting from the problem to a solution is well understood. This would be the case, for example, in a consulting company where valuing corporations as possible takeover targets is a core part of the business. In this situation, the consulting company would have a well-developed procedure for modeling the future value of a target company. Essentially the same model structure would then be reused each time a new valuation was needed. The analyst's task in this case would not be to formulate the model from scratch but to identify the data inputs for the company in question and to carry out the analysis as prescribed in the procedure. Routine modeling efforts such as this are *not* this book's major focus, although some of the tools we discuss here can be used effectively in routine modeling.

The focus of this book is on modeling new and ill-structured problems from scratch when no existing procedure is available to follow. This aspect of modeling has received very little attention in existing books or courses. Some people claim that this most creative aspect of modeling cannot be taught but can only be learned through experience. We disagree. We know that real expertise in modeling requires both breadth and depth of experience and cannot be acquired *solely* from a book or course. However, that does not mean modelers cannot improve their skills in any way other than through experience. As with most skills, learning solely through experience does lead to improvement, but the rate of improvement is often very slow and bad practices can be learned just as well as good ones. Many people learn most effectively through a combination of experience and structured study. We believe the methods presented in this book can help you dramatically improve your modeling skills if you take the time to work through the chapters slowly and carefully and if you continue to practice and develop your skills after you complete the book.

1.1 WELL-STRUCTURED VERSUS ILL-STRUCTURED PROBLEMS

Problems come to the analyst from many sources and in many forms. Most problems can be dealt with adequately without great modeling skill. Typically, these are *well-structured* problems, such as preparing a budget for the coming year in a stable business or selecting the lowest cost vendor to supply raw materials. By a well-structured problem, we mean one in which

- The objectives of the analysis are clear
- The assumptions that must be made are obvious
- All the necessary data are readily available
- The logical structure behind the analysis is well understood

By contrast, in an *ill-structured* problem, few or none of the features listed above are present. Thus, in these problems, the objectives are unclear, the assumptions one should make are ambiguous, little or no data are available, and the necessary logical structure is not well understood. Here are several examples of ill-structured problems:

- Should the American Red Cross institute a policy of paying for blood donations?
- Should Boeing's next major commercial airliner be a small supersonic jet or a slower jumbo jet?
- Should an advertiser spend more money on the creative aspects of an ad campaign or on the delivery of the ad?
- How much should a mid-career executive save out of current income toward retirement?

Why is each of these problems ill structured? The Red Cross situation illustrates a problem in which objectives are unclear. As a nonprofit organization, the organization exists to serve the public, but it cannot at the same time run an unlimited deficit. And does the quality of the blood supplied by the Red Cross figure into the analysis? The Boeing problem illustrates a situation in which the assumptions one makes may heavily influence the outcome of the analysis. Should one assume, for example, that the company can only develop one new airliner or more than one? The advertiser's problem is difficult in part because no data are available that can suggest how the quality of ads might improve with the number of drafts chosen. Finally, the retirement problem illustrates a case where the precise objective is unclear and a logical structure must be constructed within which to carry out an analysis.

1.2 MODELING VERSUS PROBLEM SOLVING

Well-structured problems typically can be *solved*. A solution to a problem consists of qualitative or quantitative information that leaves no questions about the problem unanswered. In many cases, we can actually prove the solution is the only possible one. An algebra problem provides an extreme but illustrative example. Consider this one:

Solve the following system of equations for X and Y:

$$3X + 4Y = 18$$

$$9X + Y = 21$$

The solution to this problem consists of the values $X = 2$, $Y = 3$. Not only can we easily demonstrate that these values actually do solve the problem, we can prove this is the only solution to this problem. Once we have found these values for X and Y, nothing more needs to be said about the problem.

Unlike well-structured problems, ill-structured problems cannot be solved in this same sense because it is impossible to develop a solution that leaves no outstanding questions about the problem. Even more, one cannot demonstrate that any particular set of information actually constitutes a solution, much less prove that it is unique. Many problem solvers confuse well- and ill-structured problems. But ill-structured problems demand different methods than well-structured problems. Ill-structured problems do not have solutions in the same way that well-structured ones do, nor is an effective approach to a problem that is well structured likely to work for a problem that is ill structured.

The distinction we have made here between the two types of problems is fundamental to this book. It is also helpful to make a parallel distinction between trying to *solve* a well-structured problem and trying to *gain insight into* an ill-structured one. We refer to the former activity as *problem solving* and to the latter as *modeling*.

If ill-structured problems cannot be solved, how can we say anything useful about them at all? First, ill-structured problems can be *explored*. Exploring a problem involves making assumptions, formulating hypotheses, building a sequence of models, and deriving tentative conclusions, all with an inquiring mind and in a spirit of discovery. Problem exploration is a more creative and open-ended process than problem solving. It often reveals aspects of the problem that are not obvious at first glance. These discoveries can become useful insights.

Another useful approach modelers can bring to ill-structured problems is to develop a problem *structure* within which exploration can proceed. For example, clarifying the outcome measures, decisions, and uncertainties in ill-structured problems provides a type of structure for problem exploration. In some cases, simply asking what options we have in the situation can focus the discussion. In other cases, a more formal model structure, consisting, perhaps, of a decision tree or a cash-flow model, provides the necessary structure. Ultimately, an effective modeler can develop insights into the factors that make one course of action preferred to all others.

1.3 MODELING FOR INSIGHT

We use the word *insight* for any useful information that can be developed about an ill-structured problem. Insights come from many types of analysis, but they are almost always expressed in words, not numbers. Some insights can be developed by exploring different objectives for the problem. Other insights come from exploring trade-offs between various consequences of a particular course of action. (A trade-off occurs when changing a decision improves one objective but worsens another.) Still others come from developing an understanding of the factors critical to succeeding or achieving a plan. Finally, some insights come from analyzing the risks inherent in a decision.

Although creativity is a necessary ingredient in the search for insights, some insights can be identified rather routinely using the techniques we describe in this book. Our fundamental goal is to present procedures for approaching ill-structured problems that, although not leading to solutions, can reliably lead to insights. "Modeling for insight" is the phrase we use for the process of modeling ill-structured problems in order to gain understanding.

1.4 NOVICE MODELERS AND EXPERT MODELERS

When faced with an ill-structured problem, novice modelers typically focus on acquiring *data*, whereas expert modelers typically focus on developing a *model structure*. Why do experts behave so differently from novices? Does it make a difference?

Let's consider the Red Cross problem: Should the Red Cross pay for blood donations? Novices are likely to look for data, such as the number of pints of blood now being donated per year, the average surplus (profit) the Red Cross earns on each pint, or perhaps the percent of donated blood that is contaminated. Although these types of data may be valuable in the eyes of the modeler, especially if they are easy to acquire, focusing on finding data early in a modeling effort has many drawbacks. The most obvious of these drawbacks is that data collection takes time and (often) money, which are resources that are always scarce in modeling. Even worse, actually acquiring data can prematurely narrow the focus of the modeling effort or even reduce the creativity brought to the task. Finally, premature data collection may well turn out to be wasted effort if, as frequently happens, the data that are collected do not affect the results. The essential point to keep in mind about data collection is that *without at least a preliminary model in mind, there is no way to determine what data are really needed.* One of the fundamental tools we offer in this book is a way to determine the data that will be valuable in modeling ill-structured problems.

Data play a very different role in well-structured problems. These types of problems actually have answers, which are likely to depend on the data. For example, if we change any one of the six parameters in the system of equations discussed above, then $X = 2$, $Y = 3$ will no longer be the correct solution. However, in an ill-structured problem, it is unlikely the insights we are looking for are highly sensitive to the data in the problem. And when data are critical, understanding why this is the case itself generates insights. As we have pointed out, problem-solving methods that are effective for well-structured problems are not always effective for ill-structured ones. This is true of data as well: data are critical in well-structured problems but less so in ill-structured ones.

Expert modelers, not surprisingly, take a very different approach from that of novices when faced with an ill-structured problem. Experts rarely look for data early in a modeling effort. Instead, they build a model structure as a backbone for their thinking. In other words, they focus on the *relationships* between variables rather than on the values of those variables. They use the structure they create to ask questions, such as:

- Have I formulated the right problem?
- Will the approach I'm taking work out?
- Will this model help me answer the client's questions?
- Can this approach lead to a practical course of action?

This approach offers several advantages. First, a preliminary model structure can suggest which data will be most useful. Second, the modeler's creativity is most unfettered when data issues are moved to the background. Third, the structure-first approach uses time effectively, allowing the modeler to consider important aspects of the problem in general terms first without committing to any one approach.

To help you learn this approach to modeling, we address several problems in which little or no data are given. At times, we even exaggerate the scarcity of data to inculcate the desired attitude toward data. We have found that it is a useful discipline to learn to model without any data at all. If you can build a credible model structure when no data are available, you can certainly build one when some data are at hand.

1.5 CRAFT SKILLS IN MODELING

An expert modeler possesses several different types of skills. Certainly an expert must have a command of the field of study's technical underpinnings. For example, a marketing analyst must understand marketing very well to be a good modeler in that domain. But modelers need a range of skills that go beyond their domain expertise. For one thing, they need the skill to determine which problems are suitable for modeling. Then they must be able to abstract the essential features of the real problem. Finally, they must be skilled in translating model results into a form that is relevant and useful in the real world.

We refer to all the skills an expert modeler possesses beyond technical expertise as *modeling craft*. We know from experience that craft skills are essential to success in modeling, but craft skills are difficult to articulate. Many expert modelers, like experts in any field, have difficulty explaining how they do their work. Thus, craft skills have been underappreciated and rarely taught. This book, by contrast, is devoted to helping you learn the craft skills of modeling.

You can find many books that describe how models are used in a particular domain, such as marketing or finance, and even more that cover specific modeling tools, such as optimization, simulation, or statistics. But very few books deal in any detail with the essential craft skills of modeling. This is our task.

1.6 A STRUCTURED MODELING PROCESS

We find it effective to organize modeling into four stages:

1. **Frame the Problem:** Problems do not come neatly packaged, ready for solution. Rather, they emerge out of the confusion of daily life. The first critical skill you need to develop as a modeler is the ability to identify a problem suitable for modeling, including which aspects of the problem to include and exclude. This process is *problem framing* or problem formulation.
2. **Diagram the Problem:** Once you have framed a problem, you will find it is often effective to develop a high-level understanding through visualization. We use influence diagrams for this purpose. Influence diagrams allow you to work out the connections between inputs and outputs without precisely specifying the relationships involved. This freedom is essential to creative modeling.
3. **Build a Model:** A well-structured model is the heart of modeling for insight. Building such a model requires a sound problem frame and a well-thought-out influence diagram. Then you can employ the tools of spreadsheet engineering to develop a model that will serve as an effective laboratory for finding insights.
4. **Generate Insights:** Generating insights is the goal in modeling ill-structured problems. Although most of us recognize insights when we see them, no recipe is available for generating them. Insights come out of a deliberate process of exploring the behavior and implications of a model. To find them, you must translate model results into meaningful results for the real problem represented. Insights usually are expressed in words and graphs, so as an effective modeler, you will need to develop skills in communicating with these tools as well as with numbers and formulas.

1.7 MODELING TOOLS

Within this overall process, our approach uses six powerful modeling tools:

1. **Influence Diagrams:** A graphical approach to model formulation is often more effective than an algebraic or a numerical approach, especially early in problem formulation. *Influence diagrams*, the tool we use in this book, involve mapping inputs, outputs, and the relationships that connect them.

2. **Spreadsheet Engineering:** A model is a tool, or a means to an end, not the end itself. The end result may be a forecast, a decision, or an insight. For a model to serve as an effective tool, it must be solidly built and skillfully used. *Spreadsheet engineering* refers to a variety of methods for building and using models effectively.
3. **Parameterization:** All models include numerical inputs. These inputs are referred to in different contexts as data, parameters, decision variables, assumptions, or, simply, inputs. Many insights come from varying these inputs in an organized way. To perform this task effectively, both the input values and the relationships that depend on them must be formulated appropriately. This process is called *parameterization*.
4. **Sensitivity Analysis:** When the problem at hand is ill structured, the modeling process will not be used to generate a single numerical solution. Rather, modeling will be used to better understand the problem—that is, to generate insights. Many insights can be found by varying one or more aspects of the model in a structured manner. This process is referred to as *sensitivity analysis*.
5. **Strategy Analysis:** Models are often used to evaluate alternative courses of action. Sometimes, the alternatives are set in advance, but very often the modeling process itself leads to the creation of new alternatives or new combinations of alternatives. *Strategy analysis* refers to the process within modeling of evaluating individual alternatives or combinations of alternatives and of inventing new alternatives, all in an effort to improve the overall outcome of the decision process.
6. **Iterative Modeling:** Novices often begin a modeling project by trying to build the ultimate model right from the start. This approach can lead to confusion, errors, wasted time, and sometimes, to complete failure of the modeling effort. Experts, on the other hand, build a series of increasingly complex and informative models. This process is called *iterative modeling*, and it is one of the keys to modeling success.

1.8 SUMMARY

A model is a simplified representation of reality. This book focuses on modeling ill-structured problems, for which no solution exists. Ill-structured problems cannot be solved, but they can be explored for insights. Novices and experts approach ill-structured problems in very different ways. The approach we develop in this book is designed to help you acquire the skills of expert modelers quickly.

Expert modelers possess both technical and craft skills. To help you learn the essential craft skills, we have organized the modeling process into four stages and have identified six powerful modeling tools that we use repeatedly to generate insights. In the next chapter, we discuss the modeling process and these six modeling tools in more detail.

CHAPTER 2

Foundations of Modeling for Insight

2.0 INTRODUCTION

Modeling for insight is based on the following four-stage modeling process:

1. Frame the problem
2. Diagram the problem
3. Build a model
4. Generate insights

At various points within this process, we draw on specific modeling tools. These essential tools are as follows:

- Influence diagrams
- Spreadsheet engineering
- Parameterization
- Sensitivity analysis
- Strategy analysis
- Iterative modeling

In this chapter, we discuss the four stages in the modeling process and give guidelines for navigating through it successfully. We also describe the six modeling tools, giving examples and principles where appropriate. Finally, we provide a brief overview of the essential presentation skills you will need to become an effective modeler. This chapter establishes a modeling toolkit that we use in subsequent chapters on a variety of ill-structured problems.

2.1 THE MODELING PROCESS

Most books on modeling present an elaborate flowchart for solving problems. Experienced modelers, however, rarely follow any set procedure as they go

Modeling for Insight: A Master Class for Business Analysts, by Stephen G. Powell & Robert J. Batt

about their work. In fact, Tom Willemain's research on expert modelers (Willemain, 1994, 1995) shows that experienced modelers tend to shift their attention from one issue to another. For example, they may first consider outcome measures, then whether they can get the data they need, then whether the client would implement the model, and so on. This doesn't necessarily mean good modelers do not use a problem-solving process. Experienced modelers have internalized through practice a multistage process that begins with an ill-structured problem and ends with a recommendation for action. They know implicitly the stages they have to go through to end up where they want, but they know too that those stages are not sequential and discrete, like following a recipe. Rather, they understand that they can always revisit an earlier stage, such as problem formulation, if necessary.

We find it helpful to use the following simple four-step process to organize our thinking about modeling ill-structured problems:

1. Frame the problem
2. Diagram the problem
3. Build a model
4. Generate insights

This process highlights the most creative aspects of modeling: framing the problem and generating insights. Without the ability to frame the problem and generate insights, modelers are merely technicians and their modeling efforts are likely to fail.

2.1.1 Frame the Problem

Problem framing is the first stage in our problem-solving process. But what does it mean to frame a problem? Ill-structured problems are neither neatly packaged nor easily solved. A major reason ill-structured problems are difficult to analyze is that they are vague about where the real problem lies and what might constitute a "solution." Until the problem becomes somewhat less vague, you run the risk of solving too many problems or even solving the wrong problem. So the essential task in problem framing is to bring some clarity to the problem itself.

We describe this stage as problem *framing* because the first step is to put some boundaries around the problem. Boundaries define what you will include and exclude from your model. To illustrate, imagine how you might develop a budget for a small business. What might your boundaries be? Or, to put it in other terms, what questions would you want to ask the business owner before you set out to build a budget model? Here are some examples:

- Should the budget cover next year, the next five years, or some other period of time?

- Should you model by months or quarters?
- Should you model every product individually or only product lines?
- Is it a business-as-usual budget, which would be based largely on past performance, or are new initiatives planned that will require more creativity in modeling?
- Are there any special circumstances that make past performance not representative of the budget period?
- Are there external factors that should be taken into account?

Each of these questions suggests a possible model boundary. For example, you might decide to build a model for the next year only and to assume business as usual. You might also assume that no new competitors will enter your markets and that none of the existing ones will leave. Finally, you might assume that, although sales have risen 5 percent per year for the past five years, markets are maturing and, therefore, future growth will be lower. This process of asking questions and making assumptions illustrates the process of putting boundaries around a problem. Although this process continues throughout the modeling process, it should be emphasized at the beginning of the process.

Boundaries are necessary to frame problems, but they are not sufficient. Even when we have placed some boundaries around an ill-structured problem, open questions remain that will haunt us later if we do not consider them at this stage. You should ask yourself these three questions whenever you frame a problem:

- How should I place a *value* on the different outcomes my model will produce?
- What are the critical *decisions* I should include in the model?
- Which factors are most *uncertain* and which will have the biggest impact on the outcome?

Somehow, you must discover, unearth, research, invent, or otherwise provide answers to these questions before you can start using the tools of modeling.

In the hypothetical budget exercise, you might decide that the outcome you are most concerned with is the annual profit you will make five years out. The critical decisions could involve whether to enter a certain new market and the level of sales support for your existing products. The key uncertain factors could be the success or failure of your new product (perhaps measured by its unit sales) and the cost of sales personnel.

The answers to the three questions stated above will help you frame the problem and focus your efforts. For example, having decided to focus on profit five years out, you may decide that your model can use annual, not quarterly, time periods. Given your decision to focus on the new product and sales support issues, you will ignore dozens of other decisions that may come up during the time period of your model. No one model can answer all questions.

Trying to do so is a guarantee of failure. Thus, the problem-framing process *must* limit the search for insights. Your goal is to choose effective boundaries, not to ignore them.

All models produce hypothetical forecasts of future outcomes based on three broad classes of assumptions:

- Parameters
- Decision variables
- Relationships

In the budget-forecasting example, you might believe sales will grow 3 percent per year in the future, although you cannot control the actual growth. The input *parameter* is 3 percent. You might also assume you will hire two new employees in the first year and three more in the second. These assumptions are also parameters (numbers), but because they are under the control of management, we refer to them as *decision variables*. Finally, you might make some assumptions about the *relationships* that connect your inputs to your outputs. One relationship in a budget model might involve the demand for your products. If you assume your sales will decline 10 percent for every 5 percent increase in price, you have an elastic demand relationship in the model (with an elasticity parameter of –2.0).

It is too early at the problem-framing stage to be concerned about the precise values of any of your inputs. However, it is not too early to think about what parameters might be important and roughly what their values might be. We will use sensitivity analysis during a later phase of modeling to test how much outputs change when we change parameters, decisions, and even relationships.

Ill-structured problems often come to the modeler in a complex form, encumbered by too much detail and circumstance. Often, a client will burden the modeler with as much detailed knowledge of the problem as time allows, presumably in an effort to help "clarify" the problem. But if the client needs a modeler's services, it is because he or she cannot identify the essential features of the problem well enough to deal with it; years of experience and expertise can make it impossible for the client to see past the details. It is up to the modeler to cut through these details to find the core of the problem, which is what the model will attempt to capture. We refer to this essential core of the problem as the *problem kernel*.

Think of the problem kernel as a *minimal* description of the problem you need to generate a model structure. As simplification is the essence of modeling, it is often helpful to simplify the problem description itself before beginning to model. Expert modelers are skilled at identifying the problem kernel, despite the complex problem statement that surrounds it—something novices cannot do. Thus, we recommend identifying the problem kernel as an essential step in formulating a model.

Claude Shannon, the father of information science, had this to say about why it is important to simplify when solving problems (Shannon, 1993):

> The first one [method] I might speak about is simplification. Suppose that you are given a problem to solve. I don't care what kind of problem—a machine to design, or a physical theory to develop, or a mathematical theorem to prove or something of that kind—probably a very powerful approach to this is to attempt to eliminate everything from the problem except the essentials; that is, cut it down to size. Almost every problem that you come across is befuddled with all kinds of extraneous data of one sort or another; and if you can bring this problem down into the main issues, you can see more clearly what you are trying to do and perhaps find a solution. Now in so doing you may have stripped away the problem you're after. You may have simplified it to the point that it doesn't even resemble the problem that you started with; but very often if you can solve this simple problem, you can add refinements to the solution of this until you get back to the solution of the one you started with.

Shannon's idea of abstracting to the essence of a problem is the same as our notion of a problem kernel.

Early on in this book, we present problems in a simplified form, starting with the problem kernel rather than a realistically complex problem description. This, of course, is the reverse of the situation encountered in practice. We build an initial model based on the problem kernel and then elaborate that model while considering successively more complex problem descriptions. Eventually, we reverse this process and start with a realistic problem description to give you the opportunity to practice forming your own problem kernel.

Problem framing is the critical first stage in modeling an ill-structured problem. By considering the problem from many angles, you begin to put boundaries around it and provide a structure for analyzing it. The assumptions you make during this phase will be tentative and subject to revision later. But to be effective, you must be bold about the assumptions you make at this stage. Furthermore, making thoughtful assumptions helps to clarify a vague problem. The problem kernel is the essence of the problem without the confusing detail. Identifying the problem kernel is an essential step in problem framing.

2.1.2 Diagram the Problem

The first stage in our process, problem framing, is designed to uncover what you already know about a problem, to set some boundaries, and to establish some initial assumptions for you to build on. It is exploratory by nature, not designed to lead to any final conclusions. But how do you get from this exploratory phase to the point where you can build a quantitative model?

It is usually ineffective to make the leap from problem framing to model building in one step. Many students and analysts open a blank spreadsheet and begin to populate it with data and relationships before they have a clear idea of the model structure they are trying to build. As a result, they get confused and ensnared in cycles of rework, and often they fail to generate insights. Taking time to diagram the problem once you have framed it will allow you to create an abstract but meaningful representation of the model to come. Influence diagrams serve this purpose. We discuss influence diagrams in more detail in Section 2.2.1.

2.1.3 Build a Model

The third stage in our modeling process involves actually building a spreadsheet model. Problem framing and influence diagrams have preceded this stage, so that you should have a clear sense of the overall purpose and structure of the model you intend to build before opening a spreadsheet. Even with this degree of advance planning, model building itself requires careful planning and execution if you hope to build a useful and error-free model. We call our structured approach to spreadsheet modeling *spreadsheet engineering*, and we describe it in more detail in Section 2.2.2 and Chapter 3.

2.1.4 Generate Insights

The final phase of modeling is to translate the technical, numerical results of the model-building and analysis process into conclusions and recommendations the client can understand and act on. This translation process takes a combination of quantitative and qualitative skills as well as an element of creativity. If you do it well, you will be able to generate insights that help your client better understand the problem and motivate him or her to act.

Earlier, we defined an "insight" as any kind of useful information that can be developed about an ill-structured problem. But some types of information are more powerful than others. Often, critical insights are developed through an iterative process that begins with a raw numerical result from the model and ends with a telling graph or diagram. The following example shows what we mean:

Consider a situation in which a friend asks you to build a model to help him determine how much of his take-home pay he should set aside each month in a retirement account. One way you might model this problem would be to forecast his income, savings, and assets over the period from the present to his date of retirement, and then to forecast his spending and assets from retirement to his death. This forecasting model would, of course, be built on several assumptions as to his savings rate, investment returns, longevity, and so on.

The first insight this model generates would probably be a single numerical output. You might express this insight as follows:

> If you save 10 percent of your pay, you will have $796,000 in savings on the date you retire.

Your friend, no doubt, will be very grateful for this information because if he were able to build a model that would generate a result like this, he probably would not have asked for your help. But there is more you can learn from the model and additional insights you can offer him. You can uncover these insights through two types of exploration: One is to perform sensitivity analysis, and the other is to examine additional output measures.

Sensitivity analysis might lead to Figure 2.1, which suggests an insight like this:

> If your savings rate is between 5 percent and 15 percent, you will have between $440,000 and $1,630,000 on the date you retire.

This new insight is more powerful than the previous one because it incorporates the prior one and adds a measure of the sensitivity of the outcome. Note that the graph communicates additional insights, for example, the rate at which assets at retirement increase.

Alternatively, you might express this insight in terms of the value of deviating from the base-case savings rate. Figure 2.2 shows this relationship. Here

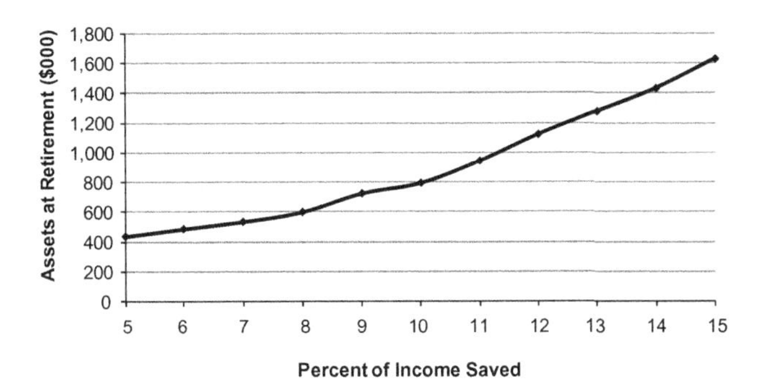

Figure 2.1. Assets at Retirement as a function of Percent of Income Saved.

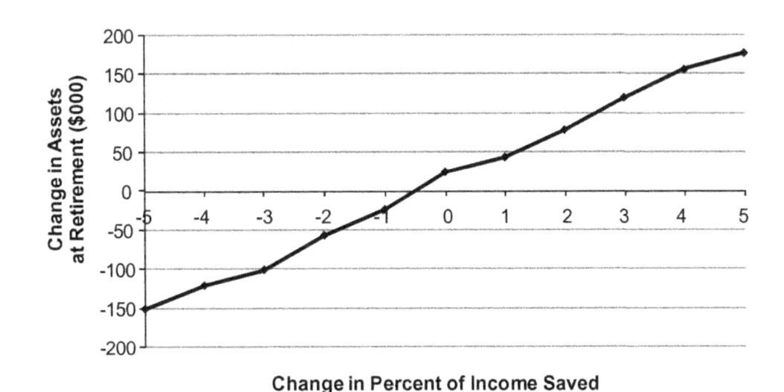

Figure 2.2. Marginal impact on Assets at Retirement of changes in Percent of Income Saved.

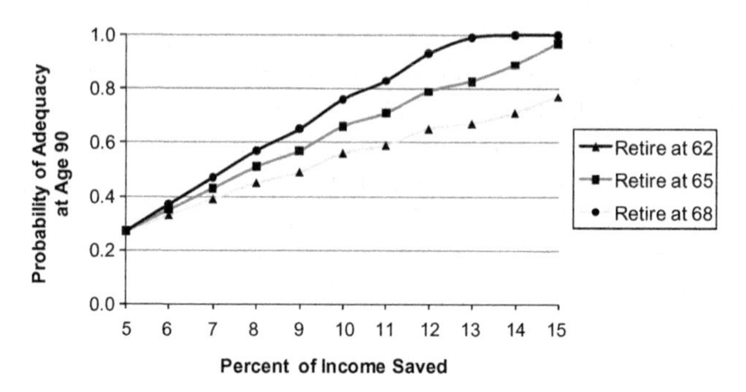

Figure 2.3. Probability of Adequacy as a function of Percent of Income Saved and Date of Retirement.

the same basic data are used in a different way to express a somewhat different point:

> Changing the base-case savings rate (10 percent) by one percentage point adds about $45,000 to your assets on the date you retire.

Another way you can explore for insights is to examine alternative outcome measures. The amount saved at retirement is one simple outcome measure, but it may not be the most useful measure for your friend. His is probably anxious to know whether a certain amount saved at retirement will be "adequate." But how do you measure adequacy? You could measure assets at death. You could measure years of comfortable retirement. You could estimate the probability of having assets left at age 90 or passing on at least $1M to heirs. These are just a few examples from an unlimited set of possibilities.

This exploration might lead to a result like Figure 2.3, which shows the probability of having adequate retirement savings through age 90 as a function of the percent of income saved and the age at retirement. This figure packs a lot of useful information into a small space in a form that is immediately clear to the client. (We explore this problem more fully in Chapter 5.)

To summarize:

- Insights are useful conclusions about the problem expressed in words.
- Insights are discovered through model exploration.
- Model exploration involves sensitivity analysis and analysis of alternative outcome measures.
- Generating insights requires a sound model, and the patience and skill to use the model to explore the problem from all angles until hidden gems emerge.

2.2 TOOLS FOR MODELING

We mentioned in Chapter 1 that modeling involves both technical and craft skills. Technical skills include, for example, knowledge of the Excel VLOOKUP function, the ability to formulate an optimization model, and the ability to interpret the results of a simulation model. Craft skills are more difficult to describe but no less essential to effective modeling. They include, for example, the ability to extract the essence of a problem, build a simple but effective model structure, and translate model results into useful insights.

Most craft skills are used in the background, out of sight. To teach them, we need to bring them into the light and to develop ways to talk about them. We have, therefore, identified six key tools for modeling ill-structured problems. We describe these tools in some detail here and show how they are used in practice throughout the book.

2.2.1 Influence Diagrams

Influence diagrams are a graphical means to display the structure of a model. They do not substitute for a quantitative model; rather, they provide a practical method for exploring alternative model structures during the early stages of problem framing. Influence diagrams are a far more effective tool than spreadsheets for creating a model structure. They encourage creativity, whereas spreadsheets encourage an inappropriate focus on details. In an influence diagram, the structure of a model is transparent; in a spreadsheet model, the structure is hidden behind the numbers.

Influence diagrams have just a few simple ingredients, each with its own symbol:

- An outcome measure (octagon)
- Intermediate variables (ellipse)
- Parameters (rectangle)
- Decision variables (parallelogram)

Consider this simple example from a well-structured situation. The problem is to determine what price to charge for a product to maximize profit. We start the influence diagram with the outcome measure, which is Profit (Figure 2.4). Next, we decompose profit into its constituent elements, which are Total Revenue and Total Cost (that is, Profit = Total Revenue − Total Cost). Total Cost, in turn, can be broken down into Fixed Cost and Total Variable Cost. Now, Total Revenue and Total Variable Cost are both a function of Quantity Sold, so we enter this variable in the diagram and connect it to both of the variables it influences. Total Variable Cost also depends on Unit Cost, so we include Unit Cost as a parameter at this point. Total Revenue is also influenced by Price, so this decision variable enters the diagram at this point. Price may

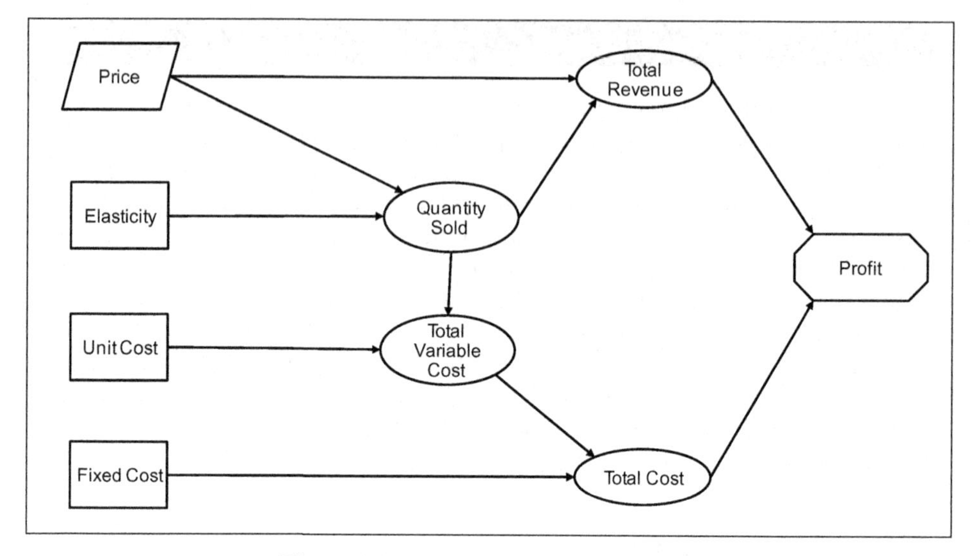

Figure 2.4. Influence diagram for Profit.

also influence Quantity Sold, so we connect Price to that variable and add another paramter, Elasticity, which influences Quantity Sold.

This diagram contains all four types of elements. The outcome measure, Profit, is shown enclosed in an octagon at the right of the diagram. The variables that influence Profit directly are some intermediate variables, shown as ellipses. Unit Cost, Elasticity, and Fixed Cost are parameters, shown as rectangles. Finally, the decision variable, Price, is shown as a parallelogram. Note how the influence diagram relates the decision variable Price to the outcome, Profit.

Here is another short problem statement, one that describes an ill-structured situation for which a model is needed.

Northlands Museum Capital Campaign

Northlands Museum is a regional museum of human and natural history located in a small city in a northern state. It has an annual budget of around $900K and an endowment of $1.2M. In recent years, the museum's board has reluctantly allowed the director to cover an operating deficit by transferring money from the endowment. The deficit has averaged $75K per year.

The museum director has recently proposed a major expansion of the museum, including building additional storage space for artifacts, upgrading exhibits, and installing air conditioning and humidity-control systems. A large capital campaign—the largest in the museum's history—will be required to fund this project. The board is concerned that the time and costs involved in the capital campaign will lead to additional requests to draw down the endowment.

The board has hired you to develop a model for analyzing the financial interrelationships among the operating budget, the capital campaign, and the endowment. The board's ultimate concern is that the endowment should not be smaller than it is now, when the campaign and the expansion project are completed in 10 or 15 years.

To the Reader: Take a few minutes to draft an influence diagram for this problem before reading on.

Our influence diagram for this problem is shown in Figure 2.5. It shows the Endowment influenced by both Interest and Deductions. Deductions, in turn, depend on the Operating Deficit and the Capital Deficit (we are assuming that surpluses in either category also flow to the endowment). The Capital Deficit itself depends on Contributions to the capital campaign as well as on the cost of the campaign (Campaign Expenses) and on the costs of the proposed expansion (Expansion Costs). Operating Deficit depends on Operating Costs and Operating Revenue.

Although you need to practice to learn how to use influence diagrams effectively, here are a few principles to follow:

- Start by defining the outcome measure.
- Decompose the outcome measure into a small set of variables that determine it directly.
- Continue to decompose each variable into its constituents, drawing each variable only once.

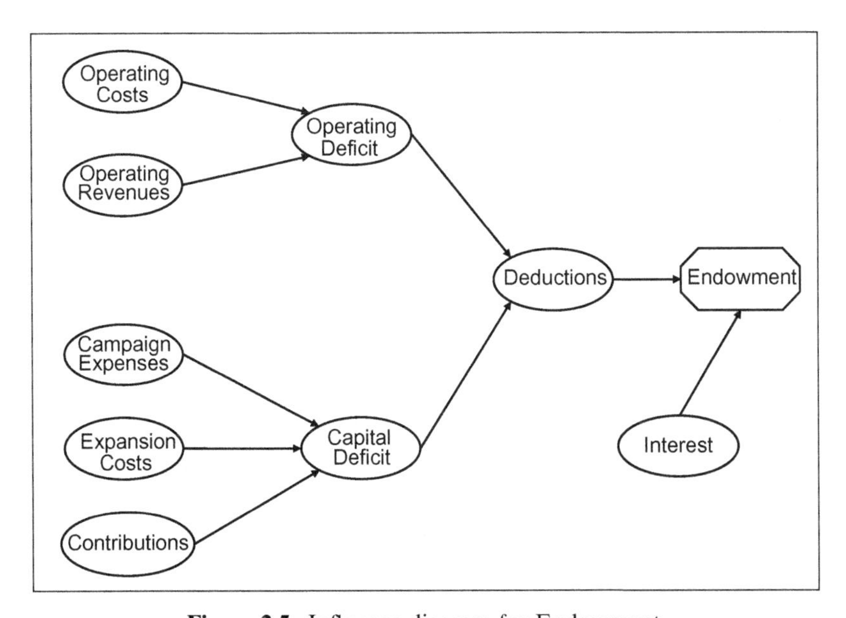

Figure 2.5. Influence diagram for Endowment.

- Link each variable to the variables it influences and is influenced by. (Variable A *influences* Variable B if Variable A is needed to *calculate* Variable B.)
- Identify parameters and decision variables.

2.2.2 Spreadsheet Engineering

Spreadsheets are the essential technical tools of business analysis. Effective modelers see to it that their spreadsheets are well structured and easy to use. Poorly structured spreadsheets not only are difficult to understand and work with, but they also are more likely to hide errors than are well-designed ones. (For more information on the risks of poorly designed spreadsheets, see the Spreadsheet Engineering Research Project website: http://mba.tuck.dartmouth.edu/spreadsheet/.)

Using spreadsheets effectively for modeling involves a set of skills we call *spreadsheet engineering*. Although problem framing is a highly creative and open-ended process, spreadsheet engineering (as the name implies) is more structured and routinized. The goal of spreadsheet engineering is to build models both *efficiently* and *effectively*. An efficient process leads to a completed model with minimum effort and waste; an effective process leads to an insightful analysis of the problem itself. Thus, it is necessary but not sufficient to build models efficiently; not doing so costs time and money. But it is still possible to build models efficiently and not be an effective modeler. Effective modelers use models to help their clients select and implement effective solutions.

We divide the process of spreadsheet engineering into four phases:

- Design
- Build
- Test
- Analyze

The essential idea in the design phase is actually to *design* your models, not simply to let them grow organically. A well-designed spreadsheet should have an easily understood logical and physical structure, with clearly defined modules. You should be able to distinguish data inputs from relationships and to document all essential assumptions.

Building a spreadsheet effectively requires you to have and use a sound design. In an effective engineering process, building the spreadsheet is mostly a mechanical task. You must pay attention to executing the design carefully without introducing flaws (or “bugs”) into the model, rather than redesigning the spreadsheet. An effective spreadsheet builder knows where most bugs originate and can therefore take special care when it is required.

Model *testing* is a crucial and often overlooked phase in spreadsheet engineering. Because all spreadsheets develop bugs as they are built—many spreadsheets in use contain bugs, in fact—you need to develop a healthy skepticism about the presence of bugs in your own work. There are many effective ways you can test a spreadsheet for bugs. Some methods are built into Excel itself (although few analysts know about them). Other methods require special add-in software. Get into the habit of thoroughly testing all your models before using them for analysis.

The final phase in spreadsheet engineering is the *analysis* phase, where the payoff to your efforts is realized. Only a well-designed, well-built, and well-tested spreadsheet should be used for analysis. The analysis phase is where you can discover insights about the problem. Analysis is one of the most creative phases of modeling, but it too has its routine aspects. For example, there are several questions you should always ask, such as how sensitive the outcome is to the input parameters and which parameters have the biggest impact on the outcome? In many cases, you can answer these questions using specific Excel tools.

We present the principles of spreadsheet engineering in more detail in Chapter 3.

2.2.3 Parameterization

The basic concept of parameterization involves separating data inputs, or parameters, from the relationships that depend on those inputs. In other words, enter an input value into the model only once and use that input in all calculations that require it. This is essential to good spreadsheet modeling, because hiding numbers in formulas leads to confusion and errors. But the concept of parameterization is broader than this.

Parameterization in the broader sense involves choosing how to represent relationships in the most effective manner. By "most effective," we mean most effective in the analysis phase of modeling. Varying parameters to see how the output changes is an essential part of analysis. But you can only vary the parameters you have chosen to include in the model. Thus, the seemingly technical question of how to parameterize a relationship can have critical implications for how useful the model is for analysis. The example below will clarify what we mean.

Imagine you are trying to model the growth in market share for a new product. You could assume some initial market share and a constant annual growth rate, as in Figure 2.6a. Or you could assume share grows steadily from time T_1 until T_2, remains level at a given steady state level S until T_3, and then declines to zero at T_4 (see Figure 2.6b).

Our point here is not that one or the other of these relationships is better, but that each supports only certain kinds of analysis. For example, in the first model, you can vary the annual rate of growth of share but you cannot vary the steady state share. Likewise, in the second model, you can vary the time

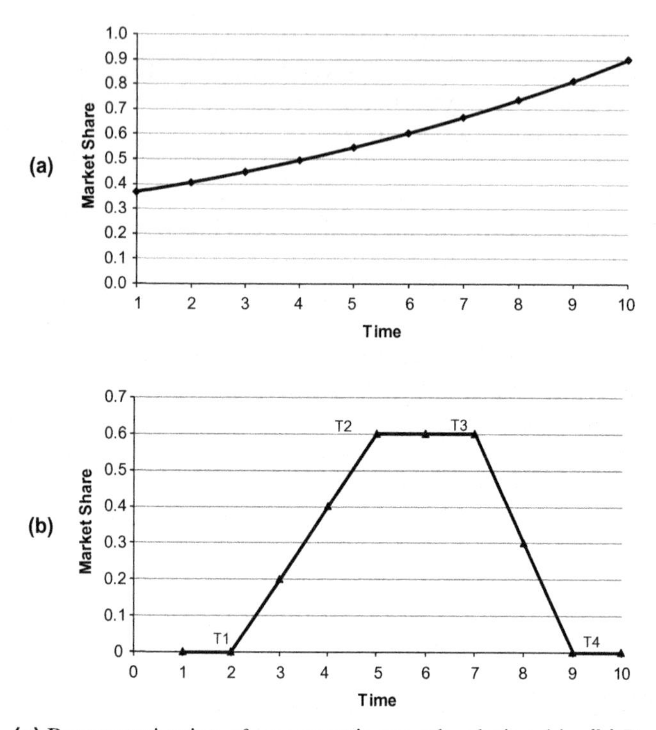

Figure 2.6. **(a)** Parameterization of a geometric growth relationship. **(b)** Parameterization of a straight-line growth relationship.

at which you first penetrate the market, but you cannot vary the initial share. Each parameterization of this relationship allows certain questions to be answered and precludes others. Designing effective parameterizations is one of the signature skills of the expert modeler.

2.2.4 Sensitivity Analysis

Sensitivity analysis is the heart of modeling for insight. It covers a range of techniques, from the most elementary notion of changing an input and observing the resulting change in the output, to sophisticated tools, such as optimization and simulation. All sensitivity analysis is based on the recognition that although a model's inputs and relationships seem to be precise, there is always some degree of uncertainty about their true value or form. If uncertainty about an input exists, it stands to reason that we investigate how sensitive the output is to that input. Doing that in an organized manner is effective sensitivity analysis.

Another form of sensitivity analysis involves identifying the best decisions in a problem. These may be categorical decisions, such as whether to buy a company, or numerical-valued decisions, such as how much to invest in a new technology. In either case, the purpose of a model is to help elucidate which decisions are better than others, and why. Insights are most likely to be found in the answers to the question: Why?

2.2.5 Strategy Analysis

The word "strategy" means different things to different people. Here we define it to mean *a related set of decisions that tell a story*.

Remember your friend who asked for your help in planning his retirement? As he sees the problem, it all boils down to one question: How much of his take-home pay should he set aside each month in a retirement account? This is a pretty straightforward decision: Determine the dollar amount (or percentage) of disposable income to save each month to maximize some as-yet-undetermined objective related to an enjoyable retirement.

A creative analyst can expand this definition of the problem by considering some of the other decisions the friend will face. One set of decisions is how much of his assets to spend each month after retirement. Another set of decisions to include might be how much long-term-care insurance to purchase and at what age. Finally, the analyst might make the date of retirement a decision as well.

Introducing additional decisions can lead to interesting insights. However, many decisions are related in the sense that certain values for the decisions either do or do not make sense together. For example, regardless of your friend's specific objectives, it probably does not make sense for him to plan to save a small amount while working, and to spend a large proportion of his assets each year after retirement. One way to reduce the problem's complexity and to simplify your recommendations to him is to select sets of decisions that naturally fit together.

One strategy for this retirement problem could be called the "Live for Now" strategy: Spend most of your income while you work and plan on a very frugal retirement. Another strategy could be called the "Save For Your Heirs" strategy: Save as much as you can while you work, delay your retirement to the latest possible date, live frugally in retirement, and purchase large amounts of long-term-care insurance as soon as possible to provide maximum protection of your assets for your heirs.

Strategies are not necessarily combinations of decisions that one would recommend or that would appeal to your client. Sometimes, they represent extreme cases, such as the "Save For Your Heirs" strategy, which could show your friend just how much he could pass on to his heirs if he did everything he could to achieve that one goal. Sometimes, strategies are primarily useful in making recommendations coherent and meaningful to clients.

2.2.6 Iterative Modeling

Expert modelers are skilled in building a series of increasingly powerful and appropriate models, rather than building one grand model from scratch. We call this *iterative modeling*. Novice modelers tend to design models that are overly complicated right from the start. Then they try to build their grand model from scratch in one frantic push, which leads to errors, wasted effort, and often to disaster. (Perhaps the most common major failing in modeling is making the model too complex and thereby not completing it in time to influence the decision.)

Effective iterative modeling starts with the simplest possible model, perhaps even an embarrassingly simple model. To identify such a model, you first need to pinpoint the problem kernel. Once you have identified it, ask yourself: What is the simplest model that begins to capture *some* of the essential elements of the problem? Notice the emphasis here is on capturing some of the *essence*, not all of the detail and complexity. Good models are radical simplifications of reality.

But effective modeling does not end with an embarrassingly simple model. How does an expert get from such an unpromising start to a worthy final product? Simply by adding more features from the real problem to the existing model, step by careful step. The trick in this process is to decide which features you will add at each stage. The goal is to add the feature that most improves your model's ability to *generate insights*. This process does not end when your model is finally "realistic." It ends when your model is just good enough for the purposes at hand, and this generally is long before it ever becomes realistic. Usually, an expert's "good enough" model is far simpler than a novice's complex model. And yet the expert can offer good insights, despite the simplicity of his or her model.

Another key to effective iterative modeling is to use each model iteration to generate tentative insights into the questions at hand. This is in contrast to an approach we often see recommended, which is to elaborate a model until it is "realistic," "sufficient," or "complete" before actually using it. This process underestimates the importance of learning during the modeling process. After all, the ultimate purpose of modeling is not to build the best model but to best educate the modeler (and, ultimately, the decision maker). Education requires learning, and learning can be facilitated by using the model frequently during the modeling process. Of course, you must keep in mind that results generated in this fashion are tentative. In later chapters, we illustrate this process and show how you can use a sequence of preliminary insights to direct the iterative modeling process.

2.3 PRESENTATION SKILLS*

A model's quantitative results almost always must be translated into terms that decision makers can understand before they take action. Therefore, it is not enough to build an excellent model and to develop correct numerical

*Professor Rob Shumsky of the Tuck School of Business at Dartmouth contributed to this section.

results; you must also express those results in managerial language. And, in most cases, you must organize your assumptions, models, results, and insights into a presentation to a specific audience. This section describes the basic skills you need to create effective business presentations.

It is important to keep in mind that the most important aspect of any presentation is the *content*, not the *style*. Edward Tufte, the most famous proponent of methods for effective presentation of quantitative information, summarized this crucial point succinctly:

> Presentations largely stand or fall on the quality, relevance, and integrity of the content. If your numbers are boring, then you've got the wrong numbers. If your words or images are not on point, making them dance in color won't make them relevant. Audience boredom is usually a content failure, not a decoration failure.
>
> —Tufte, 2003

Having said that, the best presentations have both excellent content and excellent visual support.

2.3.1 Special Challenges for Quantitative Analysts

Developing an effective presentation of quantitative analysis presents some special challenges. One challenge is that many quantitative analyses are carried out by consultants, who are outside the client organization and less knowledgeable than their clients about the industry, product, or process they are studying. This means the analysts must make special efforts to establish their credibility before their clients.

Another challenge facing quantitative analysts arises from the fact that they are often trying to convince their clients to accept a radical change from business as usual. Changes in worldview are difficult for everyone, and in an organization there will always be someone who believes his or her interests will be harmed by the change. This individual is sure to resist the recommendations, and this resistance must be anticipated when designing the presentation.

The final challenge is that quantitative analysis is often complex. The details of the analysis may be beyond the understanding of the client, but the analyst still has to find a way to convince the client of his or her results. One of the marks of an effective presentation of quantitative analysis is that it reduces the complexity in the problem to a manageable level while preserving the essential features of the problem.

2.3.2 Principles for Effective Presentations

Although every presentation is unique in some ways, there are a small number of fundamental principles that all good presentations follow. Here is a list of principles to keep in mind as you design any presentation:

- Tell a story.
- Present your insights, not your analytic process.
- Limit yourself; less is more.
- Use pictures, not words.
- Use visual aids to support, not dominate.

Tell a Story: If you have immersed yourself in the problem and found some real insights, when it comes time to create your presentation, you will have a good story to tell. People understand, enjoy, and remember stories, so design your presentation as if you were telling a story.

Present Your Insights, Not Your Analytic Process: The story you tell should help the client understand his or her problem, not recount all the work you have done to come to your insights. Figure 2.7 helps to contrast the analytic process with the presentation. On the left is a depiction of the analytic process. Try as we might to make this process rational, it inevitably includes false starts, bad ideas, and confusion. On the right in the figure is the message you want to convey after you have sifted through all your analytic work for insights. The message is straightforward, linear, and well structured. No matter how complex and challenging your analysis has been, your message should be simple, direct, and easy to follow.

Limit Yourself; Less is More: Most presentations try to do too much. If you try to convey nine or ten points to your audience, you may find they really understand none of them. However, if you limit yourself to just two or three key points, and you justify them well, your audience is more likely to understand and remember them all. In presentations, less is more.

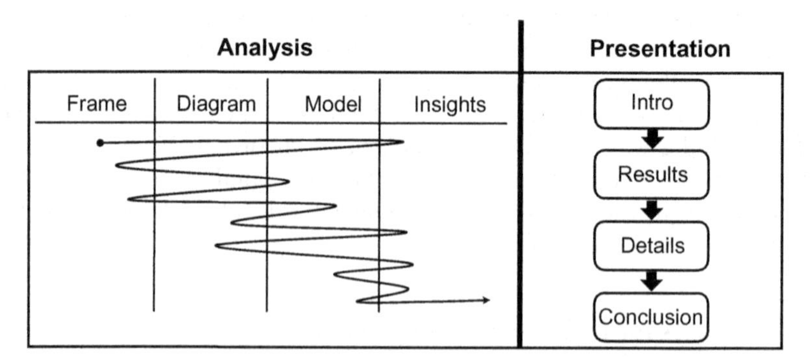

Figure 2.7. Analytic process contrasted with presentation process. (After Munter, 2006.)

Use Pictures, Not Words: Research and common sense tells us that we understand well-designed pictures better than words. Therefore, you should try when possible to replace wordy descriptions with pictures. Sometimes all that is needed is the addition of an arrow or an icon to a wordy slide. More complex qualitative relationships can be described best using diagrams. Figure 2.8 shows a variety of diagrams for organizing concepts, sequencing concepts, or subsetting a whole. Quantitative relationships can best be shown with tables or graphs. We provide more details about using tables and graphs later.

Use Visual Aids to Support, Not Dominate: Finally, beware of relying too much on PowerPoint slides in a presentation. A presentation is essentially a dialog with your audience. PowerPoint slides can be effective when used in a supporting role, but when they become the center of attention, the presenter begins to appear superfluous and the sense of dialog is lost.

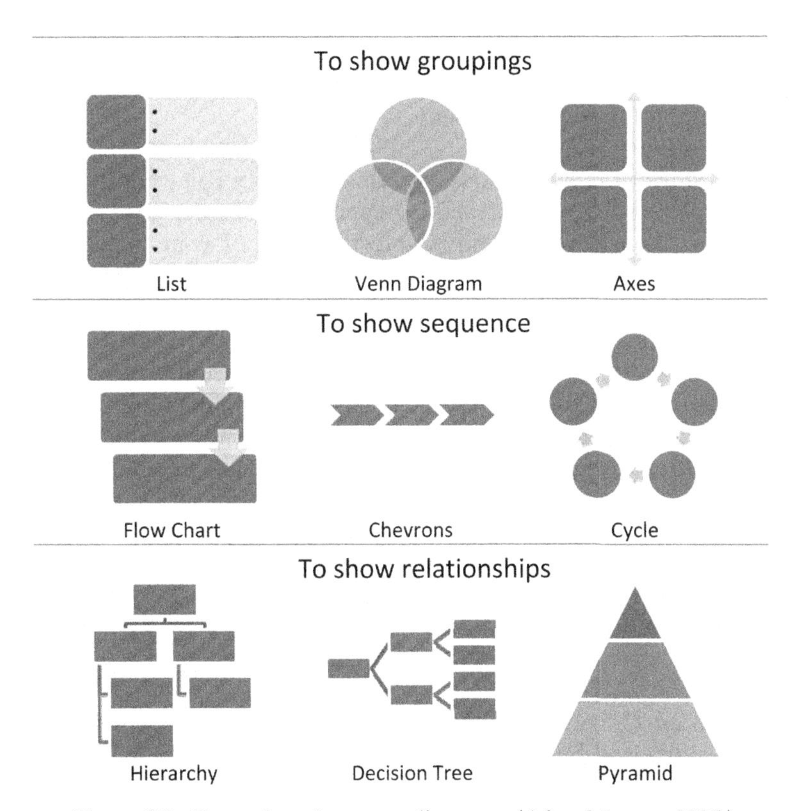

Figure 2.8. Examples of concept diagrams. (After Munter, 2006.)

2.3.3 The Process of Creating an Effective Presentation

The process of creating a presentation can be broken down into the following steps:

1. Understand your audience.
2. Outline the story.
3. Mock up slides.
4. Create charts.
5. Test the flow of ideas.
6. Practice.

Understand Your Audience: Your first step should always be to consider your audience so that you can design your presentation to meet their needs. In most business situations, your audience will include the client for your project, with whom you have discussed the project many times. But she may bring along her bosses, peers from other divisions, analysts from her own area, even outside experts. Most audiences consist of a mixture of people, with different backgrounds, job titles, power, knowledge, and points of view. Ideally, your presentation will address them all and be understandable to each of them. Practically, however, you may need to compromise and direct your presentation primarily to the most important people present.

Outline the Story: The next step is to outline the story you want to tell. There are two main alternatives for presenting your results: Either give your recommendations at the beginning and devote the body of your presentation to explaining your reasoning or give your recommendations at the end, after you have built up your reasoning. The first of these alternatives is known as the *direct* approach, whereas the second is the *indirect* approach. The indirect approach is common in academic settings, but the direct approach better fits most business situations. One of the advantages of the direct approach is that it gets your key message across at the beginning, when the audience is most attentive. It also allows you to repeat the message at the end, after you have laid out your reasoning. The direct approach is generally easier for your audience to understand, and it takes the mystery out of your presentation. The indirect approach can be advantageous when you suspect your audience may resist or not approve of your recommendations. Since you state your reasoning before your recommendations in this approach, you hope to convince your audience of the logic before they see the conclusions.

Mock Up Slides: Once you have outlined your presentation, consider where PowerPoint slides will be needed and how they should appear. Sketch your slides before you invest the time to create them in PowerPoint.

Not every point in your presentation needs a matching slide. In fact, putting everything you have to say on slides may result in a stilted presentation. When you do construct slides with words (e.g., bullet-point slides), examine each word and phrase and ask yourself, "Is this absolutely necessary to make my point?" If not, eliminate it.

Create Charts: We have already pointed out that a picture is often a more powerful means of communication than words. Therefore, pictures or charts are an important tool in designing effective presentations.

A *chart* is any visual depiction of information. Charts come in three main varieties: diagrams, tables, and graphs. *Diagrams* are used to show qualitative relationships. Influence diagrams, for example, show the relationships among inputs, outputs, and intermediate variables in a model. Figure 2.8 shows a variety of other diagrams. *Tables* and *graphs* are both used to show quantitative relationships. Tables use only numbers for this purpose, whereas graphs use geometric shapes (lines, bars, and so on).

Tables can be effective when you have just a few critical numbers to discuss and the actual numerical values are important to your audience. However, graphs are often more effective than tables. If you want to discuss a general trend, use a graph instead of a table. If you do not plan to draw attention to most of the numbers in a table, replace it with a graph or leave it out altogether.

Graphs are the most important type of chart for modelers and the most difficult to design well. Every graph is designed to make a comparison visible. The main types of comparisons are as follows:

- Showing parts of a whole
- Comparing two or more items
- Showing changes over time
- Showing a frequency distribution
- Showing correlation

Most graphs fall into one of the following five types:

- Pie chart
- Horizontal bar graph
- Vertical bar graph
- Line graph
- Scatter plot

Figure 2.9 shows which types of graphs are usually most appropriate for each type of comparison. For example, a horizontal or vertical bar graph is most appropriate for comparing several items, whereas a line graph is most appropriate for showing how a variable changes over time.

Do not make your audience guess the purpose of your charts. Provide a slide title that succinctly captures the essential point you are making. A title

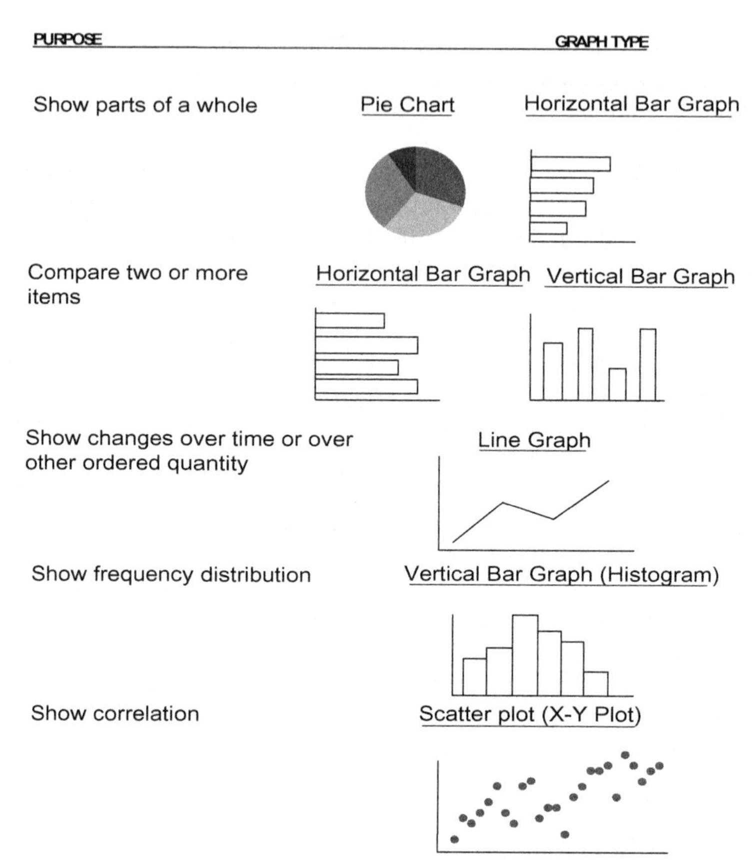

Figure 2.9. Matching comparison with chart type (After Zelazny, 2001 and Munter, 2006).

like this one: "Sales by Region for 2007," merely describes the data and does not convey the point of the chart. If the point you wish to make is that sales in the western region are falling behind, use a title such as "Sales in West Are 20 Percent Behind Average."

Finally, we suggest that you create rough sketches of your graphs before you even begin the analysis and model-building. This helps you to identify the goals of the model and to think ahead through the modeling process. When you create your final presentation, you will find these initial sketches useful even if you have learned new things from the analysis so that some of the initial sketches are obsolete.

Test the Flow of Ideas: At this point in the process, you have an outline of the story and a series of slides to back it up. Just as you refine a model by adding and correcting successive drafts, you can refine a presentation in a series of stages. Test that the presentation flows well by doing a dry run without an audience. Look for places where there are too many slides, or not enough slides. Make sure the sequence of slides is appropriate, and add summaries or reminders to the audience of where you are in the agenda. You can add animation at this point as well. Animating slides involves showing the content in stages, for example, by displaying one bullet point at a time. Complex animation can be merely distracting; once again simplicity should be the goal. Finally, determine how you will introduce each slide and how you will make the transition to the next slide.

Practice: No presentation is ready to go until you have practiced delivering it in as realistic a setting as possible. Find a room like the room you will be presenting in. Ask a few colleagues to act as your audience. Limit yourself to the actual time you will have for the presentation. And practice, practice, practice.

2.3.4 Pitfalls in Presentations

Careful observers of presentations cite just a few critical pitfalls to avoid. We have already mentioned several. Avoid telling your audience what you *did*; tell them what you *learned*. (Present your insights, not your analytic process.) Do not go into too much detail, and do not try to make too many points. (Less is more.) Do not turn your back on your audience and read your slides: Hold a dialog with them. (Tell a story.)

When appropriate, invite audience interaction. Asking the audience, "What are your questions?" is a good way to start. When you are asked a question, listen carefully to the question and let your questioner finish speaking (even if you know what he or she is going to say). If you do not fully understand the question, ask for clarification before answering. Check with your questioner after giving your answer to make sure he or she is satisfied. Be nondefensive and honest, and do not be afraid to say, "I don't know."

Beware of presenting tables of numbers. Tables are used too often and generally contain too many numbers. If the numbers are so small they cannot be read by the audience, do not use a table. If the numbers are used to establish a general trend, use a graph instead. Avoid spurious precision, which means showing more accuracy in your results than is reasonable. For example, if your forecast for sales for the year 2025 is $1,234,567.89, you are showing too much precision. A more appropriate forecast might be $1.2 M.

Finally, do not rush through a graph without explaining it carefully. Do not assume your audience will understand a quantitative chart without explanation. If you want your audience to really understand a graph, you must lead them through it step-by-step. First, explain the elements: What is on each axis, and what data are displayed? For example, you might say, "On the horizontal axis, we are measuring time in years since the start of the project, and on the vertical axis, we are measuring revenues. The three lines show projected revenues under our three scenarios." Second, explain any coding involved (e.g.,

"The green bar refers to the western region," or "The dotted line represents new product sales."). Third, state the chart's message (e.g., "This chart shows sales trending down in the west."). Usually, the message will also be the chart's title, so by describing the message in words you are repeating and amplifying what the audience has already read. Fourth, draw out the implications of the chart (e.g., "Because sales are trending down in the west, we recommend adding sales staff in that region.").

Remember that although you have been thinking deeply about your model and analysis for the past few days or weeks, many in your audience may not have thought about the problem at all. Therefore, you must design your presentation carefully to bring your audience with you, and tell your story deliberately and earnestly.

In this section we have covered only the essential aspects of presentations. Several good books are available that discuss these topics in more detail. Munter (2006) and Zelazny (2000) present good overviews of managerial presentations. Tufte (1983) and Zelazny (2001) focus specifically on graphs, whereas Few (2004) discusses both graphs and tables in great detail. Finally, Koomey (2001) is an excellent guide to all aspects of quantitative analysis and presentation.

2.4 SUMMARY

This chapter covers the foundations of modeling for insight. It is brief because you can only learn so much about modeling by reading about it. To become an effective modeler, you must do it. We described a four-stage modeling process that we use to organize our approach to ill-structured problems. These four stages are as follows:

1. Frame the problem
2. Diagram the problem
3. Build a model
4. Generate insights

Six powerful modeling tools form the foundations of the modeling craft. These essential tools are as follows:

- Influence diagrams
- Spreadsheet engineering
- Parameterization
- Sensitivity analysis
- Strategy analysis
- Iterative modeling

Two tools are especially necessary for effective modeling: influence diagrams and spreadsheet engineering. Influence diagrams are used to structure a model at a conceptual level before undertaking the task of building a numerical model. Spreadsheet engineering is a set of tools and principles designed to improve the efficiency and effectiveness of modeling. Since most business models are built in spreadsheets, having good spreadsheet-engineering skills is a prerequisite if you wish to become an expert modeler.

Finally, we presented the basic principles of effective presentations, because the results of most modeling efforts are delivered in presentations. Designing effective presentations has become another essential craft skill you need in order to become an effective modeler. Since modeling results are inherently quantitative, you must pay particular attention to designing effective tables and graphs.

REFERENCES

Few S. *Show Me the Numbers*. Analytics Press, 2004.

Koomey J. *Turning Numbers into Knowledge*. Analytics Press, 2001.

Munter M. *Guide to Managerial Communication*. 7th edition. Prentice Hall, 2006.

Shannon C. "Creative Thinking." Mathematical Sciences Research Center, AT&T, 1993. (Cited in Luenberger D. *Information Science*. Princeton University Press, 2006, p. 18.)

Spreadsheet Engineering Research Project: http://mba.tuck.dartmouth.edu/spreadsheet/index.html

Tufte E. *The Visual Display of Quantitative Information*. Graphics Press, 1983.

Tufte E. "PowerPoint is Evil," *Wired* 11.9, September 2003.

Zelazny G. *Say It With Presentations*. McGraw-Hill, 2000.

Zelazny G. *Say It with Charts*. 4th edition. McGraw-Hill, 2001.

Willemain TR. Insights on Modeling from a Dozen Experts. *Operations Research* 1994;**42**(2):213–222.

Willemain TR. Model Formulation: What Experts Think about and When. *Operations Research* 1995;**43**(6):916–932.

CHAPTER 3

Spreadsheet Engineering

3.0 WHY USE SPREADSHEETS?

Hundreds of different software programs are available for modeling, most of them highly sophisticated and specialized. However, one type of software, the electronic spreadsheet, has come to dominate modeling in many areas, including business and education. Although the spreadsheet may not be the *ideal* platform for many problems, it is *good enough* for almost all end-user modelers almost all the time. Because it is good enough, it gets used every day, and everyday use increases skill and confidence. Thus, many end users avoid using more powerful or flexible software for a particular problem because of cost or time constraints. And, anyway, they know they can always tackle the problem in Excel!

Because spreadsheets are the dominant modeling software almost everywhere, we have chosen to concentrate on spreadsheet use in this book. That is not to imply, however, that the ideas here are not equally useful when modeling with other software. In fact, the craft skills of modeling are largely independent of the particular software used. For example, iterative modeling is at least as important when using advanced software as it is when using spreadsheets because the consequences of inefficient modeling worsen the more complex the software.

Throughout this book, we assume you are an experienced user of Excel. Therefore, we do not explain the basic Excel skills, such as entering text and data, using formulas and functions, or creating charts. You will find many books and training programs that can help you acquire the basic skills; several of these are listed at the end of this chapter. When we use an advanced feature of Excel, we describe it fully.

The discipline of spreadsheet engineering is designed to take the analyst from being a competent *user of* Excel to being an effective *modeler with* Excel.

3.1 SPREADSHEET ENGINEERING

Spreadsheet engineering is broken down into four phases:

- Design
- Build
- Test
- Analyze

Just as an airplane or a bridge is designed before it is built, a spreadsheet model should be designed before any data or formulas are entered. Once designed, a spreadsheet model should be built carefully, with an eye toward avoiding bugs. Despite a careful design-and-build approach, bugs are still likely, so you will need to test thoroughly every spreadsheet model before you use it. A spreadsheet model that is carefully designed, built, and tested can safely be used for analysis. The analysis process itself can be somewhat routine, although it remains the most creative phase in the process.

In the remainder of this section, we present an overview of the essential principles and skills of spreadsheet engineering. These skills are organized into easy-to-remember rules of thumb, such as "Keep It Simple." A more complete discussion of this subject is available in Chapters 5 and 6 of *Management Science: The Art of Modeling with Spreadsheets* (Powell and Baker, 2007).

3.1.1 Design

Most Excel users do not consciously design their spreadsheets but allow them to evolve somewhat chaotically. An hour spent designing a spreadsheet so that it achieves its purpose can save many hours in rework and debugging.

Sketch the Spreadsheet. On a piece of paper or a whiteboard, sketch the physical layout of the spreadsheet. It should have enough detail that someone else could build the spreadsheet from your sketch. Decide where data, decisions, outputs, and calculations will be located. Indicate how each major formula will be calculated using either cell references [e.g., IF(C6 < G18,C6,G18)] or pseudocode [e.g., IF(Demand < Supply, Demand, Supply)]. Revise the layout and the sequence of calculations until all formulas refer to cells above and to their left, if possible.

You will notice that sketching the spreadsheet is different from sketching an influence diagram. An influence diagram depicts the model structure at a high level while ignoring the details of equations and layout. A spreadsheet sketch is more like a blueprint, showing the exact details of how the model is to be constructed.

Organize the Spreadsheet into Modules. A well-structured spreadsheet is built up from small, well-structured components, or modules, that are logically, physically, and visually distinct. Modules group similar data or calculations together, making it easy to understand the model's overall flow. They also separate components that are related but logically distinct. Where possible, modules should reference other modules above and/or to the left. The idea of modularization also applies to the use of multiple worksheets in a workbook.

Start Small. Novices tend to build their ultimate model right from the start, with all the details they feel they will ever need. Experts build a simplified version of the model first and, after testing it, expand it gradually. Thus, the ultimate model emerges through a sequence of iterations.

Isolate Input Parameters. Make sure all of the numerical inputs to your model are collected in one module; no formula should contain a number. This practice helps to clarify the numerical assumptions behind the model. It also helps avoid the common error of changing a parameter in one formula but not in others that require the same parameter value. Finally, parameterization is essential to using the sensitivity analysis tools we describe later.

Design for Use. During the design phase, think about how the spreadsheet is likely to be used. Make it easy to change parameters and observe the effect on the outcome measure by locating these modules close to each other. If a graph is a more useful way to view the results, include it in the spreadsheet in a central location. If the model is large and complex, use different worksheets for inputs and outputs.

Keep It Simple. Many spreadsheets are unnecessarily complicated, in part because of a lack of discipline during the design phase. If you carefully design the simplest possible spreadsheet first and test it before adding complications, you are likely to end up with a far simpler end product than if you had built the ultimate model from the start.

Likewise, keep formulas simple. If necessary, break down a complex calculation into a number of simpler parts, calculate each part in a separate cell, and aggregate them in a final formula. Five formulas involving 30 characters each are far easier to debug than one formula involving 150 characters.

Design for Communication. Spreadsheets almost always live on much longer than their builders anticipate. Many companies today use spreadsheets that were built years ago by someone who has left the firm. Given the long life of a typical spreadsheet, it is important for you to design a spreadsheet that easily allows an outsider to understand it. Good modularization and documentation help. So do simple formulas. An overview worksheet, with text that describes the purpose and structure of the workbook, is also a good idea.

Document Important Data and Formulas. Documentation is the most overlooked phase of spreadsheet design. Full documentation is impractical for most models, but there are several ways you can easily add vital information to your model. For example, you can document important formulas using Cell Comments. Documenting the source of data or parameters is also helpful. That way, when you read an old spreadsheet, you quickly know whether values are estimates or verifiable data.

3.1.2 Build

Your primary focus when you build a spreadsheet should be ensuring that you do not introduce bugs. Some bugs are built into a spreadsheet design right from the start because of a conceptual error, but many more are created as the spreadsheet is actually built.

Follow a Plan. A well-designed spreadsheet is easy to build without introducing bugs. If the physical layout is already determined and the formulas flow from one to another in a logical manner, the process of building the spreadsheet will be mechanical—as it should be. Your attention should be entirely focused on accuracy: implementing the plan without introducing errors.

Build One Module at a Time. Rather than building the entire model, build a module and test it before moving on. This practice helps to localize errors and makes them easier to detect.

Predict the Outcome of Each Formula. After entering a formula but before pressing the "Enter" key, mentally predict the rough magnitude of the result you expect. If you expect a value of about $20,000 and the formula gives you a value of $200,000, you have probably uncovered a bug. This approach is an effective way to catch errors as soon as you have made them, before they are hidden by the complexities of the rest of the spreadsheet.

Copy and Paste Formulas Carefully. The "Copy" and "Paste" commands offer a great deal of power, since they make it possible to copy well-designed formulas to hundreds or thousands of cells. However, these commands are also the source of many of the bugs that infest spreadsheets. Always check one or two of the cells in the "Paste" range to see that the formula has been copied as you intended.

Use Relative and Absolute Addressing to Simplify Copying. Appropriate use of relative and absolute addresses is the key to efficient copying of formulas. But mistakes in addresses are another major source of bugs. Use the "F4" key to change relative to absolute addresses efficiently. Use the [Control] + [~] key combination to display all the formulas in the "Paste" range to check that the addresses have been copied as intended.

Use the Function Wizard to Ensure Correct Syntax. Advanced Excel modelers use built-in functions skillfully. However, many functions have complex syntax that is difficult to remember. Use the function wizard (the f_x icon to the left of the formula bar) to list all the available functions and display their inputs. Remember, the answer to the question "Is there a function to do this in Excel?" is almost always "Yes."

Use Range Names to Make Formulas Easier to Read. Range names allow you to attach a verbal descriptor or name to a cell or range of cells. Excel recognizes that name in formulas, allowing you to create formulas that are much easier for you and subsequent users of the model to understand. For example, if we name cell C6 *Demand* and cell G18 *Supply*, the formula

IF(C6<G18, C6, G18)

appears instead in the more transparent form

IF(Demand<Supply, Demand, Supply)

To create range names, select Formulas—Name Manager.

Use Dummy Input Data to Make Errors Stand Out. Novices usually begin a model by entering the most realistic data available. The advantage to this approach is that results are meaningful right from the start. The disadvantage is that the numbers are hard to debug. An alternative is to use dummy inputs, like 1, 10, and 100, in the first model. These inputs lead to easy-to-debug intermediate values. For example, if the price is expected to be $35.89 and the number of units sold is 5,789, then total revenue is $207,767.21. Unless you are especially good with numbers, this is a hard number to estimate. However, with dummy inputs, we can specify a price of $10 and units sold of 1,000 and easily predict total revenue of $10,000. After the model has been built and tested with dummy inputs, it is easy to replace the dummy inputs with the real ones.

3.1.3 Test

Many spreadsheets in use in corporations have bugs. This uncomfortable fact is suspected by all spreadsheet users, although the depth of the problem is not well understood. Most users do not test their spreadsheets nearly enough because they do not know how to do so efficiently. Here are some powerful methods:

Maintain a Skeptical Attitude. The most important quality you can have as a spreadsheet model builder is *skepticism*. A modeler with a skeptical attitude knows that her models may contain bugs even after careful testing. Test, test, and test some more, even when you are convinced there cannot be any more bugs. Find a friend or coworker to test your model. Never assume a model is perfect without testing.

Check that Numerical Results Look Plausible. There are several ways to determine whether your model results are at least plausible. One way is to make rough estimates before entering a function, as we mentioned above. Another way is to replicate an entire row or column of your model using a calculator. A third option is to use extreme inputs in your model to see whether the results make sense. For example, if we set the price to zero, do we get zero profits?

Check that Formulas Are Correct. Most spreadsheet users check their models by examining their formulas cell by cell, across a row, or down a column. Of all the ways to check formulas, this is probably the least efficient. It is also a method that can tire you out and cause you to miss your own errors because you become so familiar with the formulas. A better approach is to double-click on each formula, since this highlights the cells used to calculate the formula and can help catch reference errors. Alternatively, you can display all the formulas in the spreadsheet at once, using the key combination [Control] + [~]. This is the first thing we do when a colleague sends us a model to debug.

Excel has a built-in error-checking option (see Figure 3.1). You can locate this window by selecting the Microsoft Office button—Excel Options—Formulas. Excel error checking works like the grammar checker in Microsoft Word: It highlights a cell that contains a potential error by placing a green triangle in the upper left corner of the cell. Nine different error conditions, or

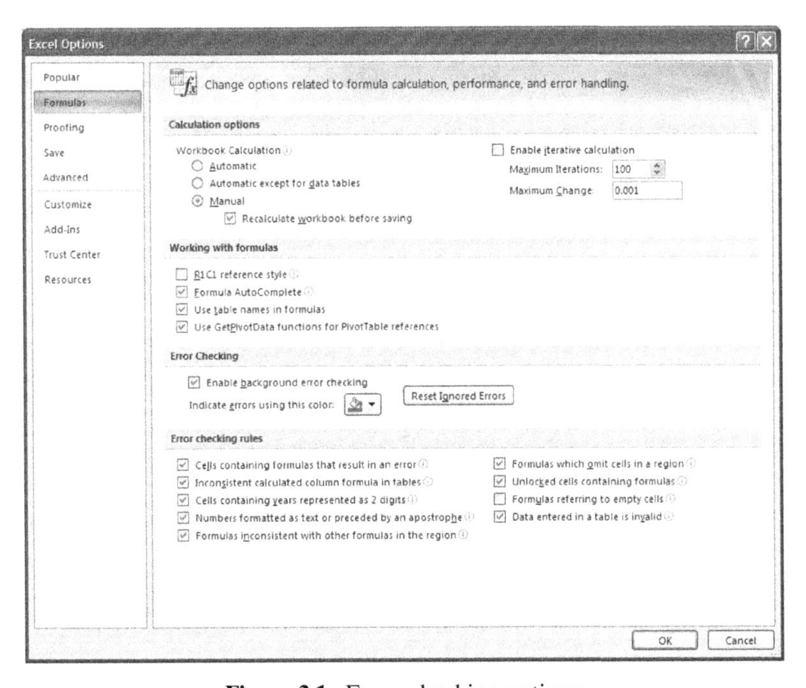

Figure 3.1. Error checking options.

"Rules," can be checked, as listed in the window. This option should always be turned on during the initial phases of model construction. (Once a model is fully debugged and passes into routine use, the error-checking option can be turned off.)

Excel also provides formula auditing tools (under Formulas—Formula Auditing), which among other things, allow you to trace the precedents and dependents of any cell. These tools are powerful and underused. With them, you can debug an individual formula by tracing its precedents. The auditing tools draw arrows from the precedent cells to the current cell. You can even display the entire logical structure of a spreadsheet, by tracing the precedents of the outcome cell and all their precedents back to the input parameters. Figure 3.2 shows what this looks like for a well-designed spreadsheet.

Test Model Performance. For many spreadsheet users, the testing and analysis phases run together. This result is not ideal, but it can be turned to an advantage. You should never trust the early results from your model. Maintain a skeptical attitude even as your analysis proceeds, since one of the most effective ways to uncover logical flaws in a model is to use it and look for surprising results. Sensitivity analysis, which we address in detail below, is a particularly powerful means of uncovering flaws.

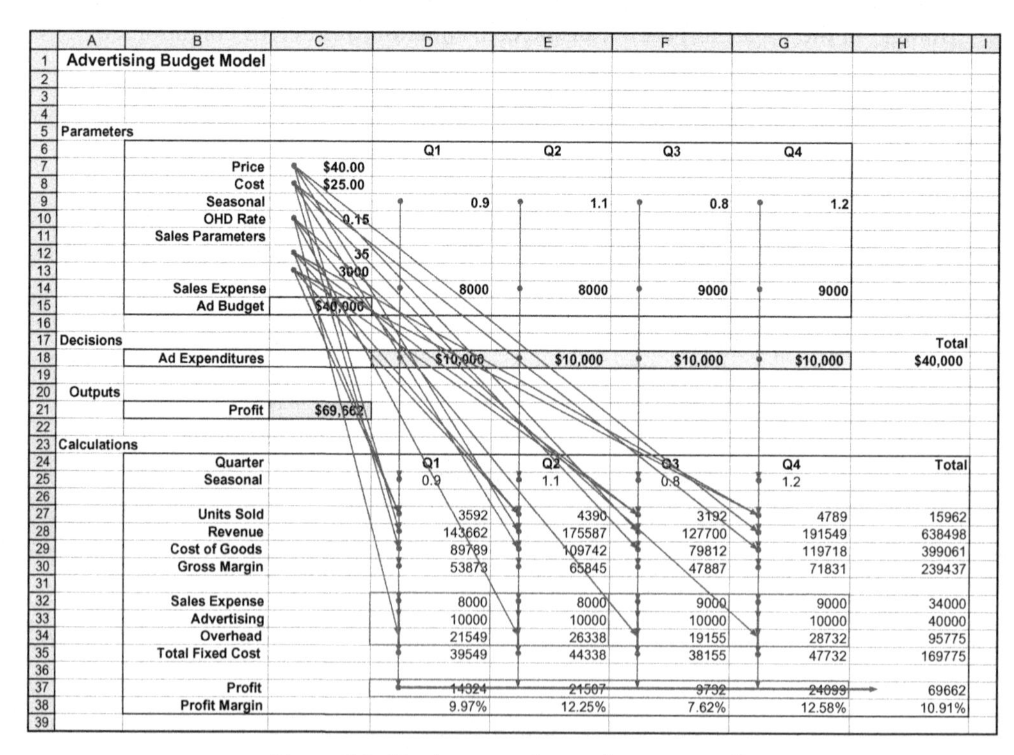

	A	B	C	D	E	F	G	H	I
1	Advertising Budget Model								
2									
3									
4									
5	Parameters								
6				Q1	Q2	Q3	Q4		
7		Price	$40.00						
8		Cost	$25.00						
9		Seasonal		0.9	1.1	0.8	1.2		
10		OHD Rate	0.15						
11		Sales Parameters							
12			35						
13			3000						
14		Sales Expense		8000	8000	9000	9000		
15		Ad Budget	$40,000						
16									
17	Decisions							Total	
18		Ad Expenditures		$10,000	$10,000	$10,000	$10,000	$40,000	
19									
20	Outputs								
21		Profit	$69,662						
22									
23	Calculations								
24		Quarter		Q1	Q2	Q3	Q4	Total	
25		Seasonal		0.9	1.1	0.8	1.2		
26									
27		Units Sold		3592	4390	3192	4789	15962	
28		Revenue		143662	175587	127700	191549	638498	
29		Cost of Goods		89789	109742	79812	119718	399061	
30		Gross Margin		53873	65845	47887	71831	239437	
31									
32		Sales Expense		8000	8000	9000	9000	34000	
33		Advertising		10000	10000	10000	10000	40000	
34		Overhead		21549	26338	19155	28732	95775	
35		Total Fixed Cost		39549	44338	38155	47732	169775	
36									
37		Profit		14324	21507	9732	24099	69662	
38		Profit Margin		9.97%	12.25%	7.62%	12.58%	10.91%	
39									

Figure 3.2. Tracing precedents of outcome cell.

3.1.4 Analyze

After you have carefully designed, built, and tested your model, it is ready to help you generate insights. Although there is no recipe for generating insights, there are several approaches to analysis that reliably provide insights in certain situations. Creativity is always a necessary ingredient in this process, but a complete set of analysis tools will automatically make your efforts more effective.

We have chosen to divide the analysis process into five stages: base case, what-if, breakeven, optimization, and simulation. These are not sequential stages, and rarely are they all used in any one modeling situation. Rather, they provide a useful framework within which to discuss specific tools and approaches.

Base Case Analysis. Every analysis starts with a base case. Although this is not a technical challenge, it does pay to consider what assumptions should go into the base case because the base case forms the starting point for all future analysis. Do you want the base case to portray an optimistic, pessimistic, or perhaps, expected outcome? The story you weave around your results could be very different depending on which alternative you choose.

What-if Analysis. Testing different model inputs to see what effect they have on the outputs takes you beyond answers to insights. What-if analysis can be as simple as testing different inputs one at a time and observing the outcome measures. But in a model with a large number of input parameters, this kind of manual what-if analysis will quickly become ineffective and tiring.

Because what-if analysis is performed on almost every model and tends to be time consuming, we have created a software toolkit to make the process easier. The Sensitivity Toolkit is a Visual Basic application that works with Excel. It is available at no charge at http://mba.tuck.dartmouth.edu/toolkit/. For more information on using the Toolkit, refer to Appendix C.

The Toolkit contains four tools:

- Data Sensitivity
- Tornado Chart
- Solver Sensitivity
- Crystal Ball Sensitivity

We describe the first two of these tools here and the last two in our discussions of optimization and simulation.

Data Sensitivity. Data Sensitivity recalculates the spreadsheet for a series of values of an input cell. It displays the results in a table and a graph on a new sheet in the workbook. In a one-way table, one input variable changes; in a two-way table, two inputs vary. (Data Sensitivity is an improved version of the built-in Excel function Data Table.)

Figure 3.3 shows a simple model of the profit of a call center. Profit is a function of the volume of calls and the associated revenue and costs. The decision variable is the Fraction of Calls Accepted. Figure 3.4 shows a sensitivity analysis of the Fraction of Calls Accepted in cell C14. We use Data Sensitivity to vary the fraction of calls from 0.0 to 1.0 in steps of 0.1. The table and chart show the resulting Profit for each of these 11 cases.

Data Sensitivity is a powerful tool for determining how sensitive the outcome is to an input over a given range. It can also be used to determine the form of the relationship between an input and the output. For example, we see in Figure 3.4 that profit first declines as we increase the fraction of calls accepted, and then increases.

Tornado Chart. In a model with a large number of inputs, it is impractical to construct a Data Sensitivity analysis for every input. Moreover, it is usually the case that many, perhaps most, input parameters have a minor impact on the outcome. Given this fact, it would be handy to have a way to tell which inputs really matter. After all, it hardly makes sense to waste time researching the value of an input that makes very little difference to the final result.

The Tornado Chart tool does just this. In its basic form, it varies each input (both up and down) by a set percentage around the base case value and recalculates the spreadsheet for both values. Then it determines the range of the output (the difference between the output value for the upside and the value for the downside). Finally, it ranks the inputs, from the biggest to the smallest ranges. This ranking creates the graph's "tornado" shape.

Figure 3.5 shows a tornado chart for the call-center model we introduced above. Five parameters are involved here, and the chart shows that two of them, Average Calls/Hour and Fixed Cost/Hour, have relatively little impact on profit.

Breakeven Analysis. What-if analysis can be used, as we have seen, to trace out the relationship between an input and the output measure. But in many cases, there is one value of the output that has special significance in the analysis. In this situation, you might want to know what value of the input would lead to that particular value of the output. Often, the special value of the output is zero. For example, you might ask what value of your product's price would cause the profit to be zero. This is commonly referred to as the *breakeven* level, although we use that term to refer to the process of finding the input to achieve any particular value of the output, not just zero—what market share will lead to profits of $1M, for instance.

Figure 3.6a shows Excel's Goal Seek tool input window (Data—Data Tools—What-If Analysis—Goal Seek). Goal Seek requires specification of the target cell (called "Set cell") and the target value (called "To value"), as well as the decision variable (called "By changing cell"). Goal Seek searches for a

	A	B	C	D
1	**Summers Group**			
2	**SGP**			
3				
4				
5	**Data**			
6		Avg Calls/Hour	500	
7		Revenue Per Call	$6	
8		Fixed Cost/Hour	$200	
9		Employee Cost/Hour	$42	
10				
11		Call Capacity (Calls/Hour)	9	
12				
13	**Decision**			
14		**Fraction of Calls Accepted**	**0.50**	
15				
16	**Ouput**			
17		**Profit (Per Hour)**	($88)	
18				
19				
20	**Calculations**			
21				
22	**Revenues**	# Calls Per Hour	250	
23		Revenue Per Hour	$1,500	
24				
25	**Costs**	Minimum # Employees	33.0	
26		Employee Variable Cost	$1,388	
27		Total Cost	$1,588	
28				

Figure 3.3. Call-center model.

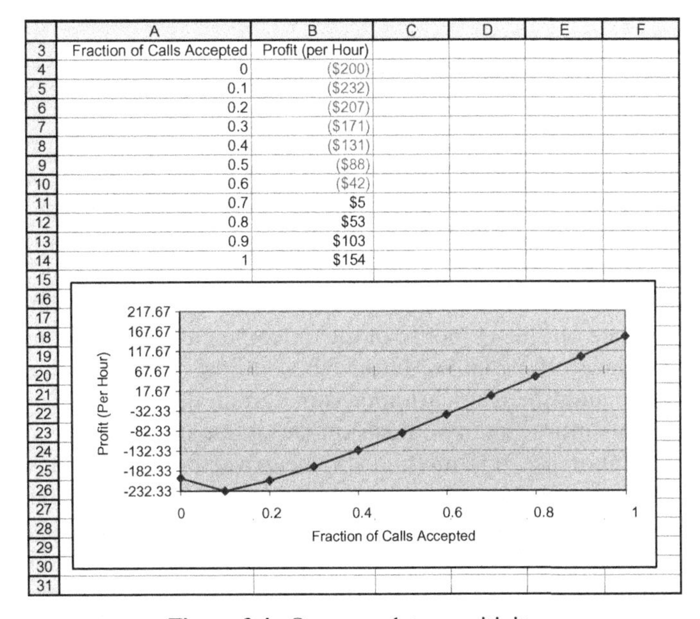

	A	B	C	D	E	F
3	Fraction of Calls Accepted	Profit (per Hour)				
4	0	($200)				
5	0.1	($232)				
6	0.2	($207)				
7	0.3	($171)				
8	0.4	($131)				
9	0.5	($88)				
10	0.6	($42)				
11	0.7	$5				
12	0.8	$53				
13	0.9	$103				
14	1	$154				

Figure 3.4. One-way data sensitivity.

value of the changing cell that drives the set cell to the target value. In this illustration, we are asking what value of Fraction of Calls Accepted (C14) will drive Profit (C17) to zero. The answer, as shown in Figure 3.6b, is 69 percent.

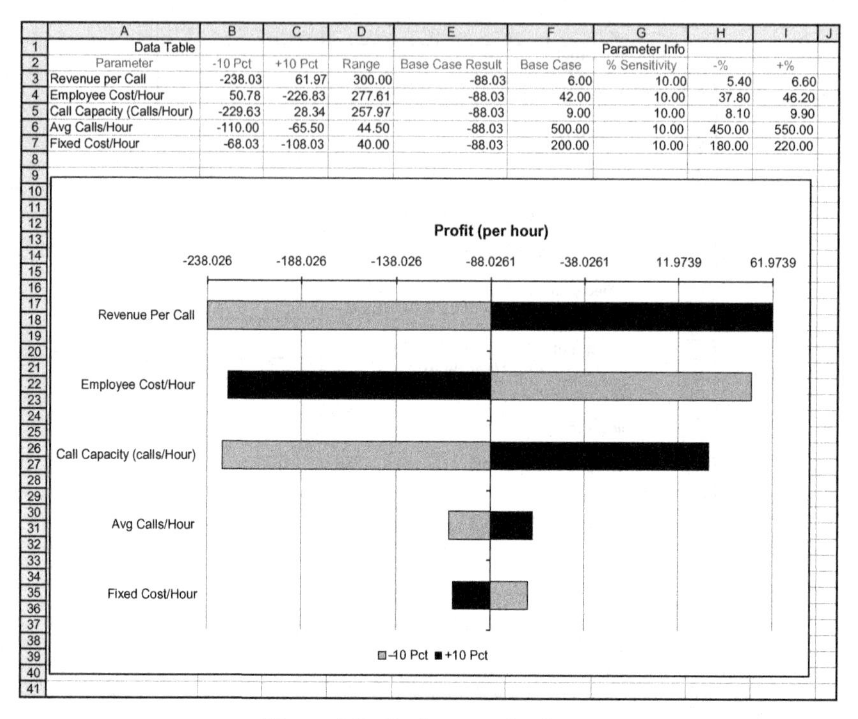

Data Table						Parameter Info		
Parameter	-10 Pct	+10 Pct	Range	Base Case Result	Base Case	% Sensitivity	-%	+%
Revenue per Call	-238.03	61.97	300.00	-88.03	6.00	10.00	5.40	6.60
Employee Cost/Hour	50.78	-226.83	277.61	-88.03	42.00	10.00	37.80	46.20
Call Capacity (Calls/Hour)	-229.63	28.34	257.97	-88.03	9.00	10.00	8.10	9.90
Avg Calls/Hour	-110.00	-65.50	44.50	-88.03	500.00	10.00	450.00	550.00
Fixed Cost/Hour	-68.03	-108.03	40.00	-88.03	200.00	10.00	180.00	220.00

Figure 3.5. Basic tornado chart.

Optimization Analysis. Optimization is the process of identifying the values of one or more decision variables that lead to the highest or lowest possible value of an outcome measure. For example, you might ask what levels of advertising each quarter will give you the highest possible profit at the end of the year.

If the problem involves only one or two decision variables, you can use Data Sensitivity to trace out the response of the outcome measure to changes in the values of the decision variables. Often, the optimal values of the decision variables can be identified, at least approximately, in this way. However, if the problem involves more than two decision variables, and possibly constraints as well, you cannot use this method. In the advertising problem shown in Figure 3.2, there are four decision variables (advertising in each quarter) and a constraint (that total advertising for the year not exceed the budget).

Excel includes an add-in called Solver that you can use for optimization. A complete treatment of Solver is beyond our scope here. See Appendix A for more information. Good references are *The Art of Modeling with Spreadsheets*, Chapters 10–13 (Powell and Baker, 2007), and *Optimization Modeling with Spreadsheets* (Baker, 2005).

Solver Sensitivity. Solver Sensitivity is a tool included in the Sensitivity Toolkit that automates sensitivity analysis for models using Solver. For example, you

(a)

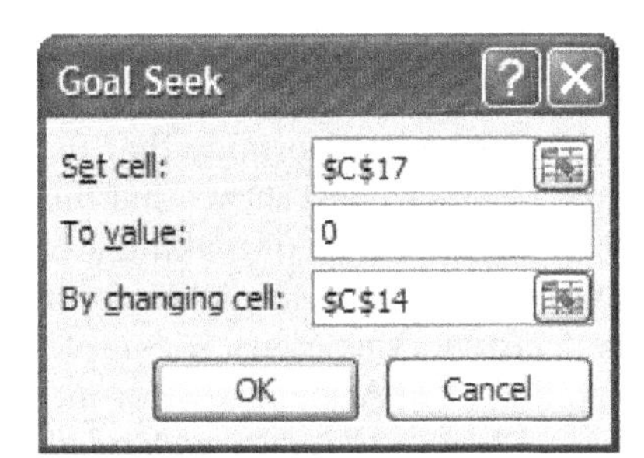

(b)

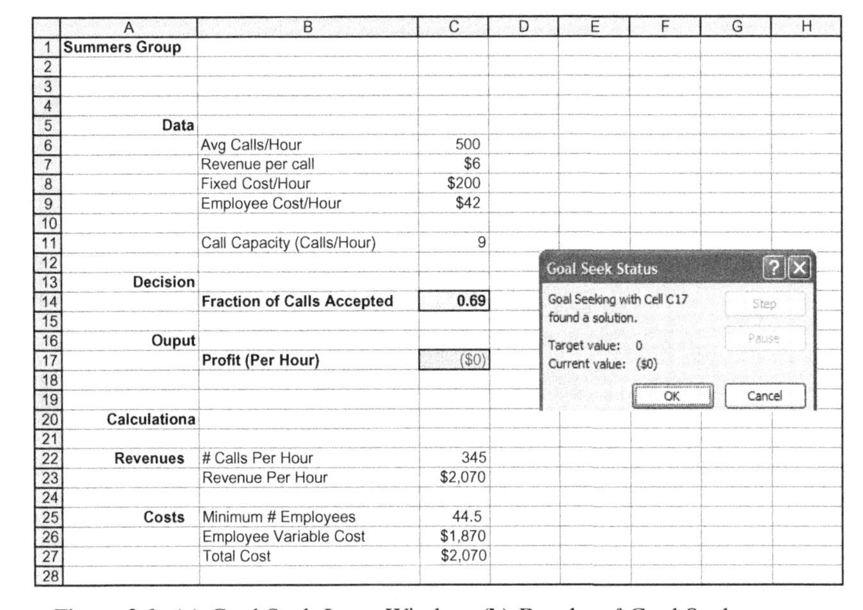

Figure 3.6. **(a)** Goal Seek Input Window. **(b)** Results of Goal Seek.

can first use Solver to find the advertising levels for each of four quarters that maximizes profit, subject to a budget constraint. Then you can use Solver Sensitivity to trace out the optimal solution to the problem as you vary the budget constraint, say, from $40K to $120K in steps of $10K.

Simulation Analysis. Most spreadsheet models are *deterministic* models, in that all the parameters and relationships are assumed to be known for sure. However, one of the major reasons to build any model is because the future is uncertain. Models that include the effects of uncertainty are called *stochastic* models.

The most effective tool for modeling uncertainty is Monte Carlo simulation, or simply "simulation" for short. In simulation analysis, fixed numerical inputs are replaced with probability distributions. The result is a probability distribution of the outcome measure. For example, in our sample model, if we assume the number of calls arriving follows a triangular distribution (Figure 3.7a), we can use simulation to determine that the Profit follows the distribution shown in Figure 3.7b. The chart shows that the mean profit is $34 per hour, with roughly an 80 percent chance of making a profit.

Several simulation add-ins are available for Excel. In this book, we use Crystal Ball (www.decisioneering.com). A complete treatment of Crystal Ball is beyond our scope here. See Appendix B for more information. Good references are *Management Science: The Art of Modeling with Spreadsheets*, Chapters 15 and 16 (Powell and Baker, 2007) , and *Introduction to Simulation and Risk Analysis*, Chapters 1–5 (Evans and Olson, 2002).

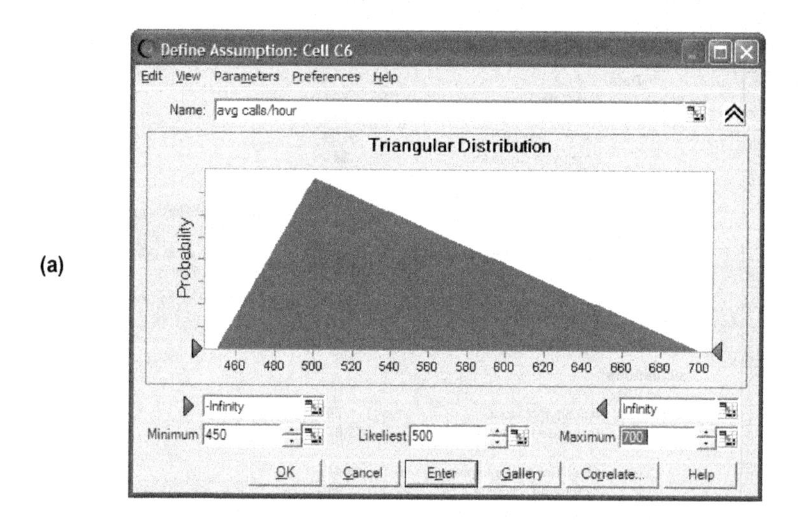

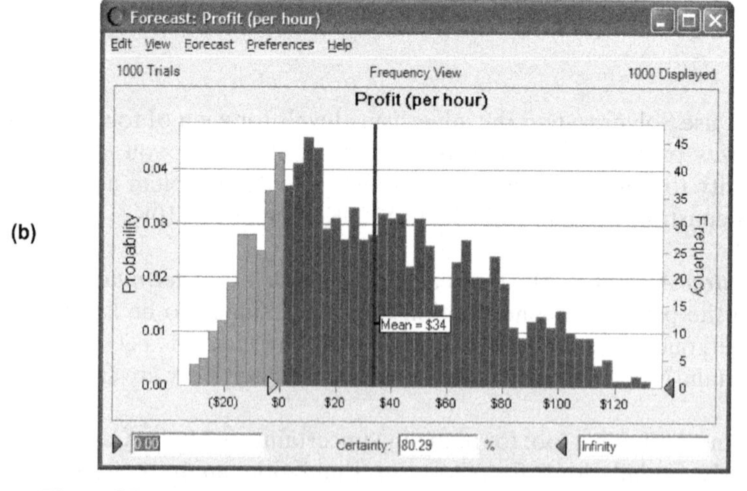

Figure 3.7. **(a)** Triangular distribution. **(b)** Distribution of Profit.

Crystal Ball Sensitivity. Crystal Ball Sensitivity is a tool included in the Sensitivity Toolkit that automates sensitivity analysis for models that use Crystal Ball. For example, Figure 3.7 shows the mean value for profit is $34/hour when the fixed cost is $200/hour. With Crystal Ball Sensitivity, you can vary the Fixed Cost/Hour, say, from $100 to $300 in steps of $25, and trace out the resulting values for mean profit per hour.

3.2 SUMMARY

Spreadsheets are the ubiquitous platform for business modeling. Careful engineering of spreadsheet models pays enormous dividends. It reduces rework and supports creative problem exploration.

Spreadsheet engineering consists of four phases:

- Design
- Build
- Test
- Analyze

The first step in creating a spreadsheet model, often overlooked, is to design the spreadsheet. Then it must be built carefully to avoid introducing bugs. The next step is to test the model, for which several powerful tools are available. Finally, the model is used for analysis. Several tools are helpful at this stage, including Goal Seek, Solver, Crystal Ball, and the Sensitivity Toolkit.

REFERENCES

Baker K. *Optimization Modeling with Spreadsheets*. Duxbury, 2005.

Evans J. and Olson D. *Introduction to Simulation and Risk Analysis*. Prentice-Hall, 2002.

Powell S. and Baker K. *Management Science: The Art of Modeling with Spreadsheets*. 2nd edition. Wiley, 2007.

Sensitivity Toolkit: http://mba.tuck.dartmouth.edu/toolkit/.

PART II

CHAPTER 4

A First Example—The Red Cross Problem

4.0 INTRODUCTION

In this chapter, we examine one ill-structured problem in detail, illustrating how we approach the four essential steps in modeling: framing the problem, diagramming the problem, building the model, and generating insights. This chapter gives you a general understanding of the approach we advocate in this book but does not teach you specific techniques or principles; those are the subjects of later chapters. Also, this narrative cannot fully capture the exploratory nature of modeling, in which we often backtrack, change our mind, uncover mistakes, or discover something new that alters earlier assumptions. The modeling process is never as tidy and straightforward as it appears in books. However, we try to convey in a realistic manner the actual process you should go through to develop a model.

4.1 THE RED CROSS PROBLEM

The Red Cross is considering a new policy under which it would pay potential blood donors for donations. You have been hired to build a model with which the Red Cross can explore the trade-offs involved in this decision.

To the Reader: We recommend that you stop reading at this point and take roughly 30 minutes to generate some initial thoughts about this problem before reading how we approach it. Get a pad of paper for taking notes and sketching pictures. Remember, the task is to begin to develop a model with which to explore trade-offs, not to "solve" the problem or get an answer. We do not expect you to develop a working model within this short time period, but you can begin to develop some ideas about how the model might be structured.

4.1.1 Frame the Problem

Many modelers, especially novices, jump right into the quantitative modeling phase without going through a problem-framing process. The danger of this approach is that the problem boundaries, options, and outcome measures remain ill defined as you build the model. With unclear problem boundaries, you are almost certain to bite off more than you can chew. With the options not carefully specified, you are unlikely to build a model with the flexibility you need to evaluate realistic alternatives. Finally, if the outcome measures are not specified, you cannot know what the model must calculate and, therefore, will not know when to stop building it.

An effective modeling process does not begin with data collection, writing down relationships between variables, or opening up a new spreadsheet. Before any of these activities comes a period of open-ended exploration of the problem. You need to understand the context and, usually, to simplify the problem as it has been presented. In short, you must identify the problem kernel. (Recall that the problem kernel is the minimum description of the problem you need to generate a model structure.) There are three important tasks at this stage: *setting problem boundaries, defining options*, and *choosing outcome measures*.

Ill-structured problems, by their nature, raise many more issues than can possibly be modeled effectively. The only way to deal with this complexity is to put some *boundaries* around the problem you are modeling. Setting boundaries to a problem not only simplifies it to some extent, it also helps to make the situation more concrete. Be aware of what is inside the boundary—and, thus, represented in the model—and what is outside and not considered. You must be able to take the model's limitations into account when translating results into recommendations for the real world.

In thinking about the Red Cross problem, you probably realized quickly how complex the real world problem is. The Red Cross is involved in lots of different activities, from blood donation to disaster relief to lifeguard training. Should we consider all aspects of the Red Cross's operations or narrow the scope somewhat? In the blood-donation market, several domestic and international organizations compete with the Red Cross. Should we consider their actions in our model? What about the economics of supply and demand? Should we worry about these effects? Lastly, the blood industry is highly regulated and the government may change the rules that affect the entire blood donation system. Do we need to model the actions of the regulatory authorities?

The real world has many more complexities than we could possibly include in our model. So we create boundaries to simplify our task. Here are the boundaries we choose in the Red Cross problem:

- We consider only the blood operations of the organization, not any of the organization's other activities, such as disaster relief.

- We ignore the actions of competitors in the market for blood donations.
- We consider U.S. operations only.
- We assume that the demand for blood is always greater than the supply.
- We assume no changes in the regulatory environment.

Another essential step to take early in a modeling effort is to define the *options* or *decisions* available to the decision maker. The primary option in this problem is to pay some amount—between $0 and a practical upper limit, perhaps $50—to each individual who offers to donate blood. But there are many ways to implement this option, and there are related options the Red Cross might pursue that could make this new policy more or less effective. For example, before they pay candidates, they could screen them (perhaps by asking them if they had engaged in risky behaviors). Or the Red Cross could pay them only after their blood was tested and found to be safe for use. Another way to limit the cost or risk of this policy would be to offer to pay only certain segments of the population or donors in certain geographical areas. They could also reduce the risks by using the paid-for blood differently than they use the blood donated for free, perhaps only using the plasma. By brainstorming in this way about the variety of ways to implement a policy, we increase the chances our analysis will uncover an effective variant of the proposed policy.

The final aspect of problem framing is to consider how we are going to measure the *outcomes* of any proposed plan of action. If our goal is to build a model with which the Red Cross can analyze trade-offs, we need to determine how to make these trade-offs concrete. An *outcome measure* allows us to quantify the result of a course of action. In this problem, we surmise that the Red Cross wants to know how the new policy will affect the operation's financial health, as well as the quality and quantity of blood it will be able to supply. We can measure the financial outcome in terms of the financial surplus (or deficit) the blood operation runs. Finally, the Red Cross certainly wants to know whether the new policy will have an impact on the quality of the blood it supplies. The quality of blood is probably measured in a complex manner, reflecting the probability of contamination by a range of agents. For a first-pass model, we choose to simplify this into a single measure: either a pint of blood is usable or it is not.

Notice that every time we make an explicit decision about problem boundaries, options, or outcome measures, we eliminate something from our analysis and get closer to identifying the problem kernel. The art of modeling involves eliminating the inessential while retaining the essential. Even an expert modeler cannot know for sure which is which, especially at the beginning of the process. So an effective modeler makes these decisions with open eyes, realizing that she may eliminate something crucial at this stage that she will want to bring back later. We recommend keeping a list of boundaries, options, and outcome measures and reviewing it periodically during the modeling process. Often, a decision made early in the process of problem framing begins to look questionable as your understanding of the problem evolves.

Here is the problem kernel we have identified for the Red Cross problem:

> The Red Cross is considering offering a fee (perhaps $10) to each individual who passes an initial screening and gives blood. This fee will be offered in all locations and for all donors, although some donors may choose to decline payment. How will this policy affect the financial aspects of the blood donation operation and the quality of blood the Red Cross delivers to hospitals?

To the Reader: Before reading on, sketch an influence diagram linking the fee to the financial surplus of the Red Cross.

4.1.2 Diagram the Problem

Although problem framing is an important first step in developing a model and generating insights, do not expect it to do more than set the stage for model building. Spreadsheet models are concrete and precise, consisting of numbers and relationships. However, our discussion to this point has been qualitative and somewhat vague. How do we get from where we are now to a working model?

We use influence diagrams to help us develop an initial problem structure. These diagrams provide a bridge between the qualitative ideas developed in the problem-framing stage and the specific details of an actual model in the model-building stage. An influence diagram is not intended to provide a detailed specification for the model. In fact, any number of models could be developed from a single influence diagram. What it does provide is a coherent view of the model's important features, including inputs and outcome measures, along with the variables and relationships that connect them.

We begin with a diagram for calculating the financial surplus (or deficit, if negative) for the Red Cross blood operation. Subsequently, we develop a diagram for the quality of the blood supply, the other major outcome measure we settled on earlier. We begin the diagram by creating an octagon with a description of the outcome measure (Surplus) and placing it on the far right side of the page. Then we consider how to calculate Surplus. More specifically, what are the highest level components of this measure? Surplus is the difference between Total Revenue and Total Cost. In the diagram, this relationship is expressed by creating the variables Total Revenue and Total Cost and connecting them with arrows to Surplus (see Figure 4.1a). The arrows signify that the variable at the tail of the arrow is needed in calculating the variable at the head.

We continue to work backward, from right to left, decomposing each of the variables in the diagram into its immediate components. For example, Total Revenue is the Price we get from selling each unit of blood, multiplied by the

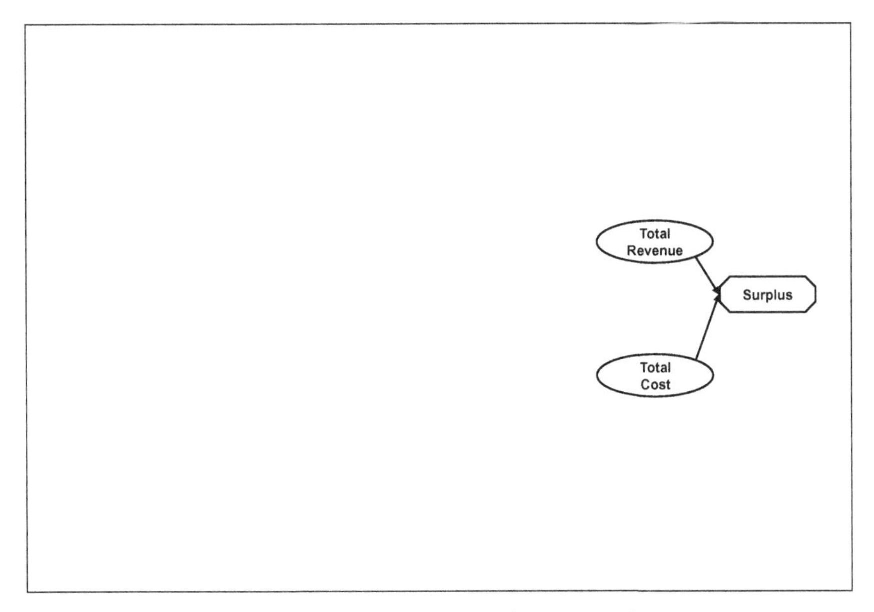

Figure 4.1. (a) Initial influence diagram for Surplus.

total units of blood. The total units of blood is the sum of the Normal Accepted units and the Paid Accepted units (see Figure 4.1b). Normal Accepted is the volume of good blood coming from unpaid donors, and Paid Accepted is the volume of good blood coming from donors who are paid.

Total Cost comes from two sources: the Normal Costs associated with Total Donors (such as salaries, testing, and handling expenses) and the additional Fee paid to each Paid Donor (see Figure 4.1c). Total Donors is the sum of Normal Donors and Paid Donors, and Normal Cost is the current cost per donor. Note that the Fee is depicted as a parallelogram, since it is a decision variable.

Since not all donated blood will pass the quality acceptance tests, Normal Accepted is some fraction of Normal Donors controlled by the Normal Acceptance Rate. Similarly, Paid Accepted is based on Paid Donors and the Paid Acceptance Rate (see Figure 4.1d). We are allowing for the possibility here that the acceptance rates for Normal and Paid Donors may differ.

We assume that there is one large pool of Potential Donors and that Normal Donors and Paid Donors both come from this pool. Of the Potential Donors, some percent (Normal Donation Rate) will become Normal Donors. Similarly, the number of Paid Donors is a function of the same Potential Donor pool and the Paid Donation Rate (see Figure 4.1e). We are assuming here that offering a fee does not change the volume of normal donations. Note how the diagramming process forces us to make assumptions and helps us to recognize them.

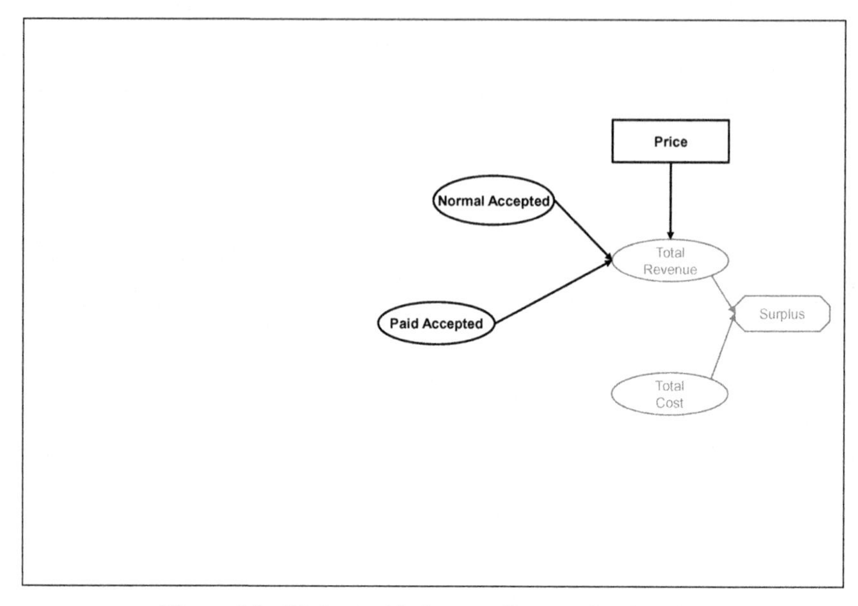

Figure 4.1. (b) Second influence diagram for Surplus.

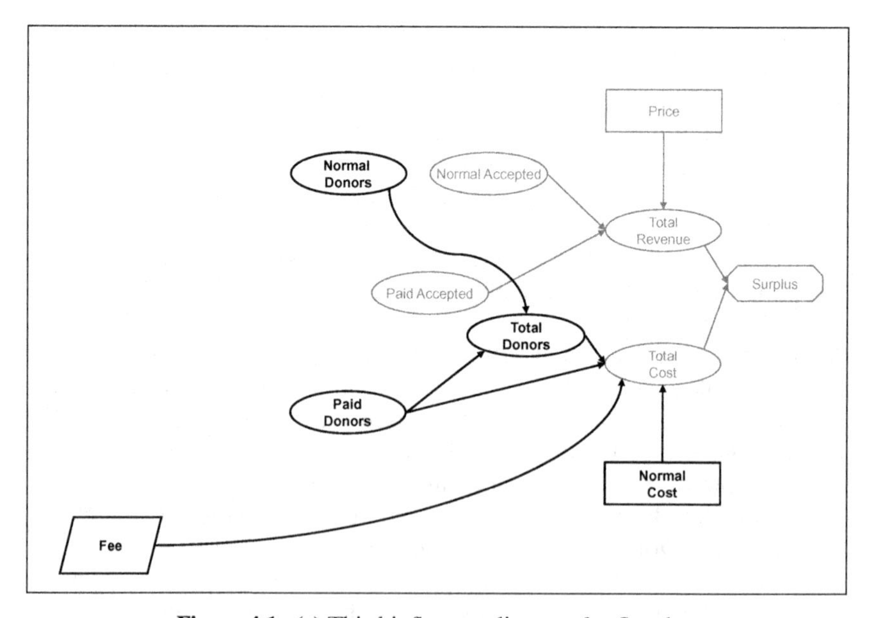

Figure 4.1. (c) Third influence diagram for Surplus.

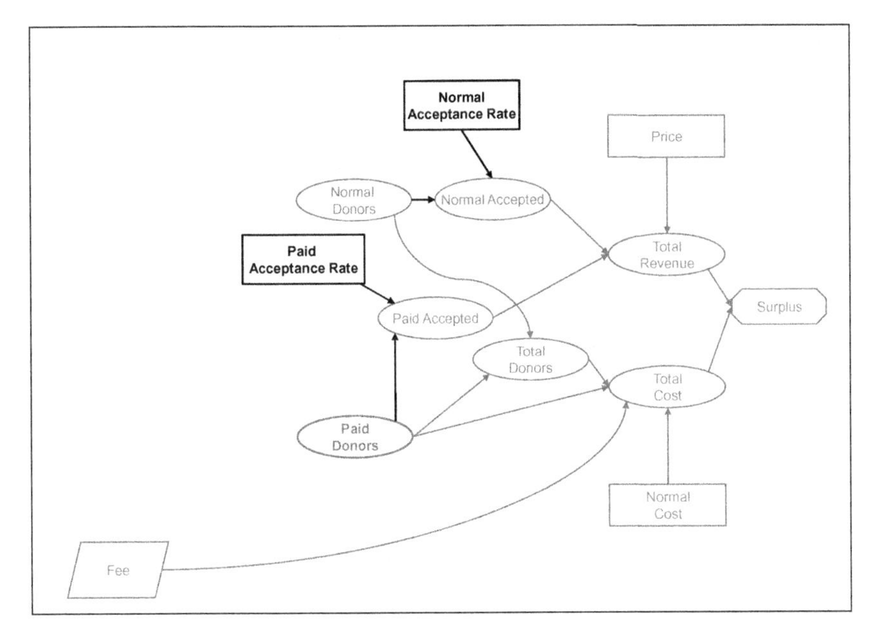

Figure 4.1. (d) Fourth influence diagram for Surplus.

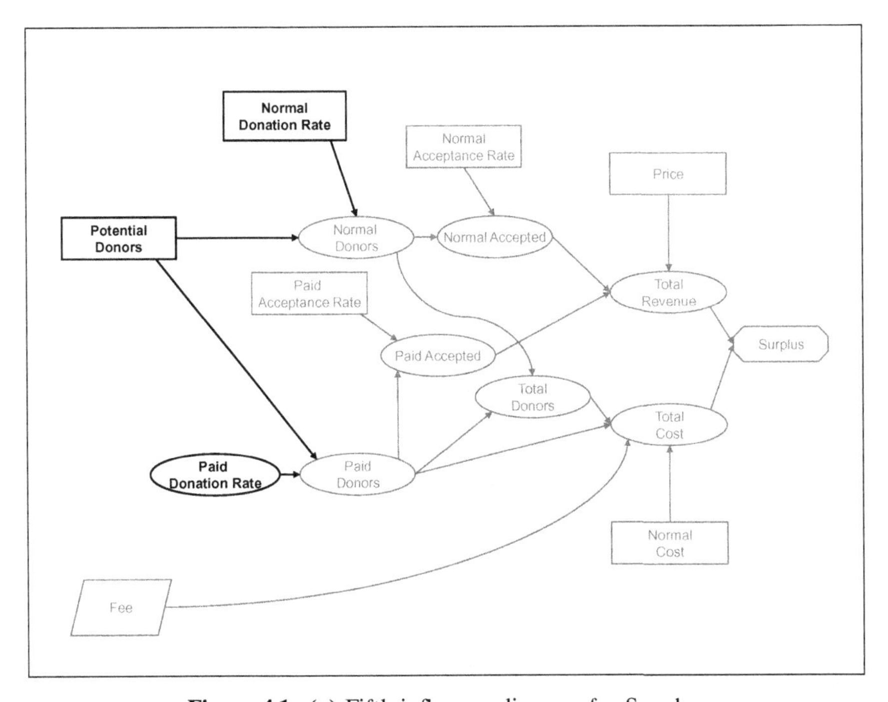

Figure 4.1. (e) Fifth influence diagram for Surplus.

Finally, we determine the Paid Donation Rate by relating it to the Fee and the Elasticity of Supply (see Figure 4.1f). We deliberately leave this last relationship ambiguous for now. We know the rate of donations may increase as the fee increases, depending on how sensitive the donor population is to fees. However, we do not need to formalize this relationship at this point. The completed influence diagram is shown in Figure 4.1g.

We do not usually include the mathematical or algebraic relationships in the influence diagram, although it is useful to consider how the calculations will actually be made. Often, you can identify a missing variable in this way. Also, we are presenting this diagram in its final version, not as it was developed the first time. In our first pass, for example, we did not distinguish Normal Accepted from Paid Accepted, or Paid Accepted from Paid Donors. The necessity of distinguishing Paid Accepted from Paid Donors only occurred to us when we realized that the fee is paid to all donors who pass initial screening, regardless of whether their blood is accepted. (Again, we made this assumption during the problem-framing stage, and we may change it later on.)

So, what have we accomplished here? We have diagramed the conceptual structure of a model for calculating the Surplus for any Fee we might choose. We have shown how all our inputs are connected to our output measure. At the highest level of abstraction, the diagram in Figure 4.1g represents a model that transforms any choice the Red Cross may make for the donation fee (including $0) into the corresponding value for the outcome measure, surplus.

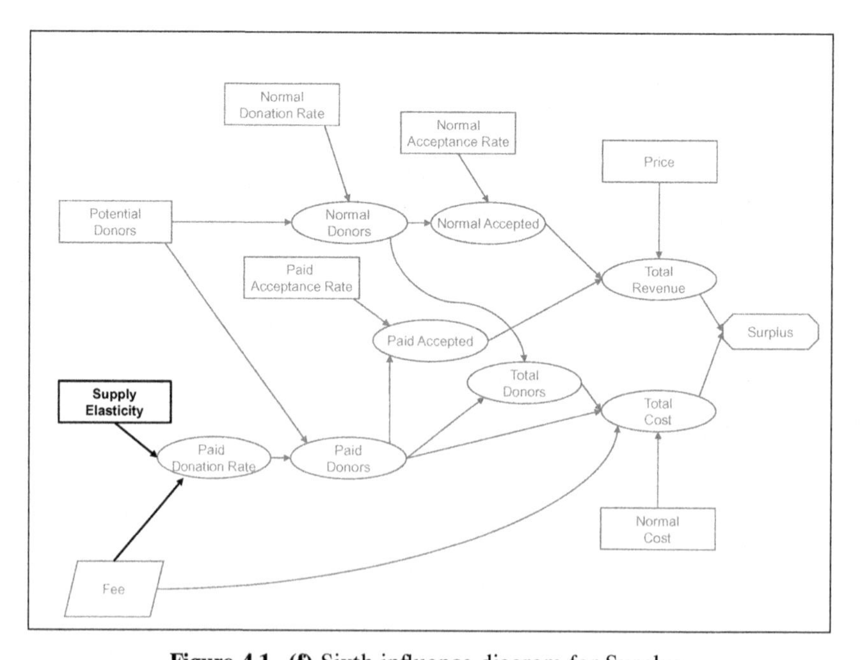

Figure 4.1. **(f)** Sixth influence diagram for Surplus.

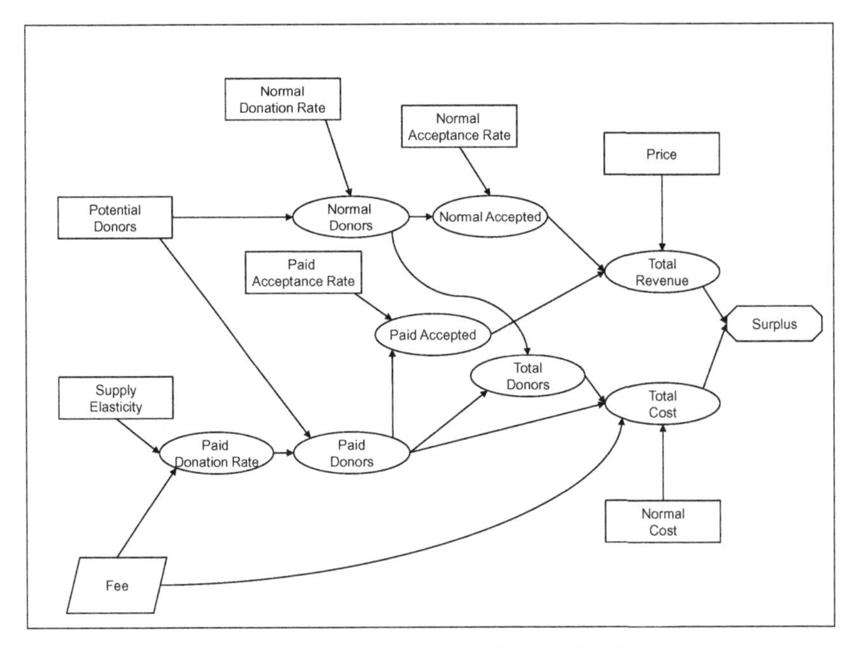

Figure 4.1. (g) Final influence diagram for Surplus.

Another accomplishment is that we have documented many of the assumptions we made along the way. All our work so far has been done without using any numbers or spending any time gathering data. Furthermore, we have deliberately kept the discussion at a high level of abstraction to avoid being distracted by irrelevant details.

A few final comments on influence diagrams are in order. The problem structure we have represented with an influence diagram can also be represented using algebra. For example, Normal Accepted = Normal Donors × Normal Acceptance Rate. (See Figure 4.2 for a complete list of the algebraic expressions in this model.)

However, the graphical approach of the influence diagram is far more flexible and intuitive than the algebraic method during the problem-formulation stage. Additionally, the influence diagram is a great tool for communicating with clients or colleagues and for helping them understand your thought process. The equations do have the advantage of providing a more compact and precise representation, and some modelers develop the algebraic version of the model after creating an influence diagram and before building a spreadsheet. However, the graphical approach is necessary for most people during the creative phases of problem formulation, whereas the algebraic approach is optional.

1. Surplus = Total Revenue – Total Cost
2. Total Revenue = Price × (Normal Accepted + Paid Accepted)
3. Total Cost = (Normal Cost × Total Donors) + (Fee × Paid Donors)
4. Total Donors = Normal Donors + Paid Donors
5. Normal Accepted = Normal Donors × Normal Acceptance Rate
6. Paid Accepted = Paid Donors × Paid Acceptance Rate
7. Normal Donors = Potential Donors × Normal Donation Rate
8. Paid Donors = Potential Donors × Paid Donation Rate
9. Paid Donation Rate = Function of Fee & Supply Elasticity

Figure 4.2. Algebraic version of model for Surplus.

4.1.3 Build the Model—Surplus

The power of an influence diagram lies in the fact that it provides a convenient pallet for sketching model structures. However, an influence diagram is not a fully articulated model. It does not contain numbers and may include only a rough idea of the precise relationships between variables. How do we take an influence diagram and use it to help us create a first model?

The goal of any model is to generate a numerical estimate of the outcome measure (or measures) in a problem. Thus, any model that produces a value for Surplus is useful. Completing a first model is a major milestone in the process of modeling for insight. Before the first model emerges, we may have lists of assumptions, ideas, data, and issues, but no answers or insights. After the first model, we have an answer for the problem, regardless of how inadequate that answer may be. And thereafter, as we elaborate on the first model, we always have an answer to give, which is whatever the latest model shows. As we demonstrate in detail in later chapters, the modeling approach we advocate involves developing a series of models, each more comprehensive than the last. Along with this series of models comes a series of quantitative answers or insights, each more reliable and insightful than the last. Iterative modeling is the process of developing this series of models in an intelligent manner.

To the Reader: Before reading on, construct a spreadsheet model for the Red Cross Surplus, based on the influence diagram in Figure 4.1g.

We continue our analysis of the Red Cross problem by showing how we translate the influence diagram of Figure 4.1g into a model.

Our first step in the translation process is to list the parameters and decision variables in a module in the spreadsheet and to assign initial values to them. The following six parameters and one decision variable can be identified directly in Figure 4.1g:

Donation Fee
Price
Potential Donors
Normal Donation Rate
Normal Acceptance Rate
Paid Acceptance Rate
Normal Cost

Since the donation fee is a decision variable, and thus under the control of the Red Cross, we can assign it any value we like for the purpose of building the first model. We will want to vary it as we come to analyzing the implications of the model, so the value we assign it initially is inconsequential. But how do we assign values to the six parameters?

If we know reasonable values for the parameters, we can certainly use them, but if not, we use plausible *placeholder values* that serve to help us complete a first model. Eventually, we can go back and replace these placeholder values with more accurate ones, but only if we determine they make a significant difference to the outcomes. Initial values for the seven inputs are given below:

Donation Fee	\$10
Price	\$50
Potential Donors	1,000,000
Normal Donation Rate	5%
Normal Acceptance Rate	95%
Paid Acceptance Rate	90%
Normal Cost	\$10

These inputs are listed in the spreadsheet in Figure 4.3 along with the corresponding placeholder values.

The same spreadsheet is shown in relationship view in Figure 4.4.

We know that the ultimate goal of this model is to calculate Surplus, so we enter a heading for that variable in row 25. Since Surplus will be determined by the Costs and Revenues generated by Normal Donations and Paid Donations (if any), we reserve column D for calculating Revenues and Costs from Normal Donations and column E for the corresponding calculations for Paid Donations.

	A	B	C	D	E	F
1	**Red Cross M1**					
2						
3						
4						
5		**Inputs**				
6			Fee	$10		
7			Baseline Fee	$20		
8			Baseline Fee Donation Rate	5%		
9			Supply Elasticity	10%		
10			Price	$50		
11			Potential Donors	1,000,000		
12			Normal Donation Rate	5%		
13			Normal Acceptance Rate	95%		
14			Paid Acceptance Rate	90%		
15			Normal Cost	$10		
16						
17		**Surplus**		**Normal**	**Paid**	
18			Donors	50,000	0	
19			Accepted	47,500	0	
20			Revenue	$2,375,000	$0	
21			Cost	$500,000	$0	
22			Surplus	$1,875,000	$0	
23						
24						
25			Total Surplus	**$1,875,000**		
26						

Figure 4.3. Surplus model—numerical view.

Using the influence diagram as a guide, we write down the sequence of intermediate variables that must be calculated to determine Surplus. These are the number of donors, the number of donations accepted, the revenue generated by accepted donations, and the costs of all donations. These variables suggest the row headings in our model:

Donors

Accepted

Revenue

Cost

Surplus

Total Surplus will be the sum of the Surplus generated by Normal and Paid Donors.

Again referring to the influence diagram, we enter formulas for the various rows. We start with Normal Donors (cell D18), which is simply:

$$\text{Potential Donors} \times \text{Normal Donation Rate} = \text{D11} * \text{D12}$$

But when we come to calculating the number of donations from Paid Donors, we run into a problem: The influence diagram is vague about how this is calculated, although it does suggest an elasticity of supply.

	A	B	C	D	E	F
1	**Red Cross M2**					
2						
3						
4						
5		**Inputs**				
6			Fee	10		
7			Baseline Fee	20		
8			Baseline Fee Donation Rate	0.05		
9			Supply Elasticity	0.1		
10			Price	50		
11			Potential Donors	1000000		
12			Normal Donation Rate	0.05		
13			Normal Acceptance Rate	0.95		
14			Paid Acceptance Rate	0.9		
15			Normal Cost	10		
16						
17		**Surplus**		**Normal**	**Paid**	
18			Donors	=Potential_Donors*Normal_donation_rate	=Potential_Donors*((((Fee-Baseline_fee)/Baseline_fee)*Supply_elasticity)+Baseline_fee_donation_rate)	
19			Accepted	=D18*Normal_rejection_rate	=E18*Paid_rejection_rate	
20			Revenue	=D19*Price	=E19*Price	
21			Cost	=D18*Normal_cost	=E18*(Normal_cost+Fee)	
22			Surplus	=D20-D21	=E20-E21	
23						
24						
25			Total Surplus	**=D22+E22**		
26						

Figure 4.4. Surplus model—relationship view.

Here we need to draw on a repertoire of useful mathematical functions for creating a plausible and flexible model for the effect of the fee on donations. The essential feature is that donations must increase as the fee increases, at least up to some point. Another important feature of such a model is that it should be easy to determine plausible values for its parameters, perhaps by asking the decision makers. The form we use here assumes that at some base fee—say, $20—paid donations will be a percentage of potential donors, say 5 percent. Then, for every 100 percent increase in the fee, the percent of donors will increase by a constant amount, say 2 percentage points. One advantage of this model is that we can ask simple questions of the decision makers and get plausible values for the three parameters we need. Before we do this, however, we complete our first model using placeholder values, which are entered into the spreadsheet in cells D7–D9. The appropriate formula for the number of Paid Donors is entered in cell E18:

Potential Donors∗((((Fee-Baseline Fee)/Baseline Fee)∗Supply Elasticity)+ Baseline Fee Donation Rate)

Now that we have linked the Fee to Paid Donations, we can complete the model fairly quickly. The amount of blood accepted is the number of donors in each category, multiplied by the corresponding acceptance rate. Revenues are the number of units accepted multiplied by the Price. Costs include the Normal Cost as well as the Fee on blood from Paid Donors. Finally, Surplus is the difference between Revenues and Costs, and Total Surplus is the sum of the Surplus from each category of donor.

We now have a completed first model! If we set the Fee at $10, our Surplus is $1,875,000. We are certainly not ready to make any final recommendations to the Red Cross based on this model, since all the numerical inputs are placeholders. But we have completed a logical structure that takes into account at least some of the essential features of the problem and gives us an unambiguous numerical estimate of one of the Red Cross's key outcome measures. More importantly, we have constructed a framework within which we can explore this problem and generate insights. We can now ask what happens to Surplus when we change the Fee and how sensitive are the results to other inputs. These questions are at the core of this problem, and, with only a few hours' work, we have built a structure that provides the foundation for developing well-reasoned answers. Problem framing, influence diagrams, and model building are powerful tools for quickly finding preliminary insights.

To the Reader: Before reading on, use the model to generate as many insights as you can about the situation facing the Red Cross.

4.1.4 Generate Insights

In the previous section, we devoted all our attention to completing a first model. We achieved this goal when we calculated a value for our outcome measure based on placeholder input values. We are not done with modeling, though; in fact, we have just begun. But where do we go from here?

One of the themes of this book is that effective modeling of ill-structured problems does not seek to find "solutions," since solutions to these problems do not exist. Rather, modeling seeks to develop *insights* that decision makers can use. It is hard to provide a recipe for generating insights. However, our analysis of the Red Cross problem should provide you with a sense of what insights look like.

One of the keys to effective modeling is learning how to *explore* a model, even a very rough first model, to see what it can tell you about the underlying problem. You should do this before modifying your first model, if only because without such exploration you have no reliable way to know what needs to be improved. You should have a list of the assumptions that went into the influence diagram, along with other options you might choose or alternative assumptions you might make. And, although you know that many of your numerical inputs are only placeholders, what you do not know yet is which assumptions are critical to your results and which placeholder values need to be replaced with more accurate values. Sensitivity analysis, which we demonstrate here and fully explore later in the book, is the key to answering these questions and, thus, the starting point for generating insights.

Now return to the first model for Surplus, Figure 4.3. The question that drives all our efforts is as follows: Should the Red Cross pay for blood donations? This model allows us to develop a tentative answer to part of that question: How does changing the Fee for donations affect the Surplus? To see what our current model suggests here, we vary the Fee from $1 to $50 and trace out the resulting values for Surplus (using Data Sensitivity). The results are shown in Figure 4.5.

We see that starting from $0, the Surplus increases as we increase the Fee, but around $23, it reaches a peak and thereafter declines. Now, we should not take these results too literally, since we are working with a very rough first

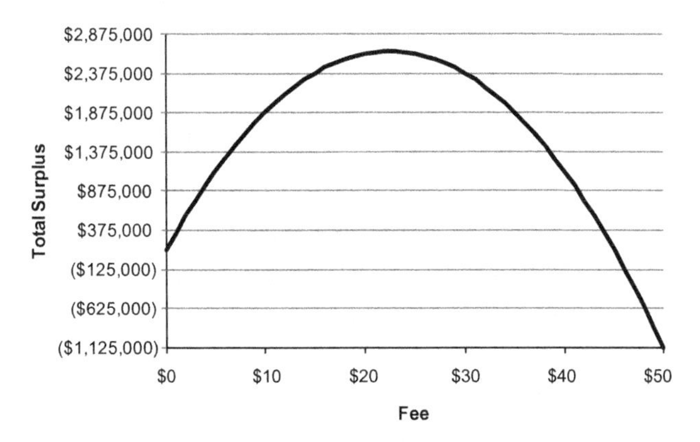

Figure 4.5. Sensitivity of Surplus to Fee (Surplus model).

model. Nevertheless, the figure does suggest that charging a fee maximizes the surplus.

Insight 1: Surplus is low for both low and high values of the fee. In between is a value of the fee that maximizes the surplus.

Another set of results comes from a tornado chart, in which we vary each input parameter by some percentage of its base-case value and compare the impact on Surplus. Figure 4.6 shows the results when each parameter varies by ±25 percent of its base value.

The chart shows the effect on Surplus in terms of the range from high to low values; the parameters are displayed from top to bottom in order of their impact. Again, although we should remember not to draw final conclusions from this chart, it does suggest some interesting preliminary insights. For example, we notice that Price has a far greater impact on Surplus than does Fee.

Insight 2: Price has a greater impact on surplus than does the fee.

Why is this? Essentially, the chart shows *absolute* changes in the Surplus that come about because of *relative* changes in the inputs. Since Price is many times larger than Fee in the base case ($50 compared with $10), a 25 percent change in Price represents a much larger absolute change than the same percentage change in Fee. So far, the result seems to be merely a consequence of

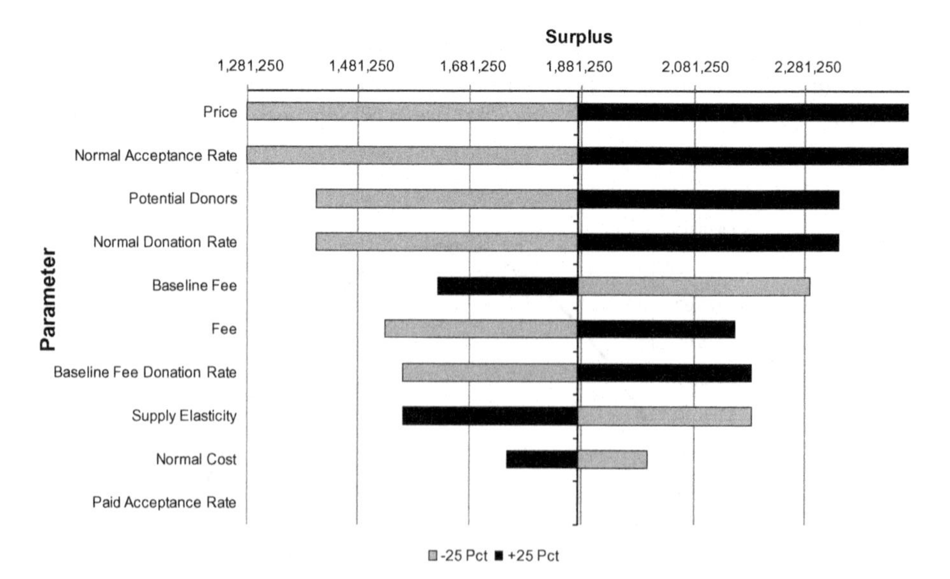

Figure 4.6. Tornado chart for Surplus.

the scale of the inputs. But it is helpful to consider here whether there is a way this result could have turned out differently. That is, under what circumstances would the Fee have a bigger impact than Price? One way would be if the Fee had a large multiplier effect on donations.

Another preliminary insight that we draw from Figure 4.6 is the notion that most of the inputs that have a significant effect on Surplus are not within the direct control of the Red Cross.

Insight 3: Most inputs that have a significant effect on surplus are beyond the control of the Red Cross.

Our focus has been on evaluating the impact of a donation fee, but Figure 4.6 suggests that Surplus is most strongly influenced by Price, Normal Acceptance Rate, Potential Donors, Normal Donation Rate, and the Baseline Fee. In fact, varying the Fee itself in a small range around $10 has relatively little effect on Surplus.

Insight 4: Variations in the fee have relatively little effect on surplus.

These results, if they hold up under additional study, should be very valuable to the decision makers, who might either lose interest in the idea of a donation fee, since it does not seem to offer a way to increase surplus materially, or gain interest, since it could be implemented without threatening the existing surplus.

4.2 BRINGING BLOOD QUALITY INTO THE ANALYSIS

During the problem-framing phase of this analysis, we realized the Red Cross would probably measure the results of the new policy in several ways. They would certainly want to know whether paying for blood donations would create an unexpected financial drain. We began to address this concern with our first model. But they would also want to know what the impact of the new policy would be on the *quality* of the blood they deliver to users. To analyze this outcome measure, we must go through the modeling process again, from influence diagram to insights.

To the Reader: Before reading on, sketch an influence diagram linking the fee to the quality of the blood supplied by the Red Cross.

4.2.1 Influence Diagram for Quality

The quality of the blood that the Red Cross supplies to the U.S. hospital system depends on whether they institute a fee and on how accurate their tests are for blood quality. It may be that paid donors' blood is much more likely to be contaminated than unpaid donors' blood, but blood tests are highly reliable, and therefore, the tests can catch almost all the new contaminated blood introduced into the system. That raises an interesting question: How accurate do the blood tests have to be to keep the overall blood quality as high as it is at present? We explore this question later as we generate insights.

Our first step toward answering this question is to draw an influence diagram for Blood Quality. Rather than review the steps in detail, we refer you to Figure 4.7 and describe some of the key relationships involved, working backward from right to left.

When we consider Blood Quality, we define "positive" to mean safe, clean blood and "negative" to mean contaminated blood. We have to keep in mind that blood tests are not perfect, and thus, some negative blood will test positive and vice versa. To measure Blood Quality, we are only concerned with blood that tests positive since any blood that tests negative will be immediately discarded. (We show this as Normal Rejects and Paid Rejects even though these values have no bearing on the outcome of Blood Quality.) We then define Blood Quality as the percent of positive-testing blood that actually is positive. (A high Blood Quality number is preferred.)

Note that some input parameters have been omitted in Figure 4.7 for visual clarity. The missing parameters are values such as Percent of Normal Actual Positive that Tests Positive. These parameters reflect the test's accuracy in detecting actual positive or negative blood. Refer to Figure 4.8 for a detailed

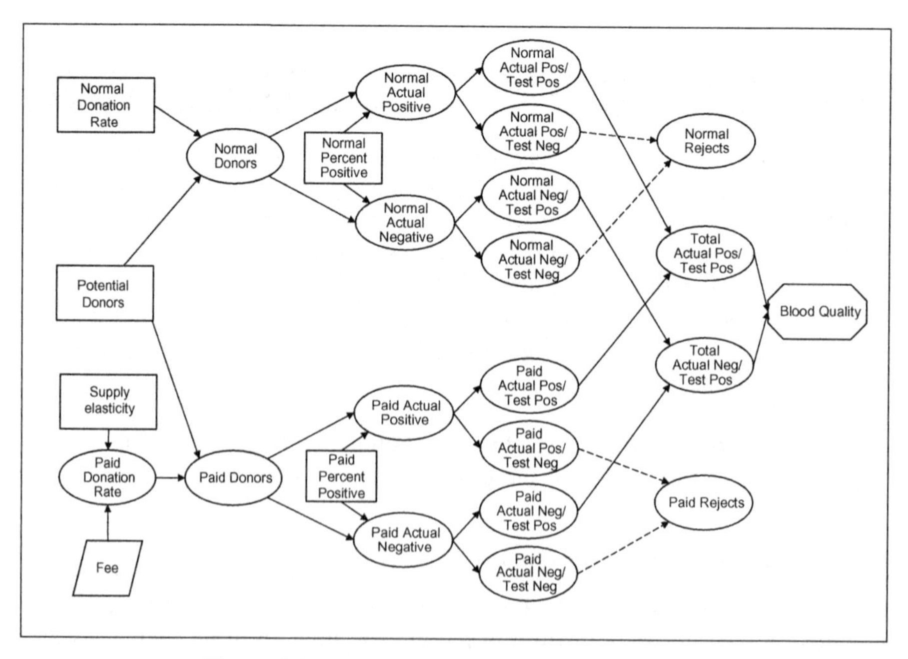

Figure 4.7. Influence diagram for Blood Quality.

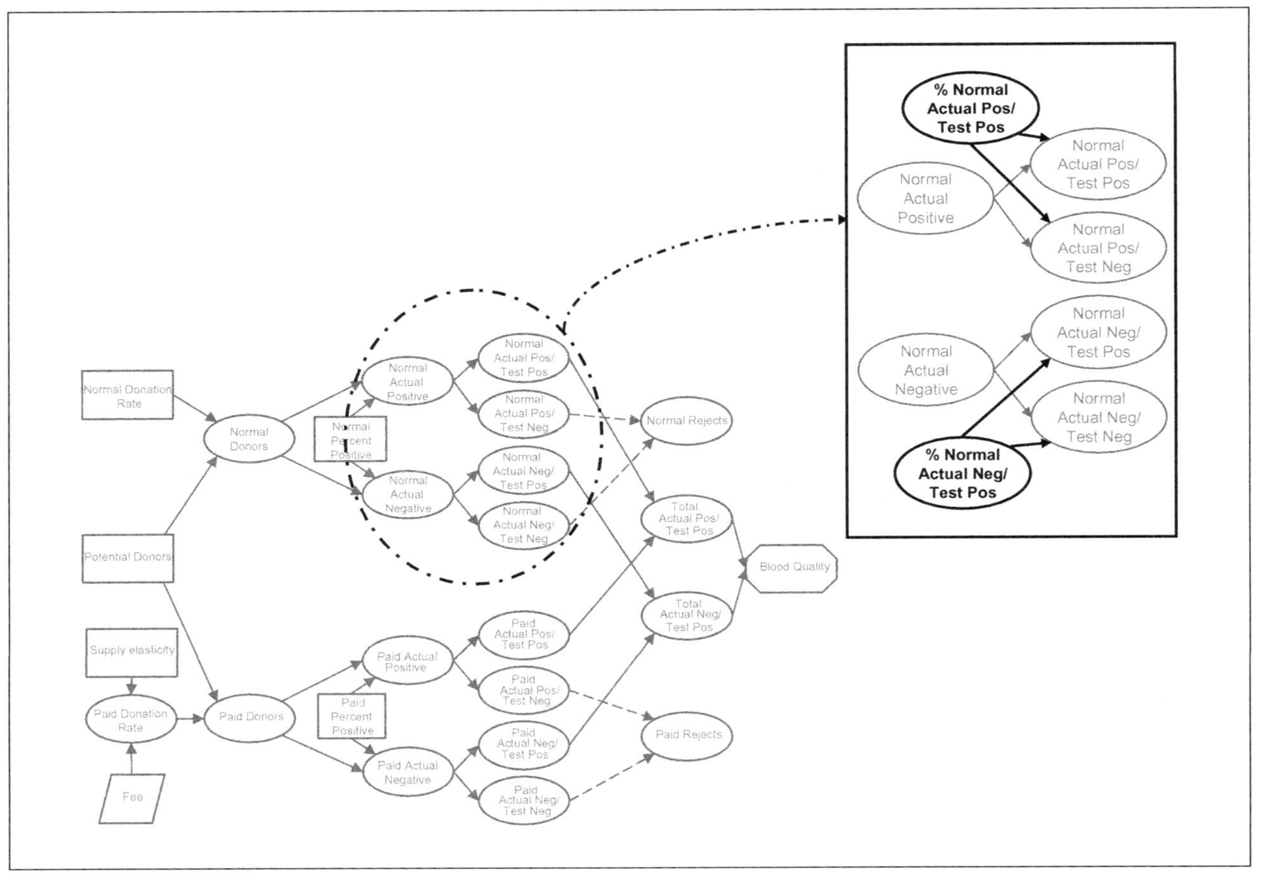

Figure 4.8. Detail of influence diagram for Blood Quality.

view of the influence diagram with some of these parameters shown. We allow here for different test accuracy rates for Normal and Paid Donors.

The Normal Percent Positive variable represents the actual percentage of Normal Donors who have safe blood. Likewise, Paid Percent Positive represents the actual percentage of Paid Donors with safe blood. By considering these variables separately, we allow for the possibility that Paid Donors have a different (lower?) actual blood safety rate.

Some portions of this influence diagram are similar to the influence diagram for Surplus (Figure 4.1g); in particular, the process whereby Normal and Paid Donors are generated is identical. However, when our outcome measure was Surplus, we transformed the number of donors into dollars, whereas when our measure is Blood Quality, we transform the same information into the percentage of units that are actually safe.

We can also express the relationships in algebraic terms, as we did for the Surplus model. Figure 4.9 lists the equations that connect the fee paid for

1. Blood Quality = Total Actual Pos Test Pos / (Total Actual Pos Test Pos + Total Actual Neg Test Pos)
2. Total Actual Neg Test Pos = Normal Actual Neg Test Pos + Paid Actual Neg Test Pos
3. Total Actual Pos Test Pos = Normal Actual Pos Test Pos + Paid Actual Pos Test Pos
4. Normal Actual Pos Test Pos = Normal Actual Positive × % Normal Actual Pos Test Pos
5. Normal Actual Neg Test Pos = Normal Actual Negative × % Normal Actual Neg Test Pos
6. Paid Actual Pos Test Pos = Paid Actual Positive × % Paid Actual Pos Test Pos
7. Paid Actual Neg Test Pos = Paid Actual Negative × % Paid Actual Neg Test Pos
8. Normal Actual Positive = Normal Donors × Normal Percent Positive
9. Normal Actual Negative = Normal Donors × (1 – Normal Percent Positive)
10. Paid Actual Positive = Paid Donors × Paid Percent Positive
11. Paid Actual Negative = Paid Donors × (1 – Paid Percent Positive)
12. Normal Donors = Potential Donors × Normal Donation Rate
13. Paid Donors = Potential Donors × Paid Donation Rate
14. Paid Donation Rate = Function of the Fee & Supply Elasticity

Figure 4.9. Algebraic version of model for Blood Quality.

donated blood to the ultimate quality of the blood supplied. Since these interrelationships are complex, the algebraic version of the model provides a useful check on the graphical view.

As we did earlier, we show finished influence diagrams here. However, we made many mistakes before settling on these versions. The process of drawing an influence diagram for an ill-structured problem is never a simple linear process. Rather, it involves mistakes, discoveries, and backtracking. In fact, we went through at least three complete diagrams before we sorted out the logic of Blood Quality to our satisfaction. Along the way, we made several discoveries about the appropriate logic and the best way to represent the relations involved. For example, in our first attempt, we overlooked the fact that testing blood only makes sense if you discard blood that seems to be contaminated. Likewise, we had to discover by trial and error that Blood Quality should be measured in terms of actual quality, not in terms of tested quality. Like any model, an influence diagram is never really done. You must decide when it is good enough, which means when most of the learning has occurred and you feel ready to tackle a spreadsheet model.

To the Reader: Before reading on, build a model for blood quality using the influence diagram in Figure 4.7.

4.2.2 Build the Model—Blood Quality

As in our earlier model, the first step is to list the parameters and the decision variable and to select realistic placeholder values. The decision variable and the first five parameters can be taken directly from the Surplus model.

Donation Fee	\$10
Baseline Fee	\$20
Baseline Fee Donation Rate	5%
Supply Elasticity	10%
Potential Donors	1,000,000
Normal Donation rate	5%

Now the model begins to diverge from the earlier one. The next parameters govern the percent of each type of blood donation that is actually positive or negative. (Positive blood is acceptable, and negative blood is unusable.)

Normal Percent Positive	95%
Paid Percent Positive	90%

Finally, we need eight parameters (not shown explicitly in the influence diagram) to represent the success rate of the blood tests, both on Normal blood and on Paid blood. Note that only four of these parameters are independent; we list all eight for clarity.

Percent Normal Actual Positive / Test Positive	98%
Percent Normal Actual Positive / Test Negative	2%
Percent Normal Actual Negative / Test Positive	5%
Percent Normal Actual Negative / Test Negative	95%
Percent Paid Actual Positive / Test Positive	93%
Percent Paid Actual Positive / Test Negative	7%
Percent Paid Actual Negative / Test Positive	15%
Percent Paid Actual Negative / Test Negative	85%

The calculations in the model can now be entered (see Figures 4.10 and 4.11). The number of donors (cells D32 and E32) are calculated just as in

	A	B	C	D	E	F	G	H	I
1	Red Cross M2								
2									
3									
4									
5		Inputs							
6			Fee	$10					
7			Baseline Fee	$20					
8			Baseline Fee Donation Rate	5%					
9			Supply Elasticity	2%					
10			Potential Donors	1,000,000					
11			Normal Donation Rate	5%					
12									
13			Normal Percent Positive	5%	"positive" indicates blood is good				
14			Paid Percent Positive	10%	"negative" indicates blood is contaminated				
15									
16			Normal						
17			Actual Positive-Test Positive Percent	98%					
18			Actual Positive-Test Negative Percent	2%					
19			Actual Negative-Test Positive Percent	5%					
20			Actual Negative-Test Negative Percent	95%					
21									
22			Paid						
23			Actual Positive-Test Positive Percent	93%					
24			Actual Positive-Test Negative Percent	7%					
25			Actual Negative-Test Positive Percent	15%					
26			Actual Negative-Test Negative Percent	85%					
27									
28									
29									
30									
31		Quality		**Normal**	**Paid**				
32			Donors	50,000	40,000				
33			Actual Positive	47500	36000				
34			Actual Negative	2500	4000				
35									
36			Normal						
37			Actual Positive-Test Positive	46550					
38			Actual Positive-Test Negative	950					
39			Actual Negative-Test Positive	125					
40			Actual Negative-Test Negative	2375					
41									
42			Paid						
43			Actual Positive-Test Positive		33480				
44			Actual Positive-Test Negative		5400				
45			Actual Negative-Test Positive		600				
46			Actual Negative-Test Negative		3400				
47									
48			Total Actual Positive	80030					
49			Total Actual Negative	725					
50									
51			Quality	**99.10%**					
52									

Figure 4.10. Blood quality model—numerical view.

	A	B	C	D	E	F
1	**Red Cross M2**					
2						
3						
4						
5		**Inputs**				
6			Fee	10		
7			Baseline Fee	20		
8			Baseline Fee Donation Rate	0.05		
9			Supply Elasticity	0.02		
10			Potential Donors	1000000		
11			Normal Donation Rate	0.05		
12						
13			Normal Percent Positive	0.05	"positive" indicates blood is good	
14			Paid Percent Positive	0.1	"negative" indicates blood is contaminated	
15						
16			Normal			
17			Actual Positive-Test Positive Percent	0.98		
18			Actual Positive-Test Negative Percent	=1-D17		
19			Actual Negative-Test Positive Percent	=1-D20		
20			Actual Negative-Test Negative Percent	0.95		
21						
22			Paid			
23			Actual Positive-Test Positive Percent	0.93		
24			Actual Positive-Test Negative Percent	=1-D23		
25			Actual Negative-Test Positive Percent	=1-D26		
26			Actual Negative-Test Negative Percent	0.85		
27						
28						
29						
30						
31		**Quality**		**Normal**	**Paid**	
32			Donors	=Potential_Donors*Normal_donation_rate	=Potential_Donors*((((Fee-Baseline_fee)/Baseline_fee)*Supply_elasticity)+Baseline_fee_donation_rate)	
33			Actual Positive	=D32*(1-Normal_Percent_Positive)	=E32*(1-Paid_Percent_Positive)	
34			Actual Negative	=D32*Normal_Percent_Positive	=E32*Paid_Percent_Positive	
35						
36			Normal			
37			Actual Positive-Test Positive	=D33*Normal_Actual_Positive_Test_Positive_Percent		
38			Actual Positive-Test Negative	=D33*Normal_Actual_Positive_Test_Negative_Percent		
39			Actual Negative-Test Positive	=D34*Normal_Actual_Negative_Test_Positive_Percent		
40			Actual Negative-Test Negative	=D34*Normal_Actual_Negative_Test_Negative_Percent		
41						
42			Paid			
43			Actual Positive-Test Positive		=E33*Paid_Actual_Positive_Test_Positive_Percent	
44			Actual Positive-Test Negative		=E33*Paid_Actual_Negative_Test_Positive_Percent	
45			Actual Negative-Test Positive		=E34*D25	
46			Actual Negative-Test Negative		=E34*Paid_Actual_Negative_Test_Negative_Percent	
47						
48			Total Actual Positive	=D37+E43		
49			Total Actual Negative	=D39+E45		
50						
51			Quality	**=D48/(D48+D49)**		
52						

Figure 4.11. Blood quality model—relationship view.

the Surplus model. Then the number of Actual Positive and Negative are calculated in D33:E34 using the appropriate parameters in D13:D14. Next, we distribute the number of Positives and Negatives into the four possible combinations of Actual and Test results, in D34:D40 for Normal Donors and D43: D46 for Paid Donors.

The influence diagram shows us how to use these eight categories to calculate Blood Quality. First, we assume that all blood that tests negative is thrown away, which leaves only blood that tests positive, some of which is actually negative. The Total Actual Positive that Tests Positive is calculated in D48 by adding the corresponding supplies from Normal and Paid Donors. Likewise, the Total Actual Negative that Tests Positive comes from both Normal and Paid Donors. Finally, the Blood Quality (percent positive) is the ratio of Total Actual Positive that Tests Positive to the sum of Total Actual Positive that Tests Positive and Total Actual Negative that Tests Positive.

Our first model for Blood Quality is complete: If we set the Donation Fee to $10, the resulting blood supply will be 99.10 percent pure. This is not a conclusion we want to show the Red Cross yet, since the parameters are only placeholders, but the accomplishment to this point is substantial nevertheless. With just a few hours work, we have built a logical structure that should stand up through subsequent iteration and elaboration. The relationships between the Donation Fee (and other inputs) and the resulting quality of the blood supply were not at all obvious when we first read the problem description. It took several attempts at creating an influence diagram to capture the logic correctly, and then it took some time to create a concrete numerical representation of this diagram in the form of a spreadsheet.

To the Reader: Before reading on, use the Blood Quality model to generate as many insights as you can about the situation facing the Red Cross.

4.2.3 Generate Insights

Our interest in Blood Quality centers around the question of what impact a Donation Fee will have on this important outcome measure. Figure 4.12 suggests what the answer might look like.

In this figure, we have varied the Fee from $0 to $50 and plotted the Blood Quality (using Data Sensitivity). The model shows that with a fee of $0, the Blood Quality measures 99.20 percent, which presumably reflects the current level of Blood Quality. (As we develop our model further, we will try to match this baseline level of Quality with the actual level of Quality in the system today.) As we increase the Fee, the Quality declines as we bring in more and more Paid Donors, with their higher level of negative blood and less effective screening. The figure shows that Quality declines to 99.02 percent for a Fee of $20. This result might appear to an outsider as a very small decline (0.18 percentage points), but we do not know whether this level of Quality is accept-

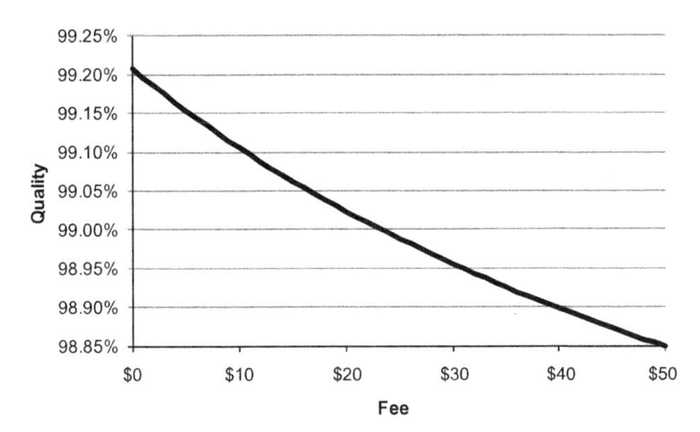

Figure 4.12. Sensitivity of Blood Quality to Fee.

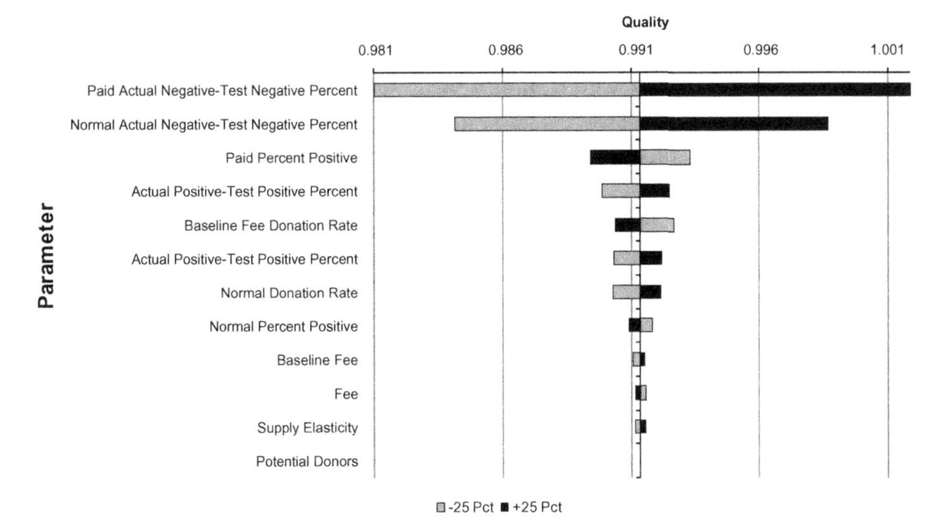

Figure 4.13. Tornado chart for Blood Quality.

able—only the decision makers know that. Our contribution is to develop a model that shows them the consequences of their actions. Although we cannot take the numerical values in Figure 4.12 as final, the figure itself does suggest an interesting preliminary insight that we can explore as we refine the model.

Insight 5: Doubling the fee has only a small effect on quality.

We complete the sensitivity analysis of the Quality model by developing a tornado chart (see Figure 4.13). This figure suggests another preliminary insight: The two most influential parameters in this case are Percent Normal

Actual Negative that Tests Negative and Percent Paid Actual Negative that Tests Negative. That is to say, the Quality of the blood supply is highly sensitive to the ability of our blood tests to identify correctly blood that is negative.

Insight 6: Quality of the delivered blood supply is highly sensitive to the accuracy of the tests.

So we see, not surprisingly, that the attractiveness (perhaps even the feasibility) of a donation fee plan may rest on our ability to detect negative blood accurately.

Insight 7: The viability of charging a fee may rest on our ability to identify contaminated blood correctly.

But recall that all the inputs to this model are placeholders, including these test percentages. Figure 4.13 gives us sound motivation for doing careful research into the actual percentages involved. Ultimately, it may also provide motivation for the Red Cross to invest in improved blood tests if it wants to implement the fee policy. This is an example of how a tornado chart can help define our real needs for data.

4.2.4 Wrap-up

We have now built two simple models and developed seven preliminary insights. We have observed, for example, that the Surplus-maximizing Fee might turn out to be low and that the Surplus itself is not terribly sensitive to the level of the Fee. We have also seen that the Blood Quality declines with the Fee but is so high currently that perhaps some decline in Quality is acceptable. Although these are just tentative ideas at this point, remember that our entire investment in modeling and analysis at this point is only a *couple of hours work*. It is a real accomplishment, given the limited investment we have made, to have several working models in hand and to have some preliminary results. We also have a sense of some areas where refinement would be worthwhile. Even more significantly, we have some ideas of the *form* our final conclusions might take (such as Figures 4.5 and 4.12), which will help us tremendously when we come to present our results.

4.3 IMPROVING AND ITERATING

We have defined iterative modeling as the process of developing a series of models in an intelligent manner so that the most important issues are dealt with first. Although we will not carry the Red Cross analysis further at this point, you should now have a sense of how developing a first model and exploring it with sensitivity analysis can help structure the modeling process.

To the Reader: Before reading on, take a few minutes to create a list of improvements to the two models we have developed here. Prioritize this list, leading with changes that will be most significant to the Red Cross.

Here are some of our ideas:

- Donor payment occurs after donated blood passes quality tests.
- Size of donation fee affects normal donation rate.
- Size of donation fee influences the number of potential donors.
- Additional costs incurred to test paid donations.

4.4 SUMMARY

In this chapter, we have used one ill-structured problem to illustrate the four steps in the modeling process:

- Frame the problem
- Diagram the problem
- Build the model
- Generate insights

The problem we began with was described in only a couple of sentences. We did not have access to the decision makers to ask them how they viewed the problem or how they wanted the analysis to proceed. We did not have access to any data. Nonetheless, we were able to develop a plausible model structure and even to formulate some interesting insights about the behavior of important outcome measures. In subsequent chapters, we elaborate on the modeling process we have illustrated here.

CHAPTER 5

The Retirement Planning Case

5.0 INTRODUCTION

This chapter is the first of three in which we work through a modeling problem in detail, illustrating how to use the concepts of the problem kernel and iterative modeling to develop a sequence of progressively more complex and useful models.

We begin with the briefest possible description of the problem and show how problem framing, problem diagramming, and model building lead to useful preliminary insights. Then we introduce an expanded description of the problem and take another pass through the same modeling process. Finally, we consider a third variant of the problem based on data collected from the client after an initial modeling effort.

5.1 RETIREMENT PLANNING (A)

Retirement Planning (A)

Bob Davidson is a 46-year-old, married, tenured professor at a small New England college. His salary is currently $116,000. His employer contributes an amount equal to 10 percent of his salary to a retirement fund, and Bob himself has been contributing $9,500 a year. The current value of his retirement fund is $167,000. Bob would like to know whether his current level of retirement savings is adequate.

To the Reader: Before reading on, frame this problem by setting boundaries for the analysis; making assumptions; listing questions; defining outputs, inputs, and decisions; and so on.

5.1.1 Frame the Problem

Where do we start with an ill-structured problem like this one? In math classes, we were taught to use all the data in the problem, but we anticipate that the

---

Modeling for Insight: A Master Class for Business Analysts, by Stephen G. Powell & Robert J. Batt

"answer" to this problem will not follow in any straightforward way from the five numbers we are given: 46, $116,000, 10 percent, $9,500, and $167,000. What form could a helpful response to Bob's request take?

The first question that occurs to us in mulling over this problem is, what does Bob mean by an "adequate" level of savings? What is adequate for one person may not be adequate for another, so one of our tasks will be to formulate ideas about how Bob might think about this question. We cannot decide on an outcome measure until we have a better understanding of the objective of adequacy. Unfortunately for us, Bob is not available for questions, so we will have to proceed on our own. How might any middle-aged and fairly well-paid person answer this question? Here are some possible answers:

- "Adequate retirement savings guarantees a comfortable retirement until both my wife and I die."
- "Adequate retirement savings leaves me a nest egg of $1.5M when I retire but allows me to maintain my current lifestyle as I continue to work."
- "Adequate retirement savings allows me to spend 30 days a year traveling internationally during retirement, while not reducing my standard of living below the level I enjoyed in my last year of working."
- "Adequate retirement savings means there is no possibility I will need financial support from my children during retirement."
- "Adequate retirement savings means my wife and I can retire at age 65 and live on 70 percent of my final salary during retirement (adjusted for inflation) for at least 20 years."

We now see one of the ways in which this problem is ill structured: There is no clear-cut result or answer we can work toward. Rather, our task involves developing a problem structure and using it to create a variety of types of information we believe Bob will find useful. Note how different this approach is from problem solving. We do not expect to give Bob a single number for an answer. Rather, we will generate several insights about retirement planning that will allow him to make an informed decision.

It is often helpful while framing a problem to ask why a specific answer is *not* sufficient for the problem at hand. For example, imagine the "right" answer to Bob's question is to save $12,000 a year. We have already discovered that such an answer is not credible because we do not know Bob's objective, but are there other reasons not to rely on it? In fact, there are. For one thing, $12,000 might be enough if Bob retires at age 65 and dies at 75, but not if he lives to be 90 years old. So, uncertainty over his date of retirement and his longevity is one reason our hypothetical answer of $12,000 is not sufficient. Other reasons quickly come to mind. Retirement savings depend not just on how much Bob saves but on the rates of return his investments earn. How much of his income he wants to consume during his working years will also influence his retirement choices. Inflation may play a key role as well. And he might be more comfortable with a level of savings that increases as his income increases than saving a fixed-dollar amount each year.

The process of challenging a simplistic answer to an ill-structured problem is designed to uncover assumptions and boundaries that make the problem more tractable. For example, we could focus our attention on Bob's savings rather than on his expenditures. This assumption means we would not attempt to determine exactly how much of his budget he can spend before and after retirement; rather, we would assume he can comfortably save somewhat more or less than he currently does and attempt to trace out the consequences of such changes on his retirement assets. Another assumption to consider is whether we should base our analysis on savings *rates* or savings *levels*. For example, we could assume that each year Bob will save a constant percentage of his gross income rather than a certain dollar amount. The advantages of this approach are that the dollar amount grows as his income grows and he only has to think about a single savings rate rather than a different savings rate for each year.

Another useful result of this thought experiment is that it highlights specific parameters and decision variables that we need values for during our analysis. We assign arbitrary (but reasonable) values to these inputs and consider whether we can then "solve" the problem. For example, can we answer Bob's question in a useful way if we assume:

- Bob will retire at 67.
- He will save 5 percent of his gross income each year until retirement.
- He will consume 70 percent of his final gross salary each year in retirement.
- His assets will earn 8 percent/year up to retirement and 5 percent after.
- He will die at 75.
- His wife will die when he would have been 80.

Note that the first three inputs (Retirement Age, Annual Savings, and Consumption Rate) are all decision variables because Bob has direct control over these values. The last three inputs (Asset Returns, Age at Death, and Age at Wife's Death) are parameters because, unfortunately, he has little or no control over the actual values. It is important to keep this distinction in mind when generating insights.

Before we move on to diagramming the problem, we must be explicit about the problem's outcome measure. As we pointed out above, there are many different outcomes that might interest Bob. However, at this early stage in problem formulation, it is best to pick one that seems to be reasonably easy to calculate. As we develop the influence diagram, we may well uncover ambiguities or contradictions in our understanding of the problem. If so, we may need to circle back to problem framing and pick a new outcome measure or adjust a boundary or assumption.

We choose Bob's Assets at Death as the outcome measure for our first influence diagram. This outcome measure is easy to calculate, and it clearly will interest Bob. For instance, if we find that his assets are negative at death, we can conclude his savings rate is too low. We also focus our attention on a single decision variable: the constant percentage of his income he should save each year until he retires. For simplicity, we assume the other two decision variables, retirement age and consumption rate after retirement, are fixed. Our goal now is to draw an influence diagram that connects the inputs to the output.

To the Reader: Before reading on, sketch an influence diagram for this problem.

5.1.2 Diagram the Problem

We start with Assets at Death as the outcome measure in Figure 5.1. To calculate Bob's assets at his death, we need to know his assets when he retires, how much his investments earn during retirement, what his consumption expenditures are during retirement, and when he dies (or, equivalently, how many years he lives in retirement). Thus, in the diagram, we show four factors influencing Assets at Death:

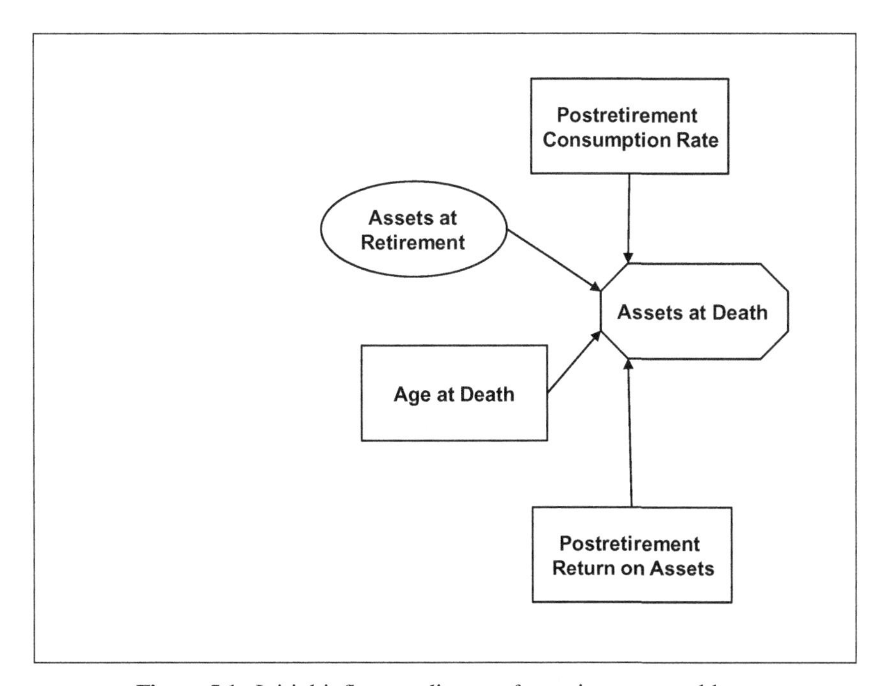

Figure 5.1. Initial influence diagram for retirement problem.

- Assets at Retirement
- Postretirement Consumption Rate
- Postretirement Return on Assets
- Age at Death

Note that at this point we leave somewhat vague exactly how we plan to calculate Assets at Death. For example, we could assume his postretirement consumption is a fixed dollar amount, a fixed percentage of his remaining assets, a variable percentage of his remaining assets, a fixed percentage of his final year's income, or something entirely different. Constructing an influence diagram should not be confused with building a spreadsheet model. A model requires precise inputs and relationships, so when we come to building a model, we will be more specific about these matters. But at this stage, we are simply trying to sort out the major factors that contribute to the output measure we have selected.

Returning to the influence diagram, we decompose Assets at Retirement into eight elements (Figure 5.2):

- Current Age
- Current Income
- Current Savings
- Preretirement Return on Assets
- Rate of Growth of Income

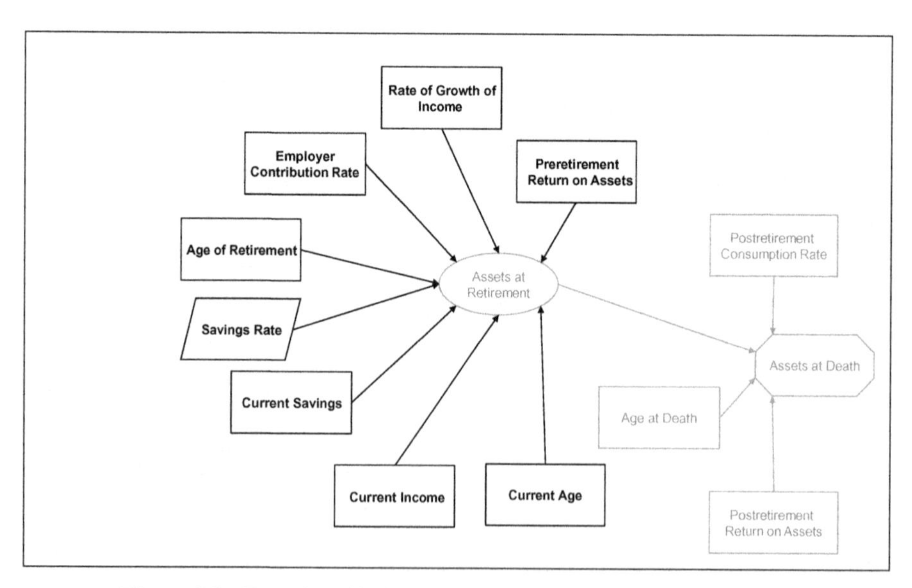

Figure 5.2. Completed influence diagram for retirement problem.

- Employer Contribution Rate
- Bob's Savings Rate
- Age of Retirement

If we know Bob's savings today and we know how much he will set aside in dollar terms each year, we can calculate how much money he has when he retires (assuming that we also know the rate of return on his assets). Also, we can determine how much he saves in dollars if we know his current income, its growth rate, and what percent of it he saves.

The influence diagram in Figure 5.2 is sufficiently detailed to provide the foundation for a reasonable first model, which we turn to next.

To the Reader: Before reading on, design and build a first model based on the influence diagram.

5.1.3 Build the Model—M1

One of the keys to success in modeling is building a first model (which we call M1) early in the process. Most modelers overestimate the amount of data collection and background research they need to do before building a model. They also underestimate the degree to which building a simple first model will allow them to uncover much of what is important in the problem. We build M1 at this point in the modeling process to get an initial idea of what savings rate might be adequate for Bob and what assumptions and relationships (if any) are missing from the influence diagram that should be included in the model.

We begin M1 by recording the five numbers given to us in the problem statement. We also record seven additional numbers that appear in our influence diagram and for which we can make reasonable guesses:

- His retirement age: 67
- His savings rate: 5 percent
- His consumption postretirement as a percent of his final salary: 70 percent
- The rates of return on his assets: 8 percent/year up to retirement and 5 percent/year thereafter
- His age at death: 75
- His age when his wife dies: 80

It is now straightforward to project Bob's income, consumption, and retirement assets over the 34 years from now to his wife's death. M1 is displayed in numerical view in Figure 5.3 and in relationship view in Figure 5.4. A summary of the key relationships follows:

	A	B	C	D	E	F	G	H
1	Retirement Planning							
2								
3								
4								
5		Parameters						
6			Current Income	$116,000				
7			Current Savings	$167,000				
8			Employer Savings Rate	10%				
9			Current Savings	$9,500				
10			Age	46				
11			Return on Assets					
12			Preretirement	8%				
13			Postretirement	5%				
14			Percent of Final Income Spent	70%				
15			Income Growth Rate	4%				
16								
17		Decision						
18			Savings Rate	5.00%				
19								
20					Savings Out of	Retirement Assets	Consumption	Retirement Assets
21		Model	Age	Income	Income	Preretirement	Postretirement	Postretirement
22			46	116,000	$17,400	$167,000		
23			47	120,640	$18,096	$198,456		
24			48	125,466	$18,820	$233,152		
25			49	130,484	$19,573	$271,377		
26			50	135,704	$20,356	$313,443		
27			51	141,132	$21,170	$359,688		
28			52	146,777	$22,017	$410,480		
29			53	152,648	$22,897	$466,215		
30			54	158,754	$23,813	$527,326		
31			55	165,104	$24,766	$594,277		
32			56	171,708	$25,756	$667,576		
33			57	178,577	$26,787	$747,768		
34			58	185,720	$27,858	$835,448		
35			59	193,149	$28,972	$931,256		
36			60	200,874	$30,131	$1,035,887		
37			61	208,909	$31,336	$1,150,095		
38			62	217,266	$32,590	$1,274,692		
39			63	225,956	$33,893	$1,410,561		
40			64	234,995	$35,249	$1,558,655		
41			65	244,395	$36,659	$1,720,007		
42			66	254,170	$38,126	$1,895,733		
43			67	264,337	$39,651	$2,087,042		
44			68				185,036	$2,006,358
45			69				185,036	$1,921,640
46			70				185,036	$1,832,686
47			71				185,036	$1,739,284
48			72				185,036	$1,641,213
49			73				185,036	$1,538,237
50			74				185,036	$1,430,113
51			75				185,036	$1,316,583
52			76				185,036	$1,197,376
53			77				185,036	$1,072,209
54			78				185,036	$940,783
55			79				185,036	$802,787
56			80				185,036	$657,890
57								

Figure 5.3. M1—numerical view.

Column D: Income each year is the previous year's income times the growth rate.

Column E: Savings for retirement each year is the savings rate, times current income, plus the employer's contribution.

Column F: Bob's retirement assets grow during his working life by the assumed rate of return, plus annual savings.

Column G: After retirement, his consumption is fixed as a percentage of his final year's salary.

Column H: His assets postretirement grow at an assumed rate of return and decline by the amount he consumes.

M1 is complete when it generates a numerical answer to the question we have posed: What will Bob's assets be at his (or his wife's) death? The answers our first model gives are reassuring: $1.317M at his death and $657,890 five years later, at his wife's death.

	A	B	C	D	E	F	G	H	I
1	Retirement Planning								
2									
3									
4									
5		Parameters							
6			Current Income	116000					
7			Current Savings	167000					
8			Employer Savings Rate	0.1					
9			Current Savings	9500					
10			Age	46					
11			Return on Assets						
12			Preretirement	0.08					
13			Postretirement	0.05					
14			Percent of Final Income Spent	0.7					
15			Income Growth Rate	0.04					
16									
17		Decision							
18			Savings Rate	0.05					
19									
20					Savings Out Of	Retirement Assets	Consumption	Retirement Assets	
21		Model	Age	Income	Income	Preretirement	Postretirement	Postretirement	
22			=D10	=D6	=($D$8+$D$18)*D22	=D7			
23			=C22+1	=D22*(1+$D$15)	=($D$8+$D$18)*D23	=F22*(1+$D$12)+E23			
24			=C23+1	=D23*(1+$D$15)	=($D$8+$D$18)*D24	=F23*(1+$D$12)+E24			
25			=C24+1	=D24*(1+$D$15)	=($D$8+$D$18)*D25	=F24*(1+$D$12)+E25			
26			=C25+1	=D25*(1+$D$15)	=($D$8+$D$18)*D26	=F25*(1+$D$12)+E26			
27			=C26+1	=D26*(1+$D$15)	=($D$8+$D$18)*D27	=F26*(1+$D$12)+E27			
28			=C27+1	=D27*(1+$D$15)	=($D$8+$D$18)*D28	=F27*(1+$D$12)+E28			
29			=C28+1	=D28*(1+$D$15)	=($D$8+$D$18)*D29	=F28*(1+$D$12)+E29			
30			=C29+1	=D29*(1+$D$15)	=($D$8+$D$18)*D30	=F29*(1+$D$12)+E30			
31			=C30+1	=D30*(1+$D$15)	=($D$8+$D$18)*D31	=F30*(1+$D$12)+E31			
32			=C31+1	=D31*(1+$D$15)	=($D$8+$D$18)*D32	=F31*(1+$D$12)+E32			
33			=C32+1	=D32*(1+$D$15)	=($D$8+$D$18)*D33	=F32*(1+$D$12)+E33			
34			=C33+1	=D33*(1+$D$15)	=($D$8+$D$18)*D34	=F33*(1+$D$12)+E34			
35			=C34+1	=D34*(1+$D$15)	=($D$8+$D$18)*D35	=F34*(1+$D$12)+E35			
36			=C35+1	=D35*(1+$D$15)	=($D$8+$D$18)*D36	=F35*(1+$D$12)+E36			
37			=C36+1	=D36*(1+$D$15)	=($D$8+$D$18)*D37	=F36*(1+$D$12)+E37			
38			=C37+1	=D37*(1+$D$15)	=($D$8+$D$18)*D38	=F37*(1+$D$12)+E38			
39			=C38+1	=D38*(1+$D$15)	=($D$8+$D$18)*D39	=F38*(1+$D$12)+E39			
40			=C39+1	=D39*(1+$D$15)	=($D$8+$D$18)*D40	=F39*(1+$D$12)+E40			
41			=C40+1	=D40*(1+$D$15)	=($D$8+$D$18)*D41	=F40*(1+$D$12)+E41			
42			=C41+1	=D41*(1+$D$15)	=($D$8+$D$18)*D42	=F41*(1+$D$12)+E42			
43			=C42+1	=D42*(1+$D$15)	=($D$8+$D$18)*D43	=F42*(1+$D$12)+E43			
44			=C43+1				=D14*D43	=F43*(1+$D$13)-G44	
45			=C44+1				=G44	=H44*(1+$D$13)-G45	
46			=C45+1				=G45	=H45*(1+$D$13)-G46	
47			=C46+1				=G46	=H46*(1+$D$13)-G47	
48			=C47+1				=G47	=H47*(1+$D$13)-G48	
49			=C48+1				=G48	=H48*(1+$D$13)-G49	
50			=C49+1				=G49	=H49*(1+$D$13)-G50	
51			=C50+1				=G50	=H50*(1+$D$13)-G51	
52			=C51+1				=G51	=H51*(1+$D$13)-G52	
53			=C52+1				=G52	=H52*(1+$D$13)-G53	
54			=C53+1				=G53	=H53*(1+$D$13)-G54	
55			=C54+1				=G54	=H54*(1+$D$13)-G55	
56			=C55+1				=G55	=H55*(1+$D$13)-G56	
57									

Figure 5.4. M1—relationship view.

To the Reader: Before reading on, use M1 to generate as many interesting insights into Bob's decision as you can.

5.1.4 Generate Insights—M1

Our modeling process could go in one of two directions at this point. We could focus on the shortcomings of M1 and improve it as a model, or we could focus on what M1 tells us about Bob's situation. It may seem dangerous to use such a simple and unsophisticated model for analysis. Nevertheless, we strongly recommend analyzing the problem before refining the model anymore. We make this suggestion because most modelers over-invest in model improvements and under-invest in model analysis. It is almost never too early to begin to use a model to learn. The iterative modeling process is designed to help you strike an appropriate balance between refining the model and using it for analysis.

What can we learn from our model? The model projects Bob's assets at age 80 will be $657,890. But this result is based on several arbitrary assumptions. We can, however, easily determine how low the savings rate can go before

Bob's final assets drop to zero. We use Goal Seek to help us find the answer; the result is 0.08 percent. So we now have a preliminary insight for Bob: His current savings rate is well above the level that would leave him with no assets at all when he dies (actually, when his wife dies, since we are assuming she dies after him).

Insight 1: Bob's current savings rate is more than adequate to provide retirement funds through age 80.

We would not share this conclusion with Bob at this point because it is far too tentative and it is based on a very preliminary model. Remember, the purpose of using M1 to generate insights is to help keep our focus on using the model to *learn*, not just to model for its own sake.

A related question Bob is sure to ask is: How sensitive are his final assets to his savings rate? Figure 5.5, which was generated using Data Sensitivity, shows that Final Assets range from $31,155 to $3,008,147 as the Savings Rate ranges from 1 percent to 20 percent. This finding suggests that Bob has a high degree of control over the amount of money left at his death.

Insight 2: Bob's assets at death are highly sensitive to his savings rate.

Whether this conclusion holds up as we further refine the model is not important at this point. What is important is that we have developed a model structure (as represented by the influence diagram and M1) as well as some preliminary insights very early in the modeling process.

Another sensitivity question interests both Bob and us: Which parameters have the biggest impact on the results? This question can be answered by varying each input around the base-case values using a tornado chart. Figure 5.6 shows a tornado chart for M1, where each of the five parameters that can change varies by 10 percent. Several aspects of these results are interesting. First, in no case do we see negative final assets. Thus, if Bob's primary goal is to ensure he does not run out of savings during retirement, this analysis sug-

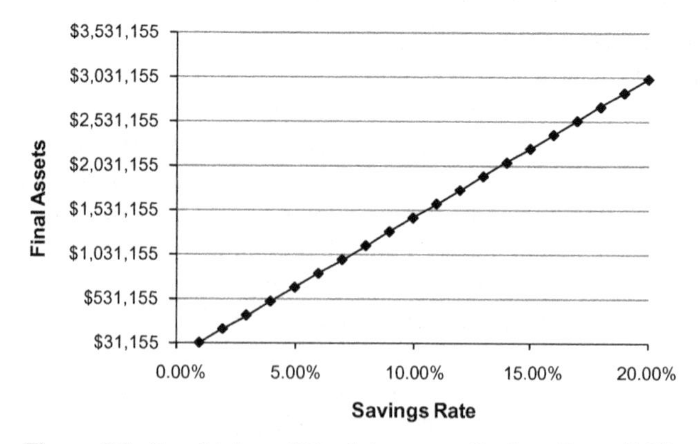

Figure 5.5. Sensitivity of Final Assets to Savings Rate (M1).

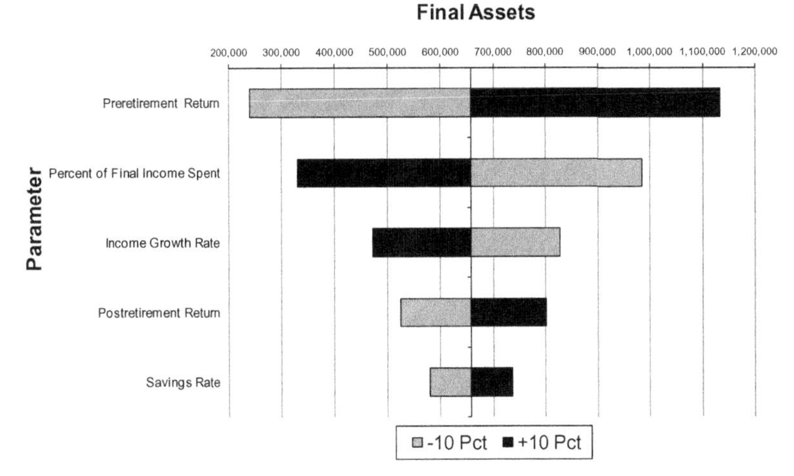

Figure 5.6. Tornado chart for retirement model (M1).

gests that modest variations in rates of return and other parameters will not threaten that goal.

Insight 3: Assuming moderate rates of return, Bob is unlikely to run out of money before he dies.

Also, we see that the parameter with the biggest impact is the rate of return on his assets before retirement. By contrast, the percent he saves has relatively little impact. (Remember that in a tornado chart, each parameter varies over the range specified with all other parameters held constant. Any conclusions we draw must be appropriate to these assumptions. For example, here we have only varied the savings rate from 4 percent to 6 percent, so to conclude that it has little impact on the outcome may be premature if values outside this range are likely.)

Insight 4: Bob's assets at death are most sensitive to the preretirement return on assets and less sensitive to his savings rate.

Having gleaned four good insights from this first model, it is now time to consider how to improve it to generate even more insights.

To the Reader: Before reading on, create a list of the elements you would add to M1 to create the next iteration of the model.

5.1.5 Improving and Iterating

The next step in the process is to refine the model. But where do we start? One place to begin is to think of alternatives to the outcome measure we have used so far. In addition to wanting to know what his assets might be at death, Bob probably would like to know:

- The number of years he can live comfortably in retirement
- The probability of running out of savings before he dies
- The maximum amount he can spend each year in retirement and not run out before a certain age
- The amount he will have saved when he retires
- Combinations of savings rates and postretirement consumption rates that guarantee sufficient assets to a certain age

An ideal model would be sufficiently detailed and flexible to answer all these and any other questions that might come up during the modeling process. Of course, building the ideal model takes infinite time and resources. So we use the iterative modeling process to develop a series of models and insights that, hopefully, will satisfy our client without requiring infinite effort.

Another good practice before deciding how to improve a model is to list its shortcomings. Our first model has:

- Fixed dates of death for Bob and his wife
- Fixed retirement date
- A single savings rate
- A single postretirement consumption rate
- Single rates of return before and after retirement
- Deterministic assumptions and outputs

To answer some of the questions we raised above, we have to make our model more complex and more flexible. We do this by developing a second model, M2.

To the Reader: Before reading on, sketch an influence diagram for the next iteration of the model, adding in whatever elements you think are most important.

5.1.6 Diagram the Problem—M2

Our primary purpose with M2 is to make the model more *flexible*. To get results quickly in M1, we fixed several parameters and decision variables that ideally should be variable, such as the date of retirement. In this next version of the model, we allow the Retirement Age and the Postretirement Consumption Rate to be decision variables (see Figure 5.7). We also add in a couple of realistic details that we left out of the first model, in particular, Postretirement Inflation and Taxes. In addition, we change the outcome measure. In M1, we

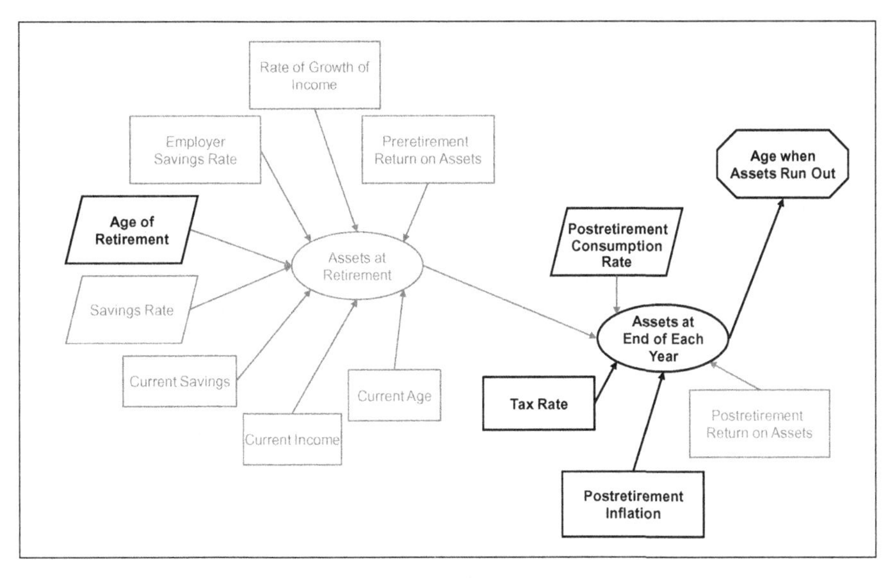

Figure 5.7. Influence diagram for M2.

determined the amount of money Bob's estate would have when his wife dies. This outcome measure has the disadvantage that it depends on an arbitrary date of death. In M2, we calculate Bob's *age* when his assets run out. This alternative outcome measure should provide somewhat better insights since it is independent of the date of death. To compute this measure we need to generate a new intermediate variable called Assets at End of Each Year. Actually, this variable is a series of values. Since we do not know when the assets will run out, we calculate the assets at the end of each year and look for the year when the assets go negative. This approach gives us the outcome measure we are looking for.

Updating the influence diagram is important because it helps us see how we need to revise the model. An up-to-date diagram also reminds us of our thought process at each step of our analysis. Lastly, the influence diagram is a good tool for communicating our work to colleagues or to the client.

To the Reader: Before reading on, design and build a model based on the influence diagram.

5.1.7 Build the Model—M2

The above changes require a more elaborate model design because a variable retirement year means that we cannot know for sure as we design the model when Bob's working life ends and his retirement begins. One solution to this design problem is to calculate all the necessary information for the maximum working life (say, 90 years) and then link the calculations for the retirement phase to the retirement date.

A summary of model M2 follows (see Figures 5.8 and 5.9):

Preretirement

Column D: Salary is the previous year's salary times the growth rate.

Column E: Employer contribution is the salary times the employer contribution rate.

Column F: Personal contribution is the salary times the personal contribution rate.

Column G: Assets at the beginning of the year are assets from the end of the previous year.

Column H: Assets at the end of the year are assets from the beginning of the year, times the quantity one plus the rate of return, plus employer contributions and personal contributions.

Postretirement

Column K: Spending in the first year of retirement is the final working year's salary times the percent of final income spent; thereafter, spending grows at the rate of inflation.

Column L: Withdrawal is the spending divided by the quantity one minus the tax rate.

Column M: Assets at the beginning of the year are assets at the end of the previous year.

Column N: Assets at the end of the year are assets from the beginning of the year, times the quantity one plus the rate of return, minus the withdrawal.

Column O: Retirement funds are depleted when end-of-year assets are negative.

Note the use of the VLOOKUP function in K26 and M26. In both cases, the VLOOKUP function is used to pick out the appropriate date to start the postretirement calculations, depending on the year of retirement. In cell K26, for example, we need to determine income in the final working year in order to calculate spending in the first year of retirement. Since the final working year changes when we change the retirement year, we use the formula

D20*VLOOKUP(J26, C26:D70, 2, 1)

to find the salary in the range D26:D70, based on matching the retirement year. This function looks for the year of retirement (J26) in the age range C26:C70. Once it finds the location of this year, it returns the matching value from the second column of the table (column D). A similar formula is used in

Retirement Planning

Parameters

Current Income	$116,000
Current Savings	$167,000
Employer Savings Rate	10%
Current Savings	$9,500
Age	46
Return on Assets	
Preretirement	8%
Postretirement	5%
Income Growth Rate (Includes Inflationa)	4%
Inflation Postretirement	2%
Tax Rate Postretirement	35%

Decision

Savings Rate	5.00%
Percent of Final Income Spent	70%
Retirement Age	65.0

Results

Year funds out	73

Tornado chart inputs

	10%	90%
Return on Assets		
Preretirement	6.00%	12.00%
Postretirement	3.00%	8.00%
Income Growth Rate	0.00%	6.00%
Inflation Postretirement	1.00%	5.00%
Tax Rate Postretirement	30.00%	40.00%
Savings Rate	10.00%	15.00%
Percent of Final Income Spent	60.00%	90.00%
Retirement Age	60	70

Model

Preretirement Age	Salary	Employer contribution	Personal Contribution	Beginning of Year Assets	End of Year Assets
46	116,000	$11,600	$5,800	$167,000	$197,760
47	120,640	$12,064	$6,032	$197,760	$231,677
48	125,466	$12,547	$6,273	$231,677	$269,031
49	130,484	$13,048	$6,524	$269,031	$310,126
50	135,704	$13,570	$6,785	$310,126	$355,291
51	141,132	$14,113	$7,057	$355,291	$404,885
52	146,777	$14,678	$7,339	$404,885	$459,292
53	152,648	$15,265	$7,632	$459,292	$518,932
54	158,754	$15,875	$7,938	$518,932	$584,260
55	165,104	$16,510	$8,255	$584,260	$655,767
56	171,708	$17,171	$8,585	$655,767	$733,984
57	178,577	$17,858	$8,929	$733,984	$819,489
58	185,720	$18,572	$9,286	$819,489	$912,907
59	193,149	$19,315	$9,657	$912,907	$1,014,911
60	200,874	$20,087	$10,044	$1,014,911	$1,126,235
61	208,909	$20,891	$10,445	$1,126,235	$1,247,671
62	217,266	$21,727	$10,863	$1,247,671	$1,380,074
63	225,956	$22,596	$11,298	$1,380,074	$1,524,374
64	234,995	$23,499	$11,750	$1,524,374	$1,681,573
65	244,395	$24,439	$12,220	$1,681,573	$1,852,758
66	254,170	$25,417	$12,709	$1,852,758	$2,039,104
67	264,337	$26,434	$13,217	$2,039,104	$2,241,883
68	274,911	$27,491	$13,746	$2,241,883	$2,462,470
69	285,907	$28,591	$14,295	$2,462,470	$2,702,353
70	297,343	$29,734	$14,867	$2,702,353	$2,963,143
71	309,237	$30,924	$15,462	$2,963,143	$3,246,580
72	321,606	$32,161	$16,080	$3,246,580	$3,554,548
73	334,471	$33,447	$16,724	$3,554,548	$3,889,082
74	347,850	$34,785	$17,392	$3,889,082	$4,252,386
75	361,764	$36,176	$18,088	$4,252,386	$4,646,842
76	376,234	$37,623	$18,812	$4,646,842	$5,075,024
77	391,283	$39,128	$19,564	$5,075,024	$5,539,718
78	406,935	$40,693	$20,347	$5,539,718	$6,043,936
79	423,212	$42,321	$21,161	$6,043,936	$6,590,933
80	440,141	$44,014	$22,007	$6,590,933	$7,184,229
81	457,746	$45,775	$22,887	$7,184,229	$7,827,629
82	476,056	$47,606	$23,803	$7,827,629	$8,525,248
83	495,098	$49,510	$24,755	$8,525,248	$9,281,532
84	514,902	$51,490	$25,745	$9,281,532	$10,101,290
85	535,498	$53,550	$26,775	$10,101,290	$10,989,718
86	556,918	$55,692	$27,846	$10,989,718	$11,952,433
87	579,195	$57,920	$28,960	$11,952,433	$12,995,507
88	602,363	$60,236	$30,118	$12,995,507	$14,125,502
89	626,457	$62,646	$31,323	$14,125,502	$15,349,511
90	651,516	$65,152	$32,576	$15,349,511	$16,675,199

Postretirement Age	Spending	Withdrawal	Beginning of Year Assets	End of Year Assets	Funds depleted? 1=Yes
65.0	171,076	263,194	1,852,758	1,682,201	0
66	174,498	268,458	1,682,201	1,497,854	0
67	177,988	273,827	1,497,854	1,298,919	0
68	181,547	279,304	1,298,919	1,084,561	0
69	185,178	284,890	1,084,561	853,900	0
70	188,882	290,588	853,900	606,007	0
71	192,660	296,399	606,007	339,908	0
72	196,513	302,327	339,908	54,576	0
73	200,443	308,374	54,576	-251,069	1
74	204,452	314,541	-251,069	-578,163	1
75	208,541	320,832	-578,163	-927,904	1
76	212,712	327,249	-927,904	-1,301,548	1
77	216,966	333,794	-1,301,548	-1,700,419	1
78	221,305	340,470	-1,700,419	-2,125,909	1
79	225,731	347,279	-2,125,909	-2,579,484	1
80	230,246	354,225	-2,579,484	-3,062,682	1
81	234,851	361,309	-3,062,682	-3,577,126	1
82	239,548	368,535	-3,577,126	-4,124,517	1
83	244,339	375,906	-4,124,517	-4,706,649	1
84	249,226	383,424	-4,706,649	-5,325,406	1
85	254,210	391,093	-5,325,406	-5,982,768	1
86	259,294	398,914	-5,982,768	-6,680,821	1
87	264,480	406,893	-6,680,821	-7,421,755	1
88	269,770	415,031	-7,421,755	-8,207,873	1
89	275,165	423,331	-8,207,873	-9,041,598	1
90	280,669	431,798	-9,041,598	-9,925,476	1
91	286,282	440,434	-9,925,476	-10,862,183	1
92	292,008	449,242	-10,862,183	-11,854,535	1
93	297,848	458,227	-11,854,535	-12,905,489	1
94	303,805	467,392	-12,905,489	-14,018,155	1
95	309,881	476,740	-14,018,155	-15,195,803	1
96	316,078	486,274	-15,195,803	-16,441,867	1
97	322,400	496,000	-16,441,867	-17,759,960	1
98	328,848	505,920	-17,759,960	-19,153,878	1
99	335,425	516,038	-19,153,878	-20,627,611	1
100	342,133	526,359	-20,627,611	-22,185,350	1

Figure 5.8. M2—numerical view.

	A	B	C	D	E	F	G	H	I	J	K	L	M	N	O	P
1	**Retirement Planning**															
2																
3																
4																
5		**Parameters**					**Results**						**Tornado Chart Inputs**			
6			Current Income	116000				Year funds out	**=INDEX(J26:J61,MATCH(1,O26:O61,0))**						0.1	
7			Current Savings	167000									Return on Assets			
8			Employer Savings Rate	0.1									Preretirement		0.06	
9			Current Savings	9500									Postretirement		0.03	
10			Age	46									Income Growth Rate		0	
11			Return on Assets										Inflation Postretirement		0.01	
12			Preretirement	0.08									Tax Rate Postretirement		0.3	
13			Postretirement	0.05												
14			Income Growth Rate (Includes Inflation)	0.04												
15			Inflation Postretirement	0.02									Savings Rate		0.1	
16			Tax Rate Postretirement	0.35									Percent of Final Income Spent		0.6	
17													Retirement Age		60	
18		**Decision**														
19			Savings Rate	0.05												
20			Percent of Final Income Spent	0.7												
21			Retirement Age	65												
22																
23		**Model**														
24			**Preretirement**		**Employer**	**Personal**	**Beginning of**	**End of**		**POSTRETIREMENT**			**Beginning of**	**End of**	**Funds Depleted?**	
25			**Age**	**Salary**	**Contribution**	**Contribution**	**Year Assets**	**Year Assets**		**Age**	**Spending**	**Withdrawal**	**Year Assets**	**Year Assets**	**1=Yes**	
26			=D10	=D6	=$D$8*D26	=$D$19*D26	=D7	=G26*(1+$D$12)+E26+F26		=D21	=D20*VLOOKUP(J26,C26:D70,2,1)	=K26/(1-$D$16)	=VLOOKUP(J26,C26:H70,6,1)	=M26*(1+$D$13)-L26	=IF(N26<0,1,0)	
27			=C26+1	=D26*(1+$D$14)	=$D$8*D27	=$D$19*D27	=H26	=G27*(1+$D$12)+E27+F27		=J26+1	=K26*(1+$D$15)	=K27/(1-$D$16)	=N26	=M27*(1+$D$13)-L27	=IF(N27<0,1,0)	
28			=C27+1	=D27*(1+$D$14)	=$D$8*D28	=$D$19*D28	=H27	=G28*(1+$D$12)+E28+F28		=J27+1	=K27*(1+$D$15)	=K28/(1-$D$16)	=N27	=M28*(1+$D$13)-L28	=IF(N28<0,1,0)	
29			=C28+1	=D28*(1+$D$14)	=$D$8*D29	=$D$19*D29	=H28	=G29*(1+$D$12)+E29+F29		=J28+1	=K28*(1+$D$15)	=K29/(1-$D$16)	=N28	=M29*(1+$D$13)-L29	=IF(N29<0,1,0)	
30			=C29+1	=D29*(1+$D$14)	=$D$8*D30	=$D$19*D30	=H29	=G30*(1+$D$12)+E30+F30		=J29+1	=K29*(1+$D$15)	=K30/(1-$D$16)	=N29	=M30*(1+$D$13)-L30	=IF(N30<0,1,0)	
31			=C30+1	=D30*(1+$D$14)	=$D$8*D31	=$D$19*D31	=H30	=G31*(1+$D$12)+E31+F31		=J30+1	=K30*(1+$D$15)	=K31/(1-$D$16)	=N30	=M31*(1+$D$13)-L31	=IF(N31<0,1,0)	
32			=C31+1	=D31*(1+$D$14)	=$D$8*D32	=$D$19*D32	=H31	=G32*(1+$D$12)+E32+F32		=J31+1	=K31*(1+$D$15)	=K32/(1-$D$16)	=N31	=M32*(1+$D$13)-L32	=IF(N32<0,1,0)	
33			=C32+1	=D32*(1+$D$14)	=$D$8*D33	=$D$19*D33	=H32	=G33*(1+$D$12)+E33+F33		=J32+1	=K32*(1+$D$15)	=K33/(1-$D$16)	=N32	=M33*(1+$D$13)-L33	=IF(N33<0,1,0)	
34			=C33+1	=D33*(1+$D$14)	=$D$8*D34	=$D$19*D34	=H33	=G34*(1+$D$12)+E34+F34		=J33+1	=K33*(1+$D$15)	=K34/(1-$D$16)	=N33	=M34*(1+$D$13)-L34	=IF(N34<0,1,0)	
35			=C34+1	=D34*(1+$D$14)	=$D$8*D35	=$D$19*D35	=H34	=G35*(1+$D$12)+E35+F35		=J34+1	=K34*(1+$D$15)	=K35/(1-$D$16)	=N34	=M35*(1+$D$13)-L35	=IF(N35<0,1,0)	
36			=C35+1	=D35*(1+$D$14)	=$D$8*D36	=$D$19*D36	=H35	=G36*(1+$D$12)+E36+F36		=J35+1	=K35*(1+$D$15)	=K36/(1-$D$16)	=N35	=M36*(1+$D$13)-L36	=IF(N36<0,1,0)	
37			=C36+1	=D36*(1+$D$14)	=$D$8*D37	=$D$19*D37	=H36	=G37*(1+$D$12)+E37+F37		=J36+1	=K36*(1+$D$15)	=K37/(1-$D$16)	=N36	=M37*(1+$D$13)-L37	=IF(N37<0,1,0)	
38			=C37+1	=D37*(1+$D$14)	=$D$8*D38	=$D$19*D38	=H37	=G38*(1+$D$12)+E38+F38		=J37+1	=K37*(1+$D$15)	=K38/(1-$D$16)	=N37	=M38*(1+$D$13)-L38	=IF(N38<0,1,0)	
39			=C38+1	=D38*(1+$D$14)	=$D$8*D39	=$D$19*D39	=H38	=G39*(1+$D$12)+E39+F39		=J38+1	=K38*(1+$D$15)	=K39/(1-$D$16)	=N38	=M39*(1+$D$13)-L39	=IF(N39<0,1,0)	
40			=C39+1	=D39*(1+$D$14)	=$D$8*D40	=$D$19*D40	=H39	=G40*(1+$D$12)+E40+F40		=J39+1	=K39*(1+$D$15)	=K40/(1-$D$16)	=N39	=M40*(1+$D$13)-L40	=IF(N40<0,1,0)	
41			=C40+1	=D40*(1+$D$14)	=$D$8*D41	=$D$19*D41	=H40	=G41*(1+$D$12)+E41+F41		=J40+1	=K40*(1+$D$15)	=K41/(1-$D$16)	=N40	=M41*(1+$D$13)-L41	=IF(N41<0,1,0)	
42			=C41+1	=D41*(1+$D$14)	=$D$8*D42	=$D$19*D42	=H41	=G42*(1+$D$12)+E42+F42		=J41+1	=K41*(1+$D$15)	=K42/(1-$D$16)	=N41	=M42*(1+$D$13)-L42	=IF(N42<0,1,0)	
43			=C42+1	=D42*(1+$D$14)	=$D$8*D43	=$D$19*D43	=H42	=G43*(1+$D$12)+E43+F43		=J42+1	=K42*(1+$D$15)	=K43/(1-$D$16)	=N42	=M43*(1+$D$13)-L43	=IF(N43<0,1,0)	
44			=C43+1	=D43*(1+$D$14)	=$D$8*D44	=$D$19*D44	=H43	=G44*(1+$D$12)+E44+F44		=J43+1	=K43*(1+$D$15)	=K44/(1-$D$16)	=N43	=M44*(1+$D$13)-L44	=IF(N44<0,1,0)	
45			=C44+1	=D44*(1+$D$14)	=$D$8*D45	=$D$19*D45	=H44	=G45*(1+$D$12)+E45+F45		=J44+1	=K44*(1+$D$15)	=K45/(1-$D$16)	=N44	=M45*(1+$D$13)-L45	=IF(N45<0,1,0)	
46			=C45+1	=D45*(1+$D$14)	=$D$8*D46	=$D$19*D46	=H45	=G46*(1+$D$12)+E46+F46		=J45+1	=K45*(1+$D$15)	=K46/(1-$D$16)	=N45	=M46*(1+$D$13)-L46	=IF(N46<0,1,0)	
47			=C46+1	=D46*(1+$D$14)	=$D$8*D47	=$D$19*D47	=H46	=G47*(1+$D$12)+E47+F47		=J46+1	=K46*(1+$D$15)	=K47/(1-$D$16)	=N46	=M47*(1+$D$13)-L47	=IF(N47<0,1,0)	
48			=C47+1	=D47*(1+$D$14)	=$D$8*D48	=$D$19*D48	=H47	=G48*(1+$D$12)+E48+F48		=J47+1	=K47*(1+$D$15)	=K48/(1-$D$16)	=N47	=M48*(1+$D$13)-L48	=IF(N48<0,1,0)	
49			=C48+1	=D48*(1+$D$14)	=$D$8*D49	=$D$19*D49	=H48	=G49*(1+$D$12)+E49+F49		=J48+1	=K48*(1+$D$15)	=K49/(1-$D$16)	=N48	=M49*(1+$D$13)-L49	=IF(N49<0,1,0)	
50			=C49+1	=D49*(1+$D$14)	=$D$8*D50	=$D$19*D50	=H49	=G50*(1+$D$12)+E50+F50		=J49+1	=K49*(1+$D$15)	=K50/(1-$D$16)	=N49	=M50*(1+$D$13)-L50	=IF(N50<0,1,0)	
51			=C50+1	=D50*(1+$D$14)	=$D$8*D51	=$D$19*D51	=H50	=G51*(1+$D$12)+E51+F51		=J50+1	=K50*(1+$D$15)	=K51/(1-$D$16)	=N50	=M51*(1+$D$13)-L51	=IF(N51<0,1,0)	
52			=C51+1	=D51*(1+$D$14)	=$D$8*D52	=$D$19*D52	=H51	=G52*(1+$D$12)+E52+F52		=J51+1	=K51*(1+$D$15)	=K52/(1-$D$16)	=N51	=M52*(1+$D$13)-L52	=IF(N52<0,1,0)	
53			=C52+1	=D52*(1+$D$14)	=$D$8*D53	=$D$19*D53	=H52	=G53*(1+$D$12)+E53+F53		=J52+1	=K52*(1+$D$15)	=K53/(1-$D$16)	=N52	=M53*(1+$D$13)-L53	=IF(N53<0,1,0)	
54			=C53+1	=D53*(1+$D$14)	=$D$8*D54	=$D$19*D54	=H53	=G54*(1+$D$12)+E54+F54		=J53+1	=K53*(1+$D$15)	=K54/(1-$D$16)	=N53	=M54*(1+$D$13)-L54	=IF(N54<0,1,0)	
55			=C54+1	=D54*(1+$D$14)	=$D$8*D55	=$D$19*D55	=H54	=G55*(1+$D$12)+E55+F55		=J54+1	=K54*(1+$D$15)	=K55/(1-$D$16)	=N54	=M55*(1+$D$13)-L55	=IF(N55<0,1,0)	
56			=C55+1	=D55*(1+$D$14)	=$D$8*D56	=$D$19*D56	=H55	=G56*(1+$D$12)+E56+F56		=J55+1	=K55*(1+$D$15)	=K56/(1-$D$16)	=N55	=M56*(1+$D$13)-L56	=IF(N56<0,1,0)	
57			=C56+1	=D56*(1+$D$14)	=$D$8*D57	=$D$19*D57	=H56	=G57*(1+$D$12)+E57+F57		=J56+1	=K56*(1+$D$15)	=K57/(1-$D$16)	=N56	=M57*(1+$D$13)-L57	=IF(N57<0,1,0)	
58			=C57+1	=D57*(1+$D$14)	=$D$8*D58	=$D$19*D58	=H57	=G58*(1+$D$12)+E58+F58		=J57+1	=K57*(1+$D$15)	=K58/(1-$D$16)	=N57	=M58*(1+$D$13)-L58	=IF(N58<0,1,0)	
59			=C58+1	=D58*(1+$D$14)	=$D$8*D59	=$D$19*D59	=H58	=G59*(1+$D$12)+E59+F59		=J58+1	=K58*(1+$D$15)	=K59/(1-$D$16)	=N58	=M59*(1+$D$13)-L59	=IF(N59<0,1,0)	
60			=C59+1	=D59*(1+$D$14)	=$D$8*D60	=$D$19*D60	=H59	=G60*(1+$D$12)+E60+F60		=J59+1	=K59*(1+$D$15)	=K60/(1-$D$16)	=N59	=M60*(1+$D$13)-L60	=IF(N60<0,1,0)	
61			=C60+1	=D60*(1+$D$14)	=$D$8*D61	=$D$19*D61	=H60	=G61*(1+$D$12)+E61+F61		=J60+1	=K60*(1+$D$15)	=K61/(1-$D$16)	=N60	=M61*(1+$D$13)-L61	=IF(N61<0,1,0)	
62			=C61+1	=D61*(1+$D$14)	=$D$8*D62	=$D$19*D62	=H61	=G62*(1+$D$12)+E62+F62								
63			=C62+1	=D62*(1+$D$14)	=$D$8*D63	=$D$19*D63	=H62	=G63*(1+$D$12)+E63+F63								
64			=C63+1	=D63*(1+$D$14)	=$D$8*D64	=$D$19*D64	=H63	=G64*(1+$D$12)+E64+F64								
65			=C64+1	=D64*(1+$D$14)	=$D$8*D65	=$D$19*D65	=H64	=G65*(1+$D$12)+E65+F65								
66			=C65+1	=D65*(1+$D$14)	=$D$8*D66	=$D$19*D66	=H65	=G66*(1+$D$12)+E66+F66								
67			=C66+1	=D66*(1+$D$14)	=$D$8*D67	=$D$19*D67	=H66	=G67*(1+$D$12)+E67+F67								
68			=C67+1	=D67*(1+$D$14)	=$D$8*D68	=$D$19*D68	=H67	=G68*(1+$D$12)+E68+F68								
69			=C68+1	=D68*(1+$D$14)	=$D$8*D69	=$D$19*D69	=H68	=G69*(1+$D$12)+E69+F69								
70			=C69+1	=D69*(1+$D$14)	=$D$8*D70	=$D$19*D70	=H69	=G70*(1+$D$12)+E70+F70								
71																

Figure 5.9. M2—relationship view.

cell M26 to determine the beginning-of-year assets for the first retirement year from the range H26:H70.

Also note the use of the INDEX and MATCH functions in cell I6. The MATCH function is nested within the INDEX function to pick out the appropriate age when assets are depleted. We use the formula

INDEX(J26:J61, MATCH(1, O26:O61, 0))

to find the age in the range J26:J61, based on the location calculated by the MATCH function. This function looks for the value 1 in the range O26:O61. Once it finds the location of this year, the INDEX function calculates the corresponding age.

The VLOOKUP, INDEX, and MATCH functions are among the more sophisticated functions in Excel, and here they enable us to build a more flexible model than we otherwise could. The benefits of this approach become apparent as we use the model to generate insights.

To the Reader: Before reading on, use M2 to generate as many interesting insights into Bob's decision as you can.

5.1.8 Generate Insights—M2

The base-case result for this model is somewhat less optimistic than our base case for M1. In that case, it seemed that if Bob died at age 75, his assets at death would be more than sufficient. Here it seems that assets will run out at age 73. Several changes account for this difference in results. First, we have allowed for inflation in postretirement spending and taxes at a hefty 35 percent of spending. Both influences reduce the lifetime of the assets available at retirement. But base-case results are often sensitive to the numerical assumptions behind them. More interesting is the sensitivity of the results to changes in assumptions. Bob's basic question can be interpreted as a sensitivity question: How much does the "adequacy" of my retirement savings change as I change my savings rate? Figure 5.10 provides a second answer to this question. As we change his savings rate from 1 percent to 20 percent of current income, the year in which his assets run out ranges from age 71 to 79. This result, if it holds up, is sobering. He is currently saving about 8 percent of his income. Even if he were to increase his savings rate to 20 percent of his income, which is probably at the upper end of what is possible, he only adds five years of "adequacy" to his retirement.

Insight 5: Increasing the savings rate to an extreme value only increases the sufficiency of funds by a few years.

Put another way (and estimating roughly the slope of the jagged line in Figure 5.10), each increase of 1 percentage point in his savings rate adds about

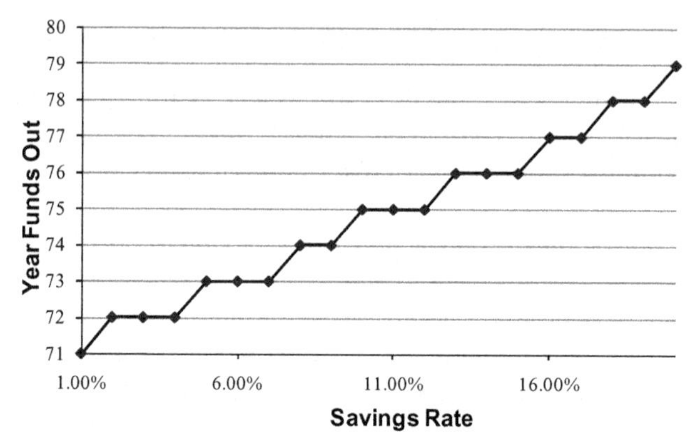

Figure 5.10. Sensitivity of Year Funds Out to Savings Rate (M2).

half a year of adequate retirement. These results suggest that it may not be easy for Bob to fund an adequate retirement if he lives more than about ten years past retirement (and all the other assumptions of our model hold).

Insight 6: Given our assumptions, it is difficult to fund retirement for more than ten years.

The tornado chart in Figure 5.11 suggests another interesting feature of this new model. By far the most important input among those tested here is the year of retirement.

Insight 7: The age when assets run out is most sensitive to the year of retirement.

With all other assumptions fixed, if Bob retires at 59, his assets will last only to age 64, whereas if he delays retirement until 71, they will last until 82. Again, these results depend on the nature of the tornado chart. A 10 percent swing in the retirement age is a much bigger absolute change than a 10 percent swing in the rates of return, inflation, or even taxes.

More realistic sensitivity analyses are possible with a tornado chart, of course. We could easily determine percent changes for each parameter separately and create a Variable Percentages Tornado Chart. Another alternative would be to use the Percentiles Tornado Chart. This requires upper and lower bounds on each input, which we interpret as the 10th and 90th percentiles of the probability distribution in the case of random variables and as likely values for the minimum and maximum in the case of decision variables (such as the retirement age). We assume the following extreme values for the key inputs:

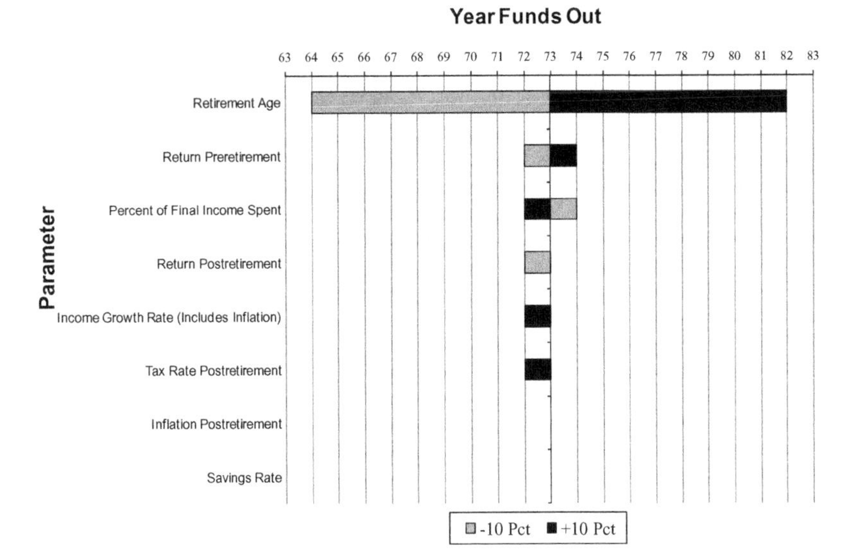

Figure 5.11. Constant percentage tornado chart (M2).

	10th Percentile	90th Percentile
Returns		
preretirement	6%	12%
postretirement	3%	8%
Income growth rate	0%	6%
Inflation postretirement	1%	5%
Tax rate	30%	40%
	Minimum	Maximum
Savings rate	10%	15%
Percent of final income spent	60%	90%
Retirement age	60	70

(Note that since we have chosen to consider savings rates between 10 and 15 percent, we have changed the base-case savings rate for this analysis to 12.5 percent. This changes the base-case outcome to 75 years.)

The results are shown in Figure 5.12. The age at which Bob chooses to retire remains the most influential factor determining the adequacy of his retirement plan. If he retires at age 70, his assets will last until he is 85, whereas if he retires at 60, they will only last to 67. The next most influential factor is the rate at which his salary grows. But here is a surprise: His assets last *longer* if his salary while working grows slowly or not at all!

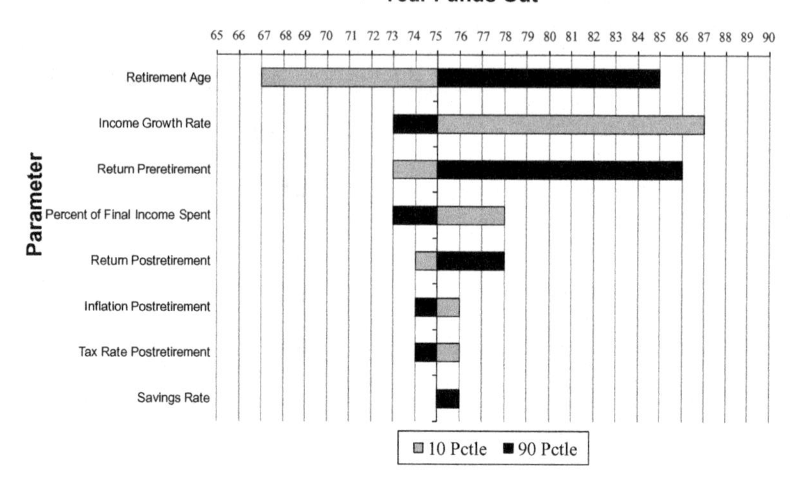

Figure 5.12. Percentiles tornado chart (M2).

Insight 8: The age when assets run out is higher if salary growth is lower.

If that result is obvious to you, you have better intuition than we do. In fact, we did not believe it until we actually put 0 percent and 6 percent salary increases in the model and verified the results. The reason is not hard to find: A low rate of increase in salary still leaves Bob with substantial assets at retirement but a significantly lower level of postretirement spending, since he spends a fixed percentage of his final salary in the first year of retirement. This should not imply that Bob should seek to reduce his income while he works, however! But it does point out a possible limitation of the model, which is that his postretirement spending is not related to some *absolute* level of consumption but to a *relative* level, which may distort the results. Some other aspects of this sensitivity analysis are worth noting as well. The rate of return on assets before retirement, not surprisingly, has a strong impact on the results, especially when that return is high. This suggests that Bob might want to consider investing in riskier assets to earn a higher return until he retires.

Insight 9: The rate of return on assets before retirement has a strong effect on the age when assets run out. Investing in riskier, higher return assets early on is beneficial.

All the remaining factors have a relatively minor impact on the outcome. The percent of his final income that he spends postretirement only changes the years of adequacy by a couple of years in either direction. Finally, the amount Bob chooses to save (at least in the range from 10 to 15 percent) has the smallest impact of all eight factors considered here. This conclusion is also somewhat startling, and it begins to suggest that we may have more than one surprise for Bob when the time comes to present our results to him.

Insight 10: Most inputs, including savings rate and postretirement spending rate, have little effect on the age when assets run out.

It is worth pausing at this point and taking stock. We set out to demonstrate how a problem as ill-structured as Retirement Planning (A) could be analyzed using the process of iterative modeling. After a modest effort, we find ourselves drawing some pretty remarkable conclusions that suggest some of the insights we may want to present to our client. We do not mean to suggest these insights are final or that our analysis is complete. Quite the opposite. There are any number of ways our analysis could still be improved. What we have demonstrated is that the tools we have used here are extremely powerful.

5.2 RETIREMENT PLANNING (B)

The problem statement we began this chapter with, Retirement Planning (A), was deliberately simplified to provide just the problem *kernel*. Recall that we defined the kernel of an ill-structured problem as a minimal description that suffices to generate a model structure. Now that we have built two such models, we can see how little information really needs to be in the kernel. In fact, we could have built essentially the same models without *any* numbers at all. We would have had to create placeholders for Bob's current salary, his assets, and his employer's contribution rate, but the structure of the model would have been the same. Building the model using these hypothetical inputs would have made it just a bit more difficult, but by no means impossible, to draw interesting conclusions.

The following version of the case, Retirement Planning (B), is the version we normally give to students in a modeling course. It is similar to the problem description one would expect to work with on a consulting assignment. It contains far more information than the first version, although none of the essential facts are changed. We present these cases in this order to highlight the importance of identifying the problem kernel before developing a first model. Having already built a simplified model structure, we now examine this fuller problem description and consider how to expand our model and analysis.

Retirement Planning (B)

Bob Davidson is a 46-year-old tenured professor at a small New England college. He has a daughter, Sue, age 6, and a wife, Margaret, age 40. Margaret is a potter, a vocation from which she earns no appreciable income.

Bob's plan is to work until he is 65. He is planning to save $100,000 in today's dollars for his daughter's college fund. (He expects her to attend college starting at age 18.) Upon retirement, he and his wife would like to be able to travel extensively (he estimates $30,000 annual travel expenses as a nice target), although he would be able to live quite modestly otherwise (his family expenses are about $5,500 per month today). He does not foresee moving from the small town where he now lives.

Bob's grandfather died at age 42, and his father died at age 58. Both died from cancer, although unrelated instances of that disease. Bob's health has been excellent; he is an active runner and skier. There are no inherited diseases in the family, with the exception of glaucoma. Bob's most recent serum cholesterol count was 190.

Bob's salary from the school where he works is currently $116,000, on which the school pays an additional 10 percent into a retirement fund. He expects his salary to grow 2 to 5 percent per year. His current tax rate is 35 percent.

In addition to the 10 percent regular contribution the school makes to Bob's retirement savings, Bob also contributes $9,500 per year (the current federal maximum allowed tax deferred amount is $12,000; this maximum is scheduled to grow to $15,000 in three years and thereafter roughly at the rate of inflation). Additional savings toward retirement above the maximum require him to invest after-tax dollars.

Bob contributes to Social Security as required by law, but his retirement plans assume essentially no benefits from Social Security.

Bob's tax-sheltered retirement funds currently amount to $167,000; they are in a 50 percent stocks/50 percent bonds fund with a long-term average annual return of 7 percent. In addition to his retirement assets, Bob's net worth consists of $200,000 equity in his home, $50,000 in short-term money market mutual funds (long-term average annual return of 3 percent), and $24,000 in a growth and income fund (long-term average annual return 6 percent) for his daughter's college tuition. Bob expects the future performance of all his investments to be consistent with historical returns. He has a term life insurance policy with a value of $600,000; this policy has no asset value but pays its face value as long as Bob continues to pay the premiums. He has no outstanding debts other than his mortgage ($200,000 principal balance outstanding, with a 7 percent interest rate).

Bob is concerned about the impact of inflation, which he expects to run at the long-term average rate of about 3 percent. He wants to plan under the assumption that all his primary future expenses (college, travel, etc.) will grow at this rate.

Should Bob die while insured, the proceeds on his life insurance go to his wife tax-free. Similarly, if he dies before retirement, his retirement assets go to his wife tax-free. Upon retirement, withdrawals from the funds are taxed as ordinary income.

Bob would like to know whether his current level of retirement savings is adequate.

To the Reader: Before reading on, frame this problem by setting boundaries for the analysis; making assumptions; listing questions; defining outputs, inputs, and decisions; and so on.

5.2.1 Frame the Problem—M3

This problem description provides a great deal of additional information on Bob's situation, but it changes only a few elements of our approach up to this point. It does help us to recognize several implicit assumptions we have made, and it induces us to draw additional boundaries. These boundaries include the following:

- We ignore potential income from Social Security.
- We ignore potential inheritances.
- We ignore the effects of a possible life insurance payout.
- We ignore the equity in his home (assuming he will use that to pay for long-term care, if needed).
- We ignore the money he has in short-term funds (assuming that it is a rainy-day fund he will continue to maintain but not add to from savings).

To the Reader: Before reading on, sketch an influence diagram for this problem.

5.2.2 Diagram the Problem—M3

With these assumptions behind us, there are four new elements to include in our influence diagram (Figure 5.13):

- Savings for college expenses
- The cap on tax-deferred savings for retirement
- Current family expenses
- Retirement travel expenses

The first two items affect the preretirement savings amount and therefore show up as inputs to Assets at Retirement. The latter two items are factors in postretirement consumption, which is an input to Assets at the End of Each Year.

To the Reader: Before reading on, design and build a model based on the influence diagram.

5.2.3 Build the Model—M3

In our new model, we assume Bob plans to save first toward his $100,000 goal for college expenses. After that goal is achieved, he will use his savings toward

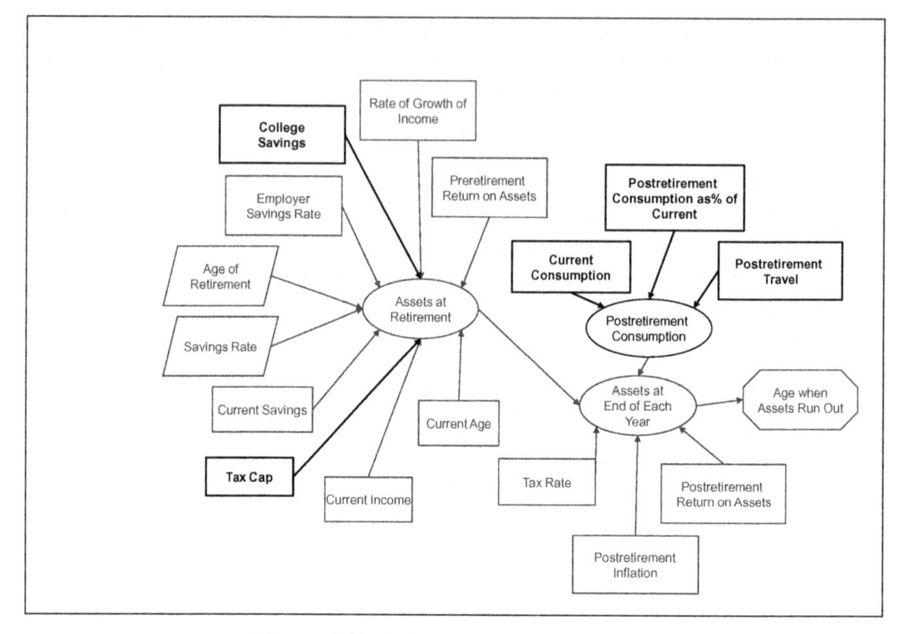

Figure 5.13. Influence diagram for M3.

retirement. The cap on tax-deferred savings requires us to compare his actual savings to the cap and calculate the taxes due on any excess. Finally, we use his current family expenses (suitably inflated) to estimate his expenses in retirement. We continue to assume that expenses after retirement are a fraction of those before retirement, but we can now base this calculation more intuitively on expenses rather than on salary. Note that we are only using current expenses to project a future consumption level. Just as in M1 and M2, we are not attempting to model his work life consumption year by year, and therefore, we do not adjust the consumption calculations in response to changes in savings or salary.

Our new model is shown in Figures 5.14 and 5.15. The changes from M2 occur in the following columns:

Column D: The cap on tax-deferred savings starts at $12,000, changes to $15,000 in three years, and thereafter grows at the rate of inflation.

Column H: The entire personal contribution goes to college savings until that reaches its target, after which it goes to the retirement fund.

Column I: The college fund grows each year with interest and contributions (if any).

Column J: The total amount actually contributed to retirement savings is the personal contribution less the contribution to the college fund and taxes (if any); the formula for age 46 is

Retirement Planning

Parameters		Results	
Current Income	$116,000	Year Funds Out	74
Current Savings	$167,000		
Employer Savings Rate	10%		
Current Savings	$9,500		
Age	46		
Return on Assets			
Preretirement	7%		
Postretirement	5%		
Income Growth Rate (Includes Inflation)	4%		
Inflation Pre- and Post- retirement	3%		
Tax Rate			
Preretirement	35%		
postretirement	25%		
Current Tax Cap	$12,000		
Future Tax Cap	$15,000		
Current Consumption Spending	$5,500		
College Fund	$24,000		
College Fund Target	$100,000		
College Fund Return	6%		
Retirement Travel Costs	$30,000		
Decisions			
Personal Savings Rate	5.00%		
Percent of pre-ret. consumption spen	70%		
Retirement age	65		

Tornado Chart Inputs	Base Case	10th Percentile	90th Percentile
Return on Assets			
Preretirement	7%	5.00%	12.00%
Postretirement	5%	3.00%	8.00%
Income Growth Rate (Includes Inflation)	4%	0.00%	6.00%
Inflation Pre- and Post- retirement	3%	1.00%	5.00%
Tax rate			
Preretirement	35%	30.00%	40.00%
Postretirement	25%	20.00%	30.00%
Current Tax Cap	$12,000	$12,000	$12,000
Future Tax Cap	$15,000	$15,000	$20,000
Current Consumption Spending	$5,500	$5,500	$5,500
College Fund	$24,000	$24,000	$24,000
College Fund Target	$100,000	$80,000	$120,000
College Fund Return	6%	4%	8%
Retirement Travel Costs	$30,000	$20,000	$45,000
Personal Savings Rate	12.50%	10.00%	15.00%
Percent of Final Income Spent	70%	60.00%	90.00%
Retirement Age	65	60	70

Model

Preretirement Age	Tax Cap	Salary	Employer contribution	Personal Contribution (before tax)	College Contribution	College Fund	Total Personal Contribution (after tax)	Beginning of Year Assets	End of Year Assets
46	$12,000	116,000	$11,600	$5,800	$5,800	$31,240	$0	$167,000	$190,290
47	$12,000	120,640	$12,064	$6,032	$6,032	$39,146	$0	$190,290	$215,674
48	$12,000	125,466	$12,547	$6,273	$6,273	$47,768	$0	$215,674	$243,318
49	$15,000	130,484	$13,048	$6,524	$6,524	$57,159	$0	$243,318	$273,399
50	$15,450	135,704	$13,570	$6,785	$6,785	$67,373	$0	$273,399	$306,107
51	$15,914	141,132	$14,113	$7,057	$7,057	$78,472	$0	$306,107	$341,648
52	$16,391	146,777	$14,678	$7,339	$7,339	$90,520	$0	$341,648	$380,241
53	$16,883	152,648	$15,265	$7,632	$7,632	$103,583	$0	$380,241	$422,122
54	$17,389	158,754	$15,875	$7,938	$0	$109,798	$7,938	$422,122	$475,484
55	$17,911	165,104	$16,510	$8,255	$0	$116,386	$8,255	$475,484	$533,534
56	$18,448	171,708	$17,171	$8,585	$0	$123,369	$8,585	$533,534	$596,637
57	$19,002	178,577	$17,858	$8,929	$0	$130,771	$8,929	$596,637	$665,188
58	$19,572	185,720	$18,572	$9,286	$0	$138,618	$9,286	$665,188	$739,609
59	$20,159	193,149	$19,315	$9,657	$0	$146,935	$9,657	$739,609	$820,354
60	$20,764	200,874	$20,087	$10,044	$0	$155,751	$10,044	$820,354	$907,910
61	$21,386	208,909	$20,891	$10,445	$0	$165,096	$10,445	$907,910	$1,002,800
62	$22,028	217,266	$21,727	$10,863	$0	$175,002	$10,863	$1,002,800	$1,105,586
63	$22,689	225,956	$22,596	$11,298	$0	$185,502	$11,298	$1,105,586	$1,216,871
64	$23,370	234,995	$23,499	$11,750	$0	$196,632	$11,750	$1,216,871	$1,337,301
65	$24,071	244,395	$24,439	$12,220	$0	$208,430	$12,220	$1,337,301	$1,467,571
66	$24,793	254,170	$25,417	$12,709	$0	$220,936	$12,709	$1,467,571	$1,608,427
67	$25,536	264,337	$26,434	$13,217	$0	$234,192	$13,217	$1,608,427	$1,760,667
68	$26,303	274,911	$27,491	$13,746	$0	$248,243	$13,746	$1,760,667	$1,925,151
69	$27,092	285,907	$28,591	$14,295	$0	$263,138	$14,295	$1,925,151	$2,102,797
70	$27,904	297,343	$29,734	$14,867	$0	$278,926	$14,867	$2,102,797	$2,294,594
71	$28,742	309,237	$30,924	$15,462	$0	$295,662	$15,462	$2,294,594	$2,501,602
72	$29,604	321,606	$32,161	$16,080	$0	$313,401	$16,080	$2,501,602	$2,724,955
73	$30,492	334,471	$33,447	$16,724	$0	$332,206	$16,724	$2,724,955	$2,965,872
74	$31,407	347,850	$34,785	$17,392	$0	$352,138	$17,392	$2,965,872	$3,225,661
75	$32,349	361,764	$36,176	$18,088	$0	$373,266	$18,088	$3,225,661	$3,505,721
76	$33,319	376,234	$37,623	$18,812	$0	$395,662	$18,812	$3,505,721	$3,807,557
77	$34,319	391,283	$39,128	$19,564	$0	$419,402	$19,564	$3,807,557	$4,132,779
78	$35,348	406,935	$40,693	$20,347	$0	$444,566	$20,347	$4,132,779	$4,483,113
79	$36,409	423,212	$42,321	$21,161	$0	$471,240	$21,161	$4,483,113	$4,860,413
80	$37,501	440,141	$44,014	$22,007	$0	$499,514	$22,007	$4,860,413	$5,266,663
81	$38,626	457,746	$45,775	$22,887	$0	$529,485	$22,887	$5,266,663	$5,703,991
82	$39,785	476,056	$47,606	$23,803	$0	$561,254	$23,803	$5,703,991	$6,174,679
83	$40,979	495,098	$49,510	$24,755	$0	$594,930	$24,755	$6,174,679	$6,681,171
84	$42,208	514,902	$51,490	$25,745	$0	$630,625	$25,745	$6,681,171	$7,226,089
85	$43,474	535,498	$53,550	$26,775	$0	$668,463	$26,775	$7,226,089	$7,812,240
86	$44,778	556,918	$55,692	$27,846	$0	$708,571	$27,846	$7,812,240	$8,442,634
87	$46,122	579,195	$57,920	$28,960	$0	$751,085	$28,960	$8,442,634	$9,120,498
88	$47,505	602,363	$60,236	$30,118	$0	$796,150	$30,118	$9,120,498	$9,849,287
89	$48,931	626,457	$62,646	$31,323	$0	$843,919	$31,323	$9,849,287	$10,632,706
90	$50,398	651,516	$65,152	$32,576	$0	$894,554	$32,576	$10,632,706	$11,474,723

Postretirement Age	Spending	Withdrawal	Beginning of Year Assets	End of Year Assets	Funds depleted? 1=Yes
65	$133,617	$178,156	$1,467,571	$1,362,794	0
66	$137,626	$183,501	$1,362,794	$1,247,432	0
67	$141,754	$189,006	$1,247,432	$1,120,798	0
68	$146,007	$194,676	$1,120,798	$982,162	0
69	$150,387	$200,516	$982,162	$830,754	0
70	$154,899	$206,532	$830,754	$665,759	0
71	$159,546	$212,728	$665,759	$486,320	0
72	$164,332	$219,110	$486,320	$291,526	0
73	$169,262	$225,683	$291,526	$80,419	0
74	$174,340	$232,453	$80,419	-$148,013	1
75	$179,570	$239,427	-$148,013	-$394,841	1
76	$184,957	$246,610	-$394,841	-$661,193	1
77	$190,506	$254,008	-$661,193	-$948,261	1
78	$196,221	$261,628	-$948,261	-$1,257,302	1
79	$202,108	$269,477	-$1,257,302	-$1,589,645	1
80	$208,171	$277,562	-$1,589,645	-$1,946,688	1
81	$214,416	$285,888	-$1,946,688	-$2,329,911	1
82	$220,849	$294,465	-$2,329,911	-$2,740,872	1
83	$227,474	$303,299	-$2,740,872	-$3,181,215	1
84	$234,299	$312,398	-$3,181,215	-$3,652,673	1
85	$241,327	$321,770	-$3,652,673	-$4,157,077	1
86	$248,567	$331,423	-$4,157,077	-$4,696,354	1
87	$256,024	$341,366	-$4,696,354	-$5,272,537	1
88	$263,705	$351,607	-$5,272,537	-$5,887,771	1
89	$271,616	$362,155	-$5,887,771	-$6,544,314	1
90	$279,765	$373,020	-$6,544,314	-$7,244,549	1
91	$288,158	$384,210	-$7,244,549	-$7,990,987	1
92	$296,802	$395,736	-$7,990,987	-$8,786,273	1
93	$305,706	$407,609	-$8,786,273	-$9,633,195	1
94	$314,878	$419,837	-$9,633,195	-$10,534,691	1
95	$324,324	$432,432	-$10,534,691	-$11,493,858	1
96	$334,054	$445,405	-$11,493,858	-$12,513,956	1
97	$344,075	$458,767	-$12,513,956	-$13,598,420	1
98	$354,398	$472,530	-$13,598,420	-$14,750,872	1
99	$365,029	$486,706	-$14,750,872	-$15,975,121	1
100	$375,980	$501,307	-$15,975,121	-$17,275,184	1

Figure 5.14. M3—numerical view.

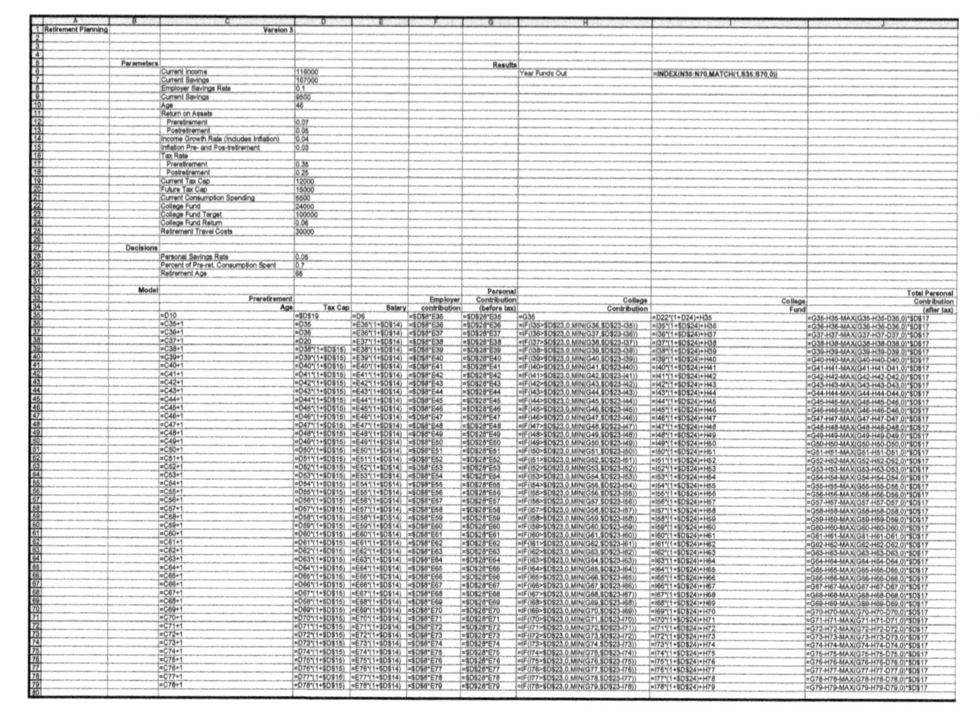

	A	B	C	D	E	F	G	H	I	J
1	Retirement Planning		Version 3							
2										
3										
4										
5		Parameters					Results			
6			Current Income	116000				Year Funds Out	=INDEX(N35:N70,MATCH(1,S35:S70,0))	
7			Current Savings	167000						
8			Employer Savings Rate	0.1						
9			Current Savings	9500						
10			Age	46						
11			Return on Assets							
12			Preretirement	0.07						
13			Postretirement	0.05						
14			Income Growth Rate (Includes Inflation)	0.04						
15			Inflation Pre- and Pos-tretirement	0.03						
16			Tax Rate							
17			Preretirement	0.35						
18			Postretirement	0.25						
19			Current Tax Cap	12000						
20			Future Tax Cap	15000						
21			Current Consumption Spending	5500						
22			College Fund	24000						
23			College Fund Target	100000						
24			College Fund Return	0.06						
25			Retirement Travel Costs	30000						
26										
27		Decisions								
28			Personal Savings Rate	0.05						
29			Percent of Pre-ret. Consumption Spent	0.7						
30			Retirement Age	65						
31										
32		Model					Personal			Total Personal
33			Preretirement			Employer	Contribution	College	College	Contribution
34			Age	Tax Cap	Salary	contribution	(before tax)	Contribution	Fund	(after tax)
35			=D10	=$D$19	=D6	=$D$8*E35	=$D$28*E35	=G35	=D22*(1+D24)+H35	=G35-H35-MAX(G35-H35-D35,0)*$D$17
36			=C35+1	=D35	=E35*(1+$D$14)	=$D$8*E36	=$D$28*E36	=IF(I35>$D$23,0,MIN(G36,$D$23-I35))	=I35*(1+$D$24)+H36	=G36-H36-MAX(G36-H36-D36,0)*$D$17
37			=C36+1	=D36	=E36*(1+$D$14)	=$D$8*E37	=$D$28*E37	=IF(I36>$D$23,0,MIN(G37,$D$23-I36))	=I36*(1+$D$24)+H37	=G37-H37-MAX(G37-H37-D37,0)*$D$17
38			=C37+1	=D20	=E37*(1+$D$14)	=$D$8*E38	=$D$28*E38	=IF(I37>$D$23,0,MIN(G38,$D$23-I37))	=I37*(1+$D$24)+H38	=G38-H38-MAX(G38-H38-D38,0)*$D$17
39			=C38+1	=D38*(1+$D$15)	=E38*(1+$D$14)	=$D$8*E39	=$D$28*E39	=IF(I38>$D$23,0,MIN(G39,$D$23-I38))	=I38*(1+$D$24)+H39	=G39-H39-MAX(G39-H39-D39,0)*$D$17
40			=C39+1	=D39*(1+$D$15)	=E39*(1+$D$14)	=$D$8*E40	=$D$28*E40	=IF(I39>$D$23,0,MIN(G40,$D$23-I39))	=I39*(1+$D$24)+H40	=G40-H40-MAX(G40-H40-D40,0)*$D$17
41			=C40+1	=D40*(1+$D$15)	=E40*(1+$D$14)	=$D$8*E41	=$D$28*E41	=IF(I40>$D$23,0,MIN(G41,$D$23-I40))	=I40*(1+$D$24)+H41	=G41-H41-MAX(G41-H41-D41,0)*$D$17
42			=C41+1	=D41*(1+$D$15)	=E41*(1+$D$14)	=$D$8*E42	=$D$28*E42	=IF(I41>$D$23,0,MIN(G42,$D$23-I41))	=I41*(1+$D$24)+H42	=G42-H42-MAX(G42-H42-D42,0)*$D$17
43			=C42+1	=D42*(1+$D$15)	=E42*(1+$D$14)	=$D$8*E43	=$D$28*E43	=IF(I42>$D$23,0,MIN(G43,$D$23-I42))	=I42*(1+$D$24)+H43	=G43-H43-MAX(G43-H43-D43,0)*$D$17
44			=C43+1	=D43*(1+$D$15)	=E43*(1+$D$14)	=$D$8*E44	=$D$28*E44	=IF(I43>$D$23,0,MIN(G44,$D$23-I43))	=I43*(1+$D$24)+H44	=G44-H44-MAX(G44-H44-D44,0)*$D$17
45			=C44+1	=D44*(1+$D$15)	=E44*(1+$D$14)	=$D$8*E45	=$D$28*E45	=IF(I44>$D$23,0,MIN(G45,$D$23-I44))	=I44*(1+$D$24)+H45	=G45-H45-MAX(G45-H45-D45,0)*$D$17
46			=C45+1	=D45*(1+$D$15)	=E45*(1+$D$14)	=$D$8*E46	=$D$28*E46	=IF(I45>$D$23,0,MIN(G46,$D$23-I45))	=I45*(1+$D$24)+H46	=G46-H46-MAX(G46-H46-D46,0)*$D$17
47			=C46+1	=D46*(1+$D$15)	=E46*(1+$D$14)	=$D$8*E47	=$D$28*E47	=IF(I46>$D$23,0,MIN(G47,$D$23-I46))	=I46*(1+$D$24)+H47	=G47-H47-MAX(G47-H47-D47,0)*$D$17
48			=C47+1	=D47*(1+$D$15)	=E47*(1+$D$14)	=$D$8*E48	=$D$28*E48	=IF(I47>$D$23,0,MIN(G48,$D$23-I47))	=I47*(1+$D$24)+H48	=G48-H48-MAX(G48-H48-D48,0)*$D$17
49			=C48+1	=D48*(1+$D$15)	=E48*(1+$D$14)	=$D$8*E49	=$D$28*E49	=IF(I48>$D$23,0,MIN(G49,$D$23-I48))	=I48*(1+$D$24)+H49	=G49-H49-MAX(G49-H49-D49,0)*$D$17
50			=C49+1	=D49*(1+$D$15)	=E49*(1+$D$14)	=$D$8*E50	=$D$28*E50	=IF(I49>$D$23,0,MIN(G50,$D$23-I49))	=I49*(1+$D$24)+H50	=G50-H50-MAX(G50-H50-D50,0)*$D$17
51			=C50+1	=D50*(1+$D$15)	=E50*(1+$D$14)	=$D$8*E51	=$D$28*E51	=IF(I50>$D$23,0,MIN(G51,$D$23-I50))	=I50*(1+$D$24)+H51	=G51-H51-MAX(G51-H51-D51,0)*$D$17
52			=C51+1	=D51*(1+$D$15)	=E51*(1+$D$14)	=$D$8*E52	=$D$28*E52	=IF(I51>$D$23,0,MIN(G52,$D$23-I51))	=I51*(1+$D$24)+H52	=G52-H52-MAX(G52-H52-D52,0)*$D$17
53			=C52+1	=D52*(1+$D$15)	=E52*(1+$D$14)	=$D$8*E53	=$D$28*E53	=IF(I52>$D$23,0,MIN(G53,$D$23-I52))	=I52*(1+$D$24)+H53	=G53-H53-MAX(G53-H53-D53,0)*$D$17
54			=C53+1	=D53*(1+$D$15)	=E53*(1+$D$14)	=$D$8*E54	=$D$28*E54	=IF(I53>$D$23,0,MIN(G54,$D$23-I53))	=I53*(1+$D$24)+H54	=G54-H54-MAX(G54-H54-D54,0)*$D$17
55			=C54+1	=D54*(1+$D$15)	=E54*(1+$D$14)	=$D$8*E55	=$D$28*E55	=IF(I54>$D$23,0,MIN(G55,$D$23-I54))	=I54*(1+$D$24)+H55	=G55-H55-MAX(G55-H55-D55,0)*$D$17
56			=C55+1	=D55*(1+$D$15)	=E55*(1+$D$14)	=$D$8*E56	=$D$28*E56	=IF(I55>$D$23,0,MIN(G56,$D$23-I55))	=I55*(1+$D$24)+H56	=G56-H56-MAX(G56-H56-D56,0)*$D$17
57			=C56+1	=D56*(1+$D$15)	=E56*(1+$D$14)	=$D$8*E57	=$D$28*E57	=IF(I56>$D$23,0,MIN(G57,$D$23-I56))	=I56*(1+$D$24)+H57	=G57-H57-MAX(G57-H57-D57,0)*$D$17
58			=C57+1	=D57*(1+$D$15)	=E57*(1+$D$14)	=$D$8*E58	=$D$28*E58	=IF(I57>$D$23,0,MIN(G58,$D$23-I57))	=I57*(1+$D$24)+H58	=G58-H58-MAX(G58-H58-D58,0)*$D$17
59			=C58+1	=D58*(1+$D$15)	=E58*(1+$D$14)	=$D$8*E59	=$D$28*E59	=IF(I58>$D$23,0,MIN(G59,$D$23-I58))	=I58*(1+$D$24)+H59	=G59-H59-MAX(G59-H59-D59,0)*$D$17
60			=C59+1	=D59*(1+$D$15)	=E59*(1+$D$14)	=$D$8*E60	=$D$28*E60	=IF(I59>$D$23,0,MIN(G60,$D$23-I59))	=I59*(1+$D$24)+H60	=G60-H60-MAX(G60-H60-D60,0)*$D$17
61			=C60+1	=D60*(1+$D$15)	=E60*(1+$D$14)	=$D$8*E61	=$D$28*E61	=IF(I60>$D$23,0,MIN(G61,$D$23-I60))	=I60*(1+$D$24)+H61	=G61-H61-MAX(G61-H61-D61,0)*$D$17
62			=C61+1	=D61*(1+$D$15)	=E61*(1+$D$14)	=$D$8*E62	=$D$28*E62	=IF(I61>$D$23,0,MIN(G62,$D$23-I61))	=I61*(1+$D$24)+H62	=G62-H62-MAX(G62-H62-D62,0)*$D$17
63			=C62+1	=D62*(1+$D$15)	=E62*(1+$D$14)	=$D$8*E63	=$D$28*E63	=IF(I62>$D$23,0,MIN(G63,$D$23-I62))	=I62*(1+$D$24)+H63	=G63-H63-MAX(G63-H63-D63,0)*$D$17
64			=C63+1	=D63*(1+$D$15)	=E63*(1+$D$14)	=$D$8*E64	=$D$28*E64	=IF(I63>$D$23,0,MIN(G64,$D$23-I63))	=I63*(1+$D$24)+H64	=G64-H64-MAX(G64-H64-D64,0)*$D$17
65			=C64+1	=D64*(1+$D$15)	=E64*(1+$D$14)	=$D$8*E65	=$D$28*E65	=IF(I64>$D$23,0,MIN(G65,$D$23-I64))	=I64*(1+$D$24)+H65	=G65-H65-MAX(G65-H65-D65,0)*$D$17
66			=C65+1	=D65*(1+$D$15)	=E65*(1+$D$14)	=$D$8*E66	=$D$28*E66	=IF(I65>$D$23,0,MIN(G66,$D$23-I65))	=I65*(1+$D$24)+H66	=G66-H66-MAX(G66-H66-D66,0)*$D$17
67			=C66+1	=D66*(1+$D$15)	=E66*(1+$D$14)	=$D$8*E67	=$D$28*E67	=IF(I66>$D$23,0,MIN(G67,$D$23-I66))	=I66*(1+$D$24)+H67	=G67-H67-MAX(G67-H67-D67,0)*$D$17
68			=C67+1	=D67*(1+$D$15)	=E67*(1+$D$14)	=$D$8*E68	=$D$28*E68	=IF(I67>$D$23,0,MIN(G68,$D$23-I67))	=I67*(1+$D$24)+H68	=G68-H68-MAX(G68-H68-D68,0)*$D$17
69			=C68+1	=D68*(1+$D$15)	=E68*(1+$D$14)	=$D$8*E69	=$D$28*E69	=IF(I68>$D$23,0,MIN(G69,$D$23-I68))	=I68*(1+$D$24)+H69	=G69-H69-MAX(G69-H69-D69,0)*$D$17
70			=C69+1	=D69*(1+$D$15)	=E69*(1+$D$14)	=$D$8*E70	=$D$28*E70	=IF(I69>$D$23,0,MIN(G70,$D$23-I69))	=I69*(1+$D$24)+H70	=G70-H70-MAX(G70-H70-D70,0)*$D$17
71			=C70+1	=D70*(1+$D$15)	=E70*(1+$D$14)	=$D$8*E71	=$D$28*E71	=IF(I70>$D$23,0,MIN(G71,$D$23-I70))	=I70*(1+$D$24)+H71	=G71-H71-MAX(G71-H71-D71,0)*$D$17
72			=C71+1	=D71*(1+$D$15)	=E71*(1+$D$14)	=$D$8*E72	=$D$28*E72	=IF(I71>$D$23,0,MIN(G72,$D$23-I71))	=I71*(1+$D$24)+H72	=G72-H72-MAX(G72-H72-D72,0)*$D$17
73			=C72+1	=D72*(1+$D$15)	=E72*(1+$D$14)	=$D$8*E73	=$D$28*E73	=IF(I72>$D$23,0,MIN(G73,$D$23-I72))	=I72*(1+$D$24)+H73	=G73-H73-MAX(G73-H73-D73,0)*$D$17
74			=C73+1	=D73*(1+$D$15)	=E73*(1+$D$14)	=$D$8*E74	=$D$28*E74	=IF(I73>$D$23,0,MIN(G74,$D$23-I73))	=I73*(1+$D$24)+H74	=G74-H74-MAX(G74-H74-D74,0)*$D$17
75			=C74+1	=D74*(1+$D$15)	=E74*(1+$D$14)	=$D$8*E75	=$D$28*E75	=IF(I74>$D$23,0,MIN(G75,$D$23-I74))	=I74*(1+$D$24)+H75	=G75-H75-MAX(G75-H75-D75,0)*$D$17
76			=C75+1	=D75*(1+$D$15)	=E75*(1+$D$14)	=$D$8*E76	=$D$28*E76	=IF(I75>$D$23,0,MIN(G76,$D$23-I75))	=I75*(1+$D$24)+H76	=G76-H76-MAX(G76-H76-D76,0)*$D$17
77			=C76+1	=D76*(1+$D$15)	=E76*(1+$D$14)	=$D$8*E77	=$D$28*E77	=IF(I76>$D$23,0,MIN(G77,$D$23-I76))	=I76*(1+$D$24)+H77	=G77-H77-MAX(G77-H77-D77,0)*$D$17
78			=C77+1	=D77*(1+$D$15)	=E77*(1+$D$14)	=$D$8*E78	=$D$28*E78	=IF(I77>$D$23,0,MIN(G78,$D$23-I77))	=I77*(1+$D$24)+H78	=G78-H78-MAX(G78-H78-D78,0)*$D$17
79			=C78+1	=D78*(1+$D$15)	=E78*(1+$D$14)	=$D$8*E79	=$D$28*E79	=IF(I78>$D$23,0,MIN(G79,$D$23-I78))	=I78*(1+$D$24)+H79	=G79-H79-MAX(G79-H79-D79,0)*$D$17
80										

Figure 5.15. M3—relationship view.

$$G35\text{-}H35\text{-}MAX(G35\text{-}H35\text{-}D35, 0)*\$D\$17$$

Column O: Spending postretirement is calculated by taking current annual spending (\$5,500 × 12 = \$66,000), inflating it to the retirement year, multiplying by the spending fraction, and adding the inflated travel costs.

M3 is an expanded and refined version of M2, based on the new information contained in Retirement Planning (B) and an additional set of assumptions. The structural changes are the three mentioned above: saving for college, the tax cap, and the calculation of retirement spending. In addition, we adjust several of the parameters from the previous model. For example, Bob expects his retirement assets to earn 7 percent, not 8 percent, so that input is lower in this model. His current tax rate is 35 percent, so we assume his postretirement tax rate will be somewhat lower at 25 percent.

To the Reader: Before reading on, use M3 to generate as many interesting insights into Bob's decision as you can.

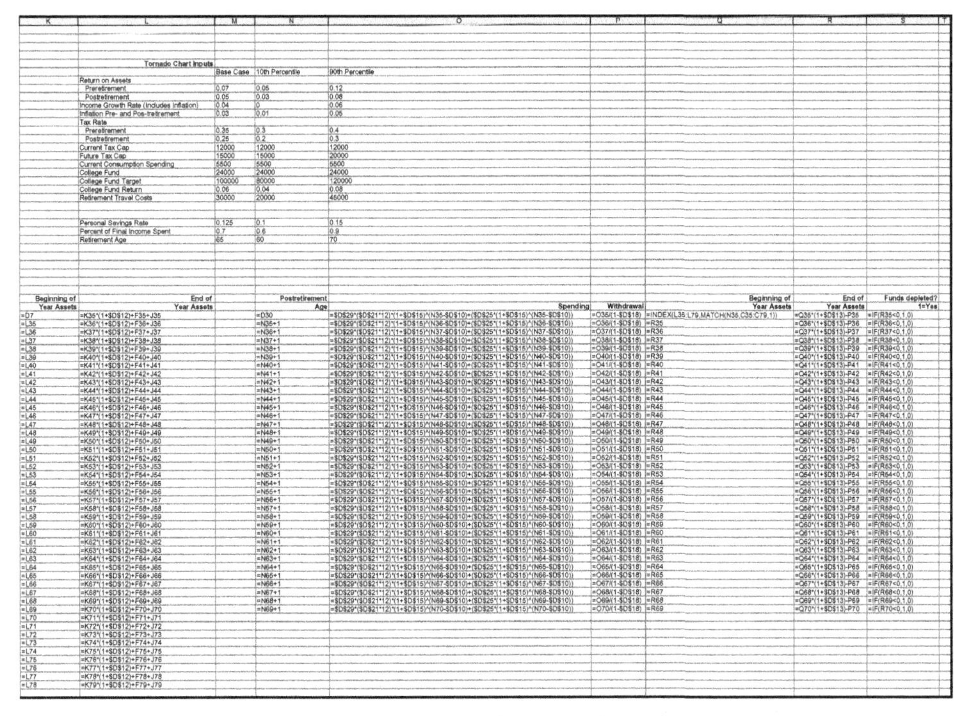

Tornado Chart Inputs	Base Case	10th Percentile	90th Percentile
Return on Assets			
Preretirement	0.07	0.05	0.12
Postretirement	0.05	0.03	0.08
Income Growth Rate (includes Inflation)	0.04	0	0.06
Inflation Pre- and Pos-tretirement	0.03	0.01	0.05
Tax Rate			
Preretirement	0.35	0.3	0.4
Postretirement	0.25	0.2	0.3
Current Tax Cap	12000	12000	12000
Future Tax Cap	15000	15000	20000
Current Consumption Spending	5500	5500	5500
College Fund	24000	24000	24000
College Fund Target	100000	80000	120000
College Fund Return	0.06	0.04	0.08
Retirement Travel Costs	30000	20000	45000
Personal Savings Rate	0.125	0.1	0.15
Percent of Final Income Spent	0.7	0.6	0.9
Retirement Age	65	60	70

Beginning of Year Assets	End of Year Assets	Postretirement Age	Spending	Withdrawal	Beginning of Year Assets	End of Year Assets	Funds depleted? 1=Yes
=D7	=K35*(1+$D$12)+F35+J35	=D30	=$D$29*($D$21*12)*(1+$D$15)^(N35-$D$10)+($D$25*(1+$D$15)^(N35-$D$10))	=O35/(1-$D$18)	=INDEX(L35:L79,MATCH(N35,C35:C79,1))	=Q35*(1+$D$13)-P35	=IF(R35<0,1,0)
=L35	=K36*(1+$D$12)+F36+J36	=N35+1	=$D$29*($D$21*12)*(1+$D$15)^(N36-$D$10)+($D$25*(1+$D$15)^(N36-$D$10))	=O36/(1-$D$18)	=R35	=Q36*(1+$D$13)-P36	=IF(R36<0,1,0)
=L36	=K37*(1+$D$12)+F37+J37	=N36+1	=$D$29*($D$21*12)*(1+$D$15)^(N37-$D$10)+($D$25*(1+$D$15)^(N37-$D$10))	=O37/(1-$D$18)	=R36	=Q37*(1+$D$13)-P37	=IF(R37<0,1,0)
=L37	=K38*(1+$D$12)+F38+J38	=N37+1	=$D$29*($D$21*12)*(1+$D$15)^(N38-$D$10)+($D$25*(1+$D$15)^(N38-$D$10))	=O38/(1-$D$18)	=R37	=Q38*(1+$D$13)-P38	=IF(R38<0,1,0)
=L38	=K39*(1+$D$12)+F39+J39	=N38+1	=$D$29*($D$21*12)*(1+$D$15)^(N39-$D$10)+($D$25*(1+$D$15)^(N39-$D$10))	=O39/(1-$D$18)	=R38	=Q39*(1+$D$13)-P39	=IF(R39<0,1,0)
=L39	=K40*(1+$D$12)+F40+J40	=N39+1	=$D$29*($D$21*12)*(1+$D$15)^(N40-$D$10)+($D$25*(1+$D$15)^(N40-$D$10))	=O40/(1-$D$18)	=R39	=Q40*(1+$D$13)-P40	=IF(R40<0,1,0)
=L40	=K41*(1+$D$12)+F41+J41	=N40+1	=$D$29*($D$21*12)*(1+$D$15)^(N41-$D$10)+($D$25*(1+$D$15)^(N41-$D$10))	=O41/(1-$D$18)	=R40	=Q41*(1+$D$13)-P41	=IF(R41<0,1,0)
=L41	=K42*(1+$D$12)+F42+J42	=N41+1	=$D$29*($D$21*12)*(1+$D$15)^(N42-$D$10)+($D$25*(1+$D$15)^(N42-$D$10))	=O42/(1-$D$18)	=R41	=Q42*(1+$D$13)-P42	=IF(R42<0,1,0)
=L42	=K43*(1+$D$12)+F43+J43	=N42+1	=$D$29*($D$21*12)*(1+$D$15)^(N43-$D$10)+($D$25*(1+$D$15)^(N43-$D$10))	=O43/(1-$D$18)	=R42	=Q43*(1+$D$13)-P43	=IF(R43<0,1,0)
=L43	=K44*(1+$D$12)+F44+J44	=N43+1	=$D$29*($D$21*12)*(1+$D$15)^(N44-$D$10)+($D$25*(1+$D$15)^(N44-$D$10))	=O44/(1-$D$18)	=R43	=Q44*(1+$D$13)-P44	=IF(R44<0,1,0)
=L44	=K45*(1+$D$12)+F45+J45	=N44+1	=$D$29*($D$21*12)*(1+$D$15)^(N45-$D$10)+($D$25*(1+$D$15)^(N45-$D$10))	=O45/(1-$D$18)	=R44	=Q45*(1+$D$13)-P45	=IF(R45<0,1,0)
=L45	=K46*(1+$D$12)+F46+J46	=N45+1	=$D$29*($D$21*12)*(1+$D$15)^(N46-$D$10)+($D$25*(1+$D$15)^(N46-$D$10))	=O46/(1-$D$18)	=R45	=Q46*(1+$D$13)-P46	=IF(R46<0,1,0)
=L46	=K47*(1+$D$12)+F47+J47	=N46+1	=$D$29*($D$21*12)*(1+$D$15)^(N47-$D$10)+($D$25*(1+$D$15)^(N47-$D$10))	=O47/(1-$D$18)	=R46	=Q47*(1+$D$13)-P47	=IF(R47<0,1,0)
=L47	=K48*(1+$D$12)+F48+J48	=N47+1	=$D$29*($D$21*12)*(1+$D$15)^(N48-$D$10)+($D$25*(1+$D$15)^(N48-$D$10))	=O48/(1-$D$18)	=R47	=Q48*(1+$D$13)-P48	=IF(R48<0,1,0)
=L48	=K49*(1+$D$12)+F49+J49	=N48+1	=$D$29*($D$21*12)*(1+$D$15)^(N49-$D$10)+($D$25*(1+$D$15)^(N49-$D$10))	=O49/(1-$D$18)	=R48	=Q49*(1+$D$13)-P49	=IF(R49<0,1,0)
=L49	=K50*(1+$D$12)+F50+J50	=N49+1	=$D$29*($D$21*12)*(1+$D$15)^(N50-$D$10)+($D$25*(1+$D$15)^(N50-$D$10))	=O50/(1-$D$18)	=R49	=Q50*(1+$D$13)-P50	=IF(R50<0,1,0)
=L50	=K51*(1+$D$12)+F51+J51	=N50+1	=$D$29*($D$21*12)*(1+$D$15)^(N51-$D$10)+($D$25*(1+$D$15)^(N51-$D$10))	=O51/(1-$D$18)	=R50	=Q51*(1+$D$13)-P51	=IF(R51<0,1,0)
=L51	=K52*(1+$D$12)+F52+J52	=N51+1	=$D$29*($D$21*12)*(1+$D$15)^(N52-$D$10)+($D$25*(1+$D$15)^(N52-$D$10))	=O52/(1-$D$18)	=R51	=Q52*(1+$D$13)-P52	=IF(R52<0,1,0)
=L52	=K53*(1+$D$12)+F53+J53	=N52+1	=$D$29*($D$21*12)*(1+$D$15)^(N53-$D$10)+($D$25*(1+$D$15)^(N53-$D$10))	=O53/(1-$D$18)	=R52	=Q53*(1+$D$13)-P53	=IF(R53<0,1,0)
=L53	=K54*(1+$D$12)+F54+J54	=N53+1	=$D$29*($D$21*12)*(1+$D$15)^(N54-$D$10)+($D$25*(1+$D$15)^(N54-$D$10))	=O54/(1-$D$18)	=R53	=Q54*(1+$D$13)-P54	=IF(R54<0,1,0)
=L54	=K55*(1+$D$12)+F55+J55	=N54+1	=$D$29*($D$21*12)*(1+$D$15)^(N55-$D$10)+($D$25*(1+$D$15)^(N55-$D$10))	=O55/(1-$D$18)	=R54	=Q55*(1+$D$13)-P55	=IF(R55<0,1,0)
=L55	=K56*(1+$D$12)+F56+J56	=N55+1	=$D$29*($D$21*12)*(1+$D$15)^(N56-$D$10)+($D$25*(1+$D$15)^(N56-$D$10))	=O56/(1-$D$18)	=R55	=Q56*(1+$D$13)-P56	=IF(R56<0,1,0)
=L56	=K57*(1+$D$12)+F57+J57	=N56+1	=$D$29*($D$21*12)*(1+$D$15)^(N57-$D$10)+($D$25*(1+$D$15)^(N57-$D$10))	=O57/(1-$D$18)	=R56	=Q57*(1+$D$13)-P57	=IF(R57<0,1,0)
=L57	=K58*(1+$D$12)+F58+J58	=N57+1	=$D$29*($D$21*12)*(1+$D$15)^(N58-$D$10)+($D$25*(1+$D$15)^(N58-$D$10))	=O58/(1-$D$18)	=R57	=Q58*(1+$D$13)-P58	=IF(R58<0,1,0)
=L58	=K59*(1+$D$12)+F59+J59	=N58+1	=$D$29*($D$21*12)*(1+$D$15)^(N59-$D$10)+($D$25*(1+$D$15)^(N59-$D$10))	=O59/(1-$D$18)	=R58	=Q59*(1+$D$13)-P59	=IF(R59<0,1,0)
=L59	=K60*(1+$D$12)+F60+J60	=N59+1	=$D$29*($D$21*12)*(1+$D$15)^(N60-$D$10)+($D$25*(1+$D$15)^(N60-$D$10))	=O60/(1-$D$18)	=R59	=Q60*(1+$D$13)-P60	=IF(R60<0,1,0)
=L60	=K61*(1+$D$12)+F61+J61	=N60+1	=$D$29*($D$21*12)*(1+$D$15)^(N61-$D$10)+($D$25*(1+$D$15)^(N61-$D$10))	=O61/(1-$D$18)	=R60	=Q61*(1+$D$13)-P61	=IF(R61<0,1,0)
=L61	=K62*(1+$D$12)+F62+J62	=N61+1	=$D$29*($D$21*12)*(1+$D$15)^(N62-$D$10)+($D$25*(1+$D$15)^(N62-$D$10))	=O62/(1-$D$18)	=R61	=Q62*(1+$D$13)-P62	=IF(R62<0,1,0)
=L62	=K63*(1+$D$12)+F63+J63	=N62+1	=$D$29*($D$21*12)*(1+$D$15)^(N63-$D$10)+($D$25*(1+$D$15)^(N63-$D$10))	=O63/(1-$D$18)	=R62	=Q63*(1+$D$13)-P63	=IF(R63<0,1,0)
=L63	=K64*(1+$D$12)+F64+J64	=N63+1	=$D$29*($D$21*12)*(1+$D$15)^(N64-$D$10)+($D$25*(1+$D$15)^(N64-$D$10))	=O64/(1-$D$18)	=R63	=Q64*(1+$D$13)-P64	=IF(R64<0,1,0)
=L64	=K65*(1+$D$12)+F65+J65	=N64+1	=$D$29*($D$21*12)*(1+$D$15)^(N65-$D$10)+($D$25*(1+$D$15)^(N65-$D$10))	=O65/(1-$D$18)	=R64	=Q65*(1+$D$13)-P65	=IF(R65<0,1,0)
=L65	=K66*(1+$D$12)+F66+J66	=N65+1	=$D$29*($D$21*12)*(1+$D$15)^(N66-$D$10)+($D$25*(1+$D$15)^(N66-$D$10))	=O66/(1-$D$18)	=R65	=Q66*(1+$D$13)-P66	=IF(R66<0,1,0)
=L66	=K67*(1+$D$12)+F67+J67	=N66+1	=$D$29*($D$21*12)*(1+$D$15)^(N67-$D$10)+($D$25*(1+$D$15)^(N67-$D$10))	=O67/(1-$D$18)	=R66	=Q67*(1+$D$13)-P67	=IF(R67<0,1,0)
=L67	=K68*(1+$D$12)+F68+J68	=N67+1	=$D$29*($D$21*12)*(1+$D$15)^(N68-$D$10)+($D$25*(1+$D$15)^(N68-$D$10))	=O68/(1-$D$18)	=R67	=Q68*(1+$D$13)-P68	=IF(R68<0,1,0)
=L68	=K69*(1+$D$12)+F69+J69	=N68+1	=$D$29*($D$21*12)*(1+$D$15)^(N69-$D$10)+($D$25*(1+$D$15)^(N69-$D$10))	=O69/(1-$D$18)	=R68	=Q69*(1+$D$13)-P69	=IF(R69<0,1,0)
=L69	=K70*(1+$D$12)+F70+J70	=N69+1	=$D$29*($D$21*12)*(1+$D$15)^(N70-$D$10)+($D$25*(1+$D$15)^(N70-$D$10))	=O70/(1-$D$18)	=R69	=Q70*(1+$D$13)-P70	=IF(R70<0,1,0)
=L70	=K71*(1+$D$12)+F71+J71						
=L71	=K72*(1+$D$12)+F72+J72						
=L72	=K73*(1+$D$12)+F73+J73						
=L73	=K74*(1+$D$12)+F74+J74						
=L74	=K75*(1+$D$12)+F75+J75						
=L75	=K76*(1+$D$12)+F76+J76						
=L76	=K77*(1+$D$12)+F77+J77						
=L77	=K78*(1+$D$12)+F78+J78						
=L78	=K79*(1+$D$12)+F79+J79						

Figure 5.15. *(continued)*

5.2.4 Generate Insights—M3

You might expect after making such significant changes to the model that our conclusions and insights would also change. But this is not the case. In fact, when we use essentially the same input parameters and the same values for the decision variables in M3 as in M2, the year funds run out changes by only one year.

Insight 11: Given the M3 base-case assumptions, assets run out at age 75. This is within one year of the base-case result of M2.

Moreover, the sensitivity results do not change much either. Figure 5.16 shows the sensitivity of the output to the savings rate. Whether Bob saves 1 percent or 20 percent of his income, his assets will run out between ages 73 and 79. (The comparable results for M2 are ages 71 and 79.)

Insight 12: The age when assets run out is relatively insensitive to the savings rate. Assets will run out somewhere between ages 73 and 79 for all reasonable values of the savings rate.

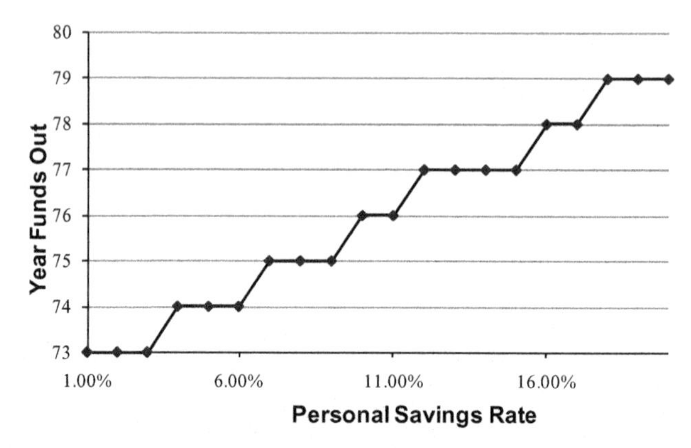

Figure 5.16. Sensitivity of Year Funds Out to Savings Rate (M3).

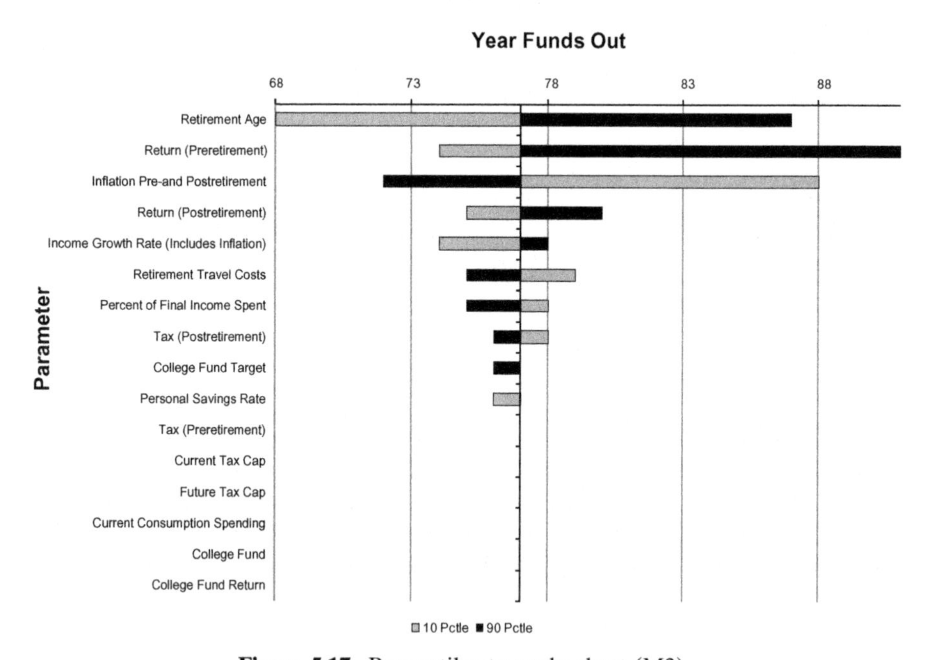

Figure 5.17. Percentiles tornado chart (M3).

A Percentiles Tornado Chart analysis (Figure 5.17) shows once again that the most important input is the Retirement Age. After that, we find the return on assets preretirement and the inflation rate are also important. Inflation plays a bigger role in this model because it affects all the years, not just the years after retirement.

Insight 13: The age of retirement continues to be the most influential input and inflation has more effect than in M2.

The rate at which income grows, which was an important factor in the previous model, is no longer important. That is because in this new version, we calculate postretirement spending not as a fraction of the final year's income but as a fraction of current spending inflated. Thus, the growth of income has a reduced effect.

Insight 14: Income growth has only a moderate effect since it is no longer used to calculate postretirement consumption.

It is time to stop and reflect on what we have accomplished to this point. We have built a sequence of three deterministic models, starting with the simplest credible model based on the minimal description of the problem. We then elaborated on that model, first by refining it to make it more flexible for analysis and then by taking into account the increased realism the more complete problem description provided. As we progressed from one model to the next, we performed basic sensitivity tests at each stage and stopped long enough to formulate preliminary insights for our client. We did not refine the model simply to make it more "realistic." Rather, we added refinements cautiously, only when we felt they were essential. And by formulating a sequence of insights, we were able to test the extent to which our insights did or did not change as the model changed.

You may have begun to feel that the story was getting repetitive toward the end of this process. We agree. When we modified M2 to create M3, it may have seemed that we made large changes. It certainly took several hours of careful spreadsheet work. But the conclusions hardly changed at all from the earlier model. No doubt the last model forms a more solid basis for making actual recommendations to our client, since it incorporates more of the essential features of his situation. But, in fact, if we had had only enough time to build M2, we would have made essentially the same recommendations. This is the most important lesson of this case: Iterative modeling leads to a sequence of insights that become more and more predictable. If, as you refine your model and test its implications, you find the same result coming out time after time, you can begin to trust that it is a sound conclusion. And you can be confident that more refinements to the model are not likely to change your conclusions substantially. Iterative modeling should give you a sense of "zeroing in" on certain insights with each successive model.

5.3 RETIREMENT PLANNING (C)

In Chapter 2, we discussed strategy analysis as a tool for generating insights. A strategy is a set of values for parameters and decision variables that make sense together or that tell a story. Up to this point in the Retirement Planning case, we have considered Bob's choices (his savings rate, his retirement age, and his postretirement spending) as independent decisions. But, clearly, he could retire sooner if he saved more of his income, and he would have to retire later if he wanted to spend more in retirement. In other words, some choices naturally make sense together and some do not. In this final section, we show how strategy analysis can be applied in the Retirement Planning case.

We start by providing additional information:

Retirement Planning (C)

Based on your initial analysis, Bob realizes his current savings plan is inadequate for achieving his goals. He has decided that two alternative approaches are worth considering, each one making a different trade-off between current lifestyle and level of investment performance risk. His "Big Saver" strategy is to push his tax-sheltered savings to the limits allowed by law and shift the funds to a more aggressive 60/40 stocks versus bonds allocation, plus save an additional $15,000 after tax in a growth and income fund. His "Go for Upside" strategy is to invest his entire retirement fund in stocks and save $3,000 per year after tax. The advantage of the Go for Upside strategy is that he will not have to reduce dramatically his expenditures during the years leading up to retirement.

The specific parameters that define each of the three strategies are summarized below:

Decision/Parameter	Base Case	Big Saver	Go for Upside
Retirement age	65	65	65
Target college fund	$100	$100	$100
Annual pretax contribution	5%	max	max
After-tax savings	$0	$15,000	$3,000
Retirement spending percent	70%	70%	70%
Fund allocation: % stocks	50%	60%	100%
Fund allocation: % bonds	50%	40%	0%

To facilitate comparison of risk and return among the strategies, Bob has assessed his beliefs about the uncertainty range for the critical uncertain inputs. He has defined "low-case" and "high-case" inputs that are his best guess of the ranges of possible values for each of the six uncertain inputs (see table below):

Uncertainty Assessment	Low	Base	High
Salary growth	2%	3%	5%
Inflation impact on expenses	2%	3%	4%
Return on stocks	5%	10%	15%
Return on bonds	2%	4%	6%
Return on growth and income fund	4%	6%	8%
Fund return postretirement	3%	4%	5%

To the Reader: Revise the influence diagram for M3 to take the new information in the case into account and modify M3 appropriately.

5.3.1 Diagram the Problem and Build the Model—M4

These new assumptions require us to make some minor changes to our influence diagram and model, although the overall structure remains the same. The additions to the influence diagram (Figure 5.18) are as follows:

- Returns on various asset classes
- Investment portfolio weights among those assets classes
- Post-tax savings contribution

We create three new models based on M3 in order to keep the logic of the three scenarios distinct. The only changes required in the base-case model are to include investment returns from stocks and bonds and to average these returns when calculating the annual return on retirement assets. In the Big Saver model, we have to distinguish the pretax contribution, which is set at the pretax cap, and the aftertax contribution, which is $15,000 each year. We assume college savings is taken from the pretax contribution until the college

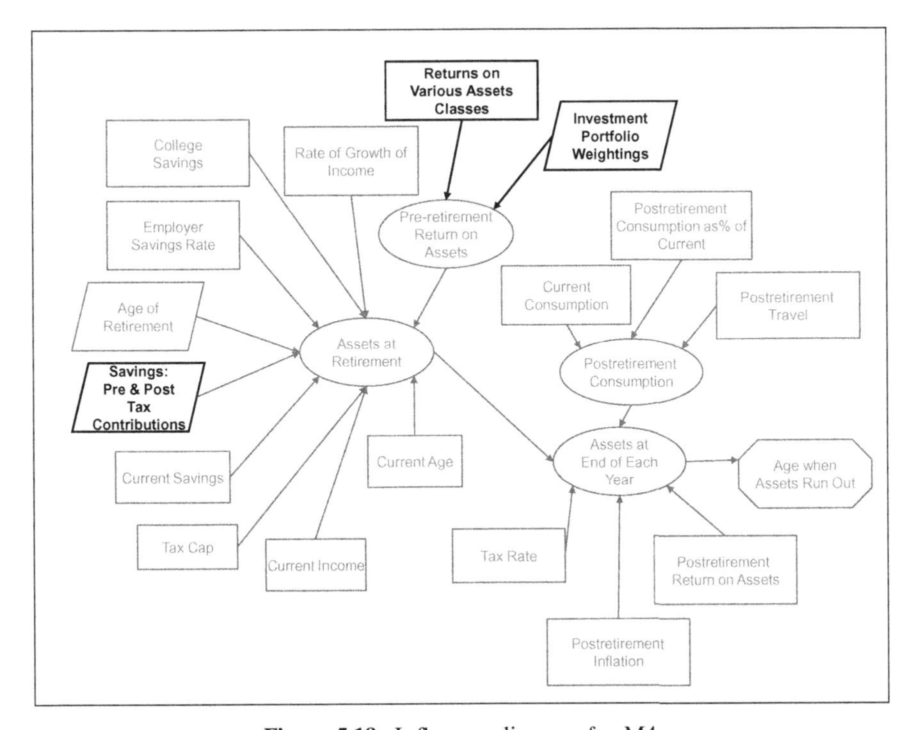

Figure 5.18. Influence diagram for M4.

fund target is reached. We also have to calculate the growth in the after-tax fund at the rate of return on the growth and income fund. Finally, the Go for Upside model is structurally identical to the Big Saver model, but the investment allocations are different and the amount saved after tax is lower.

To assess the range of outcomes possible under each of the strategies, we need to make our model stochastic by creating probability distributions for the six elements for which Bob has provided uncertainty assessments (see the table above). We use Crystal Ball to model each uncertain input with a uniform distribution between Bob's low and high estimates.

To the Reader: Use the model you have built to generate as many interesting insights into Bob's decision as you can.

5.3.2 Generate Insights—M4

Figure 5.19 shows that in the base-case scenario, on average, Bob's retirement funds run out when he is 73.6 years old. There is only a 31 percent chance that his funds will be adequate past age 75. This is very similar to the somewhat chastening conclusion our previous analysis suggested.

Insight 15: The base case continues to suggest that assets will run out at about age 74, and there is less than a 50 percent chance of them lasting past age 75.

Figure 5.20 presents a very different picture. If Bob follows the Big Saver strategy and saves $15,000 over the pretax cap (and invests somewhat more aggressively), he can expect his funds to run out on average when he is 79.2, and there is a 92 percent chance his funds will be adequate to age 75. Whether or not he is motivated to pursue this aggressive savings strategy, our analysis helps him to understand the quantitative benefits: He can postpone the expected day of reckoning by about five years and be secure in the knowledge that his funds are highly unlikely to run out early.

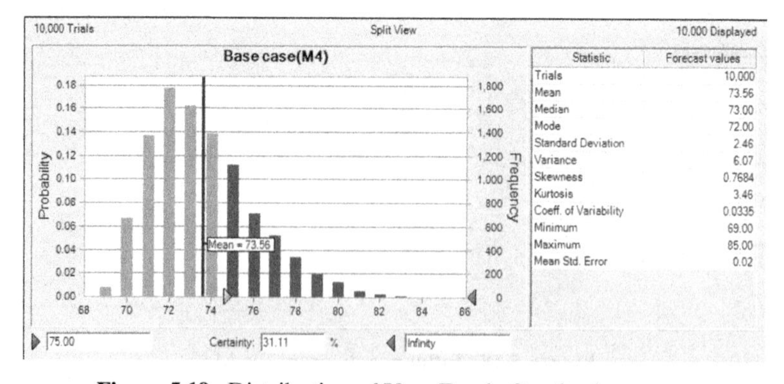

Figure 5.19. Distribution of Year Funds Out for base case.

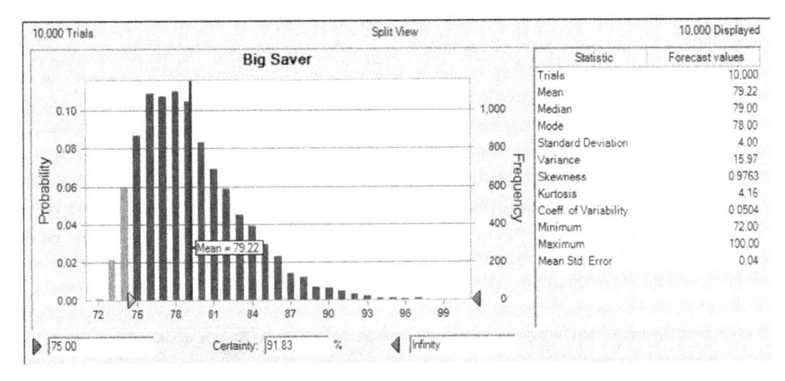

Figure 5.20. Distribution of Year Funds Out for Big Saver case.

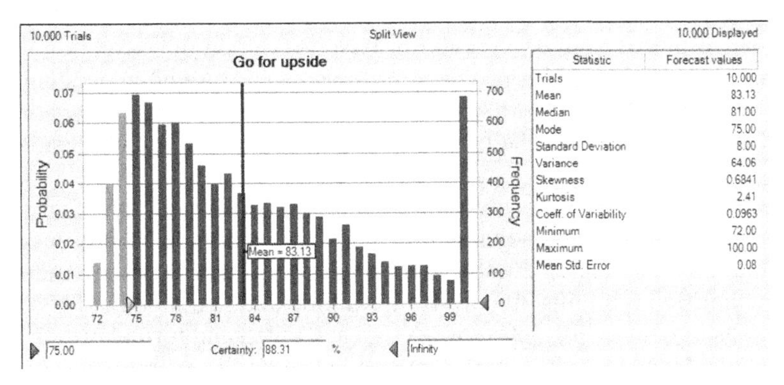

Figure 5.21. Distribution of Year Funds Out for Go for Upside case.

Insight 16: The Big Saver strategy pushes out the expected age of bankruptcy by five years and dramatically improves the probability of having sufficient funds past age 75.

Figure 5.21 shows the comparable results for the Go for Upside strategy. This strategy involves the most aggressive investment strategy but a much less onerous savings plan. How do these choices pan out? Our results suggest that this strategy may be even more attractive than the Big Saver plan: On average, his funds will last until he is 83.1, and there is an 88 percent chance they will last at least to age 75. (The spike on the right end of the distribution is because our model only calculates savings to age 100. Any trial in which assets last to at least 100 returns a result of 100. Thus, the true expected year in which assets run out is somewhat higher than 83.1.) This outcome is better on average than under the Big Saver strategy, with only a slightly increased risk of his funds running out early. And, as Bob is well aware, under this strategy, he will have significantly more money available during his working life for consumption.

Insight 17: The Go for Upside strategy provides a higher expected age of bankruptcy than the other two strategies, is highly probable to provide funds past age 75, and allows for higher consumption during Bob's working years.

Which strategy Bob chooses is, of course, up to him. We have done our job as analysts by providing him with information and insights that help him make a decision he understands, is comfortable with, and is confident is the best, given his circumstances.

5.4 PRESENTATION OF RESULTS

We have explored Bob's situation in substantial depth, and the time has come to organize our results into a form he can understand and act on. With four models and 17 insights, we have no shortage of material to work with. However, all of our results depend to a greater or lesser extent on our assumptions and the way we have framed his problem. We must make sure he understands the approach we have taken and appreciates that we cannot hope to predict his future. We can, however, give him information that will help him decide how he might shape his future.

The first step in creating any presentation is to consider the audience. The problem statement suggests that Bob is likely to understand his current financial situation well, but he does not know how to think about retirement. We assume that Bob is intelligent and strongly motivated to understand his retirement decisions better. Therefore we design a detailed and formal presentation that highlights the impacts his decisions will have on the retirement outcomes he values.

In most presentations we prefer the direct approach, in which we first present the conclusions and then the logic that justifies them. In this case, however, we take a more indirect approach, since there is no single result we can cite as our conclusion and we need to explain our approach before we can present our results.

When we look back at our results, we see they naturally divide into two parts: deterministic (M1 through M3) and stochastic (M4). Since deterministic results are easier to explain, we start with them. The stochastic results can then be presented as an elaboration of the deterministic results. Within the deterministic results, we present a base case and sensitivity analyses of key parameters. Within the stochastic results, we also present a base case, largely to introduce Bob to uncertainty analysis, but our focus is on the insights we derive from the two scenarios: Big Saver and Go for Upside.

Our story starts with an introduction that explains the approach we took and how to interpret our results. Then we use a base case from M3 to give Bob a concrete foundation from which to explore the impact of changes in his decisions. Next, we present sensitivity results for the savings rate, retirement date, and asset returns to show how the decisions he makes can influence the outcomes. In the final part of the presentation, we show the variability in results he faces, both in the base case and under the two strategies we formulated (Big Saver and Go for Upside).

Here is an outline of our story in a bit more depth:

- Challenges of planning retirement
- Our approach to the problem
- How we visualize the problem (i.e., the influence diagram)
- A base case
- Sensitivities
    - Savings rate
    - Retirement date
    - Return on assets
    - Postretirement consumption
- The effects of uncertainty
- The "Big Saver" strategy
- The "Go for Upside" strategy

Our opening slide (Figure 5.22) sets the stage by describing the decisions Bob must make with respect to retirement, the uncertainties he faces, and the multitude of outcome measures he could adopt. The next slide (Figure 5.23) summarizes our approach, which is to use a series of simple models to explore possible outcomes, not to predict. Then we display our influence diagram

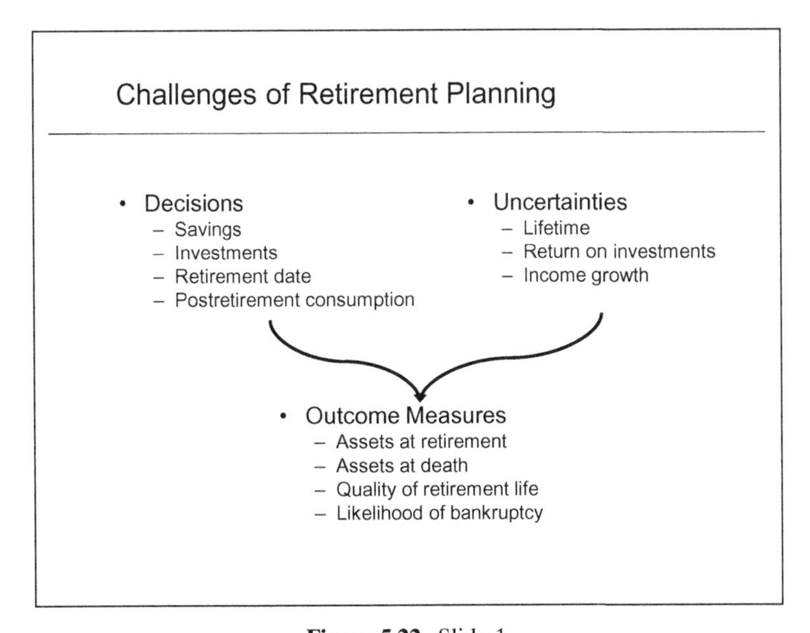

Figure 5.22. Slide 1.

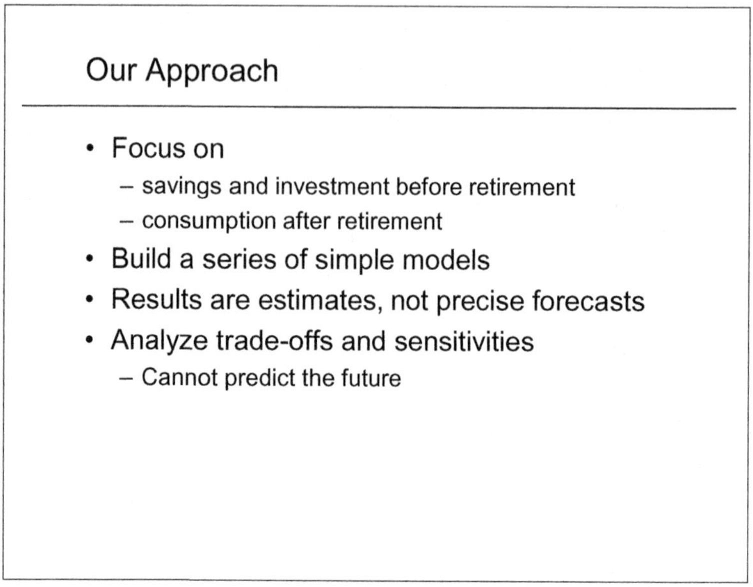

Our Approach
• Focus on
– savings and investment before retirement
– consumption after retirement
• Build a series of simple models
• Results are estimates, not precise forecasts
• Analyze trade-offs and sensitivities
– Cannot predict the future

Figure 5.23. Slide 2.

Influence Diagram
Rate of Growth of Income
College Savings
Preretirement Return on Assets
Postretirement Consumption as% of Current
Employer Savings Rate
Current Consumption
Postretirement Travel
Age of Retirement
Assets at Retirement
Postretirement Consumption
Savings Rate
Current Age
Assets at End of Each Year
Age when Assets Run Out
Current Savings
Tax Cap
Current Income
Tax Rate
Postretirement Return on Assets
Postretirement Inflation

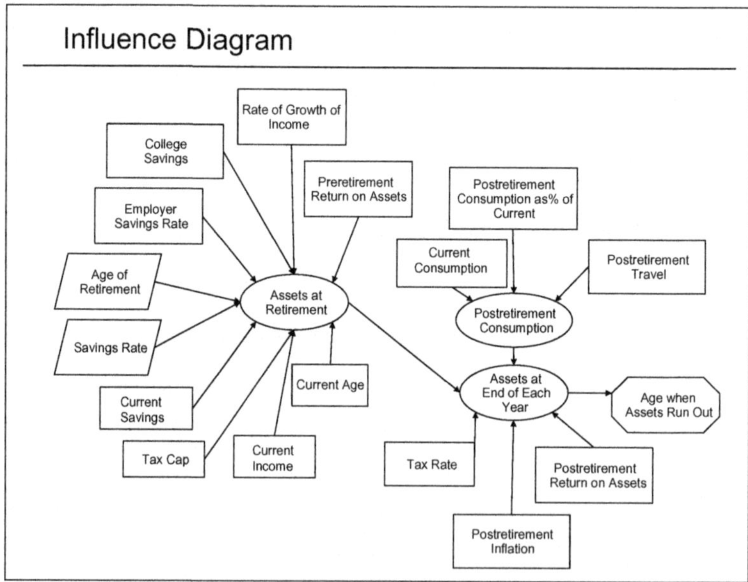

Figure 5.24. Slide 3.

(Figure 5.24) as a way to orient Bob to the critical variables in the problem and how they are interrelated. Although we are familiar with our influence diagram, it might look confusing to Bob if presented all at once. Therefore, we use the Animation controls in PowerPoint to allow us to display the influence

Assets will run out at age 74 under conservative assumptions

	Base Case
Savings Rate	5%
Retirement Age	65
Asset Return (preretirement)	7%
Asset Return (postretirement)	5%
Consumption Fraction	70%
Asset Exhaustion Age	**74**

Figure 5.25. Slide 4.

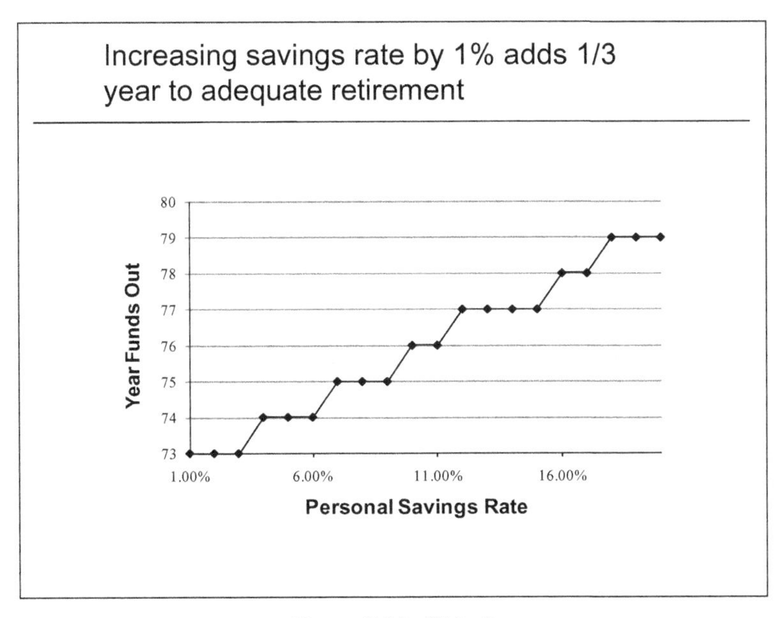

Figure 5.26. Slide 5.

diagram in a sequence of steps. We start with the outcome measure and work our way backward through the diagram, just like we did when we created it.

The next seven slides (Figures 5.25–5.31) form the core of the presentation. Figure 5.25 describes a base case, which, we explain to Bob, is just a reasonable

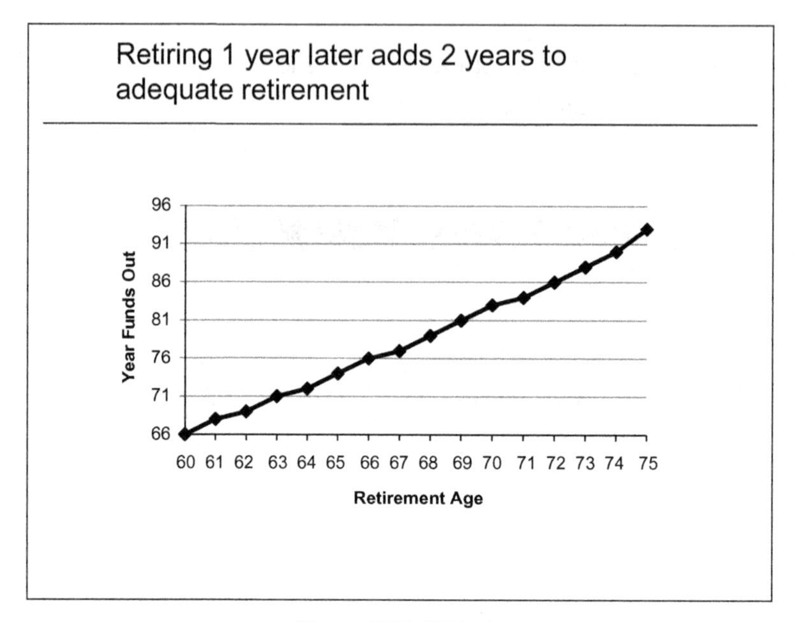

Figure 5.27. Slide 6.

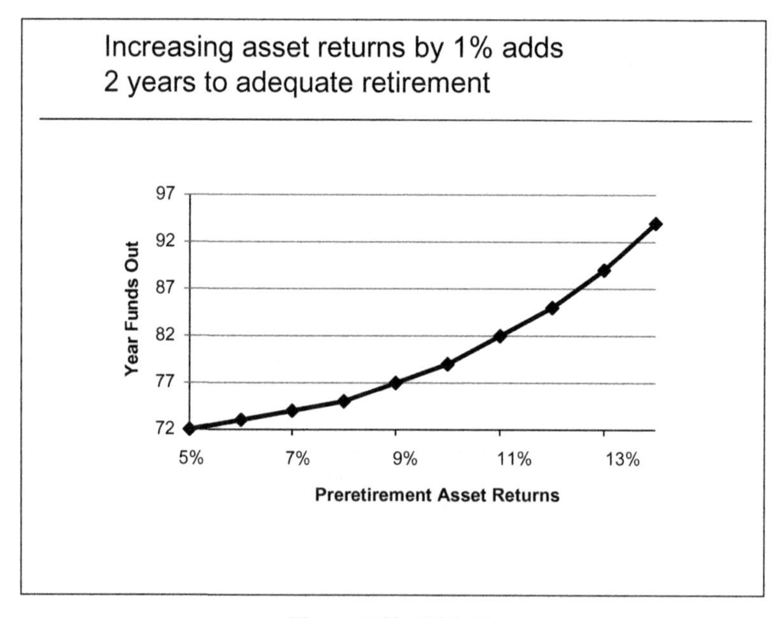

Figure 5.28. Slide 7.

starting place, not an outcome he can count on. The really interesting results are in the following four figures, which show how his choices influence the year his assets run out. Figure 5.26 shows that, contrary to what he might expect, changes to his savings rate have little effect on the adequacy of his funds. By contrast, Figure 5.27 shows him that changing his retirement date can have a

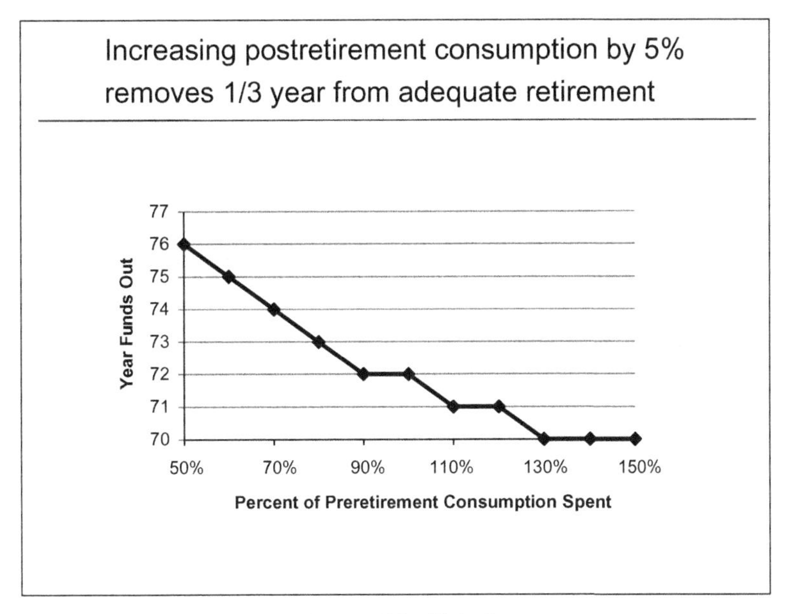

Figure 5.29. Slide 8.

Retirement age and asset return have the greatest effect on asset sufficiency

	Sensitivity
Savings Rate	Moderate
Retirement Age	High
Asset Return (preretirement)	High
Consumption Fraction	Low

Figure 5.30. Slide 9.

big impact on the year his assets run out. Delaying retirement from age 65 to age 70, for example, adds ten years to the adequacy of his retirement funds. Figure 5.28 shows the sensitivity of the results to the return his assets earn. This chart suggests that every additional percentage point he earns on his assets adds roughly two years to the adequacy of his assets. Figure 5.29 shows

Assets can last much longer with minor changes in decisions

	Base Case	Better Case
Savings Rate	5%	10%
Retirement Age	65	68
Asset Return (preretirement)	7%	9%
Asset Return (postretirement)	5%	5%
Consumption Fraction	70%	70%
Asset Exhaustion Age	**74**	**87**

Figure 5.31. Slide 10.

that increasing his postretirement consumption by 5% removes a third of a year from the adequacy of his assets. We summarize these four sensitivities in a table in Figure 5.30 to highlight which two inputs really matter. Finally, we complete our presentation of deterministic results by offering a case (Figure 5.31) we think might appeal to Bob: Save an additional 5% (10% vs. 5%), retire 3 years later (68 vs. 65), and earn 2% more on your assets (9% vs. 7%), and then he can expect to have adequate assets to age 87.

At this point in the presentation, we shift gears a bit and address the uncertainty he faces. We explain in general terms what simulation is and how we developed the results he is about to see. Then we show him Figure 5.32, a base-case distribution for the year funds run out. After explaining what a distribution chart shows, we concentrate on two aspects: the mean (74 years) and the probability that his assets will last to at least age 75 (31 percent, in this case).

This slide sets up the next three slides. First we explain the assumptions behind the two scenarios, Big Saver and Go for Upside (Figure 5.33). In Figure 5.34, we show the distribution under the Big Saver strategy. We explain the assumptions behind this strategy and show that by saving the maximum, he can increase the average number of years to 79 and the probability of exceeding age 75 to 92 percent. We next show Figure 5.35, which shows comparable results for the Go for Upside strategy: an average of 83 years and a 88 percent probability of assets exceeding age 75. We use Figure 5.36 to make the point that both of these strategies produce outcomes much better than the base case, but that the Go for Upside strategy also allows for higher consumption during Bob's working life. We conclude with a summary slide (Figure 5.37) that

Assets run out too soon in base case

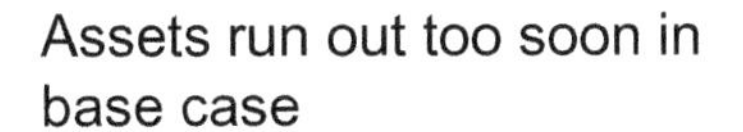

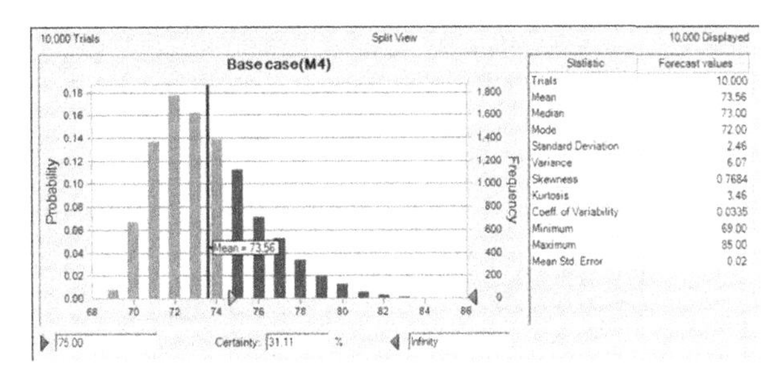

- Mean age assets exhausted: 74
- Probability assets last to age 75: 31%

Figure 5.32. Slide 11.

Two Strategies

	Big Saver	Go for Upside
Savings (tax-shielded)	Max	Max
Savings (taxed)	$15,000	$3,000
Retirement Age	65	65
Asset Allocation: Stocks	60%	100%
Asset Allocation: Bonds	40%	0%
Consumption Fraction	70%	70%

Figure 5.33. Slide 12.

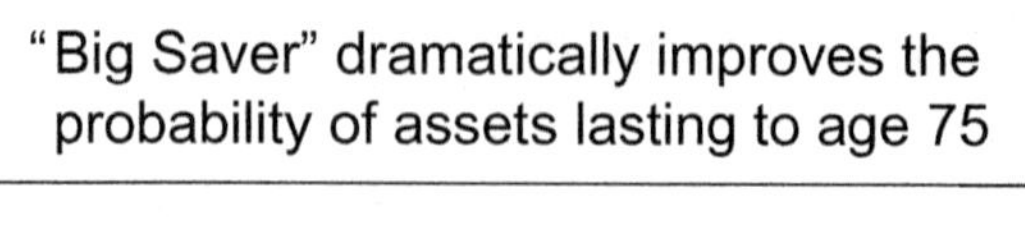

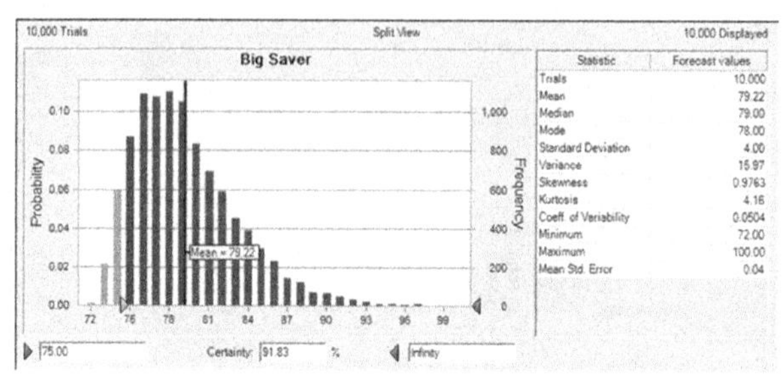

- Mean age assets exhausted: 79
- Probability assets last to age 75: 92%

Figure 5.34. Slide 13.

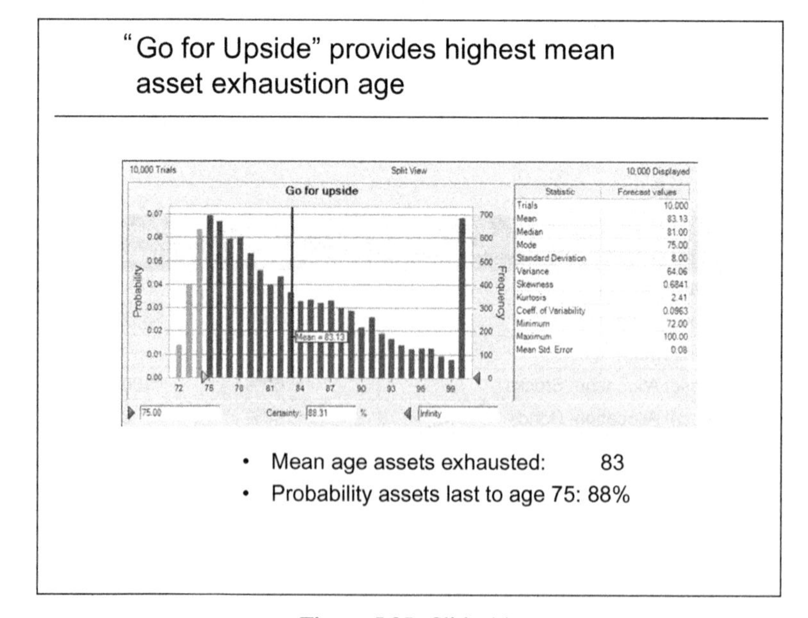

Figure 5.35. Slide 14.

Both strategies produce good outcomes

- "Go for Upside" allows for considerably more consumption during working life

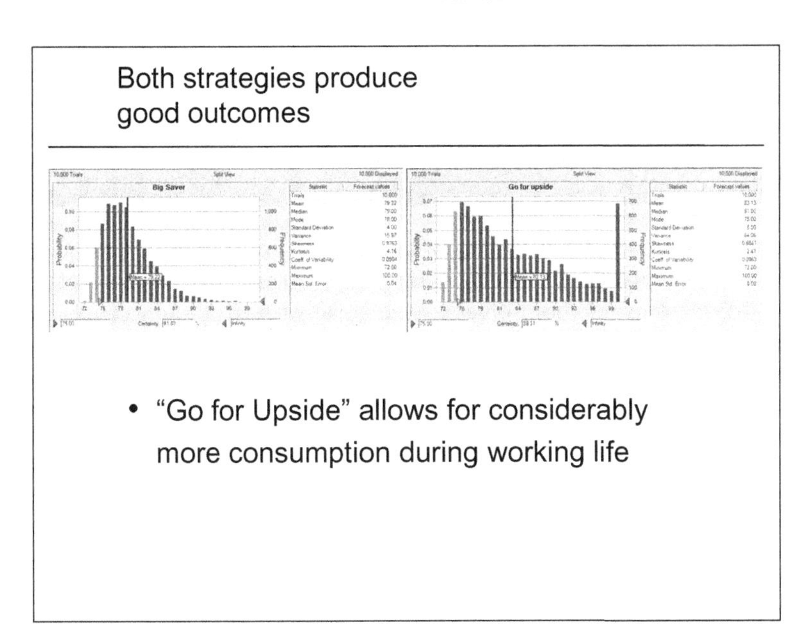

Figure 5.36. Slide 15.

Summary

- Assets could run out early with current plans
- Assets could be adequate into your 80s with minor changes
  - Asset returns
  - Retirement age
- Consider a higher return strategy

Figure 5.37. Slide 16.

includes our recommendation to consider switching to a higher asset return strategy.

5.5 SUMMARY

In this chapter, we built a sequence of models for a realistic problem situation, starting with the problem kernel. As explained earlier, ill-structured problems usually do not present themselves in the form of a problem kernel. Rather, they arrive encrusted with both relevant and irrelevant detail. Your first job often is to bore through the details to find what is essential. This is the problem kernel. We began with the kernel in this chapter in order to demonstrate how you can build an effective first model, and even derive useful insights from it, by carefully framing the problem, diagramming the problem, and building a simple model.

After building a first model, we restrained ourselves from improving it until we had put it through its analytic paces. That involved establishing a base-case result and performing various kinds of sensitivity analyses using Goal Seek, Data Sensitivity, and Tornado Chart. These analyses suggested insights that provided the foundation for future analysis.

We carefully selected what elements to add to each model so as to maximize our learning and reduce the chance of errors. For example, we only added uncertainty to the model (M4) once we felt we had learned all we could from the previous deterministic models.

Rather than focusing on building a sequence of more complex models for its own sake, we focused on the sequence of insights we were able to derive from our models. We observed a "zeroing-in" effect as later insights reinforced insights from previous models. It is this sequence of results that ultimately led to the final insights we presented to our client.

CHAPTER 6

Technology Option

6.0 INTRODUCTION

In this chapter, we tackle another ill-structured problem, this time one related to a straightforward business problem: Should we undertake research and development (R&D) that could reduce the production cost of a new product? Just as we did in Chapters 4 and 5, we will use the problem kernel and iterative modeling to develop a sequence of progressively more complex and insightful models.

We begin with the briefest possible description of the problem and then use problem framing, problem diagramming, and model building to generate preliminary insights. Next, we introduce an expanded description of the problem and take another pass through the same modeling process. Finally, we consider a variant of the problem based on data gathered from the client after an initial modeling effort.

6.1 TECHNOLOGY OPTION (A)

Technology Option (A)

Ultra is a recently developed specialty chemical with high market potential. It currently costs \$90 per barrel to produce and would sell for about \$60 per barrel. An R&D project has been proposed that will take two years, cost \$15M, and has a 10 percent chance to reduce the costs by 50 percent. If this project succeeds, it will take another year and \$250M to build a manufacturing plant with a capacity of 30M barrels per year.

Do you recommend this project?

To the Reader: Before reading on, frame this problem by setting boundaries for the analysis; making assumptions; listing questions; defining outputs, inputs, and decisions; and so on.

6.1.1 Frame the Problem

The initial statement of this problem is fairly straightforward. Given the limited amount of detail we have to work with, however, a realistic goal is to develop a rough estimate of the financial return the company could expect if it undertook this project. In the spirit of iterative modeling, we attempt to build the simplest model that answers the question: Should the company take on this project?

The first boundary we draw is that we are only concerned with the direct financial results of this project. Thus, we do not consider the possibility that the R&D will lead to technology breakthroughs useful in other projects, or that this product could help the company break into a lucrative market segment even if its direct return was negative.

Our next step in framing the problem is to select an outcome measure. Because this problem asks for a financial recommendation, we want to develop a concrete measure of the dollar impact of taking on this project. Since the cash flows involved will occur over time, it makes sense to evaluate the project on a net present value (NPV) basis. Thus, we will determine the net cash flows (inflows less outflows) each year and discount them to the present. The problem statement does not specify the appropriate discount rate or time horizon for this evaluation, but we can choose plausible numbers for these inputs and test their impact using sensitivity analysis.

Having decided on the NPV of cash flows as the outcome measure, we next consider accounting issues. In a full-fledged analysis of this type of investment, we would need to deal with a variety of accounting issues, such as taxes and depreciation. The problem description gives us no specific information about how depreciation and taxes would be handled in this case, so we ignore these issues temporarily. This decision may seen drastic, but remember that our goal at the moment is not to build the *ultimate* model but only a plausible *first* model. Furthermore, ignoring depreciation probably underestimates the true NPV, while ignoring taxes overestimates it, so these two assumptions are partially offsetting.

Another issue we must deal with is the uncertainty surrounding the success of the R&D project itself. Apparently, the project will either succeed in achieving a 50 percent reduction in costs or fail completely. This is certainly a simplification of the results of any real R&D project, but the problem kernel is always a simplification. If the company undertakes the project and it fails, it is out only \$15M. (They would not build the \$250M plant knowing with certainty that they would lose money on it.) If the R&D project succeeds, they may ultimately make an overall profit or loss; at this point, we do not know which one. One way to combine these two outcomes is by calculating the *expected value*. In other words, if they have a 10 percent chance of receiving \$50M and a 90 percent chance of receiving nothing, we value that uncertain amount as if it were worth \$5M ($= 10\% \times 50 + 90\% \times 0$).

The three main elements of our problem frame are as follows:

- Outcomes measured with the NPV of cash flows
- Depreciation and taxes ignored
- Uncertainty reflected in expected values

To the Reader: Before reading on, sketch an influence diagram for this problem.

6.1.2 Diagram the Problem

The next step is creating an influence diagram. Figure 6.1a shows that we start with the outcome measure, the NPV of Cash Flows, and decompose it into Yearly Profit, Plant Cost, and R&D Cost. Note that we have not specified the time unit here, leaving that detail to the model-building process.

Yearly Profit itself is decomposed into Revenue and Cost (Figure 6.1b). Both are determined in part by Unit Sales, which, for simplicity, we model as always equal to capacity (we assume demand is larger than capacity). The parameter Unit Variable Cost and the decision variable Price complete this portion of the model.

The final element of this diagram is the success or failure of the R&D effort. We have drawn this as a random variable with two concentric rectangles in Figure 6.1c. R&D Success affects NPV Cash Flows by way of reducing Unit Variable Costs and determining whether or not they build the plant. However,

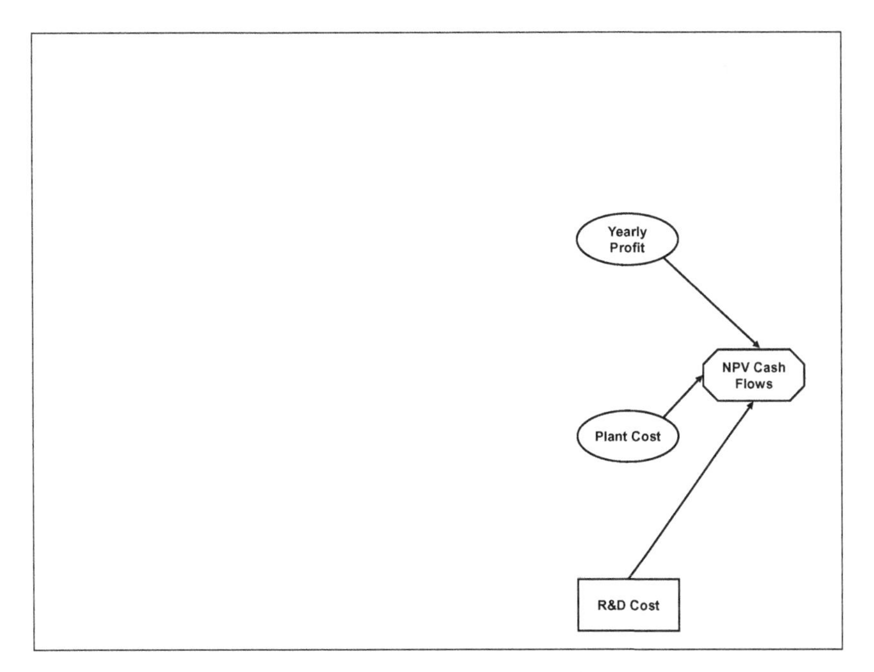

Figure 6.1. **(a)** Initial influence diagram for M1.

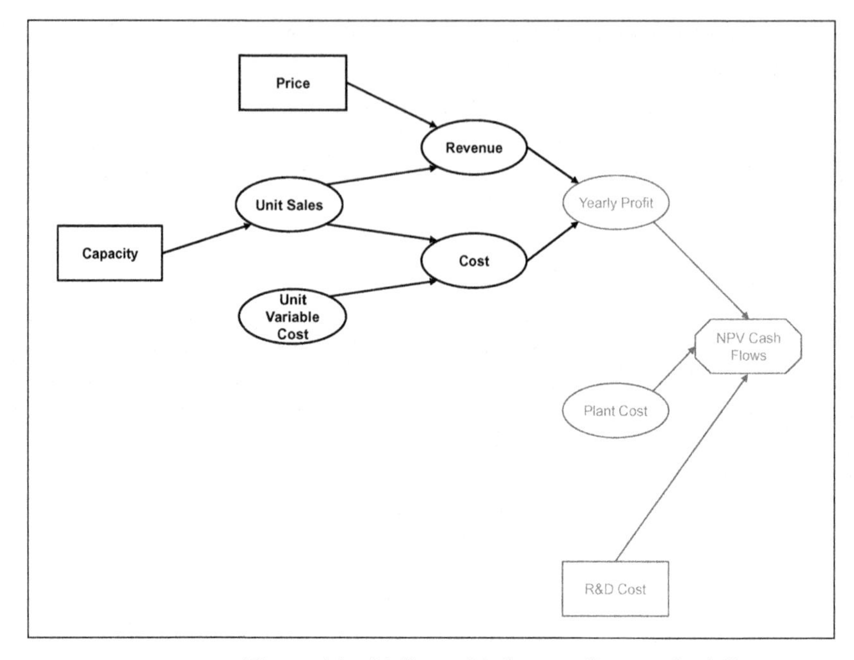

Figure 6.1. (b) Second influence diagram for M1.

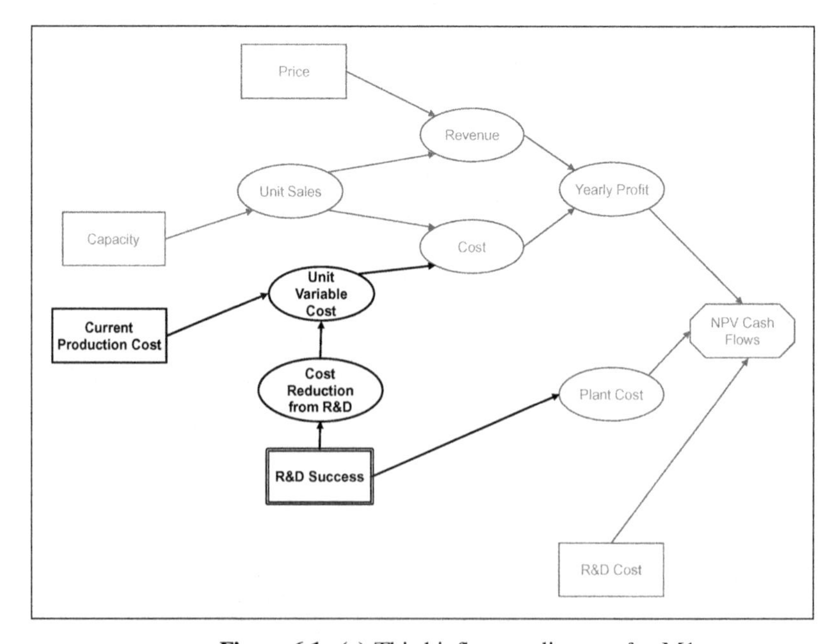

Figure 6.1. (c) Third influence diagram for M1.

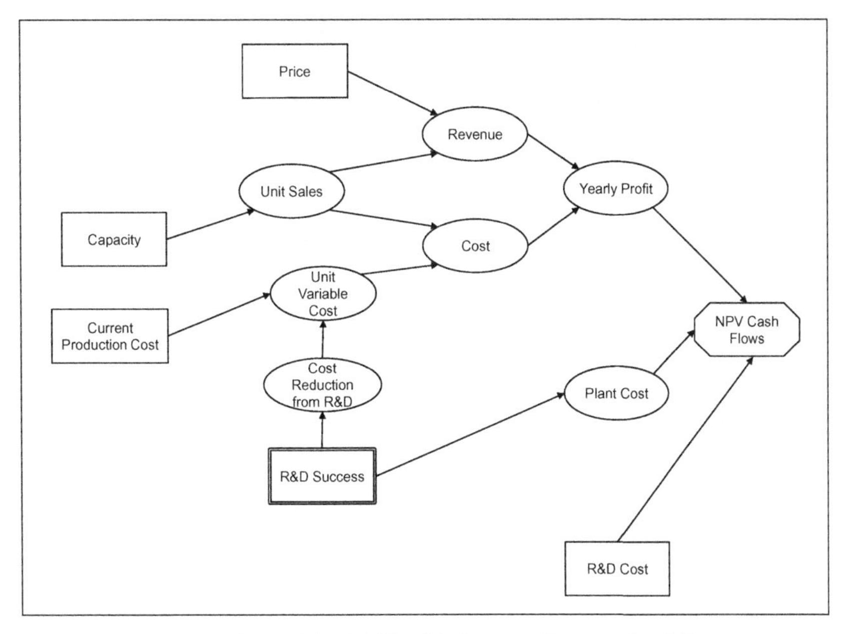

Figure 6.1. (d) Final influence diagram for M1.

	A	B	C	D	E	F	G	H	I	J	K	L
1	**Technology Option M1**											
2												
3												
4												
5		**Parameters**										
6			Production Cost	$90								
7			Price	$60								
8			Plant Cost	$250,000,000								
9			Plant Capacity	30,000,000								
10			Discount Rate	10%								
11			Cost of R&D	$15,000,000								
12			Cost Reduction From R&d	50%								
13			Chance of Success	10%								
14												
15		**Model**						**Year**				
16				1	2	3	4	5	6	7	8	
17			R&D Cost	$15,000,000								
18			Plant Cost			$250,000,000						
19			Revenues				$1,800,000,000	$1,800,000,000	$1,800,000,000	$1,800,000,000	$1,800,000,000	
20			Costs				$1,350,000,000	$1,350,000,000	$1,350,000,000	$1,350,000,000	$1,350,000,000	
21			Profit	-$15,000,000		-$250,000,000	$450,000,000	$450,000,000	$450,000,000	$450,000,000	$450,000,000	
22												
23		**Results**										
24			NPV	$117,350,368								
25												

Figure 6.2. M1—numerical view.

following our assumption about expected values, we simply assume they pay the plant cost and receive the Unit Cost savings with a probability of 10 percent.

The completed influence diagram is shown in Figure 6.1d.

To the Reader: Before reading on, design and build a first model based on the influence diagram.

6.1.3 Build the Model—M1

Once we have framed the problem and drawn an influence diagram, a working model almost builds itself. In Figures 6.2 and 6.3, we show our first model. The

	A	B	C	D	E	F	G	H	I	J	K	L
1	**Technology Option M1**											
2												
3												
4												
5		**Parameters**										
6			Production Cost	90								
7			Price	60								
8			Plant Cost	250000000								
9			Plant Capacity	30000000								
10			Discount Rate	0.1								
11			Cost of R&D	15000000								
12			Cost Reduction From R&D	0.5								
13			Chance of Success	0.1								
14												
15		**Model**						**Year**				
16				1	2	3	4	5	6	7	8	
17			R&D Cost	=D11								
18			Plant Cost			=D8						
19			Revenues				=$D$7*$D$9	=$D$7*$D$9	=$D$7*$D$9	=$D$7*$D$9	=$D$7*$D$9	
20			Costs				=$D$6*(1-$D$12)*$D$9	=$D$6*(1-$D$12)*$D$9	=$D$6*(1-$D$12)*$D$9	=$D$6*(1-$D$12)*$D$9	=$D$6*(1-$D$12)*$D$9	
21			Profit	=-D17		=-F18	=G19-G20	=H19-H20	=I19-I20	=J19-J20	=K19-K20	
22												
23		**Results**										
24			NPV	=D21+D13*NPV(D10,E21:K21)								
25												

Figure 6.3. M1—relationship view.

parameters and decision variables are isolated in D6:D13. The model itself extends over eight years. The components of the NPV are calculated as follows:

Row 17: The R&D cost is paid in year one.

Row 18: The cost of the plant is paid in year three.

Row 19: Revenues are calculated as price times plant capacity (assuming demand always exceeds capacity).

Row 20: Variable costs are the current production cost times (1 – the cost reduction from R&D) times plant capacity (assuming sales and production are equal each year).

Row 21: Yearly profit is revenue less variable cost.

In cell D24, we calculate the outcome measure by subtracting the R&D cost from the probability-weighted NPV of cash flows. (The NPV of cash flows in years two through eight is scaled back by the probability of the R&D effort's success.) We now have our first result: The R&D plan has an expected NPV of $117M. It looks as if this might be an attractive project.

This first model (M1) is very simple, but it does provide an answer to the question we face. It probably took an hour or two of work to come up with this result. Now we are in a position to use the model to generate insights.

To the Reader: Before reading on, use M1 to generate as many interesting insights into this situation as you can.

6.1.4 Generate Insights—M1

Before we consider improving our model, it is important to take some time to explore it. The conclusions we draw will be preliminary because the model is so limited at this point. But this process will begin to suggest which parameters really matter and which parts of the model we should expand or improve first.

Figure 6.4 answers one question that is surely on the minds of our clients: How much does the R&D program have to reduce costs before the project becomes profitable? By performing data sensitivity on the cost reduction parameter, we see that roughly a 37 percent reduction is sufficient. We also see that the NPV from this project is linear in the cost reduction, so the NPV more than doubles if we can achieve a 65 percent cost reduction instead of the 50 percent reduction assumed in the base case.

Insight 1: The NPV of the project is linearly related to the cost reduction. We only need a 37 percent cost reduction to break even on the project.

Figure 6.5 shows the results of another sensitivity analysis. Here we vary the chance of R&D success from 1 percent to 25 percent, while continuing to

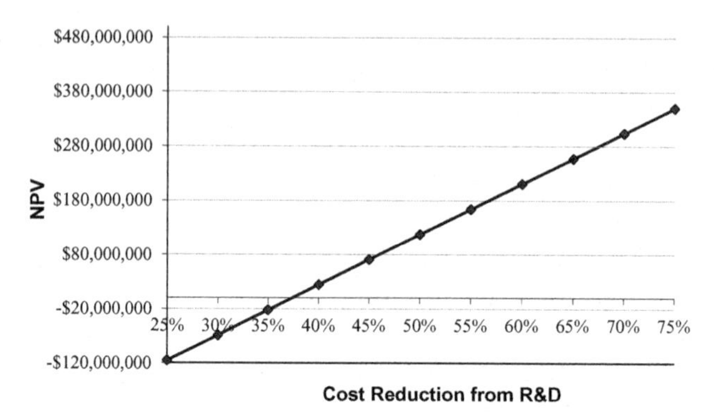

Figure 6.4. Sensitivity to Cost Reduction (M1).

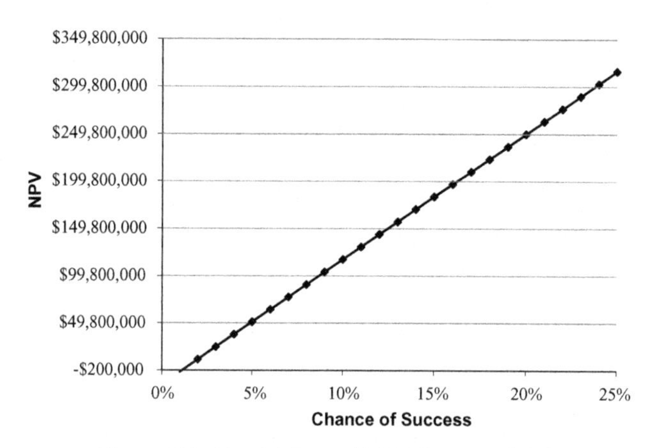

Figure 6.5. Sensitivity to R&D Success (M1).

assume a 50 percent cost reduction. This graph suggests that the project returns a positive NPV almost regardless of the chances of success.

Insight 2: Although the NPV is sensitive to the probability of R&D success, it remains positive for all reasonable values of the success probability.

This conclusion is somewhat startling. How can the attractiveness of the project be so insensitive to the chances of success? The answer, of course, lies in the numbers. We can understand this result better if we focus on the cost and return from building the plant. At the beginning of year three, the company invests $250M and receives a profit of $450M in each of the following five years. The NPV at the beginning of year three of these cash flows is about $1.5 *billion*. To receive this NPV, the company has to invest only $15M now in R&D, which is about 1/100th of the subsequent profit. So it does make sense that

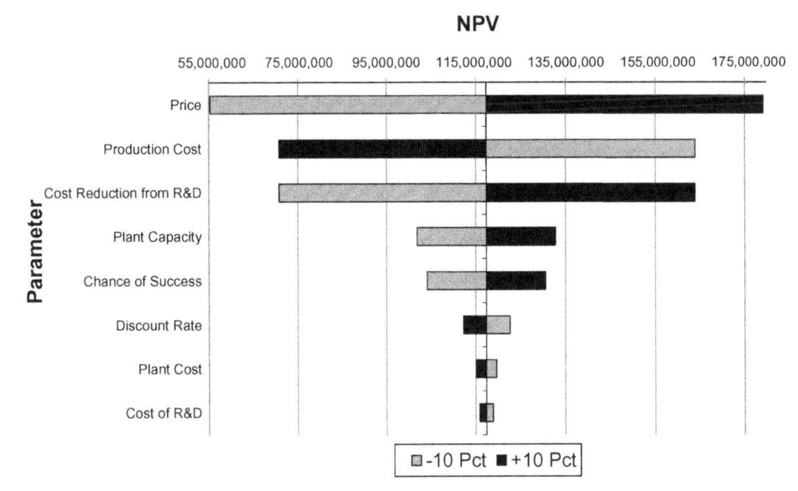

Figure 6.6. Tornado chart for NPV (M1).

even with a success probability around one percent the company can still break even on this project.

In a final sensitivity analysis, we create a constant-percentage Tornado Chart for this first model. The results are shown in Figure 6.6. Three parameters stand out here as having the biggest impact on the NPV:

- Price
- Production Cost
- Cost Reduction from R&D

Perhaps more surprising, the cost of the R&D itself and the cost of the plant have little impact, at least when considering ±10 percent changes from the base-case values.

Insight 3: The NPV is insensitive to moderate changes in R&D cost and plant cost.

Another parameter that has a small impact is the probability of success; this is related to our earlier finding that the project breaks even with a success probability as low as one percent.

6.2 TECHNOLOGY OPTION (B)

We now introduce the complete description of the Ultra problem from which we extracted the kernel in the previous section.

Technology Option (B)

SuperChem first conceived Ultra, a proprietary specialty chemical, in early 2001. Ultra is designed for use in the intermediate stages of plastic manufacturing and is targeted to replace a competing product with 2003 sales of over $2B worldwide. The advantages of Ultra over the competition are proven, and market research indicates customers are willing to pay a 10–25 percent premium for Ultra. The future unit price (per barrel) is uncertain given the volatility of key chemicals used in the existing manufacturing process; the market price of the current chemical technology is near its ten-year average of $55 per barrel, with a low of $45 and a high of $75.

Although SuperChem's engineers are excited by Ultra's market potential, they cannot profitably manufacture it today. Using current approaches in a small-scale pilot plant, the team demonstrated they could produce Ultra for $90 per barrel but could not reduce the cost further. Last month, management recommended terminating the Ultra program unless new options could be found for achieving profitability. Historically, SuperChem's R&D program has a return in excess of five times R&D spending, and they typically expect new opportunities to meet or exceed this performance hurdle.

To save the program, the engineers are proposing a two-year R&D study they believe has a 10 percent chance of reducing unit costs by 50 percent. This $15M effort will upgrade the pilot plant to test several new manufacturing approaches that will work together to bring down the unit cost. The team agrees that the Ultra program should be shut down if this effort is unsuccessful.

They also note that given the rate of advance of next-generation technology, the effective product life for Ultra is four to seven years (from launch) before new technology comes out and replaces Ultra.

If the R&D effort works, the current business plan calls for construction of a full-scale manufacturing plant at a cost of $250M to be on-line within a year. The proposed plant has a design capacity of 30M barrels per year and could be expanded if sales exceed capacity. The plant has negligible fixed costs and is designed for flexible production; SuperChem would only produce Ultra if the market price exceeds the unit cost.

To the Reader: Before reading on, consider how the original problem frame should be adjusted for the new information contained in the (B) case.

6.2.1 Frame the Problem—M2

What is new in this full problem description that was not reflected in the problem kernel? We are given somewhat more information on the current market price and on the premium buyers might pay for Ultra. We are also given a range for the product lifetime. Perhaps most important, we learn that R&D projects are not merely expected to break even but to return a present value (PV) of profits at least five times initial costs. In other words, the present value of revenues and costs from the product itself should be at least five times the investment in R&D. This is a potentially important change to the outcome measure we used in the first model.

This problem description also reveals a new and potentially significant *real option* for SuperChem. (A real option is an alternative one can take or not take in the future, such as closing down a plant, which could increase the value of an asset.) The company has the ability to shut down the plant and not produce product in any year in which the price is too low. This option could increase the value of the project, and we probably should include it at some point in our evaluation.

At this stage in our modeling, we need to prioritize the changes we want to make to the model. If we introduce too many changes to the model at one time, we will have a hard time understanding which changes make a difference and why. On the other hand, if we make too many small changes, we will learn at too slow a pace.

Here is a list of the potential changes to the model we see at this point:

- Add taxes and depreciation on the plant investment.
- Change the outcome measure to PV as a multiple of costs.
- Include the ranges and uncertainty in inputs.
- Include the option to shut down the plant.

In this case, we will make the easiest changes first. Our plan is as follows. In M2, we add in taxes and depreciation and change the outcome measure, while keeping the model otherwise essentially unchanged from M1. In M3, we introduce uncertainty in the inputs and in the outcome of R&D. This change requires the use of Monte Carlo simulation, which represents a substantial increase in complexity. Not only is the model itself more complex, but the data inputs are more complex, since they involve probability distributions. Furthermore, we cannot evaluate a single case without running repeated simulation trials, which means that analysis is also more complex. Finally, we build M4, in which we model the option to shut down production when costs exceed revenues.

To the Reader: Before reading on, sketch an influence diagram for this new problem.

6.2.2 Diagram the Problem—M2

We need to return at this point to the influence diagram and elaborate it to reflect the new features in our model. In this version of the model, we incorporate three substantive changes:

- Include taxes and depreciation
- Allow sales to differ from plant capacity
- Change the outcome measure to the PV multiple (i.e., the present value of profits should be at least five times initial project costs)

In addition, we introduce new parameters suggested in the problem description. For example, we know the price is around $55 and customers will pay a premium of about 10 percent. We also know the product lifetime will be around five years, so we introduce a parameter for the project life in order to vary it during sensitivity analysis (Figure 6.7).

To the Reader: Before reading on, design and build a model based on the influence diagram in Figure 6.7.

6.2.3 Build the Model—M2

M2 has very much the same structure as M1. We have added six new inputs:

- Price Premium
- Tax Rate
- Project Life
- Demand
- Depreciation Rate
- PV Multiple

Some values for these inputs were suggested in the problem description. Where input values were not specified directly in the problem, we enter

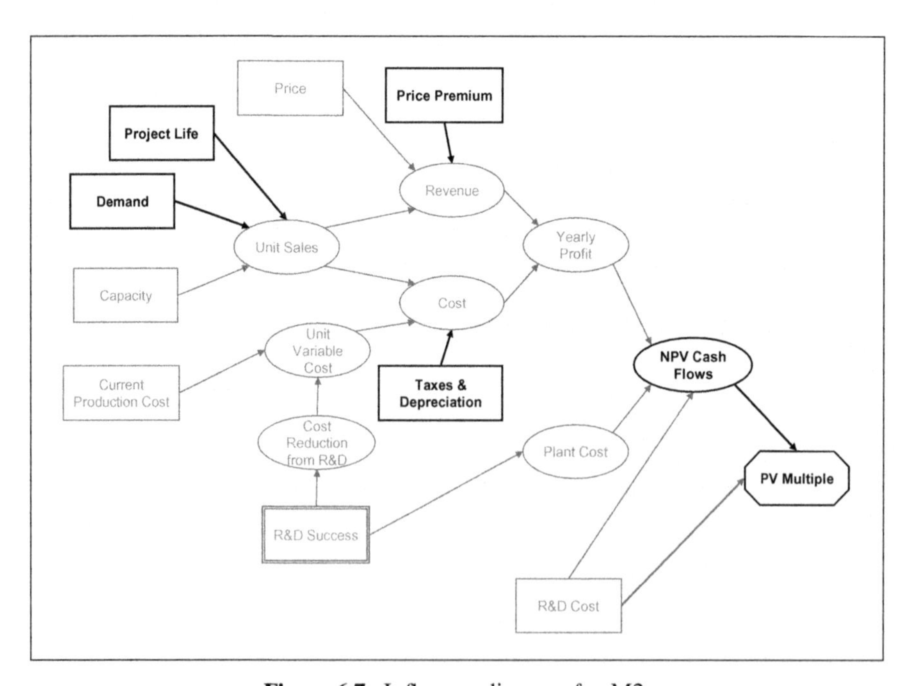

Figure 6.7. Influence diagram for M2.

reasonable values (see Figures 6.8 and 6.9). We also extend the time horizon to ten years to accommodate the maximum product life of seven years.

The core of the model occupies rows 23–30:

Row 23: The R&D cost is paid in year one.

Row 24: The cost of the plant is paid in year three.

Row 25: Revenues are calculated as price times the price premium times the minimum of plant capacity and sales.

Row 26: Variable costs are the current production cost times (1 – the cost reduction from R&D) times the minimum of plant capacity and sales.

Row 27: Depreciation is the depreciation rate times the cost of the plant.

Row 28: Taxable income is revenue minus cost minus depreciation.

Row 29: Taxes are the tax rate times taxable income.

Row 30: Cash flow is taxable income less taxes plus depreciation (note that the project lifetime is used to set profits to zero when the lifetime is exceeded).

Finally, we collect the results in cells D33 through D35. In cell D33, we calculate the PV of the cash flows from the project (excluding the costs of R&D). In cell D34, we calculate the ratio of project PV to R&D costs; this is the new outcome measure.

This model is a modest extension of M1. It has the same physical layout and shares many of the same assumptions. It is a refinement, not a new approach. And it also gives an unambiguous answer to our question: Under these new assumptions, we should *not* recommend the project because it does not return five times its costs. However, with a multiple of 4.07, it is not far from breakeven.

To the Reader: Before reading on, use M2 to generate as many interesting insights into the situation as you can.

6.2.4 Generate Insights—M2

Sensitivity analysis helps determine the changes in parameters required to achieve the critical PV multiple of five times cost. For example, we vary the cost reduction from R&D from 25 percent to 75 percent, as in Figure 6.10, and we see that a cost reduction of 54 percent is sufficient to meet the target. Similarly, we vary the probability of R&D success from 1 percent to 25 percent, as in Figure 6.11, and we see that a success probability of about 13 percent is enough to meet the target. Together, these results suggest that although the project does not meet the target in the base case, it would not take a very large improvement in one or more of the parameters to make it attractive. This insight is very useful.

	A	B	C	D	E	F	G	H	I	J	K	L	M	N
1	**Technology Option M2**													
2														
3														
4														
5		**Parameters**												
6			Production Cost	$90										
7			Current Price	$55										
8			Price Premium	10%										
9			Plant Cost	$250,000,000										
10			Plant Capacity	30,000,000										
11			Discount Rate	10%										
12			Cost of R&D	$15,000,000										
13			Cost Reduction from R&D	50%										
14			Chance of Success	10%										
15			Tax Rate	35%										
16			Project Life	5										
17			Annual Sales	25,000,000										
18			Depreciation Rate	10%										
19			Required PV Multiple	5										
20														
21		**Model**						**Year**						
22				1	2	3	4	5	6	7	8	9	10	
23			R&D Cost	$15,000,000										
24			Plant Cost			$250,000,000								
25			Revenues				$1,512,500,000	$1,512,500,000	$1,512,500,000	$1,512,500,000	$1,512,500,000	$1,512,500,000	$1,512,500,000	
26			Costs				$1,125,000,000	$1,125,000,000	$1,125,000,000	$1,125,000,000	$1,125,000,000	$1,125,000,000	$1,125,000,000	
27			Depreciation				$25,000,000	$25,000,000	$25,000,000	$25,000,000	$25,000,000	$25,000,000	$25,000,000	
28			Taxable Income				$362,500,000	$362,500,000	$362,500,000	$362,500,000	$362,500,000	$362,500,000	$362,500,000	
29			Taxes				$126,875,000	$126,875,000	$126,875,000	$126,875,000	$126,875,000	$126,875,000	$126,875,000	
30			Profit	-$15,000,000	0	-$250,000,000	$260,625,000	$260,625,000	$260,625,000	$260,625,000	$260,625,000	$0	$0	
31														
32		**Results**												
33			PV for Product	$60,989,570										
34			PV Multiple	**4.07**										
35			Recommend?	NO										
36														

Figure 6.8. M2—numerical view.

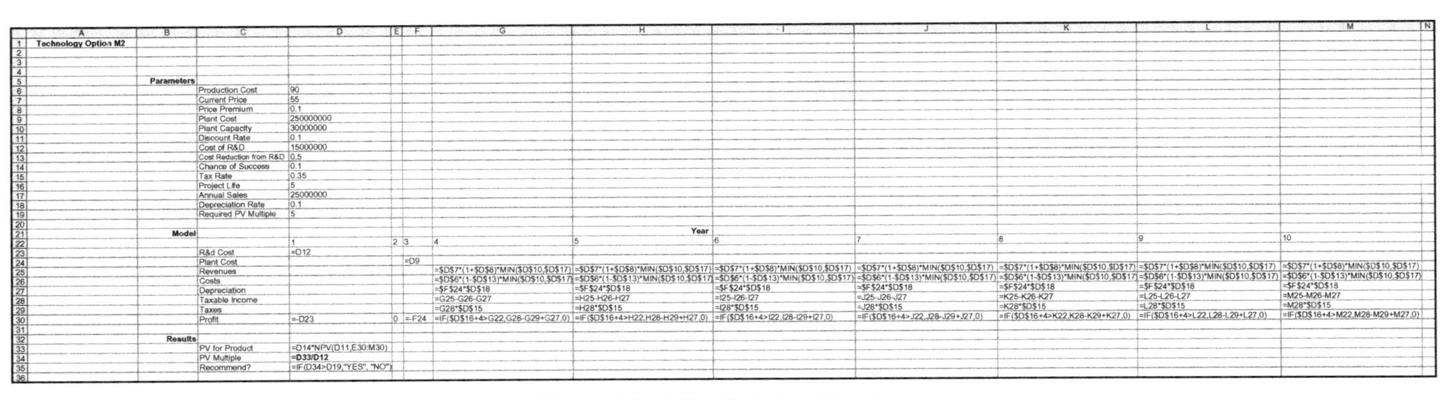

	A	B	C	D	E	F	G	H	I	J	K	L	M	N
1	**Technology Option M2**													
2														
3														
4														
5		**Parameters**												
6			Production Cost	90										
7			Current Price	55										
8			Price Premium	0.1										
9			Plant Cost	250000000										
10			Plant Capacity	30000000										
11			Discount Rate	0.1										
12			Cost of R&D	15000000										
13			Cost Reduction from R&D	0.5										
14			Chance of Success	0.1										
15			Tax Rate	0.35										
16			Project Life	5										
17			Annual Sales	25000000										
18			Depreciation Rate	0.1										
19			Required PV Multiple	5										
20														
21		**Model**						**Year**						
22				1	2	3	4	5	6	7	8	9	10	
23			R&d Cost	=D12										
24			Plant Cost			=D9								
25			Revenues				=$D$7*(1+$D$8)*MIN($D$10,$D$17)	=$D$7*(1+$D$8)*MIN($D$10,$D$17)	=$D$7*(1+$D$8)*MIN($D$10,$D$17)	=$D$7*(1+$D$8)*MIN($D$10,$D$17)	=$D$7*(1+$D$8)*MIN($D$10,$D$17)	=$D$7*(1+$D$8)*MIN($D$10,$D$17)	=$D$7*(1+$D$8)*MIN($D$10,$D$17)	
26			Costs				=$D$6*(1-$D$13)*MIN($D$10,$D$17)	=$D$6*(1-$D$13)*MIN($D$10,$D$17)	=$D$6*(1-$D$13)*MIN($D$10,$D$17)	=$D$6*(1-$D$13)*MIN($D$10,$D$17)	=$D$6*(1-$D$13)*MIN($D$10,$D$17)	=$D$6*(1-$D$13)*MIN($D$10,$D$17)	=$D$6*(1-$D$13)*MIN($D$10,$D$17)	
27			Depreciation				=$F$24*$D$18	=$F$24*$D$18	=$F$24*$D$18	=$F$24*$D$18	=$F$24*$D$18	=$F$24*$D$18	=$F$24*$D$18	
28			Taxable Income				=G25-G26-G27	=H25-H26-H27	=I25-I26-I27	=J25-J26-J27	=K25-K26-K27	=L25-L26-L27	=M25-M26-M27	
29			Taxes				=G28*$D$15	=H28*$D$15	=I28*$D$15	=J28*$D$15	=K28*$D$15	=L28*$D$15	=M28*$D$15	
30			Profit	=-D23	0	=-F24	=IF($D$16+4>G22,G28-G29+G27,0)	=IF($D$16+4>H22,H28-H29+H27,0)	=IF($D$16+4>I22,I28-I29+I27,0)	=IF($D$16+4>J22,J28-J29+J27,0)	=IF($D$16+4>K22,K28-K29+K27,0)	=IF($D$16+4>L22,L28-L29+L27,0)	=IF($D$16+4>M22,M28-M29+M27,0)	
31														
32		**Results**												
33			PV for Product	=D14*NPV(D11,E30:M30)										
34			PV Multiple	**=D33/D12**										
35			Recommend?	=IF(D34>D19,"YES", "NO")										
36														

Figure 6.9. M2—Relationship view.

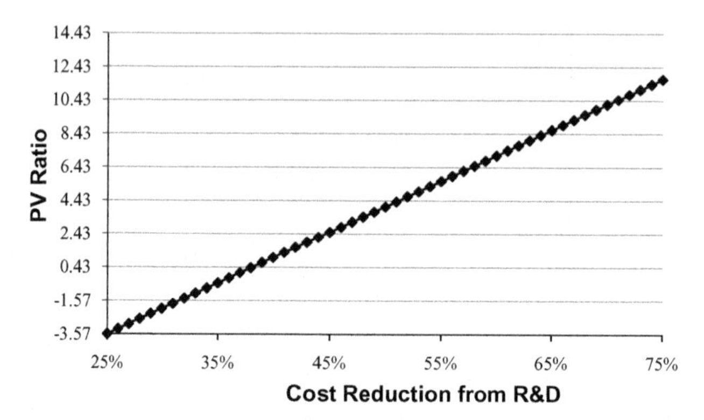

Figure 6.10. Sensitivity to Cost Reduction (M2).

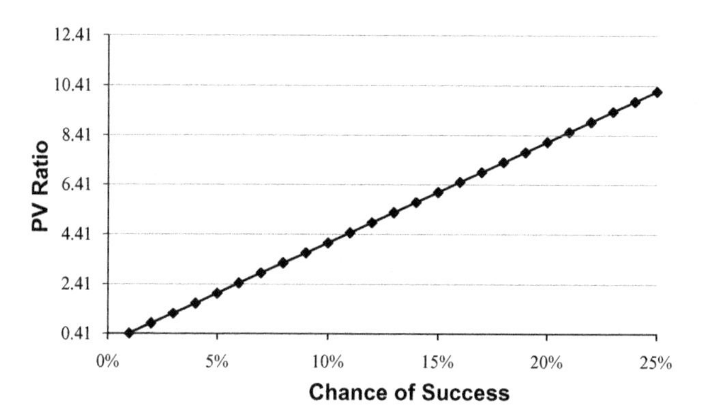

Figure 6.11. Sensitivity to R&D Success (M2).

Insight 4: In the base case, the project is very close to meeting the target.

Finally, we create a tornado chart to determine which of the inputs in our new model have the biggest impact on the outcome measure. Figure 6.12 shows the results: Current Price, Cost Reduction from R&D, and Production Cost dominate the results. These same three parameters were most important in M1. Thus, despite the changes we have made in the model, the parameters with the biggest impact on the outcome remain the same.

Insight 5: Current price, cost reduction from R&D, and production cost are the dominant inputs in M2, just as they were in M1.

There are some changes, however, in this tornado chart. Plant capacity ranked high in our first model, but it has essentially no impact in the second model, whereas annual sales does have a significant impact. This is because in

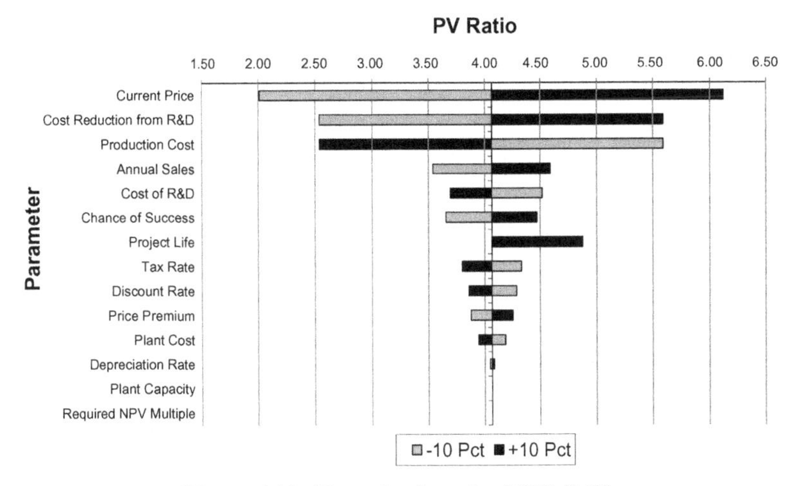

Figure 6.12. Tornado chart for NPV (M2).

this model we assumed that sales are less than capacity, and within the ranges of variation we test here, changes to capacity have no impact on the outcome.

Insight 6: Plant capacity has no effect on the PV multiple because we have assumed demand is always less than capacity.

6.3 ADDITIONAL REFINEMENTS

During the iterative modeling process, we produce a carefully designed sequence of models and a matching series of results. Rather than focus on the numerical results themselves, which, after all, may change in the next model, we focus on the general patterns we see and on the insights we develop. These are more likely to be stable over the entire process and to represent information of value to our clients.

In this case, M1 established the important fact that the expected NPV could be positive for the proposed project. If the estimated NPV had been a large negative number, we might have decided to stop the analysis at that point and recommend against the project. But given the initially positive estimate, we decided to refine the model to see how more realistic assumptions would influence the projection.

M2 included taxes, depreciation, and the client's actual objective. The results of that model suggested that perhaps the project was not attractive, since the expected PV multiple was only 4.07 times the investment, whereas the company expects a multiple of five or more. Sensitivity analysis, however, showed the breakeven multiple could be reached with only modest improvements in one or another of the input parameters. This suggests pursuing the analysis further.

To the Reader: Before reading on, consider how the original problem frame should be adjusted to include uncertainty in the analysis.

6.3.1 Frame the Problem—M3

The key feature we address in our next model is uncertainty. Since our first model, we have faced uncertainty in the outcome of the R&D process. We have handled this so far by using expected values. But expected values somewhat distort the true nature of the situation. With a 10 percent chance of R&D success, the company loses its $15M initial investment 90 percent of the time for a PV multiple of zero, whereas during the remaining 10 percent of the time, it will succeed at R&D and generate a PV multiple of about 40. This extreme range of outcomes is not apparent in the overall expected PV ratio of 4.07.

In addition to uncertainty in the outcome of the R&D, there is, undoubtedly, uncertainty in many of our model's other parameters. We have not addressed this uncertainty directly, although we have performed sensitivity analysis and created a tornado chart to explore the impacts of variations in the parameters. At this stage in the analysis, we are ready to incorporate both the uncertainty in the R&D and the uncertainty in parameters.

To the Reader: Before reading on, sketch an influence diagram for this new problem.

6.3.2 Diagram the Problem—M3

Since we are not changing any of the model's structure, the only change to the influence diagram is indicating which parameters will now be modeled as random variables. As we did with "R&D Success," we use a double rectangle to indicate an input that is subject to randomness (Figure 6.13). The randomized parameters are as follows:

- Project Life
- Demand
- Price Premium
- Cost Reduction from R&D
- Taxes
- Plant Cost
- R&D Cost

To the Reader: Before reading on, design and build a model based on the influence diagram in Figure 6.13.

6.3.3 Build the Model—M3

The problem statement suggests plausible ranges for some parameters. For example, it states that the price premium could range between 10 and 25

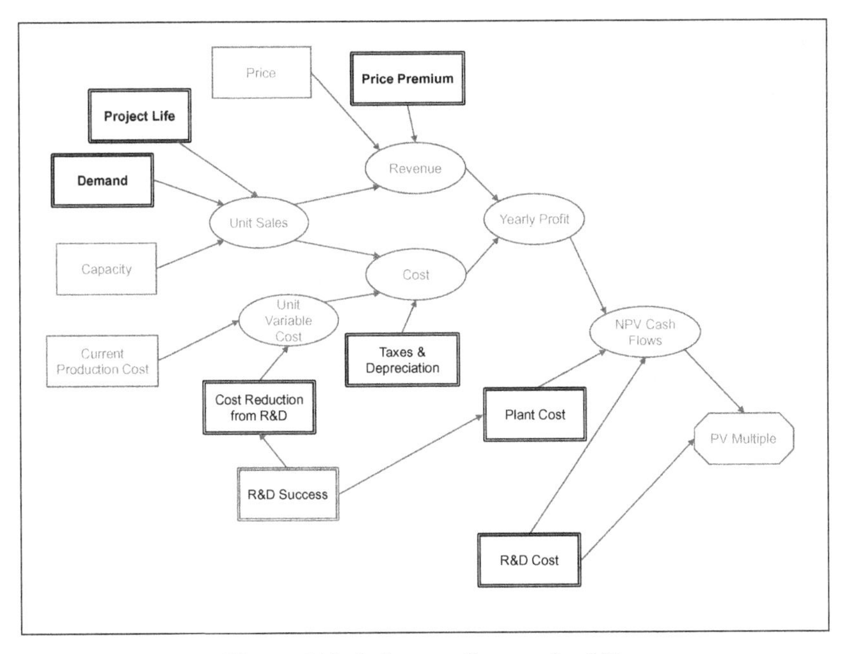

Figure 6.13. Influence diagram for M3.

percent. It also states that prices historically have ranged from a low of $45 to a high of $75. Finally, it states that the product life could be four to seven years. We could assume this information indicates the true uncertainty of these parameters, but it would probably be worth our effort to ask the client at this point for a more careful assessment of the uncertainties in the case. If we did this, we not only could determine the ranges for these three parameters, but we could also ask the client about uncertainties in the other parameters. Just as with any other decision we make during modeling, we need to decide before we contact the client whether this is the best use of our time at this point. There is no way to know for sure, but our reasoning here is that uncertainty and risk are already key parts of the problem. Getting better information about all the uncertainties in the situation will probably allow us to build a more appropriate model and provide more useful information to the client.

(We will not address the issue here of how to approach decision makers for information relating to uncertainties. Powerful techniques have been developed that are both practical and unbiased. A good reference on this subject is Clemen and Reilly, *Making Hard Decisions*.)

The results of our inquiries with the client are summarized in the following table, which gives low, base, and high values for the seven parameters we believe are most uncertain.

Parameter	Low	Base	High
Price Premium	10%	15%	25%
Plant Cost	$220M	$250M	$350M
Cost of R&D	$10M	$15M	$20M
Cost Reduction from R&D	40%	50%	60%
Tax Rate	30%	35%	40%
Project Life	4	5	7
Annual Sales	$25M	$30M	$35M

The client also informed us that prices are highly volatile year-to-year. A lognormal distribution is often appropriate for a parameter such as a price that cannot go negative. In this case, we use a lognormal distribution with a mean of $55 and a matching standard deviation. This distribution is shown in Figure 6.14. We sample independently each year from this distribution to model the high year-to-year volatility.

It is fairly straightforward to modify M2 to incorporate the uncertainty in these parameters. Figure 6.15 shows our new model, M3. In columns F–H, we have entered the three values for each of the uncertain parameters. Then, instead of having a single number for each uncertain parameter in column D, we have a probability distribution. The distribution we use here is only one of many we could have used given the information provided to us by the client. Here we assume the low value occurs with probability 0.25, the base value with probability 0.50, and the high value with probability 0.25. An example of the resulting distribution is shown in Figure 6.16 for the price premium.

There is one more change in this model. Instead of calculating the expected PV, we use the actual R&D outcome (success or failure) on each simulation

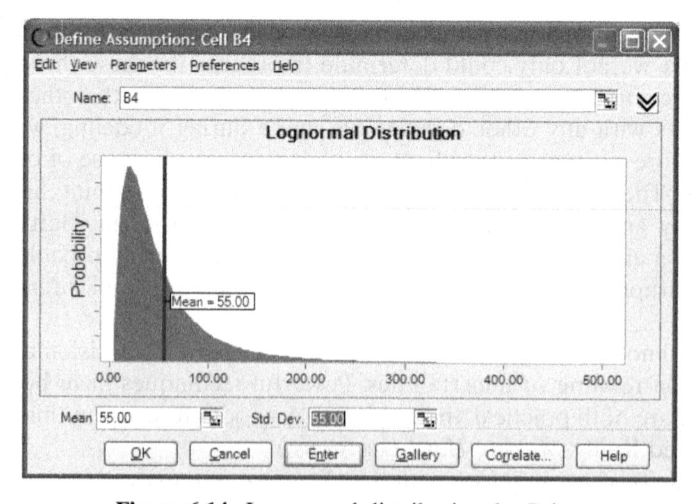

Figure 6.14. Lognormal distribution for Price.

	A	B	C	D	E	F	G	H	I	J	K	L	M	N
1	**Technology Option M3**													
2														
3														
4														
5		**Parameters**				**Low**	**Base**	**High**						
6			Production Cost	$90			$90							
7			Current Price-mean	$55			$55							
8			Current Price-SD	55			$10							
9			Price Premium	25%		10.00%	15.00%	25.00%						
10			Plant Cost	250,000,000		$220,000,000	$250,000,000	$350,000,000						
11			Plant Capacity	30,000,000			30,000,000							
12			Discount Rate	10%			10%							
13			Cost of R&D	10,000,000		$10,000,000	$15,000,000	$20,000,000						
14			Cost Reduction from R&D	60%		40.00%	50.00%	60.00%						
15			Chance of Success	10%			10%							
16			Tax Rate	40%		30%	35%	40%						
17			Project Life	7		4	5	7						
18			Annual Sales	35,000,000		25,000,000	30,000,000	35,000,000						
19			Depreciation Rate	10%			10%							
20			Required PV Multiple	5			5							
21														
22			R&D Success?	0										
23		**Model**						**Year**						
24				1	2	3	4	5	6	7	8	9	10	
25			R&D Cost	$10,000,000										
26			Plant Cost			$250,000,000								
27			Revenues				$10,959,754,457	$2,727,637,054	$1,044,193,005	$486,583,465	$1,773,586,705	$1,466,762,572	$1,773,586,705	
28			Costs				$1,080,000,000	$1,080,000,000	$1,080,000,000	$1,080,000,000	$1,080,000,000	$1,080,000,000	$1,080,000,000	
29			Depreciation				$25,000,000	$25,000,000	$25,000,000	$25,000,000	$25,000,000	$25,000,000	$25,000,000	
30			Taxable Income				$9,854,754,457	$1,622,637,054	-$60,806,995	-$618,416,535	$668,586,705	$361,762,572	$668,586,705	
31			Taxes				$3,941,901,783	$649,054,822	-$24,322,798	-$247,366,614	$267,434,682	$144,705,029	$267,434,682	
32			Profit		0	-$250,000,000	$5,937,852,674	$998,582,233	-$11,484,197	-$346,049,921	$426,152,023	$242,057,543	$426,152,023	
33														
34		**Results**												
35			PV for Product	$0										
36			PV Multiple	**0.00**										
37														

Figure 6.15. M3—numerical view.

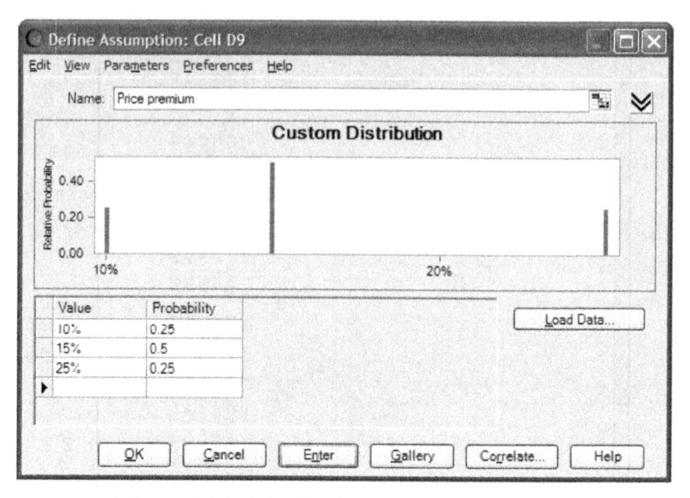

Figure 6.16. Distribution for Price Premium.

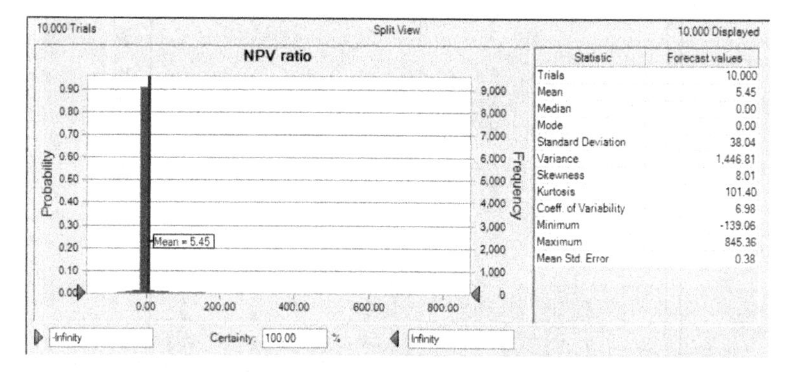

Figure 6.17. Frequency distribution and statistics for PV Ratio (M3).

trial. To model this type of outcome, we use a Bernoulli trial, implemented with a Yes/No distribution in Crystal Ball. The probability of R&D success is given in cell D15 (10 percent). This distribution is located in cell C22, and it returns the value 1 when R&D is successful and 0 otherwise. We use this result in cell D35, where we multiply it by the PV of profit. This essentially cancels out any profit when R&D fails.

Now we can explore all possible combinations of input parameters and determine the distribution of the resulting PV multiple. To do this requires Monte Carlo simulation. The results of 10,000 trials are shown in Figure 6.17. The mean PV ratio is now 5.45, which is slightly *above* the minimum requirement.

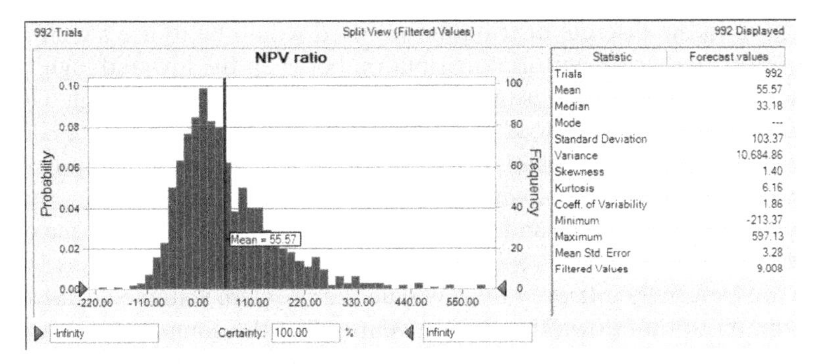

Figure 6.18. Frequency distribution and statistics for PV Ratio (M3 with zero outcomes eliminated).

To the Reader: Before reading on, use M3 to generate as many interesting insights into the situation as you can.

6.3.4 Generate Insights—M3

What can we learn from this new model? As we have observed, the mean PV ratio is now slightly above the minimum required by the company. But the PV ratio is *zero* 90 percent of the time, in all cases in which R&D fails, so the average of 5.45 must mean that the ratio is well above five some of the time. The worst outcome is a PV ratio of –139, whereas the best outcome is a ratio of 845. So even when R&D succeeds, the outcome is highly variable. This is shown more clearly in Figure 6.18, in which we have filtered out all the outcomes in which R&D fails. This figure makes clear that we can fail to achieve the PV goal even when R&D succeeds.

Insight 7: The project can fail to meet the PV multiple target even when R&D is successful.

This is an entirely new idea, one that we could not have uncovered with a purely deterministic analysis. It must be the case that we can succeed at R&D and yet the values of the other inputs are such that the PV ratio is below five. But there is also an upside to this project; in fact, there are combinations of parameters in which we not only exceed our breakeven target of five, we exceed it by a factor of about 170.

Insight 8: It is possible to exceed the PV multiple target by a factor of 170.

There is much more we can do with this latest model beyond this base case. For example, we could test the sensitivity of our results to any of our assumptions by running another simulation. We could also test the impact of the probability distribution we chose to represent our client's uncertainty. One alternative to the discrete distribution we used would be to use a continuous distribution, perhaps a uniform distribution between the low and high values or a triangular distribution using all three values: low, base, and high. Using a continuous distribution instead of a discrete one would give us a more continuous frequency distribution for the PV ratio, but it seems unlikely it would substantially alter the two main conclusions we have drawn so far—that the average value exceeds five and the range from the minimum to the maximum outcome is huge.

We are especially intrigued by one issue the problem statement raises. That is the option to shut down the plant in years when the company would take a loss. Our intuition suggests this real option must raise the value of the project, but we cannot guess how much the value will increase. It also might have an impact on the project's risks. Following these hunches, we move on to our next model.

To the Reader: Before reading on, modify the influence diagram in Figure 6.13 to include the real option to shut the plant.

6.3.5 Diagram the Problem—M4

Our fourth model is identical to the previous one except for the option to shut down. The calculated value that influences the decision to shut down is yearly profit. Any year in which profit would be negative, the factory shuts down and cash flow for that year goes to zero. We show this on the influence diagram in Figure 6.19.

To the Reader: Before reading on, design and build a model based on the influence diagram in Figure 6.19.

6.3.6 Build the Model—M4

We implement this option in the model by setting profit to zero in any year in which the company otherwise would have made a loss. This is carried out in row 34, where we calculate the nominal profit as in the earlier model, and in row 35, in which we use an IF statement in each year to set negative profits to zero.

6.3.7 Generate Insights—M4

The results from this new model are shown in Figure 6.20. A quick glance shows that including the option to shut down improves the average PV ratio from 5.45 to 10.98. Perhaps more importantly, it also radically changes the *shape* of the distribution. Whereas the minimum ratio was –139 in the previous model, shutting down the plant when profits are negative improves the worst-case outcome to –21. Furthermore, the upside outcome has also improved: The maximum PV ratio is 1,335 for a 58 percent improvement.

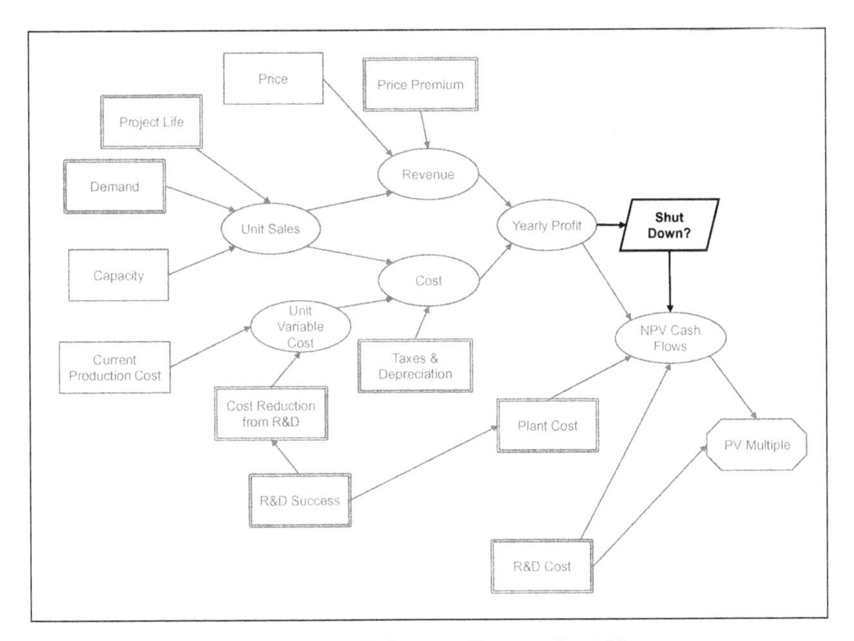

Figure 6.19. Influence diagram for M4.

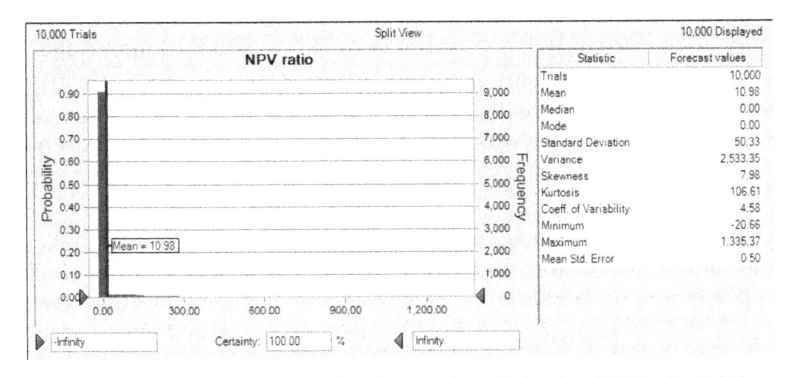

Figure 6.20. Frequency distribution and statistics for PV Ratio (M4).

Insight 9: The shut-down option dramatically reduces the downside risk and improves the upside potential.

Our hunch has paid off handsomely. We thought the option might improve the prospects of the project and that has turned out to be true. Moreover, we can quantify just how much of an improvement is involved: The average PV multiple improves by about 100 percent (from 5.45 to 10.98), and the worst case has improved even more. To be more precise, a PV ratio of –139 on an investment of \$15M implies that the project loses \$2.1 *billion*. We faced this situation when analyzing the problem without the option. But with the option to shut down included in the analysis, the worst-case PV ratio is –21, representing a much smaller loss of \$315M.

6.4 PRESENTATION OF RESULTS

The time has come to organize our models and insights into a presentation to SuperChem management. To do this well, we need to select our most telling insights and organize them into a logical story. But first, as always when designing a presentation, we must consider our audience. Since we have no specific information on our audience in this case, we will use our imagination. Most senior managers at SuperChem are likely to know about Ultra and even know some of the details about its costs and potential. They may have been in previous meetings where it was discussed. We imagine that we are an internal analysis team that has been asked to develop a framework for studying the decision whether to fund R&D on reducing the manufacturing costs for Ultra. Our study has been exploratory, so the data we have used and the models we have built are somewhat rough. Nonetheless, a clear picture has emerged that we can share with management. We need to walk a fine line between overselling our results by claiming more accuracy and certainty than is warranted, and underselling them by undercutting our credibility with too many caveats.

The focus of our study has been on a single decision: to undertake R&D on Ultra or not. In this situation, a direct presentation is more effective than an indirect one. Recall that in a direct presentation, we present our conclusions first and then back them up with analysis, whereas in an indirect presentation, we present our conclusions last, after building up the evidence that supports them. Since we are an independent analysis group, we do not have a stake in this decision. Hence, our audience likely understands that we do not have a bias regarding the decision. This suggests that a direct approach, which gets their attention right at the start, will work best. If, on the other hand, we worked for the R&D division or for the division that would market Ultra, we might choose an indirect approach to gain credibility for our analytic work before we put forth our recommendation.

What story do we want to tell here? Our analysis suggests that Ultra may well be an excellent investment, despite the high likelihood of failure in R&D. The \$15M investment in R&D is modest, certainly when compared with the potential revenue. In the best case, the present value of revenues could be more than 1000 times this investment, which is a blockbuster result if there ever was one. But these results emerged after we built a sequence of four models, M1 through M4. Should we include all four models and their associated results in our presentation, or should we present only selected results? The choice is obvious here: Showing our audience all the steps we took to reach our conclusions will burden them and not add insight. We could present results only from M4, since that model is the most complete of the four. However, the value added by the shutdown option is so significant that we will isolate it by first presenting the results from M3, then showing the effects of the option (M4).

Here is an outline of the presentation:

- The Ultra R&D Decision
- *Go Ahead with $15M R&D*
- Assumptions and Models
- Logic of Analysis (Influence Diagram)
- Tree of Outcomes
- Base Case: NPV and Risks
- Shutdown Option: NPV and Risks
- Summary

Our PowerPoint slides are shown in Figures 6.21– 6.30. The first slide introduces the problem and states some of the pros and cons of developing Ultra (Figure 6.21). The second slide states our conclusion: Investing $15M in R&D is well justified, despite the low probability of success, because the downside is acceptable and the upside is huge (Figure 6.22).

Having stated our conclusion, we now develop the reasoning behind the conclusion. In slide 3 (Figure 6.23), we lay out the essential elements of our

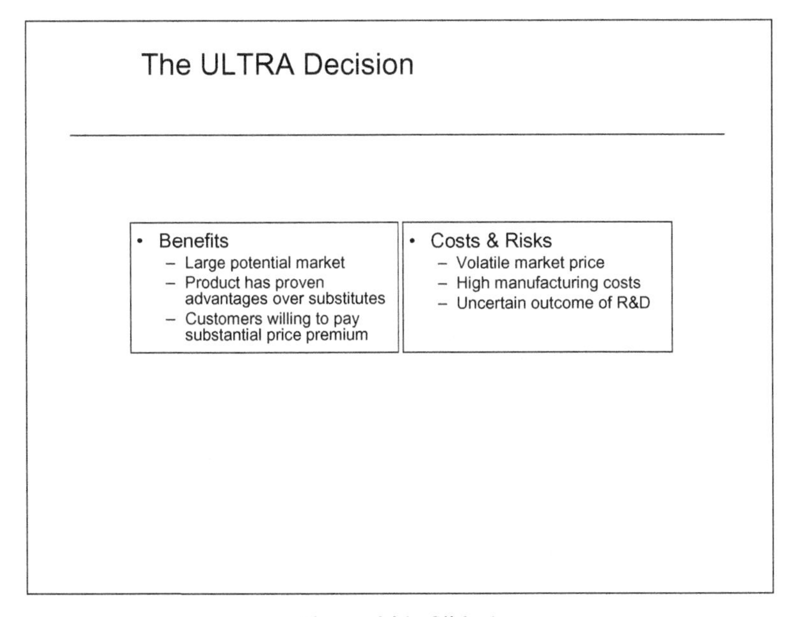

Figure 6.21. Slide 1.

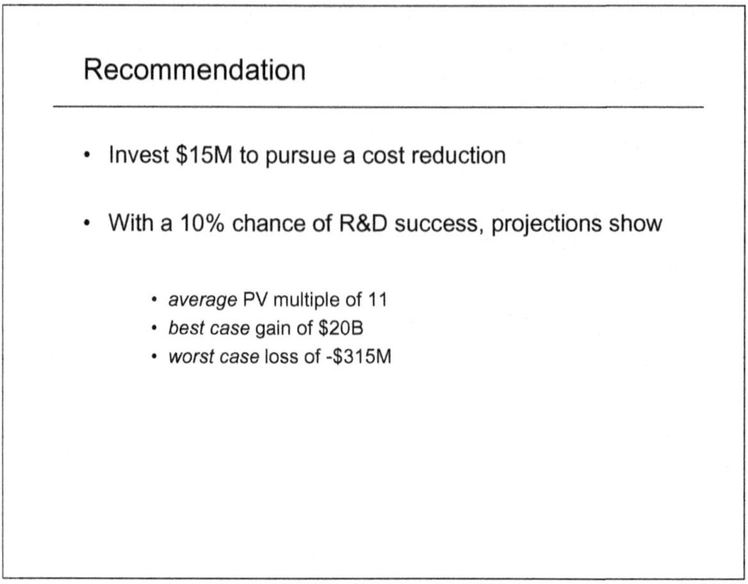

Recommendation

- Invest $15M to pursue a cost reduction
- With a 10% chance of R&D success, projections show
  - *average* PV multiple of 11
  - *best case* gain of $20B
  - *worst case* loss of -$315M

Figure 6.22. Slide 2.

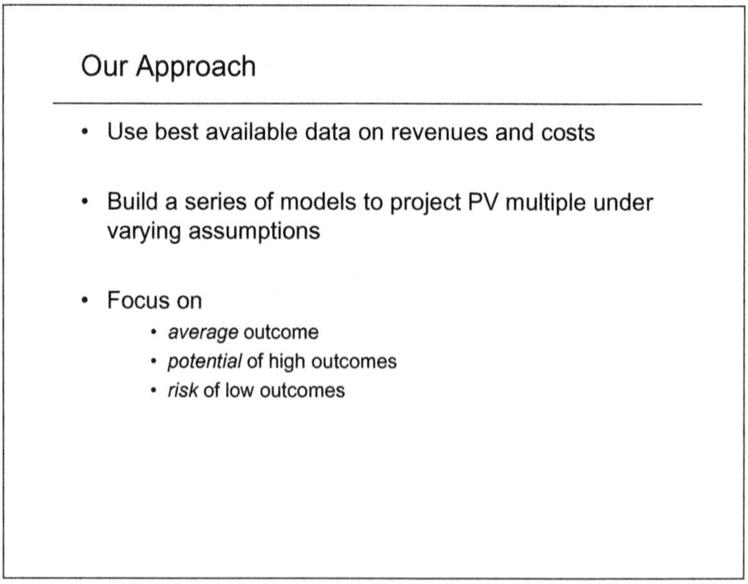

Our Approach

- Use best available data on revenues and costs
- Build a series of models to project PV multiple under varying assumptions
- Focus on
  - *average* outcome
  - *potential* of high outcomes
  - *risk* of low outcomes

Figure 6.23. Slide 3.

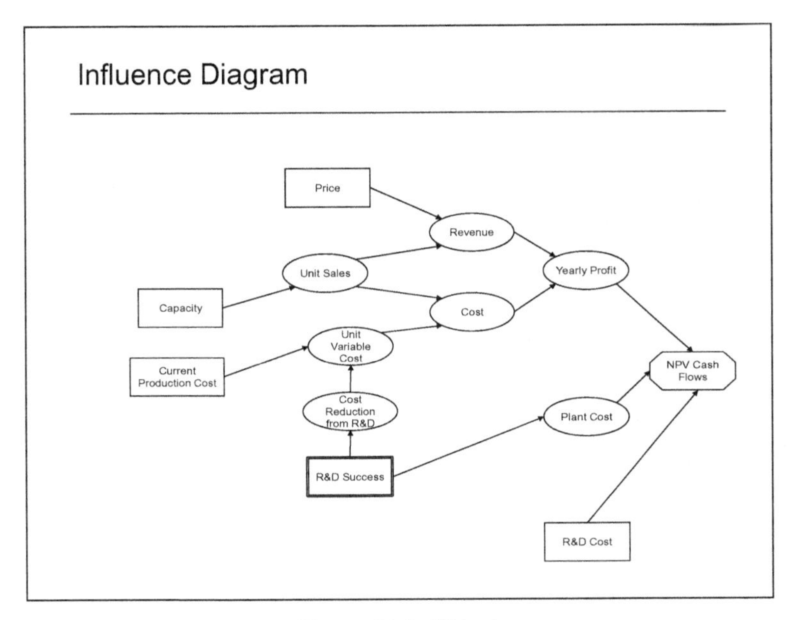

Figure 6.24. Slide 4.

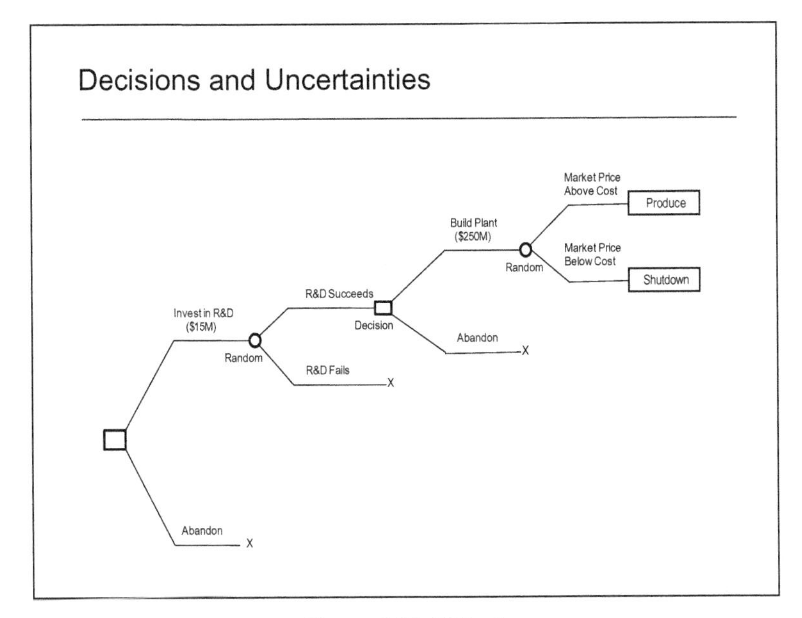

Figure 6.25. Slide 5.

Project meets PV hurdle even without shutdown option

	Base Case
Probability of failure	90%
PV multiple:	
mean	5.5
maximum	845
minimum	-139

Figure 6.26. Slide 6.

When R&D succeeds, the average PV multiple is 56

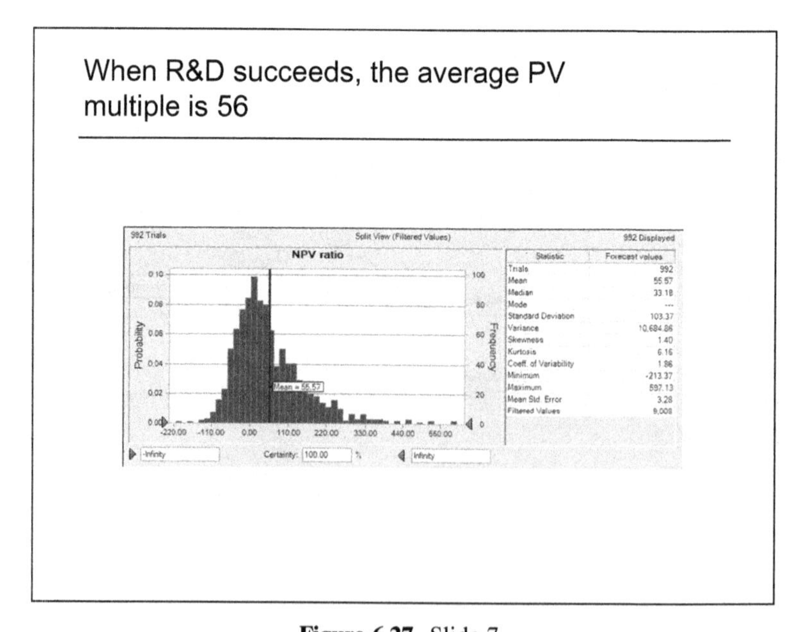

Figure 6.27. Slide 7.

Shutdown option improves mean PV and reduces risk

	Shutdown Option	Base Case
Probability of failure	90%	90%
PV multiple:		
mean	11.0	5.5
maximum	1,335	845
minimum	-21	-139

Figure 6.28. Slide 8.

With the shutdown option, when R&D succeeds the average PV multiple is 114

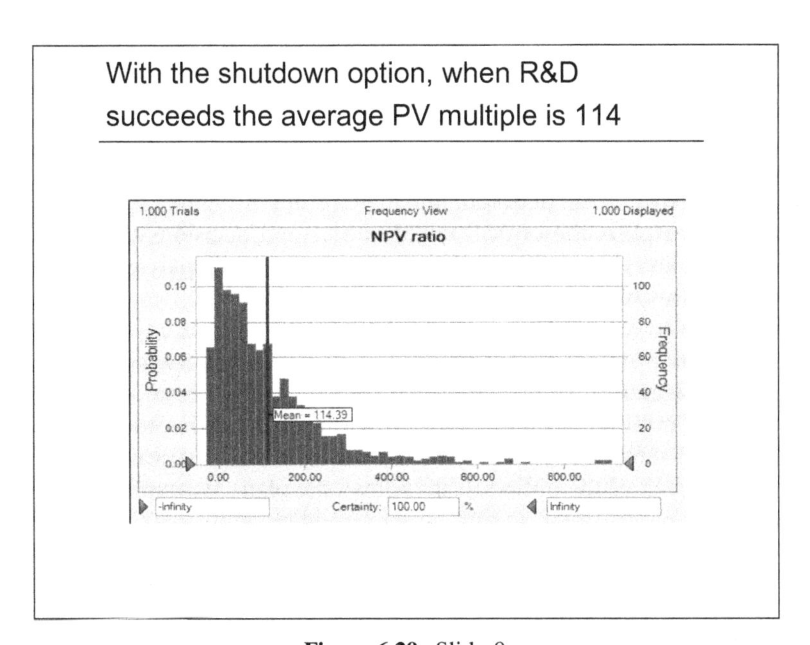

Figure 6.29. Slide 9.

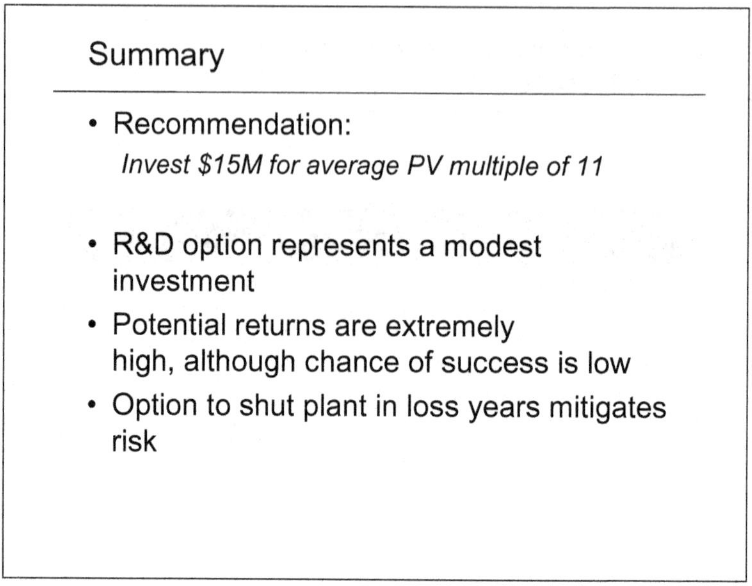

Summary

- Recommendation:
  Invest $15M for average PV multiple of 11
- R&D option represents a modest investment
- Potential returns are extremely high, although chance of success is low
- Option to shut plant in loss years mitigates risk

Figure 6.30. Slide 10.

approach: we used the best available data, built a series of models, and analyzed both averages and extreme results. Slide 4 (Figure 6.24) shows the basic influence diagram we used to link the parameters in the problem to the PV outcome measure. (When we present this slide, we use the Animation feature in PowerPoint to show how we developed the influence diagram step-by-step from the objective to the decision, much as we did in Figures 6.1a–d.) Slide 5 (Figure 6.25) displays the critical uncertainties and decisions that make this problem complex. This decision tree is a simple way to display graphically the essential features of our models.

We now get to results. As stated, we first show a base case that includes all elements of the problem except the option to shut down the plant in years a loss would be suffered. Slide 6 (Figure 6.26) shows these results: a just barely acceptable PV multiple and a substantial downside offsetting the upside. Slide 7 (Figure 6.27) shows the same results in the form of a distribution.

Next we introduce the option to shut down the plant. We explain this option verbally in the presentation so our audience understands how this case differs from the last. Then in Slides 8 and 9 (Figures 6.28 and 6.29), we report the results: a higher PV multiple and a much better worst-case outcome. Note that Slides 6 through 9 use the same format to convey parallel results for different cases. We do this to make it easy for the audience to compare the two cases.

We conclude with a summary slide (Figure 6.30) that repeats in slightly different words the message first introduced in the second slide: Our recommendation is to invest, because the upfront cost is low and the possible loss is low compared with the potential gain.

6.5 SUMMARY

In this chapter, we have taken a typical ill-structured business problem, involving a decision to undertake R&D, and have modeled it using the tools of iterative modeling. We began the analysis with the problem kernel, as we have in previous chapters, to demonstrate how to build a first model using just a fraction of the information usually conveyed in a problem statement. This first model, M1, radically simplified the problem, but it generated several interesting findings nevertheless.

We then introduced the full problem statement and developed a list of potential refinements to our first model. Experienced modelers develop a good sense of which refinements will be most productive and what order to introduce them into the analysis. (However, all modelers run into dead ends from time to time.) In this case, we laid out a plan to develop three refinements of the first model. The first step was to clean up the accounting in the initial model and to introduce the client's outcome measure (the PV multiple), which was not included in M1. This version of the model represented a modest improvement over the first one. The changes were easy to make, which was the major reason why we undertook them first. M2, the resulting model, came closer to representing a realistic base case than M1, which was based only on the problem kernel. Although the structure of the model did not change significantly at this stage, the results were closer to reality.

Our third model, M3, was a more significant undertaking. At this stage, we tackled the issue of uncertainty, thinking that uncertainty would probably be an essential feature of R&D planning, one that our analysis must address. We expended the resources to gather data from the client about the uncertainty in parameters, and we used this data to build a Monte Carlo simulation model. The added complexity of this approach was rewarded when our results allowed us to quantify the extreme risk our client faced in this case.

Our last model refinement, M4, also built on the previous model. We used Monte Carlo simulation again and essentially the same model. The only change was that we included the option to shut down the plant in any year in which profits would be negative. The change to the model only involved adding a single row, so we knew the time required to build this model version would be minimal. Our intuition suggested that the option might substantially improve the results from the previous model. Our hunch panned out. Not only did the average result improve substantially, but also the option substantially reduced the risk of loss.

Our client originally asked a simple Yes/No question, but we have provided much more than a Yes/No answer. By carefully analyzing the uncertainties and options in the problem, we have been able to quantify the range of outcomes they face and have been to able to provide a convincing rational for taking this project on.

The major artificial aspect of this chapter is that we began with the problem kernel and only later introduced the full problem statement. Our purpose in this approach was to show how a well-thought-out problem kernel can provide all that is needed for a first model. Once you have built and analyzed a first model, the task ahead is far simpler than it may have seemed originally. As we demonstrated here, building a series of successively more realistic and insightful models is not difficult if you carefully search the problem statement for important issues and then introduce them into the model in a well-thought-out sequence.

REFERENCE

Clemen RT, Reilly T. *Making Hard Decisions: Introduction to Decision Analysis*. 2nd edition. Pacific Grove, CA: Duxbury, 2005.)

PART III

CHAPTER 7

MediDevice

7.0 INTRODUCTION

In this chapter, we tackle a business-planning problem in a somewhat different manner than in previous chapters. Here we start with the complete problem description and simplify it by identifying the problem kernel, which is a process that is closer to what experts use in actual practice. With the kernel isolated from the details of the problem description, we formulate influence diagrams and models in the same fashion we demonstrated earlier. We also introduce new information into the problem and show how this can be incorporated effectively into the existing model.

7.1 MEDIDEVICE CASE (A)

MediDevice (A)

MediDevice Inc., a medical device manufacturer, has developed a new blood analyzer for rapid blood testing in doctors' offices. The much anticipated device will receive final regulatory approval in time for a January 2010 launch. MediDevice has the next 12 months to work out the final details for its commercial plans. This device is revolutionary and has worldwide appeal. Target markets for 2010 are the United States, Europe, and Japan.

Demand for this product is driven by the fraction of doctors' offices adopting the technology. Forecasts indicate 10 percent of doctors' offices in any country will potentially adopt this type of device. The time it takes to achieve this level of technology adoption is expected to be two to seven years, depending on several factors including the effectiveness of the sales force in each country.

Each office will buy only one analyzer, but once an office buys an analyzer, MediDevice receives recurring revenue from each office from royalties on analyzer supplies (sold by a third party). These annual royalty revenues will continue until MediDevice's analyzer is displaced by next-generation technology.

Modeling for Insight: A Master Class for Business Analysts, by Stephen G. Powell & Robert J. Batt

MediDevice expects to be first to market, but a major Japanese competitor is working on a next-generation device. When this competitive device hits the market, it will most likely displace the MediDevice analyzer from a portion of MediDevice's installed base of doctors' offices. However, some of the installed base will continue in use until third-generation devices ultimately take over the entire market.

Opinions are divided about the commercial strategy MediDevice should use. It can build its own U.S. sales force and sell directly to doctors, but the company does not have time to put together a worldwide sales team. Alternatively, MediDevice can license the device to a competitor to take advantage of its worldwide sales force. A licensing arrangement guarantees instant access to doctors and will probably lead to faster market penetration but limits MediDevice revenues to a negotiated license fee on each device. On the other hand, building an internal sales force will increase fixed costs, but it will allow MediDevice to keep all the sales revenue for itself. (In either plan, MediDevice earns the same recurring royalty revenue from each office once the device is sold.)

MediDevice will be the sole manufacturer and distributor. Its engineers have assessed the capital required to build manufacturing capacity for either a U.S.-only or a worldwide launch.

MediDevice has identified two alternative commercialization plans:

- In the Current Plan, MediDevice will build an internal sales force to sell to the U.S. market only.
- In the Partner Plan, MediDevice will license the device to the competitor and go for a full worldwide launch.

The MediDevice board has commissioned you to perform a cash flow analysis to compare the value of the two plans and make a recommendation. The board's assessments of the key parameters are summarized in the table below:

MediDevice Assessment	Low	Base	High
Market Size—U.S. (# of Dr. Offices)		600,000	
Market Size—Europe (# of Dr. Offices)		450,000	
Market Size—Japan (# of Dr. Offices)		140,000	
Peak market penetration of technology (% of Dr. Offices)	5%	10%	15%
Years to peak adoption (including launch year)	3	5	8
Direct sales revenue per unit ($000s)	$8	$10	$12
Annual recurring royalty revenue per office ($000s)	$0.5	$1	$2
Year when competitor launches next-generation device	2012	2015	2019
Percent of offices currently using the MediDevice analyzer that will switch to the competitive product when it launches	50%	70%	80%
Estimated year when third-generation devices completely take over the market	2018	2019	2023
Annual cost to run in-house MediDevice U.S. sales force for Current Plan ($M)	$10	$15	$20

MediDevice Assessment	Low	Base	High
License fee for Partner Plan (% of sales revenues paid to MediDevice)	40%	50%	60%
Improved time-to-peak adoption in Partner Plan (years)	0	1	2
Total capital cost for U.S. manufacturing ($M, all spent in 2009)	$20	$25	$35
Additional capital cost for international manufacturing ($M, all spent in 2009)	$15	$20	$35
Variable manufacturing cost per unit ($000s)	$3	$4	$5

To the Reader: Before reading on, frame this problem by setting boundaries for the analysis; making assumptions; listing questions; defining outputs, inputs, and decisions; and so on.

7.1.1 Frame the Problem

The first conclusion we draw from the problem statement is that the goal does not seem to be to decide *whether* to market this new product, but *how*. This new device is a breakthrough product that is expected to penetrate 10 percent of all doctors' offices over the next few years. It seems clear that it will be profitable. The real question is how best to market it.

These observations suggest that we focus initially on the key decision: whether MediDevice should license the product for marketing to the whole world by a competitor, or build up its own sales force and market it only in the United States. Within the broad outlines of those choices may lie other options. We want to give some thought to crafting strategies that maximize profit while limiting risks.

These decisions involve a trade-off that may be key to modeling and analyzing this problem successfully. If MediDevice licenses the device, it can expect quick access to the entire market at low initial cost. However, the company will realize only a fraction of the revenues generated by the device. If, on the other hand, MediDevice develops its own sales force, it will only be able to market the product in the United States but will capture all the revenues.

Having identified some key trade-offs that our analysis will explore, we also notice in the problem description several details that relate to *timing*: timing of competitor entry and timing (and level) of market penetration.

MediDevice asked us to compare the two alternatives on the basis of cash flows, which potentially extend over the next 15 years. Differences in the timing of the cash flows could be critical, so we need to take the time value of money into account. The usual way to do this is to compute the net present value (NPV). So we choose the net present value of cash flows as our outcome measure. Here is our problem kernel:

Our client has developed a new medical device that will be sold to doctors' offices. It will generate revenue both from initial sales and from recurring usage of consumable supplies. The volume of sales and the rate at which market share grows will depend on the number of doctors' offices, marketing efforts, and perhaps other factors. The key decision is whether to sell the device through a licensing agreement or through direct sales. Our outcome measure is the net present value of cash flows.

To the Reader: Before reading on, sketch an influence diagram for this problem.

7.1.2 Diagram the Problem—M1

We start with a generic cash flow influence diagram as in Figure 7.1, which shows that the Cash Flows can be calculated as Net Income after Tax, plus Depreciation, less Capital Expenditures. Net Income, in turn, is calculated as EBITDA (earnings before interest, taxes, depreciation, and amortization), less Taxes and Depreciation.

A more detailed influence diagram is shown in Figure 7.2. Here we decompose the factors influencing the NPV of cash flows to the point where we begin to see the influence of the licensing decision on costs and revenues. This

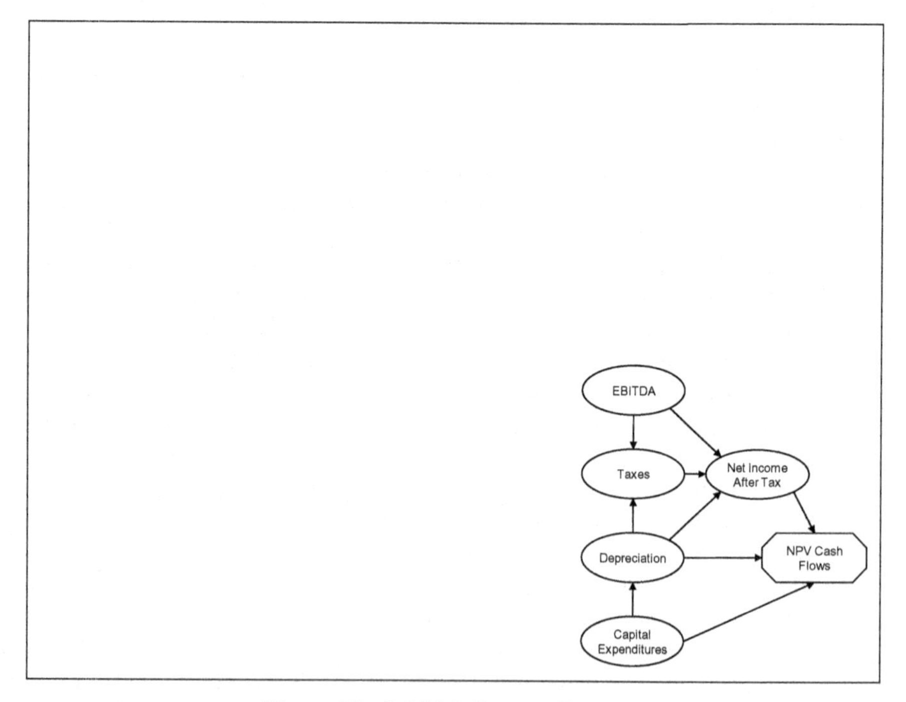

Figure 7.1. Initial influence diagram.

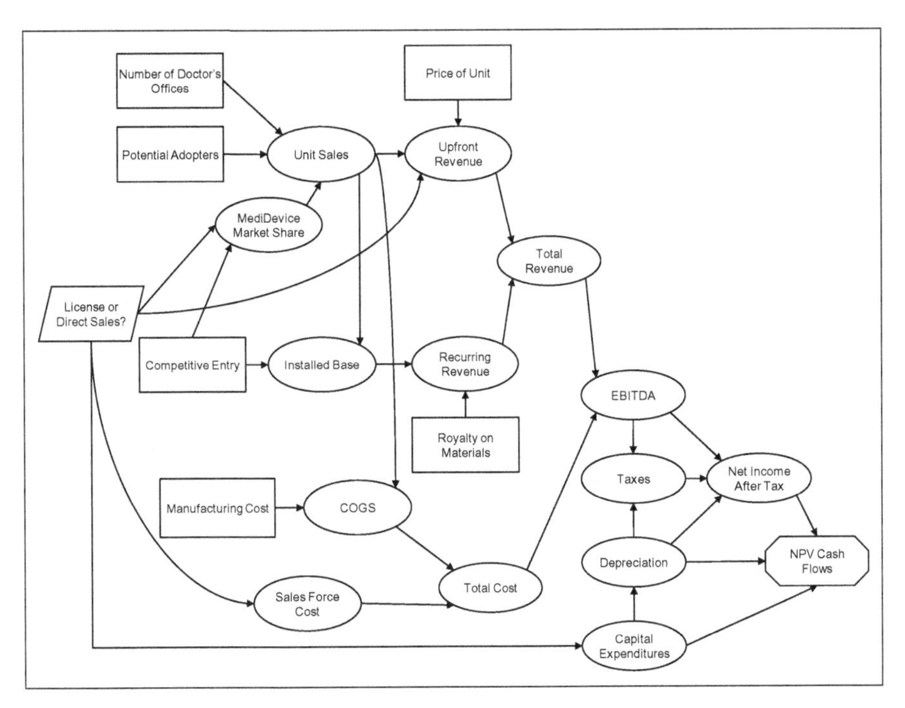

Figure 7.2. Completed influence diagram for M1.

diagram includes the distinction between revenues from initial sales and use of consumables by the installed base of devices. It also shows how unit sales are influenced by the total size of the market (the number of doctors' offices), MediDevice's share of that market, and its marketing decisions.

It is important to develop an intuition for when an influence diagram is sufficiently detailed. Remember, an influence diagram is not a model; it cannot give you answers to your questions. Its sole purpose is to help you articulate the assumptions and relationships in your model. If you develop it beyond the point of usefulness, you will be wasting your time. On the other hand, if you do not develop it enough, you will find yourself confused when building your first model.

Much of what is recorded in the influence diagram in Figure 7.2 is standard accounting and needs no more elaboration. But there are several places in which we have been vague, and it is worth developing these areas in more detail before we build our model. We focus on two related issues: how Competitive Entry affects the Installed Base and how Unit Sales are determined.

We know from the problem statement that a second-generation product is expected in about 2015 and that about 70 percent of the installed base of MediDevice customers will switch to that new product. Then, in about 2019, a third-generation product will arrive and take all the remaining MediDevice customers. Thus, up until 2015, we track the total number of machines sold and calculate MediDevice's recurring revenues based on that total. After the second-generation machine arrives, MediDevice's installed base drops by 70 percent, but the remaining 30 percent continues to generate royalty revenues. After the third-generation machine arrives, all of MediDevice's revenues drop to zero.

Our approach to modeling unit sales each year is based on the following decomposition:

$$\text{Unit Sales} = \text{Total Number of Doctors' Offices} \times \text{Market Penetration Level} \times \text{MediDevice Market Share}$$

We first project the total number of doctors' offices by region (United States, Europe, Japan). Then we model the growth of potential adopters over time. Here we assume that advertising and word of mouth lead doctors to learn about this technology at a rate that increases the number of potential adopters in a linear fashion. This relationship is based on four inputs:

- Launch Year
- Peak Year
- Initial Level
- Peak Level

Thus, if the launch year is 2010, and the peak year 2014, and the initial and peak penetration levels are 0 percent and 10 percent, respectively, the penetration level starts at zero in 2010 and grows linearly until it reaches its peak of 10 percent in 2014 (see Figure 7.3). Thereafter, it remains constant. The licensing option complicates this model slightly since one of the impacts of licensing is to increase the rate of adoption. This has the effect of moving the peak year earlier.

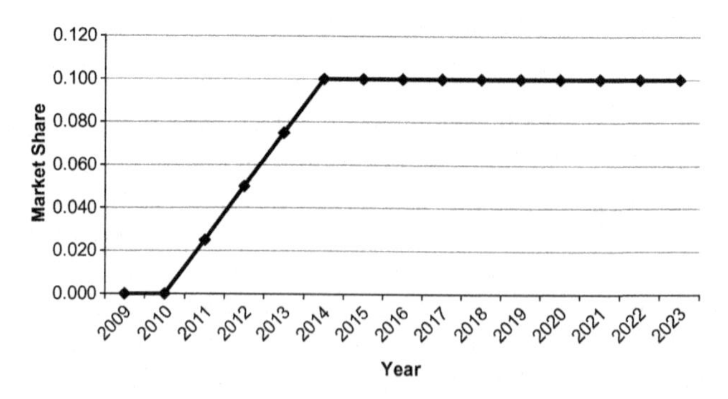

Figure 7.3. Market Share model.

MediDevice's market share will be 100 percent until the second-generation technology arrives, at which point it drops to 30 percent. We assume this means that 30 percent of the installed base of MediDevice customers continue to use their machines and generate royalty income, and 30 percent of new sales continue to go to MediDevice. After the third-generation device arrives, the MediDevice installed base will drop to zero and no new sales will occur.

To the Reader: Before reading on, design and build a first model based on the influence diagram in Figure 7.2.

7.1.3 Build the Model—M1

Figures 7.4 and 7.5 show our first model, M1. All the parameters are taken directly from the problem statement and need no elaboration. (Note that although some inputs, such as price, are technically under our control, we list them as parameters, not decision variables, here because we are concentrating on the choice between licensing and direct sales in this model.) However, one design detail is worth mentioning. As we saw in the influence diagram, the licensing decision has implications for several aspects of the model. In particular, if MediDevice decides to license, unit revenues drop, potential adopters increase sooner, and sales force cost drops. One way to compare the direct sales option with the licensing option would be to build two separate models, perhaps on two worksheets in the same workbook. The problem with this approach is that a high percentage of the content of these models would be the same, and not only would we duplicate effort in creating two sheets, but also we would run a severe risk that at some point we would make a change to one model but not to the other.

A better approach is to use a switching variable to turn on the licensing option. Thus, in cell C7, we enter the value 0 to represent the base case in which MediDevice uses its own direct sales force and sells only in the United States. When we enter 1 in C7, we are turning on the licensing option, and we must then change all the aspects of the model that change under licensing. We show how this is done in more detail below.

The core of the model is in rows 50–70, where we calculate revenue. We begin by projecting the number of doctors' offices in the three regions: the United States, Europe, and Japan. Note that we have projected the number of doctors' offices to remain constant over the next 15 years. By explicitly showing in the model the number of doctor's offices for each year, we create flexibility that allows us to come back and try other assumptions later. We jot down a note to reconsider this assumption in the second model.

In row 55, we calculate market penetration: the percentage of all doctors' offices that are potential adopters each year. The IF statement here is complex, but we can write this in pseudo-code more simply as follows:

	A	B	C	D	E	F	G	H	I	J	K	L	M	N	O	P	Q
1	MediDevice	M1															
2																	
3																	
4																	
5																	
6	Decisions				NPV	194.4		Benchmarks			% changes						
7		Licensing (0/1)?	0						direct sales	194	0.000						
8									licensing	212	-0.08						
9	Parameters																
10		Number of Doctor's Offices			BASE												
11		US	600		600												
12		Europe	450		450												
13		Japan	140		140												
14																	
15		First Year	2009		2009												
16		Tax Rate	40%		0.4												
17		Discount Rate	10%		0.1												
18																	
19		Market Penetration															
20		Launch Year	2010		2010												
21		Peak Year - Base	2014		2014												
22		Initial Level	0%		0												
23		Peak Level	10%		0.1												
24		Licensing Benefit	1		1												
25		Peak Year - Actual	2014		2014												
26																	
27		Second Generation															
28		Entry Year	2015		2015												
29		Share of Installed Base Captured	70%		0.7												
30																	
31		Third Generation															
32		Entry Year	2019		2019												
33		Share of Installed Base Captured	100%		1												
34																	
35		Prices															
36		Device	10		10												
37		Materials	1		1												
38																	
39		Costs															
40		Unit Manufacturing	4		4												
41		Sales Force	15		15												
42		US Plant Cost	25		25												
43		ROW Plant Cost	20		20												
44		Depreciation Lifetime	10		10												
45		Licensing Fee	50%		0.5												
46																	
47			2009	2010	2011	2012	2013	2014	2015	2016	2017	2018	2019	2020	2021	2022	2023
48	Revenue																
49		Number of Doctor's Offices															
50		US	600	600	600	600	600	600	600	600	600	600	600	600	600	600	600
51		Europe	450	450	450	450	450	450	450	450	450	450	450	450	450	450	450
52		Japan	140	140	140	140	140	140	140	140	140	140	140	140	140	140	140
53		Total	1190	1190	1190	1190	1190	1190	1190	1190	1190	1190	1190	1190	1190	1190	1190
54																	
55		Market Penetration	0.000	0.000	0.025	0.050	0.075	0.100	0.100	0.100	0.100	0.100	0.100	0.100	0.100	0.100	0.100
56																	
57		Potential Users - Total	0	0	15	30	45	60	60	60	60	60	60	60	60	60	60
58		Potential Users - New	0	0	15	15	15	15	0	0	0	0	0	0	0	0	0
59		MediDevice Share	1.00	1.00	1.00	1.00	1.00	1.00	0.30	0.30	0.30	0.30	0.00	0.00	0.00	0.00	0.00
60																	
61		Unit Sales	0	0	15	15	15	15	0	0	0	0	0	0	0	0	0
62		Cumulative Unit Sales	0	0	15	30	45	60	60	60	60	60	60	60	60	60	60
63		MediDevice Installed Base	0	0	15	30	45	60	18	18	18	18	0	0	0	0	0
64																	
65		Revenues															
66		New Devices															
67		Direct Sales	0	0	150	150	150	150	0	0	0	0	0	0	0	0	0
68		Licensing	0	0	0	0	0	0	0	0	0	0	0	0	0	0	0
69		Materials	0	0	15	30	45	60	18	18	18	18	0	0	0	0	0
70		Total	0	0	165	180	195	210	18	18	18	18	0	0	0	0	0
71																	
72	Operating Cost																
73		COGS	0	0	60	60	60	60	0	0	0	0	0	0	0	0	0
74		Sales Force Cost	0	0	15	15	15	15	0	0	0	0	0	0	0	0	0
75		Total Cost	0	0	75	75	75	75	0	0	0	0	0	0	0	0	0
76																	
77	Capital Expenditure																
78		Manufacturing Plant Cost	25														
79		Depreciation	2.5	2.5	2.5	2.5	2.5	2.5	2.5	2.5	2.5	2.5	0	0	0	0	0
80																	
81	Income Statement																
82		Revenue	0.0	0.0	165.0	180.0	195.0	210.0	18.0	18.0	18.0	18.0	0.0	0.0	0.0	0.0	0.0
83		Cost	0.0	0.0	75.0	75.0	75.0	75.0	0.0	0.0	0.0	0.0	0.0	0.0	0.0	0.0	0.0
84		EBITDA	0.0	0.0	90.0	105.0	120.0	135.0	18.0	18.0	18.0	18.0	0.0	0.0	0.0	0.0	0.0
85		Depreciation	2.5	2.5	2.5	2.5	2.5	2.5	2.5	2.5	2.5	2.5	0.0	0.0	0.0	0.0	0.0
86		Taxes	-1.0	-1.0	35.0	41.0	47.0	53.0	6.2	6.2	6.2	6.2	0.0	0.0	0.0	0.0	0.0
87		NIAT	-1.5	-1.5	52.5	61.5	70.5	79.5	9.3	9.3	9.3	9.3	0.0	0.0	0.0	0.0	0.0
88																	
89	Cash Flow																
90		NIAT	-1.5	-1.5	52.5	61.5	70.5	79.5	9.3	9.3	9.3	9.3	0.0	0.0	0.0	0.0	0.0
91		Depreciation	2.5	2.5	2.5	2.5	2.5	2.5	2.5	2.5	2.5	2.5	0.0	0.0	0.0	0.0	0.0
92		CAPX	25.0	0.0	0.0	0.0	0.0	0.0	0.0	0.0	0.0	0.0	0.0	0.0	0.0	0.0	0.0
93		Cash flow	-24.0	1.0	55.0	64.0	73.0	82.0	11.8	11.8	11.8	11.8	0.0	0.0	0.0	0.0	0.0
94																	
95	NPV	194.4															

Figure 7.4. M1—numerical view.

IF(Current Year < Launch Year)

  0,

ELSE IF(Current Year = Launch Year) AND

  (Current Year ≠ Peak Year)

  Launch Level,

	A	B	C	D	E	F	G
1	MediDevice	M1					
2							
3							
4							
5							
6	Decisions				NPV	=B95	
7		Licensing (0/1)?	0				
8							
9	Parameters						
10		Number of Doctor's Offices			BASE		
11		US	600		600		
12		Europe	450		450		
13		Japan	140		140		
14							
15		First Year	2009		2009		
16		Tax Rate	0.4		0.4		
17		Discount Rate	0.1		0.1		
18							
19		Market Penetration					
20		Launch Year	2010		2010		
21		Peak Year - Base	2014		2014		
22		Initial Level	0		0		
23		Peak Level	0.1		0.1		
24		Licensing Benefit	1		1		
25		Peak Year - Actual	=MAX(C21-C7*C24,C20)		2014		
26							
27		Second Generation					
28		Entry Year	2015		2015		
29		Share of Installed Base Captured	0.7		0.7		
30							
31		Third Generation					
32		Entry Year	2019		2019		
33		Share of Installed Base Captured	1		1		
34							
35		Prices					
36		Device	10		10		
37		Materials	1		1		
38							
39		Costs					
40		Unit Manufacturing	4		4		
41		Sales Force	15		15		
42		US Plant Cost	25		25		
43		ROW Plant Cost	20		20		
44		Depreciation Lifetime	10		10		
45		Licensing Fee	0.5		0.5		
46							
47			**=C15**	**=C47+1**	**=D47+1**	**=E47+1**	
48	Revenue						
49		Number of Doctor's Offices					
50		US	=C11	=C50	=D50	=E50	
51		Europe	=C12	=C51	=D51	=E51	
52		Japan	=C13	=C52	=D52	=E52	
53		Total	=SUM(C50:C52)	=SUM(D50:D52)	=SUM(E50:E52)	=SUM(F50:F52)	
54							
55		Market Penetration	=IF(C47<$C$20,0,IF(AND(C47=$C$20,C47<>$C$25),$C$22,IF(C47<$C$25,($C$23-$C$22)/($C$25-$C$20)+B55,$C$23)))	=IF(D47<$C$20,0,IF	=IF(E47<$C$20,0,IF	=IF(F47<$C$20,0,IF	
56							
57		Potential Users - Total	=C55*(C50+$C$7*(C51+C52))	=D55*(D50+$C$7*(	=E55*(E50+$C$7*(	=F55*(F50+$C$7*(F	
58		Potential Users - New	=0	=D57-C57	=E57-D57	=F57-E57	
59		MediDevice Share	=IF(AND(C47<$C$28, C47<$C$32),1,IF(AND(C47>=$C$28, C47<$C$32),1-$C$29, 1-$C$33))	=IF(AND(D47<$C$2	=IF(AND(E47<$C$2	=IF(AND(F47<$C$2	
60							
61		Unit Sales	=+C59*C58	=+D59*D58	=+E59*E58	=+F59*F58	
62		Cumulative Unit Sales	0	=C62+D61	=D62+E61	=E62+F61	
63		MediDevice Installed Base	=C62*C59	=D62*D59	=E62*E59	=F62*F59	
64							
65		Revenues					
66		New Devices					
67		Direct Sales	=(1-$C$7)*C61*$C$36	=(1-$C$7)*D61*C	=(1-$C$7)*E61*C	=(1-$C$7)*F61*C	
68		Licensing	=$C$7*$C$45*C61*$C$36	=$C$7*$C$45*D61*	=$C$7*$C$45*E61*	=$C$7*$C$45*F61*	
69		Materials	=C63*$C$37	=D63*$C$37	=E63*$C$37	=F63*$C$37	
70		Total	=SUM(C67:C69)	=SUM(D67:D69)	=SUM(E67:E69)	=SUM(F67:F69)	
71							
72	Operating Cost						
73		COGS	=C61*$C$40	=D61*$C$40	=E61*$C$40	=F61*$C$40	
74		Sales Force Cost	=IF(C61>0,$C$41,0)*(1-$C$7)	=IF(D61>0,$C$41,0	=IF(E61>0,$C$41,0	=IF(F61>0,$C$41,0	
75		Total Cost	=C73+C74	=D73+D74	=E73+E74	=F73+F74	
76							
77	Capital Expenditure						
78		Manufacturing Plant Cost	=C42+C43*C7				
79		Depreciation	=IF(C47<$C$15+$C$44,$C$78/$C$44,0)	=IF(D47<$C$15+$C	=IF(E47<$C$15+$C	=IF(F47<$C$15+$C	
80							
81	Income Statement						
82		Revenue	=C70	=D70	=E70	=F70	
83		Cost	=C75	=D75	=E75	=F75	
84		EBITDA	=C82-C83	=D82-D83	=E82-E83	=F82-F83	
85		Depreciation	=C79	=D79	=E79	=F79	
86		Taxes	=(C84-C85)*$C$16	=(D84-D85)*$C$16	=(E84-E85)*$C$16	=(F84-F85)*$C$16	
87		NIAT	=C84-C85-C86	=D84-D85-D86	=E84-E85-E86	=F84-F85-F86	
88							
89	Cash Flow						
90		NIAT	=C87	=D87	=E87	=F87	
91		Depreciation	=C79	=D79	=E79	=F79	
92		CAPX	=C78	=D78	=E78	=F78	
93		Cash Flow	=C90+C91-C92	=D90+D91-D92	=E90+E91-E92	=F90+F91-F92	
94							
95	NPV	**=C93+NPV(C17,D93:Q93)**					
96							

Figure 7.5. M1—relationship view (columns G through Q omitted for clarity).

ELSE IF(Current Year < Peak Year)
    Interpolate,
ELSE
    Peak Level.

The interpolation we use here is a linear function between the initial level and the peak level. The annual increment is calculated by taking the ratio of the change in the levels, divided by the change in the dates:

$$(\text{Peak Level} - \text{Initial Level}) \div (\text{Peak Year} - \text{Launch Year})$$

We then add last year's penetration rate to determine this year's penetration level.

In rows 57–59, we calculate more of the variables needed to determine unit sales. In row 57, we determine the number of potential adopters by multiplying the total number of doctors' offices by the market penetration. Recall, however, that MediDevice's market is only the United States when it chooses to use direct sales and the entire world when it uses licensing. This is the first time in the model that we use the switch variable in C7 to represent licensing.

The formula can be written as follows:

Number of Potential Adopters
= Market Penetration × Number of Doctors' Offices
= Market Penetration × [U.S. Doctors + License Switch × (Europe Doctors + Japan Doctors)]

When the license switch is set to 0, the market is just U.S. doctors; when the switch is set to 1, meaning MediDevice chooses licensing, the market expands to include Europe and Japan.

In row 58, we calculate the change in the number of potential adopters each year. This is the group that MediDevice can potentially sell to each year. (Notice another assumption here, that MediDevice sells its device to all the new adopters each year. Jot down a note to reconsider this assumption in a subsequent model.)

In row 59, we calculate MediDevice's market share. As we determined when we created the influence diagram, MediDevice's share will be 100 percent until the second-generation technology arrives, at which point it will drop to 30 percent and, finally, to zero when the third generation arrives. This is accomplished through an IF statement expressed in pseudo-code as

```
IF(Current Year < Entry Year Second Generation) AND
  (Current Year < Entry Year Third Generation)
      100%,
ELSE IF(Current Year ≥ Entry Year Second Generation) AND
        (Current Year < Entry Year Third Generation)
      30%,
ELSE
      0%.
```

Now we are ready to calculate MediDevice's Unit Sales and Installed Base (from which we will calculate recurring revenues). Unit Sales (row 61) are New Potential Users times MediDevice's Market Share. Cumulative Unit Sales (row 62) are last year's Cumulative Unit Sales plus this year's sales. MediDevice's Installed Base (row 63) is Cumulative Unit Sales times Market Share. Note we are assuming that when competitors enter, they take market share equally from both the installed base and the current unit sales. Is this an assumption we want to revisit? Would it be more realistic to assume the effects of new technology differ on the installed base and current sales? Would it be more realistic to assume these affects occur gradually over time, rather than all in one year? Add these considerations to your list of potential refinements to the model.

We can now calculate revenues from sales of new devices and royalties from previously sold machines (rows 67–70). The logic is simple: Revenues equal Unit Sales times Price. But there is a complication here: Licensing changes both Unit Sales and Price. We incorporate these changes by using the switch input in C7 for licensing. Also, we calculate Revenue under the direct sales option on a separate line from Revenue under licensing, in order to keep the formulas down to a reasonable length.

Revenue from direct sales is calculated using this formula:

(1 – $C$7)*(C61*$C$36)

C61*$C$36 is unit sales times price. The factor (1 – $C$7) zeros this quantity out when we choose licensing.

The formula for revenues under licensing is similar:

$C$7*$C$45*(C61*$C$36)

Again, C61*C$36 is unit sales times price. $C$45 is the license fee, or the fraction of total sales that MediDevice receives. The factor $C$7 zeros this out when we choose direct sales.

Since only one of the rows 67 and 68 is non-zero, we can add the two rows to calculate total revenues. The other component of revenues, royalties on previously sold machines, is calculated by multiplying the installed base by the relevant price.

The remainder of the model is simpler. We highlight just a few important points here. The sales force cost in row 74 uses the licensing switch to turn these costs off when that option is selected. (Also, we assume here that MediDevice will only incur costs in years when sales are positive.) We also use the licensing switch in the formula in cell C78 to add the incremental cost of manufacturing for Europe and Japan under the licensing option to the manufacturing cost for the United States.

To the Reader: Before reading on, use M1 to generate as many interesting insights into the situation as you can.

7.1.4 Generate Insights—M1

This first model answers our most basic question: Should MediDevice choose direct sales or licensing? Under the direct sales approach, we estimate an overall NPV of $194M—well worth doing even though MediDevice is selling only in the U.S. market. Under licensing, however, the NPV is $212M, an improvement of 9 percent. Licensing is more profitable than direct sales, even though it reduces our revenue per unit by 50 percent. Its positive impacts on the penetration rate and the savings it offers in sales costs more than compensate for the lost revenues.

Insight 1: Licensing seems to be a better choice than direct sales.

For a first model, this one is fairly complex. It has one decision variable and 24 parameters. Before we contemplate changing the model to take other factors into account, we should continue to explore the model we have.

Intuition suggests that market penetration of the new device is a critical factor, so we explore the sensitivity of our results to the peak penetration level, which is 10 percent in the base case. Figure 7.6 shows that licensing dominates direct sales over the entire range of values for peak penetration (5–20 percent). Both NPVs are positive over this range, so the device looks profitable for a wide range of assumptions for market penetration.

Insight 2: Licensing dominates direct sales over a wide range of values of the peak penetration level.

How important is the licensing benefit to the advantage of licensing over direct sales? Recall that the licensing benefit is the number of years that licensing brings forward the peak penetration year. It is set to one year in the base case. Figure 7.7 shows the NPV under licensing as the licensing benefit varies from zero to five years. If the licensing benefit is zero, and licensing has *no* effect on penetration, we see that licensing has a slightly lower NPV than direct sales ($183M versus $194M). So our conclusion that licensing is more profitable than direct sales is highly dependent on our assumption that it moves adoption forward.

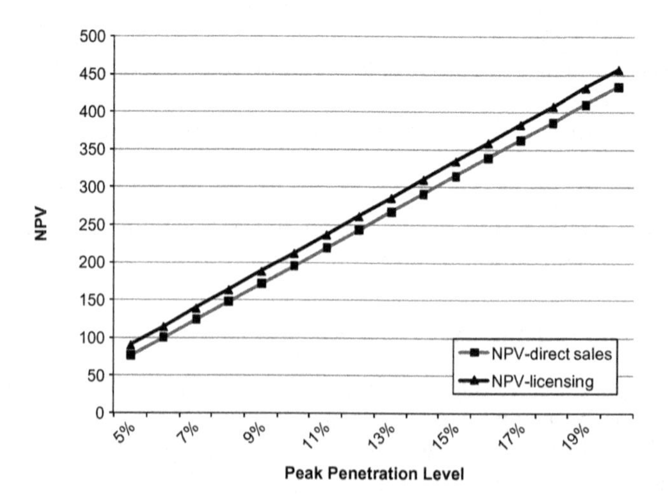

Figure 7.6. Sensitivity to Peak Penetration Level (M1).

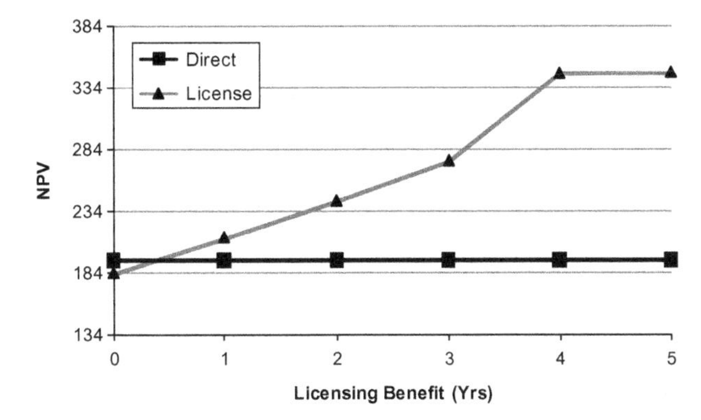

Figure 7.7. Sensitivity to Licensing Benefit (M1).

Insight 3: The benefits of licensing are highly sensitive to the effect it has on the rate of adoption.

If this effect is taken away, direct sales has a higher NPV. By the same token, if licensing has a more powerful effect on penetration, the NPV from licensing is much higher than in the base case. If the licensing benefit is four years, the NPV increases to $345M, which is an increase of 78 percent over the base case.

How important is our assumption that the licensing agreement gives MediDevice 50 percent of revenue? One way to answer this question is to ask how low the fee would have to go before licensing offers the same NPV as direct sales. In other words, how *unattractive* would the licensing arrangement have to be before MediDevice would be indifferent to a choice between it and direct sales? We use Goal Seek to show that a licensing fee of 46.8 percent drives the NPV of licensing to $194M, which is the same as direct sales. Thus, our conclusion that licensing is more profitable is highly sensitive to this input parameter.

Insight 4: The relative advantage of licensing disappears if the fee falls much below 50 percent.

Finally, we consider which of our 24 inputs matter most in determining the overall NPV. A tornado chart can begin to answer this question. However, if we highlight all 24 parameters and request the usual ±10 percentage variation, we get an unpleasant surprise: The Tornado Chart tool gives us the message "*Run time error 13.*" What has happened? The program varied each parameter by ±10 percent as we requested, but one or more of the resulting input values caused an error in the spreadsheet. If we examine the spreadsheet at this point, we see that the NPV cell contains #VALUE! and the input for the launch year in C20 is 1804.5. Obviously, it makes no sense to launch before the War of

1812, but that is what we have asked Excel to do. The problem originates in row 55, where we calculate market share. The formula here is a nested IF statement, which was not set up to calculate market share when all the years of the model fall after the launch date. When this cell returns an error, the rest of the model does as well. A simple solution is to exclude the date inputs from the tornado chart. These inputs are better handled using Data Sensitivity since they actually can change only in units of one year.

Another point to note is that there are really two versions of our model: one for the direct sales option and one for licensing. Some parameters may be more important in one case than another. For example, the number of doctors in Europe will certainly not be important in the direct sales case since, in that case, MediDevice will only be marketing to the United States. To deal with this, we generate two tornado charts, one for each case. Figure 7.8 shows the tornado chart for the direct sales case, and Figure 7.9 shows the chart for the licensing case.

Figure 7.8 shows that three parameters are particularly important in the direct sales case: the price of new units (Device Price), the number of U.S. doctors' offices (US Doctors), and the Peak Penetration Level. We should also note that a ±10 percent swing in any of these parameters does not create a case where the NPV is negative. As we suspected from the start, this new device is highly likely to be profitable.

Figure 7.9 shows more parameters having a significant impact on the NPV since, in the licensing case, more parameters play a role. Device Price still ranks highest, but equally important is the Licensing Fee. (Recall that we discovered earlier that the choice of direct sales or licensing was highly sensitive to this

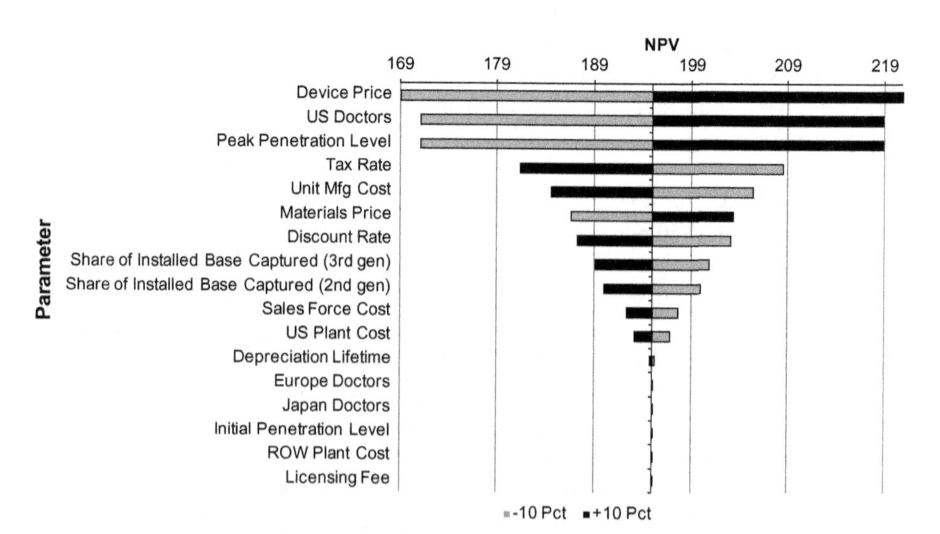

Figure 7.8. Tornado chart—Direct sales option (M1).

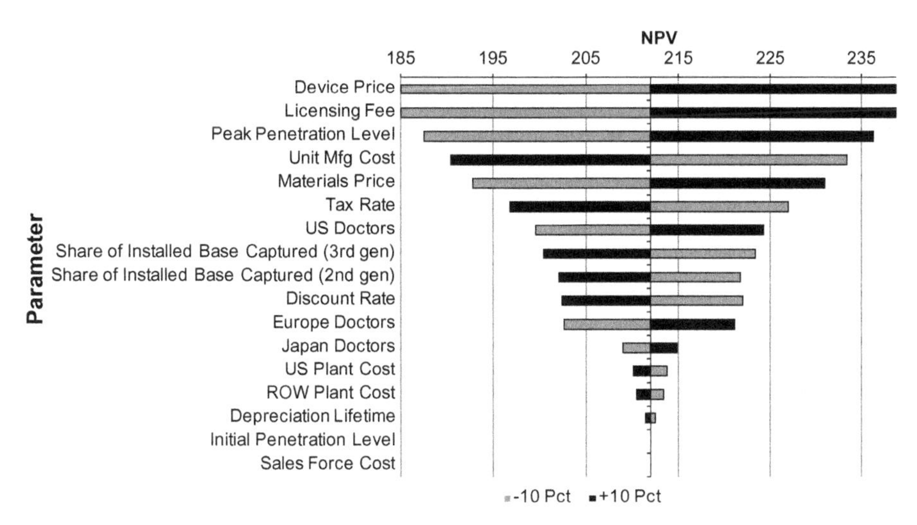

Figure 7.9. Tornado chart—Licensing option (M1).

parameter, which reflects the same effect.) Third is the Peak Penetration Level, followed by Unit Manufacturing Cost, Materials Price, and a host of lesser factors. Note that US Doctors has been bumped well down the list. It seems reasonable for the size of the domestic market to be less critical when we are also selling to the rest of the world.

Insight 5: The price of new units is the most important parameter in both the direct sales and the licensing scenarios.

Our first model has been very productive. We managed to include most of the information in the problem statement, and the conclusions have been strong: Licensing seems to be the best choice (at the moment), but that choice is sensitive to a number of our assumptions. As we think about improving our model, what direction should we take?

To the Reader: Before reading on, create a list of the factors you would add to M1 to create the next iteration of the model.

7.2 REVISING THE MODEL

7.2.1 Frame the Problem—M2

We have already noted several areas where we might refine the model. First, the problem statement mentions nothing about possible growth in the number of doctors' offices, but it seems reasonable to assume this number will grow somewhat over the next 15 years, perhaps at different rates in the three geographic regions. Would changing our assumption of no growth change our results substantially? We can guess that our estimates of the NPV under either direct sales or licensing would increase if we included growth in the number of doctors' offices, since that change by itself will lead to an increase in sales and revenues. But will it change the *relative* attractiveness of direct sales versus licensing? Certainly it could if growth was markedly higher in Europe and Japan, which provide revenues only under a licensing agreement. But significantly different growth rates in these regions seem unlikely. Our best guess is that growth in the number of doctors' offices will not significantly change the relative NPVs under our two marketing choices. Given this hunch, it does not seem vital to include this effect in the model at this early stage in the modeling process, so we leave it outside the boundaries of the model.

Another area we noted as questionable was our assumption that MediDevice sells devices to all of the new adopters every year. We could modify this aspect of the model and create a module in which potential adopters become purchasers at a slower rate. But, again, this change would not seem to shift the choice between direct sales and licensing very much. So we leave this consideration outside the model boundary as well.

A third area we noted was the process by which competitors take share from MediDevice. In M1, we made the simplest possible assumption: that each competitor takes a fixed percentage of both the share of new sales and the installed base in the year of entry. It is probably more realistic to assume the process of capturing share evolves over several years, and it is possible that a competitor could initially capture a higher percentage of new sales than of installed base (or vice versa). But changing this aspect of the model would not radically alter the results, and we have no specific information on which to base a more detailed model. So, once again, we draw a boundary and choose to keep the model simple.

There is one important set of information in the problem statement that we have not employed in our model to this point. That is the given ranges for the input parameters. (Usually, we would have to ask the client a detailed series of questions to develop information like this. We are fortunate to have ranges given to us in the initial problem statement.) Since we have data about input uncertainty, we can use it to estimate uncertainty in the NPV of cash flows. Perhaps this issue is worth the effort of additional modeling. If the distributions of the NPV under direct sales and licensing are substantially different, we might have a stronger foundation for making our recomendation. Three factors suggest that we pursue this issue in our second model. The first is that we have the necessary information. The second is that we cannot easily rule out the possibility that the change will matter to the decision. The third is that only minor effort is required.

To the Reader: Before reading on, sketch an influence diagram for the new version of the model.

7.2.2 Diagram the Problem—M2

It is easy to update our influence diagram to show the use of ranges for uncertain inputs. The only change is that we indicate with a double line the seven elements that are now random inputs (Figure 7.10).

To the Reader: Before reading on, design and build a model based on the influence diagram.

7.2.3 Build the Model—M2

Adapting M1 to use uncertain inputs is likewise quite straightforward. Since we are only given upper and lower values, we use the same discrete distribution for each uncertain parameter. Specifically, we assume that the base-case value has a probability of 0.50 of occurring, whereas each of the lower and upper values has a probability of 0.25. Under this assumption, we need only make minor modifications to M1 for simulation.

This approach is a "quick and dirty" way to add variability to a model when we have limited information about the actual variability of the situation. We can even see how sensitive the model is to changes in variability by changing the probabilities of the three outcomes. Increasing the probability of the extreme cases increases the overall variability.

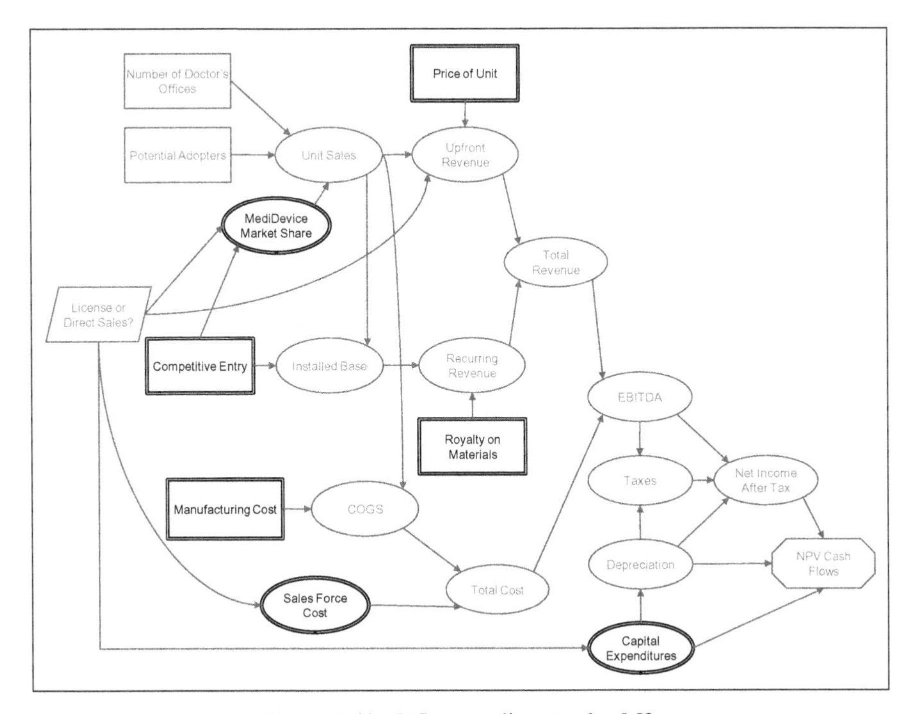

Figure 7.10. Influence diagram for M2.

M2 is shown in Figure 7.11. The only changes are to the parameters. In columns E, F, and G, we input the three values for each of the uncertain inputs: low, base, and high. Then, in column D, we have the Crystal Ball Custom distribution shown in Figure 7.12. This generates the values 1, 2, and 3 with probabilities 0.25, 0.50, and 0.25, respectively. The same distribution appears in every shaded cell (D21:D45), although the samples returned by each distribution are independent. Finally, in column C, we use the INDEX function to read

	A	B	C	D	E	F	G	H	I	J	K	L	M	N	O	P	Q	R
1	MediDevice	M2																
2																		
3																		
4																		
5																		
6	Decisions				NPV	268.1		Benchmarks			% changes							
7		Licensing (0/1)?	0						direct sale	194	0.38							
8									licensing	212	0.27							
9	Parameters																	
10		Number of Doctor's Offices																
11		US	600															
12		Europe	450															
13		Japan	140															
14																		
15		First Year	2009															
16		Tax Rate	40%															
17		Discount Rate	10%															
18																		
19		Market Penetration																
20		Launch Year	2010															
21		Peak Year - Base	2017	3	2012	2014	2017											
22		Initial Level	0%															
23		Peak Level	0.10	2	5%	10%	15%											
24		Licensing Benefit	1	2	0	1	2											
25		Peak Year - Actual	2017															
26																		
27		Second Generation																
28		Entry Year	2019	3	2013	2015	2019											
29		Share of Installed Base Capt	0.80	3	50%	70%	80%											
30																		
31		Third Generation																
32		Entry Year	2019	2	2018	2019	2023											
33		Share of Installed Base Capt	100%															
34																		
35		Prices																
36		Device	10	2	8	10	12											
37		Materials	2	3	0.5	1	2											
38																		
39		Costs																
40		Unit Manufacturing	5	3	3	4	5											
41		Sales Force	10	1	10	15	20											
42		US Plant Cost	20	1	20	25	35											
43		ROW Plant Cost	35	3	15	20	35											
44		Depreciation Lifetime	10															
45		Licensing Fee	0.60	3	40%	50%	60%											
46																		
47			2009	2010	2011	2012	2013	2014	2015	2016	2017	2018	2019	2020	2021	2022	2023	
48	Revenue																	
49		Number of Doctor's Offices																
50		US	600	600	600	600	600	600	600	600	600	600	600	600	600	600	600	
51		Europe	450	450	450	450	450	450	450	450	450	450	450	450	450	450	450	
52		Japan	140	140	140	140	140	140	140	140	140	140	140	140	140	140	140	
53		Total	1190	1190	1190	1190	1190	1190	1190	1190	1190	1190	1190	1190	1190	1190	1190	
54																		
55		Market Penetration	0.00	0.00	0.01	0.03	0.04	0.06	0.07	0.09	0.10	0.10	0.10	0.10	0.10	0.10	0.10	
56																		
57		Potential Users - Total	0	0	8.57143	17.1429	25.7143	34.2857	42.8571	51.4286	60	60	60	60	60	60	60	
58		Potential Users - New	0	0	8.57143	8.57143	8.57143	8.57143	8.57143	8.57143	8.57143	0	0	0	0	0	0	
59		MediDevice Share	1.00	1.00	1.00	1.00	1.00	1.00	1.00	1.00	1.00	1.00	0.00	0.00	0.00	0.00	0.00	
60																		
61		Unit Sales	0	0	8.57143	8.57143	8.57143	8.57143	8.57143	8.57143	8.57143	0	0	0	0	0	0	
62		Cumulative Unit Sales	0	0	8.57143	17.1429	25.7143	34.2857	42.8571	51.4286	60	60	60	60	60	60	60	
63		MediDevice Installed Base	0	0	8.57143	17.1429	25.7143	34.2857	42.8571	51.4286	60	60	0	0	0	0	0	
64																		
65		Revenues																
66		New Devices																
67		Direct Sales	0	0	85.7143	85.7143	85.7143	85.7143	85.7143	85.7143	85.7143	0	0	0	0	0	0	
68		Licensing	0	0	0	0	0	0	0	0	0	0	0	0	0	0	0	
69		Materials	0	0	17.1429	34.2857	51.4286	68.5714	85.7143	102.857	120	120	0	0	0	0	0	
70		Total	0	0	102.857	120	137.143	154.286	171.429	188.571	205.714	120	0	0	0	0	0	
71																		
72	Operating Cost																	
73		COGS	0	0	43	43	43	43	43	43	43	0	0	0	0	0	0	
74		Sales Force Cost	0	0	10	10	10	10	10	10	10	0	0	0	0	0	0	
75		Total Cost	0	0	53	53	53	53	53	53	53	0	0	0	0	0	0	
76																		
77	Capital Expenditure																	
78		Manufacturing Plant Cost	20															
79		Depreciation	2	2	2	2	2	2	2	2	2	2	0	0	0	0	0	
80																		
81	Income Statement																	
82		Revenue	0.0	0.0	102.9	120.0	137.1	154.3	171.4	188.6	205.7	120.0	0.0	0.0	0.0	0.0	0.0	
83		Cost	0.0	0.0	52.9	52.9	52.9	52.9	52.9	52.9	52.9	0.0	0.0	0.0	0.0	0.0	0.0	
84		EBITDA	0.0	0.0	50.0	67.1	84.3	101.4	118.6	135.7	152.9	120.0	0.0	0.0	0.0	0.0	0.0	
85		Depreciation	2.0	2.0	2.0	2.0	2.0	2.0	2.0	2.0	2.0	2.0	0.0	0.0	0.0	0.0	0.0	
86		Taxes	-0.8	-0.8	19.2	26.1	32.9	39.8	46.6	53.5	60.3	47.2	0.0	0.0	0.0	0.0	0.0	
87		NIAT	-1.2	-1.2	28.8	39.1	49.4	59.7	69.9	80.2	90.5	70.8	0.0	0.0	0.0	0.0	0.0	
88																		
89	Cash Flow																	
90		NIAT	-1.2	-1.2	28.8	39.1	49.4	59.7	69.9	80.2	90.5	70.8	0.0	0.0	0.0	0.0	0.0	
91		Depreciation	2.0	2.0	2.0	2.0	2.0	2.0	2.0	2.0	2.0	2.0	0.0	0.0	0.0	0.0	0.0	
92		CAPX	20.0	0.0	0.0	0.0	0.0	0.0	0.0	0.0	0.0	0.0	0.0	0.0	0.0	0.0	0.0	
93		Cash flow	-19.2	0.8	30.8	41.1	51.4	61.7	71.9	82.2	92.5	72.8	0.0	0.0	0.0	0.0	0.0	
94																		
95	NPV	268.1																
96																		

Figure 7.11. M2—numerical view.

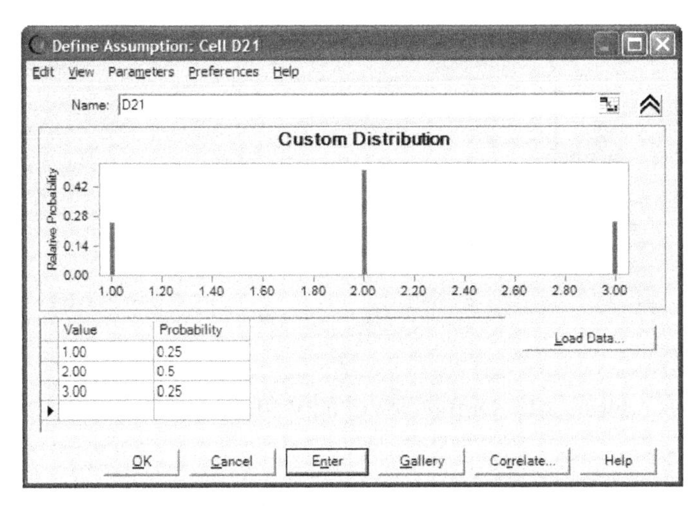

Figure 7.12. Custom distribution to generate random input parameters.

which of the three inputs to use based on the random outcome in column D. For example, to generate the parameter Peak Year-Base in cell C21, we use the formula

INDEX(E21:G21, D21)

This picks the first, second, or third value in the range E21:G21, depending on the (random) value in D21.

To the Reader: Before reading on, use M2 to generate as many interesting insights into the situation as you can.

7.2.4 Generate Insights—M2

M2 is now ready for simulation. The results are shown in Figures 7.13 and 7.14. The mean NPV is 25 percent higher under licensing than under direct sales: \$243M versus \$194M. This result makes licensing look considerably more attractive than it did with the deterministic model. Uncertainty seems to favor licensing, probably because the upside potential is captured by licensing better than the direct sales option.

We gain a bit more insight into this choice by comparing the ranges of outcomes. Under direct sales, the NPV ranges from –\$34M to \$786M, whereas under licensing, the range is from –\$69M to \$1,505M. Both options can lead to a negative NPV; in fact, licensing has a somewhat lower worst-case outcome and a slightly higher chance of the NPV being negative. But the licensing option offers a much higher maximum outcome than does direct sales: \$1,505M against \$786M. However, because the maximum outcome is unlikely to occur,

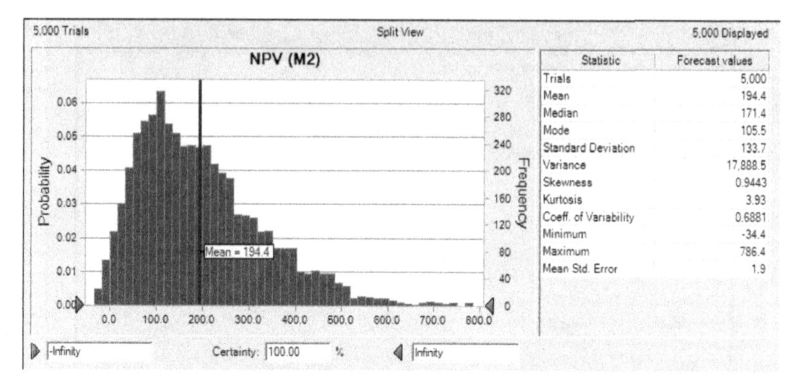

Figure 7.13. Distribution of NPV—Direct sales (M2).

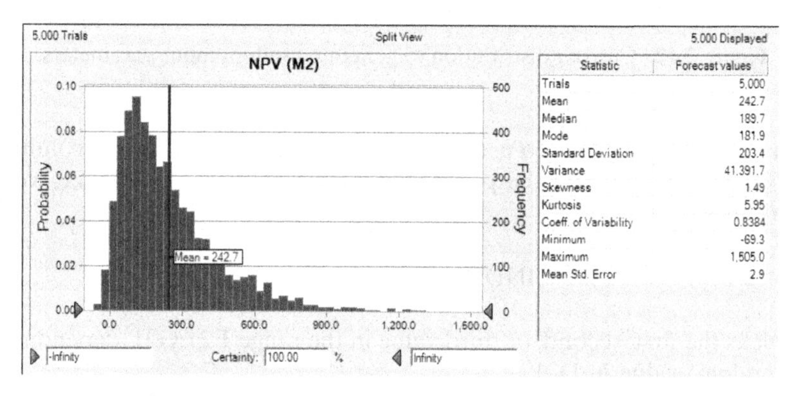

Figure 7.14. Distribution of NPV—Licensing (M2).

it is not a realistic measure of upside potential. A better measure is a high percentile, such as the 90th percentile, above which 10 percent of the outcomes lie. The 90th percentile for the direct sales option is $381M, whereas the 90th percentile for the licensing option is $527M (Figure 7.15).

Insight 6: Not only is the average outcome better under licensing, but there is a substantially higher chance of extremely good outcomes.

Uncertainty analysis has added a new dimension to our understanding of the choices MediDevice faces. Our recommendation, to adopt licensing over direct sales, is the same as it was previously, but our reasoning is deeper and therefore sounder. It was worth the effort.

New information and new options often become available during analysis. A sign of an effective modeling process is the ability to incorporate this new information easily. The next section presents new information about the MediDevice case, which tests our ability to expand and adapt our analysis.

NPV Percentiles (M2)

Percentile	Direct Sales ($M)	Licensing ($M)
0%	–34	–69
10%	46	42
20%	78	77
30%	106	112
40%	137	149
50%	171	190
60%	207	241
70%	246	298
80%	301	379
90%	381	527
100%	786	1,505

Figure 7.15. NPV percentiles (M2).

7.3 MEDIDEVICE CASE (B)

MediDevice (B)

OfficeSupply Co. is the company that will sell analyzer supplies to MediDevice's installed base of doctor's offices. OfficeSupply is now offering a different royalty deal that gives MediDevice an incentive to sell more analyzers.

OfficeSupply is offering to pay an annual royalty per office, based on the number of doctors' offices adopting the device, as follows:

- $500 per office if the installed base is fewer than 50,000 offices.
- Between 50,000 and 100,000 offices, the royalty increases linearly from $500 to $2,000.
- $2,000 per office if the installed base is 100,000 offices or more.

OfficeSupply states that it prefers this new deal to the flat royalty of $1,000 per office [the original deal described in MediDevice (A)], but it will still accept the original deal if MediDevice is not interested in renegotiating.

Note: As in MediDevice (A), assume the royalty is a recurring annual payment per office that ends if MediDevice's analyzer is displaced in the future by substitute technology.

As so often happens during the analysis of a business problem, a new wrinkle has been introduced. The supplier of the materials used in the device has offered a new contractual arrangement. Should MediDevice accept it, or stick with the current agreement?

New information tests our ability to refine our existing model. If we have built a logically sound and flexible model and have implemented it in a well-designed spreadsheet, we find that much of the model structure remains unchanged as we introduce the new elements. If not, we find ourselves radically redesigning the model.

The previous royalty agreement called for an annual payment of $1,000 for each machine regardless of the number of machines in use. Now the supplier is offering a sliding schedule of payments that increase as the installed base increases. Up to 50,000 machines, the annual payment will only be $500, half the current rate. But if MediDevice can sell more than 100,000 machines, the annual payment will be $2,000, twice the current rate. It seems the supplier is trying to hedge against the possibility that only a modest number of machines will be sold. This makes perfect sense for OfficeSupply, but what will the impact be on MediDevice? In particular, should MediDevice accept this new arrangement, and if so, does the arrangement alter the decision to adopt the licensing option?

To the Reader: Before reading on, sketch an influence diagram for this new version of the problem.

7.3.1 Diagram the Problem—M3

The influence diagram requires only a minor change to represent this new option. Since we already calculate the installed base of MediDevice machines, we can replace the fixed annual royalty with a function that represents the new sliding scale royalty. We show this by connecting the Installed Base circle to the Royalty on Materials circle (Figure 7.16). We leave the details of this relationship to the model-building step. For simplicity, we assume constant values for our input parameters and ignore uncertainty. Thus, M3 is based directly on M1.

To the Reader: Before reading on, design and build a model based on the influence diagram in Figure 7.16.

7.3.2 Build the Model—M3

We make our modeling task easier if we incorporate the new royalty plan as another option independent of the marketing choice between direct sales and licensing. We do this by defining four strategies:

- Direct sales with current royalty plan
- Direct sales with ramp-up royalty plan
- Licensing with current royalty plan
- Licensing with ramp-up royalty plan

For each of these strategies, we want the model to incorporate the appropriate assumptions automatically. We do this by using a strategy table in cells E8:G11 in our new model (Figure 7.17). The first column lists the strategy

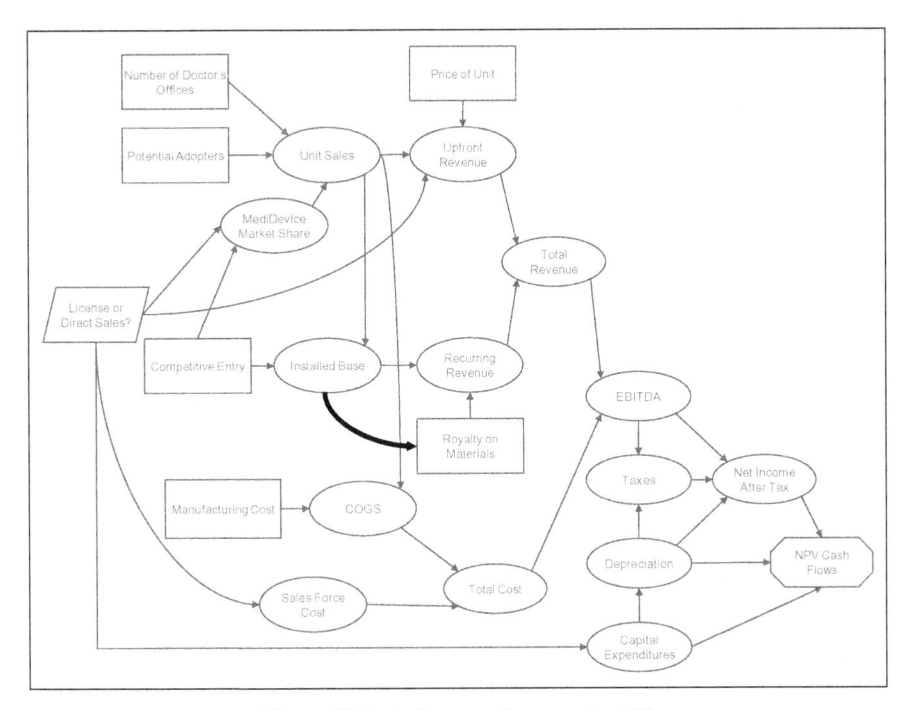

Figure 7.16. Influence diagram for M3.

number. The second and third columns specify the appropriate choice of marketing and royalty plans using 0/1 codes much as in our previous models (licensing = 1, ramp-up = 1).

These inputs are brought into our model in the following way: First, in cell B6, we choose one of the four strategies by entering its number. Then, in cells C7 and C8, we use a VLOOKUP function to copy the appropriate 0/1 indicator for the marketing and royalty options. For example, in cell C7, the formula is

VLOOKUP(B6, E8:G11, 2)

In this function, VLOOKUP searches for the value in B6 in column E and returns the corresponding value from the second column of the range: column F.

Once the codes are entered in C7 and C8, it only remains to modify the model for the royalty option, since the licensing option is already included in the model. The royalty is either the fixed value of $1,000 under the current royalty plan (ramp-up = 0) or the sliding-scale royalty under the proposed plan (ramp-up = 1). Royalty revenues from materials are calculated in row 78 using a complex formula. The nested IF functions can be written in pseudocode as follows:

	A	B	C	D	E	F	G	H	I	J	K
1	**MediDevice**	**M3 - B Case**									
2											
3											
4											
5											
6	**Strategies**	1				**Strategy Table**					
7		Licensing (0/1)?	0		**Number**	**Licensing**	**Ramp Up**				
8		Ramp Up (0/1)?	0		1	0	0				
9					2	0	1				
10					3	1	0				
11					4	1	1				
12	**Parameters**										
13		Number of Doctor's Offices			**NPV**	194.4		Benchmarks			% changes
14		US	600						1	194	0.00
15		Europe	450						2	160	0.22
16		Japan	140						3	212	-1.00
17									4	287	-1.00
18		First Year	2009								
19		Tax Rate	40%								
20		Discount Rate	10%								
21											
22		Market Penetration									
23		Launch Year	2010								
24		Peak Year - Base	2014								
25		Initial Level	0%								
26		Peak Level	10%								
27		Licensing Benefit	1								
28		Peak Year - Actual	2014								
29											
30		Second Generation									
31		Entry Year	2015								
32		Share of Installed Base Captured	70%								
33											
34		Third Generation									
35		Entry Year	2019								
36		Share of Installed Base Captured	100%								
37											
38		Prices									
39		Device	10								
40		Materials									
41		Base	1								
42		Min IB	50								
43		Max IB	100								
44		Min Royalty	0.5								
45		Max Royalty	2								
46		Slope	0.03								
47											
48		Costs									
49		Unit Manufacturing	4								
50		Sales Force	15								
51		US Plant Cost	25								
52		ROW Plant Cost	20								
53		Depreciation Lifetime	10								
54		Licensing Fee	50%								
55											

	A	B	C	D	E	F	G	H	I	J	K	L	M	N	O	P	Q
56			**2009**	**2010**	**2011**	**2012**	**2013**	**2014**	**2015**	**2016**	**2017**	**2018**	**2019**	**2020**	**2021**	**2022**	**2023**
57	**Revenue**																
58		Number of Doctor's Offices															
59		US	600	600	600	600	600	600	600	600	600	600	600	600	600	600	600
60		Europe	450	450	450	450	450	450	450	450	450	450	450	450	450	450	450
61		Japan	140	140	140	140	140	140	140	140	140	140	140	140	140	140	140
62		Total	1190	1190	1190	1190	1190	1190	1190	1190	1190	1190	1190	1190	1190	1190	1190
63																	
64		Market Penetration	0.000	0.000	0.025	0.050	0.075	0.100	0.100	0.100	0.100	0.100	0.100	0.100	0.100	0.100	0.100
65																	
66		Potential Users - Total	0	0	15	30	45	60	60	60	60	60	60	60	60	60	60
67		Potential Users - New	0	0	15	15	15	15	0	0	0	0	0	0	0	0	0
68		MediDevice Share	1.00	1.00	1.00	1.00	1.00	1.00	0.30	0.30	0.30	0.30	0.00	0.00	0.00	0.00	0.00
69																	
70		Unit Sales	0	0	15	15	15	15	0	0	0	0	0	0	0	0	0
71		Cumulative Unit Sales	0	0	15	30	45	60	60	60	60	60	60	60	60	60	60
72		MediDevice Installed Base	0	0	15	30	45	60	18	18	18	18	0	0	0	0	0
73																	
74		Revenues															
75		New Devices															
76		Direct Sales	0	0	150	150	150	150	0	0	0	0	0	0	0	0	0
77		Licensing	0	0	0	0	0	0	0	0	0	0	0	0	0	0	0
78		Materials	0	0	15.0	30.0	45.0	60.0	18.0	18.0	18.0	18.0	0	0	0	0	0
79		Total	0	0	165.0	180.0	195.0	210.0	18.0	18.0	18.0	18.0	0	0	0	0	0
80																	
81	**Operating Cost**																
82		COGS	0	0	60	60	60	60	0	0	0	0	0	0	0	0	0
83		Sales Force Cost	0	0	15	15	15	15	0	0	0	0	0	0	0	0	0
84		Total Cost	0	0	75	75	75	75	0	0	0	0	0	0	0	0	0
85																	
86	**Capital Expenditure**																
87		Manufacturing Plant Cost	25														
88		Depreciation	2.5	2.5	2.5	2.5	2.5	2.5	2.5	2.5	2.5	2.5	0	0	0	0	0
89																	
90	**Income Statement**																
91		Revenue	0.0	0.0	165.0	180.0	195.0	210.0	18.0	18.0	18.0	18.0	0.0	0.0	0.0	0.0	0.0
92		Cost	0.0	0.0	75.0	75.0	75.0	75.0	0.0	0.0	0.0	0.0	0.0	0.0	0.0	0.0	0.0
93		EBITDA	0.0	0.0	90.0	105.0	120.0	135.0	18.0	18.0	18.0	18.0	0.0	0.0	0.0	0.0	0.0
94		Depreciation	2.5	2.5	2.5	2.5	2.5	2.5	2.5	2.5	2.5	2.5	0.0	0.0	0.0	0.0	0.0
95		Taxes	-1.0	-1.0	35.0	41.0	47.0	53.0	6.2	6.2	6.2	6.2	0.0	0.0	0.0	0.0	0.0
96		NIAT	-1.5	-1.5	52.5	61.5	70.5	79.5	9.3	9.3	9.3	9.3	0.0	0.0	0.0	0.0	0.0
97																	
98	**Cash Flow**																
99		NIAT	-1.5	-1.5	52.5	61.5	70.5	79.5	9.3	9.3	9.3	9.3	0.0	0.0	0.0	0.0	0.0
100		Depreciation	2.5	2.5	2.5	2.5	2.5	2.5	2.5	2.5	2.5	2.5	0.0	0.0	0.0	0.0	0.0
101		CAPX	25.0	0.0	0.0	0.0	0.0	0.0	0.0	0.0	0.0	0.0	0.0	0.0	0.0	0.0	0.0
102		Cash flow	-24.0	1.0	55.0	64.0	73.0	82.0	11.8	11.8	11.8	11.8	0.0	0.0	0.0	0.0	0.0
103																	
104	**NPV**	**194.4**															
105																	

Figure 7.17. M3—numerical view.

IF(Ramp-Up = 0)

$1,000,

ELSE IF(Installed Base < 50,000)

$500,

ELSE IF(Installed Base < 100,000)
 Straight-line interpolation between \$500 and \$2,000,
ELSE
 \$2,000.

To the Reader: Before reading on, use M3 to generate as many interesting insights into the situation as you can.

7.3.3 Generate Insights—M3

Results from this model are interesting (Figure 7.18). The new royalty plan *reduces* the NPV under the direct sales marketing plan, but it *increases* the NPV under the licensing plan. Under direct marketing, the NPV drops from \$194M to \$160M; under the licensing plan, the NPV increases from \$212M to \$287M. Why is this? The new royalty plan rewards MediDevice for selling more machines, which it does in the licensing case. So the NPV drops in the direct sales case because the installed base is too low to benefit from the higher payments. The NPV increases in the licensing case because the installed base is high enough that the higher royalty payments kick in. So the new option actually increases the attractiveness of the marketing plan we would have recommended anyway.

Insight 7: The new royalty plan makes the licensing option even more attractive.

Having completed our deterministic analysis, similar to M1, we are ready to consider uncertainty as we did in M2.

To the Reader: Before reading on, modify M3 to allow for simulation and use it to refine the insights gathered from M2 and M3.

7.3.4 Build the Model and Generate Insights—M4

In M4, we introduce ranges for the same 13 uncertain parameters as in M2, so all we have to do is rerun the simulations. Again we use the distribution shown in Figure 7.13 to capture the uncertainty. The results are summarized in terms of the distribution of NPVs (Figures 7.19–7.22).

		Sales Method	
		Direct	License
Royalty plan	Fixed	194	212
	Sliding	160	287

Figure 7.18. NPV of the four strategies (M3).

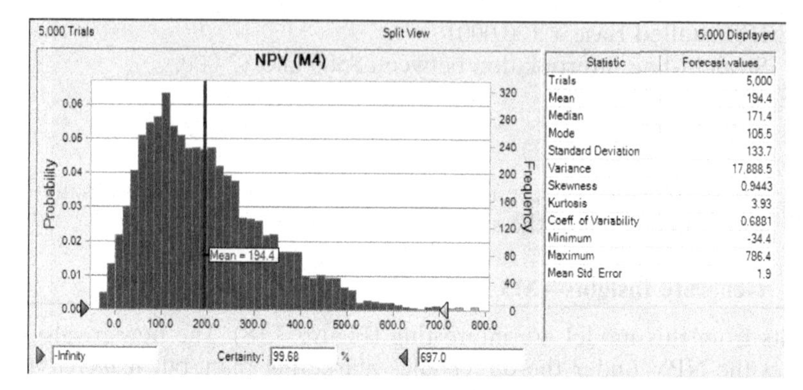

Figure 7.19. Distribution of NPV—Direct sales with flat royalty plan (M4).

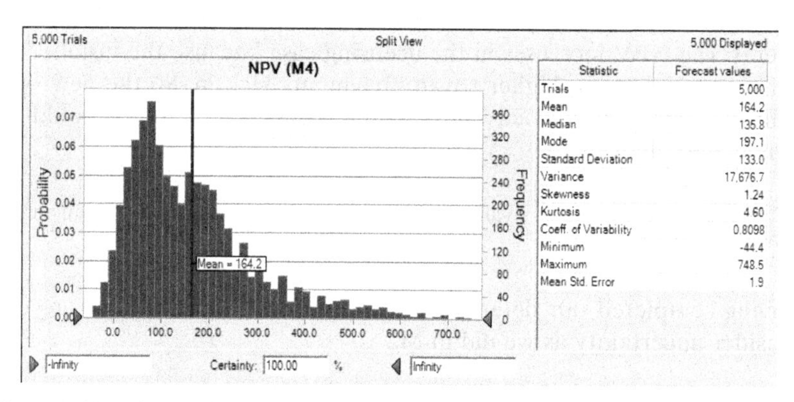

Figure 7.20. Distribution of NPV—Direct sales with ramp-up royalty plan (M4).

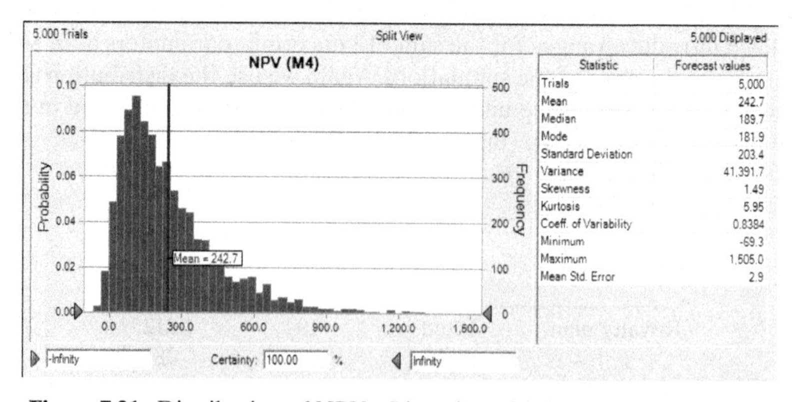

Figure 7.21. Distribution of NPV—Licensing with flat royalty plan (M4).

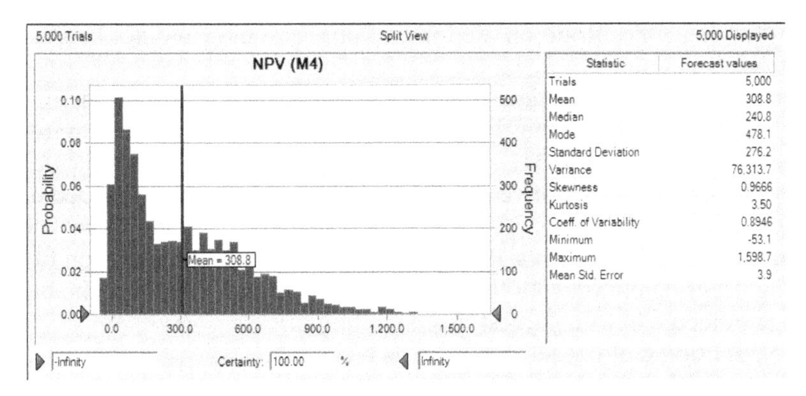

Figure 7.22. Distribution of NPV—Licensing with ramp-up royalty plan (M4).

The mean values tell essentially the same story as in our deterministic model. For the direct sales option, the mean NPV under the current royalty plan is $194M, decreasing to $164M under the new royalty plan. In the licensing case, the NPV for the current royalty plan is $243M, increasing to $309M under the new royalty plan. So it still seems that the best choice is licensing with the new royalty plan. (The worst choice, it would seem, is direct sales under the new royalty plan.)

Insight 8: Even when taking uncertainty into account, the licensing option with the new royalty plan is best.

7.4 PRESENTATION OF RESULTS

The time has come to wrap up the analysis phase of our work on this case and to develop a presentation of our results for MediDevice management. We have plenty of model results and insights to work with. The challenge is to select the most useful results and to present them in a way that allows management to understand what they should do and why.

We assume our audience consists of senior executives at MediDevice. Although they have various levels of expertise and knowledge of the details of this problem, we assume they all understand the company and the role this device plays in its overall strategy. Finally, we assume we represent an internal consulting group, which brings analytic expertise to the problem but has no stake in the outcome of the decisions on how the product is sold or which royalty agreement is selected.

We use a direct approach in this presentation because we have clear recommendations to offer on the two decisions of interest and we have no reason to expect resistance to our findings. A direct approach allows us to state our recommendations at the start of the presentation and then to develop the reasoning behind those findings. When the audience knows our recommendations, they can concentrate on the logic of our reasoning and not be guessing about our conclusions.

The story we tell centers around the two choices management must make: how to distribute the device and whether to accept the new royalty arrangement. Recall that M1 showed us that licensing seemed to offer the higher profit, and this tentative conclusion held up when we introduced uncertainty into the model. When we introduced the new royalty option, we had not two but four possible choices. Perhaps the most important conclusion of all is that the licensing option with the new royalty deal offers not only the highest average outcome but the largest upside potential.

A basic outline of our story takes this form:

- Introduction
- Options
  - Direct sales or lease
  - Current royalty or new royalty
- Influence diagram
- Direct sales/leasing decision
  - Base-case results
  - Sensitivities
    - Penetration level
    - License benefit
  - Uncertainty analysis
- New royalty option
  - Base-case results
  - Sensitivities
    - Penetration level
  - Uncertainty analysis

Note that we have chosen to discuss the first decision, direct sales versus leasing, in complete detail before discussing the royalty decision. We use the same approach in discussing both decisions: Establish a base case using a deterministic model, perform sensitivity analyses, and then present an uncertainty analysis. This type of parallelism helps our audience follow the presentation more easily.

Our first slide (Figure 7.23) introduces the problem. The second slide (Figure 7.24) lists the two decisions and the options available under each one. The third slide (Figure 7.25), which shows our influence diagram, allows us to walk the audience through the basic structure of our models at a conceptual level. We have two goals with this slide: one is to gain credibility for our analysis, and the other is to solidify the mental models of our audience so they can follow our train of thought later as we explain our results. As we have mentioned in prior chapters, we would use PowerPoint animation to display the influence diagram section by section.

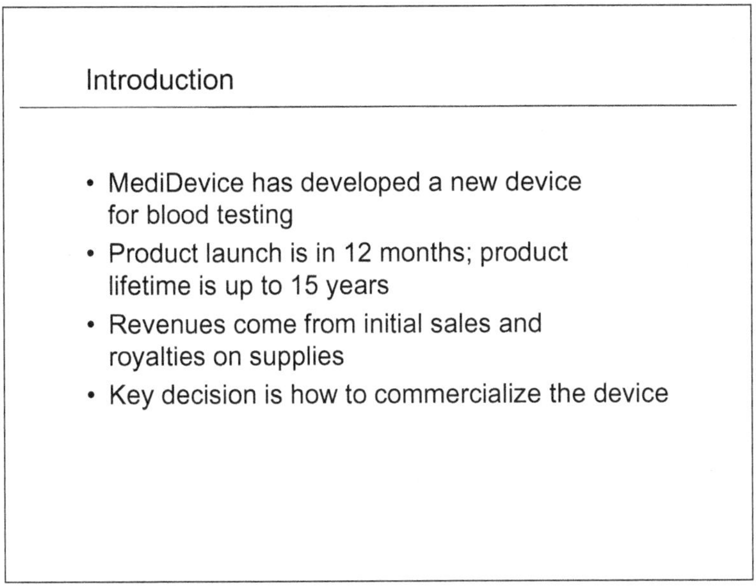

Introduction
• MediDevice has developed a new device for blood testing
• Product launch is in 12 months; product lifetime is up to 15 years
• Revenues come from initial sales and royalties on supplies
• Key decision is how to commercialize the device

Figure 7.23. Slide 1.

MediDevice must choose a sales method and a royalty plan
Sales Method
• Build a direct sales force to market in U.S. only
OR
• License to a competitor for worldwide sales
Royalty Plan
• A flat royalty rate
OR
• A royalty rate that rises with installed base of machines

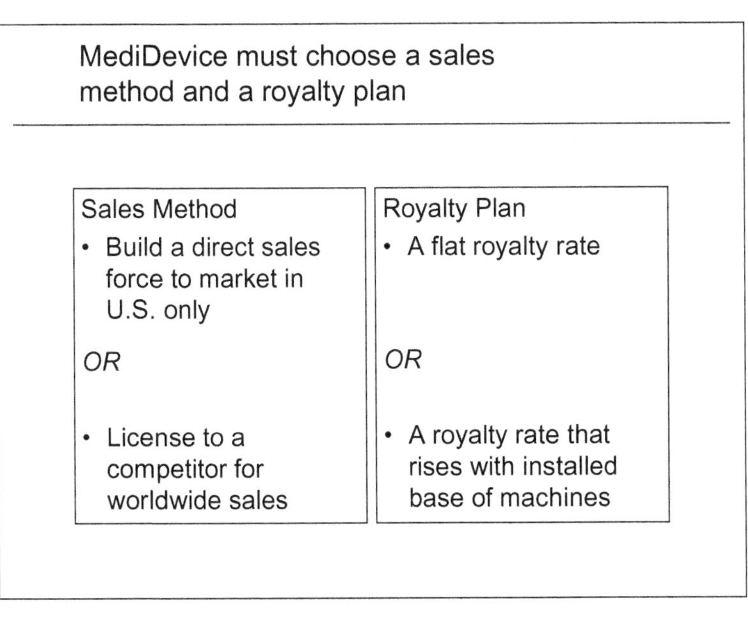

Figure 7.24. Slide 2.

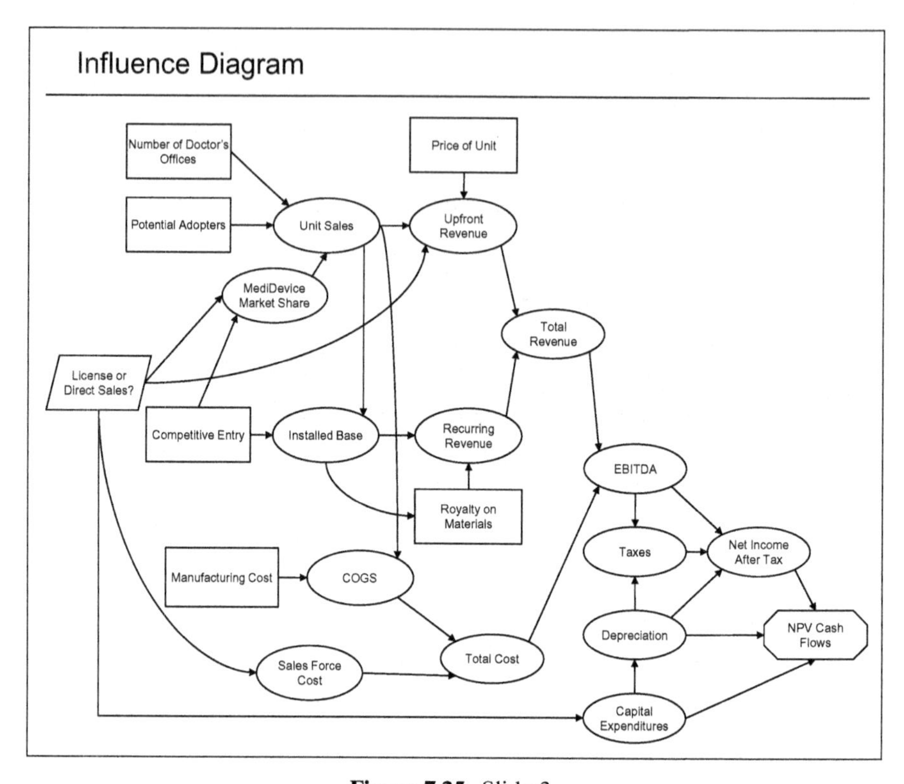

Figure 7.25. Slide 3.

We then introduce base-case results, based on M1, that show licensing is a better option than direct sales (Figure 7.26). We stress that these results are just a base case, which means that if we change assumptions, we will almost certainly change the result. This sets us up to introduce the sensitivity analyses that appear on the following two slides. Figure 7.27 shows how the NPVs for the two options change as the peak penetration rate changes, and it shows as well that licensing is always better than direct sales. Figure 7.28 shows how changing the licensing benefit affects the NPV. The message here is that as long as licensing moves the peak adoption rate ahead by at least a year, licensing is the preferred choice.

At this point, we leave the deterministic world behind and introduce uncertainty analysis. We take a few minutes to discuss the relevance of uncertainty in our assumptions and how we model it. Depending on the sophistication of our audience, we may even explain what a distribution is by comparing it to a histogram. Then we display Figure 7.29, which shows the distributions of the NPV for the direct sales and licensing options. We have highlighted two points of interest on this slide: the mean values and the 90th percentiles. This allows us to point out that both the mean and the upper tail of the distribution are better under licensing than under direct sales.

Licensing for U.S. & international sales yields a larger NPV

(NPV)	Direct Sales	Licensing
Flat Royalty	$192	$212

Figure 7.26. Slide 4.

Licensing is preferred regardless of the market penetration level

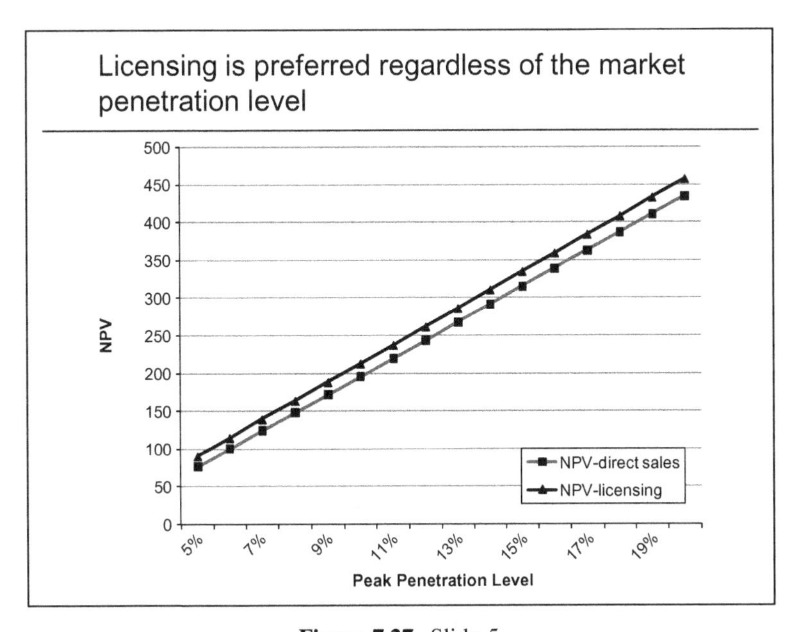

Figure 7.27. Slide 5.

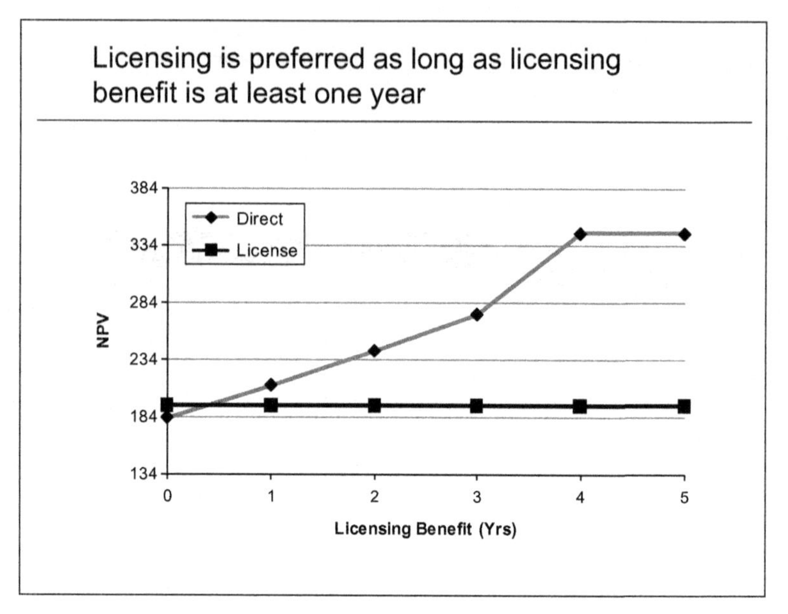

Figure 7.28. Slide 6.

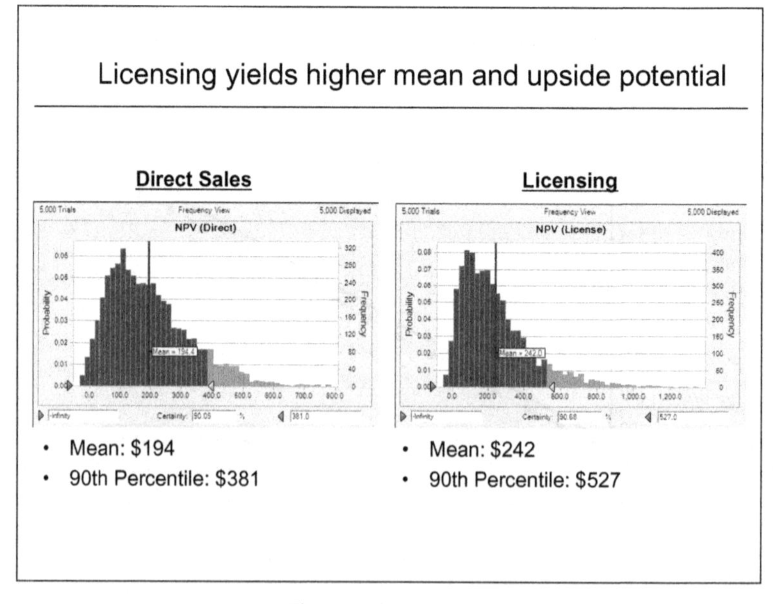

Figure 7.29. Slide 7.

Sliding royalty plan is beneficial with licensed sales strategy

(Mean NPV)	Direct Sales	Licensing
Flat Royalty	$194	$242
Sliding Royalty	$164	**$309**

Figure 7.30. Slide 8.

Sliding royalty plan is best as long as peak market penetration is above 7%

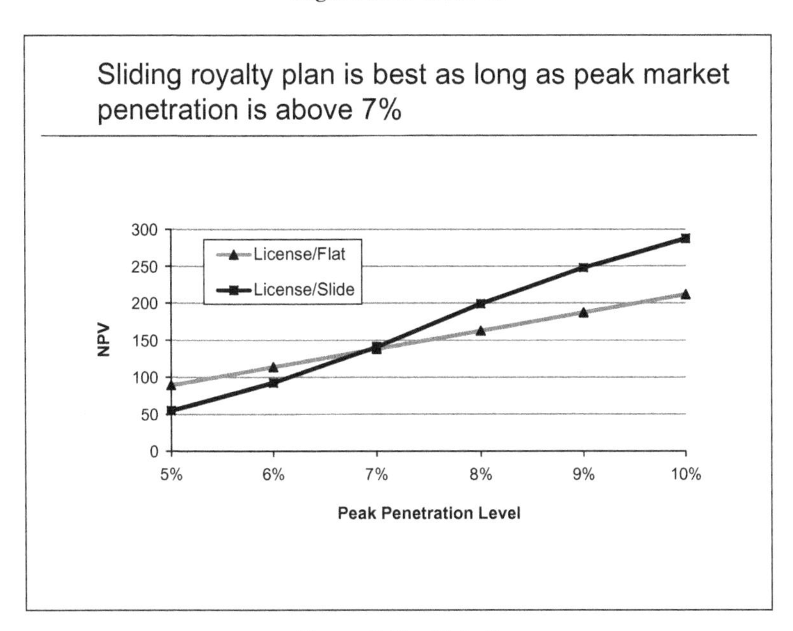

Figure 7.31. Slide 9.

We now consider the royalty options. Our next slide, Figure 7.30, compares the four combinations of decisions in terms of mean NPV. This slide shows that the new royalty plan is most advantageous under a licensing agreement, which was already shown to be the preferred commercialization option. In the next slide (Figure 7.31), we show how sensitive this conclusion is to variations in the peak penetration level. Finally, we show in Figure 7.32 the distribution of NPV under the new royalty/licensing combination in comparison with the distribution under the flat royalty plan. This shows how beneficial the new royalty is in terms of both the mean outcome and the 90th percentile. We close with a summary slide (Figure 7.33) that recapitulates our findings.

Sliding royalty plan provides higher mean and upside potential

License & Flat Royalty | **License & Sliding Royalty**

- Mean: $242
- 90th Percentile: $527

- Mean: $309
- 90th Percentile: $697

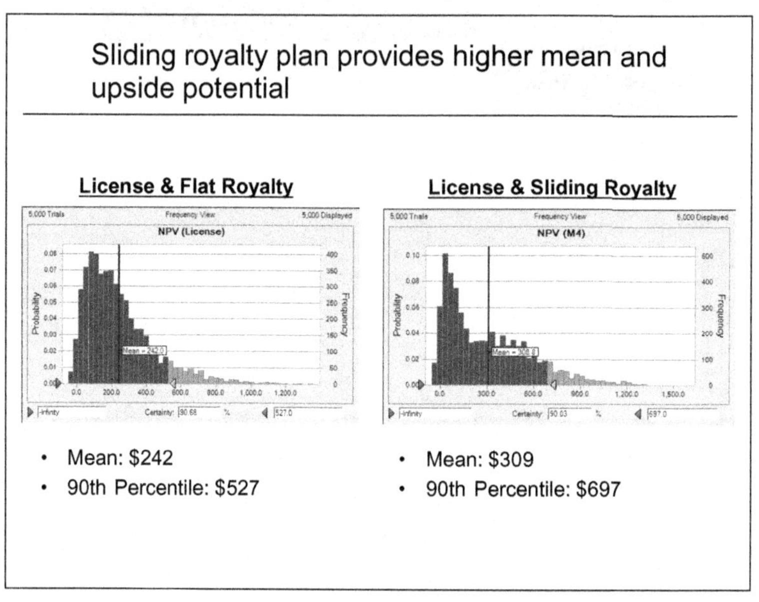

Figure 7.32. Slide 10.

Summary

- Key decisions for MediDevice
  - Direct sales or licensing
  - Flat royalty or sliding-scale royalty
- Licensing offers an NPV 25% higher than direct sales under current royalty
- New royalty offers a further 35% increase in NPV
- These conclusions are robust to reasonable changes in assumptions

Figure 7.33. Slide 11.

7.4 SUMMARY

This case illustrated several key features of our approach to modeling. Influence diagrams were essential in helping us lay out the accounting framework and behavioral processes that govern sales. We used parameterization to create a flexible function to represent market-share dynamics. Careful spreadsheet engineering was vital throughout and paid off when we were able to incorporate the new royalty option easily into our existing models. Sensitivity analysis and simulation helped us uncover essential insights into which of the available choices were best and *why*.

CHAPTER 8

Draft Commercials

8.0 INTRODUCTION

Many aspects of business are governed more by habit than by analysis. One example is the common practice in many companies of relying on a single advertising agency to create commercials. Considerable evidence indicates that the most important factor in a commercial's success is its originality and impact. Nevertheless, many companies devote more effort and money to selecting the media and timing of their campaigns than to the creative process of commercial design. The case in this chapter offers an opportunity to examine this practice and to determine whether an alternative approach of commissioning multiple competing draft advertisements might be more effective.

This case also illustrates the use of prototyping to show a client how modeling can shed light on a problem. In that sense, it is similar to the National Leasing case that appears in Chapter 10. A prototype is a stripped-down, inexpensive model that shows modeling *works*. Usually, it is used to convince managers to fund a subsequent, full-scale effort to build a version of the model that can be used day-to-day.

In this case, we find that a generic model, developed despite a lack of information specific to the case, captures the fundamental forces that determine whether one draft commercial, or multiple draft commercials, will lead to the most effective overall campaign. By showing that we have captured these fundamentals, we make it more likely that the client will accept this new approach to advertising.

8.1 DRAFT COMMERCIALS CASE

Draft Commercials

Our client directs advertising for a large corporation that currently relies on a single outside advertising agency. For years, the corporation has followed the same strategy: The ad agency creates a draft commercial and, after testing and getting the client's approval, completes production and arranges for it to be aired.

The client's budget is divided between creating and airing commercials. Typically, about 5 percent of the budget is devoted to creating commercials and 95 percent to airing them. Lately, the client has become dissatisfied with the quality of the ads being created. He believes an ad campaign's ultimate profitability is more strongly influenced by the commercial's content than by either the total sum spent on airing or the media used (assuming reasonable levels of expenditure). Thus, he is considering increasing the percentage of his budget devoted to the first, more creative part of the process.

One way to do this would be to commission two or more ad agencies to develop a draft commercial independently. He could then select the one he determines would be most effective in promoting sales. Of course, since his budget is essentially fixed, the more money he spends creating draft commercials, the less he has to spend airing commercials. Note that he will have to pay up front for all of the draft commercials before he has a chance to evaluate them.

The standard technique for evaluating a draft commercial involves a screening before a test audience and later asking viewers what they remember about the advertisement; this is known as "next day recall." Ads with higher next day recall are generally those with higher effectiveness in the marketplace, but the correlation is far from perfect. A standard method for assessing the effectiveness of a commercial after it has aired is to survey those who watched the show and estimate "retained impressions." Retained impressions is a measure of the number of viewers who can recall the ad's essential features. Ads with higher retained impressions generally are more effective in generating sales, but again, the correlation is not perfect. Both the commercial's effectiveness (the number of retained impressions it creates) and its exposure (the number of times it is aired) influence sales.

The client would like to know if he should pursue this new strategy, and if so, how many draft commercials he should commission. We have offered to develop a prototype model that will assist him in this task.

Data from published reports in *Ad Week* on the number of retained impressions per dollar spent have been collected on 204 television advertisements in the following categories:

- Beer
- Coffee
- Cereals
- Detergents
- Bath soaps
- Cold remedies
- Airlines

These data are displayed in Figure 8.1 in the form of a histogram.

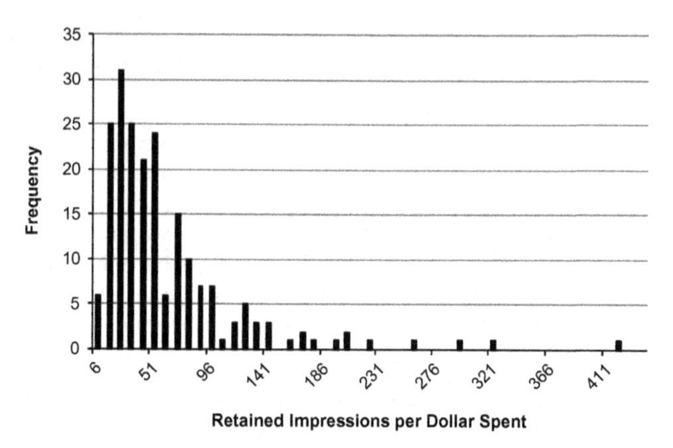

Figure 8.1. Retained impressions per dollar spent on airing.

To the Reader: Before reading on, frame this problem by setting boundaries for the analysis; making assumptions; listing questions; defining outputs, inputs, and decisions; and so on.

8.2 FRAME THE PROBLEM

This problem is presented in an abstract form, with few details and little data. Thus, it is easier to identify the problem kernel, but we have to be creative when it comes to data. The essential challenge here is to evaluate a new strategy for advertising, one in which more than one draft commercial is purchased. Paying for more than one draft commercial obviously costs more than paying for a single draft, leaving less money for purchasing air time (assuming the total budget is fixed). However, buying multiple drafts might result in a more effective ad, compensating for the added cost. Our task is to explore this possibility in a structured fashion so that we can intelligently describe the circumstances under which multiple drafts are advantageous. Here is a statement of our problem kernel:

> Commissioning multiple draft commercials may improve the quality of the final advertisement chosen. Under what circumstances will this improvement more than outweigh the added cost?

The case itself gives limited information on the process the company uses to develop an advertisement. Before building a model, it is a good idea to attempt to lay out the process in some detail and, where details seem important, to check with the client.

Our client is considering purchasing more than one draft commercial, presumably from different ad agencies. The drafts must be paid for completely up front before the quality of the ad can be determined (although an arrangement that rewards the agencies for the quality of their drafts might be worth considering at some point). Under the current plan, the draft ad is shown to a consumer panel and their reactions are captured in a measure called next day recall. Although this is not stated explicitly, we assume the company can modify the ad (or perhaps cancel it) at this stage; otherwise, there would be little point in paying for the panel.

If multiple drafts are available, we assume that each one is shown to a panel of viewers. In some way, the company then chooses the "best" of the drafts. This process could rely entirely on the next day recall data provided by the panels. It is more likely that the decision is based on the next day recall data and on the judgment of management. For one thing, we know that next day recall does not perfectly predict the ad's actual impact in terms of retained impressions. Thus, some error is involved in estimating retained impressions by next day recall. We may eventually want to examine this issue in our models.

We also assume that having chosen (and perhaps modified) one of the drafts, the company then creates an advertising campaign that incorporates a choice of media (television, radio, print), markets (e.g., East Coast daytime television, Midwest Sunday papers), and timing (how many airings in a given market over how many weeks). One important assumption we make at this point is that the company's current marketing process will not change if the company adopts the new multiple-draft strategy. This simplifies the problem. Besides, we have no information on these details and no reason to expect that simply buying more drafts will allow the corporation to develop better campaigns. This assumption focuses the analysis on the effects multiple drafts have on ad quality.

What should be the output measure for our analysis? The ultimate goal of advertising is to increase sales, but we have no information on the effect of advertising on sales. The information we do have relates to retained impressions, which is the measurable impact of advertisements. (It is generally accepted that more retained impressions lead to higher sales, although the relationship is complex.) Thus, we use retained impressions as the measure of success and assume the company's goal is to choose the number of draft commercials that maximizes retained impressions.

We might also consider the effect (if any) the new policy has on the ad agencies themselves. By assuming the company can commission as many ads as it chooses from different agencies, we are assuming there are as many agencies as required and that all would agree to compete with other agencies in this manner. We also might consider the possibility that the ads produced under this scheme might differ in some ways from those produced under the current exclusive arrangement. An agency might devote more or less effort to a draft ad if it knows it is competing against several others. Since we have no evidence on this question, we assume the quality of the individual ads does not change under the new policy.

To the Reader: Before reading on, sketch an influence diagram for this problem.

8.3 DIAGRAM THE PROBLEM

In our influence diagram (Figure 8.2), we begin with the objective, Retained Impressions. Retained Impressions can be decomposed in many different ways. However, we have assumed the effectiveness of the advertising campaign will not be changed by the number of draft ads created. This suggests that each dollar spent on airing an ad has the same effectiveness (in terms of gaining Retained Impressions). Thus, the total number of Retained Impressions is the product of Retained Impressions per Dollar Spent on Airing (RI/$) and Dollars Spent on Airing ($Spent).

$Spent is what's left over in the budget after paying for the draft ads. Thus, three factors influence $Spent: Total Budget, Cost per Draft, and Number of Drafts, as follows:

$$\$\text{Spent} = \text{Total Budget} - (\text{Cost per Draft} \times \text{Number of Drafts})$$

The final influence required in our model is the link between the number of drafts and the RI/$. This is the critical relationship in the analysis. The rationale for commissioning more than one draft is to increase the effectiveness of the final ad chosen, which is to say, to increase RI/$. But how much of an increase does each additional draft ad buy us?

We know that the process used to choose the best ad involves using a panel to provide an evaluation of next day recall (NDR). So if the client were to commission five draft ads, he would have five different estimates of NDR. A simple assumption is that the company would choose the ad with the highest

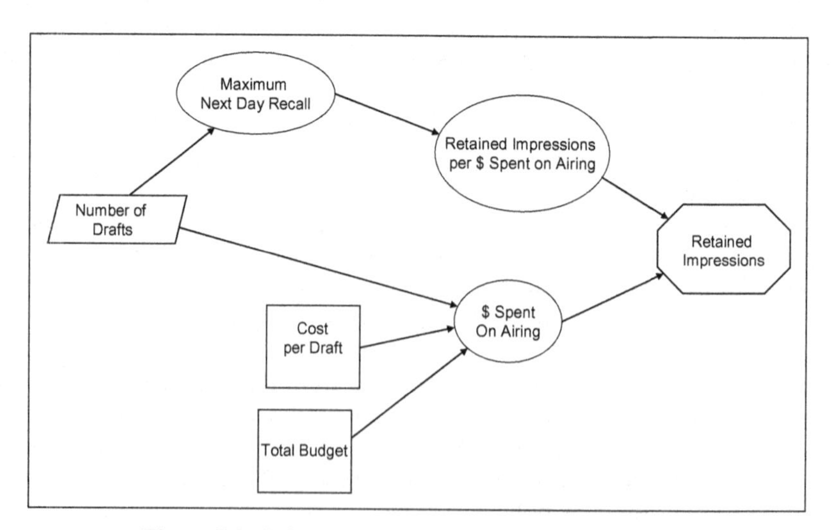

Figure 8.2. Influence diagram for Draft Commercials.

NDR. Therefore, we need to specify two relationships: the link between the number of drafts and the maximum NDR (MaxNDR), and that between MaxNDR and RI/$.

An analogy may help you to understand the form of the relationship between the number of draft commercials and the highest NDR among those drafts. Individual draft ads must vary in quality or there would be no point in considering buying multiple drafts. We measure the quality of a draft by its NDR. Since we cannot predict the NDR of a draft before it is evaluated, we can think of NDR as a random number drawn from some distribution. When we pick the highest NDR from among five drafts, we are in effect drawing five random samples from a distribution and choosing the highest. So the relationship between the number of drafts and the highest NDR is akin to the relationship between the number of random samples and the *best* of those samples.

We approach this question by performing a thought experiment with a simple model. Imagine a bag filled with 101 balls numbered from 0 to 100. You want to draw numbers from the bag and keep track of the *highest number* you draw. That will be your score in this experiment. Reach into the bag and randomly draw your first number. Although the probability of drawing any specific number is equally likely (1 out of 101), the expected value of the first draw is 50 (the probability-weighted average of all possible outcomes). Suppose you do pull out the number 50; your best score is now 50. Record that value and return the ball to the bag.

What is the probability that your score will *improve* with another pull from the bag? Fifty of the balls have values higher than 50, and 50 balls have values less than 50. So, about half the time, the second draw will improve your score, and half the time it will not. Let us assume you reach into the bag and pull out the number 60. Your overall score improves to 60 because that is the higher of your two draws. Return the ball to the bag.

Now consider the probability that drawing a third ball will improve your score. The probability has gone down to about 40 percent since there are now only 40 balls in the bag with a value higher than your current best value (60). The value of an additional draw has decreased! There is about a 60 percent chance that an additional draw will be of no value to you since it will not increase your overall score.

If you continue with this experiment, you realize that every time you draw a high-numbered ball, your score increases, but you simultaneously reduce the probability that the next draw will be of any value to you. Thus, although each additional draw has value, the value of each draw is less than that of the previous draw. There are diminishing returns to additional draws.

The situation in the commercials case is similar. Buying draft commercials is like pulling balls from a hypothetical bag. The quality of the commercials is presumably randomly distributed, and purchasing each draft is like pulling another number from the bag. So we can conjecture that the NDR of the best ad increases with the number of drafts, but with diminishing returns. This conjecture does not fully specify the relationship between the number of drafts and the NDR of the best ad, but it does give some sense of the shape it must have. We apply this idea in several alternative ways when we build our models.

The final relationship to consider is that between MaxNDR and RI/$. What we know about this relationship suggests that the two factors are highly correlated, but there is noise in the relationship. In other words, the higher the MaxNDR, the higher the RI/$, but the relationship is subject to some variability.

To the Reader: Before reading on, design and build a first model based on the influence diagram.

8.4 M1 MODEL AND ANALYSIS

To build a numerical model along the lines of the influence diagram in Figure 8.2 requires some assumptions about the size of the budget and the cost of a draft ad. Since we are building a prototype and we lack real data, we use representative numbers. Thus, we assume the total budget is 100 and the cost of an individual draft ad is 5. This allows up to 20 draft commercials. (We have even left the units vague: This model applies just as well to budgets of $100,000 or $10,000,000.)

Next, we specify the quantitative relationship between the chosen draft's NDR and RI/$. We expect retained impressions to increase with next day recall, but it is doubtful the relationship is strictly proportional. There probably are some ads that gain a high NDR and are expected to show a high RI/$ but in reality fail to live up to this expectation. This could be because the focus group that generated the NDR did not represent the broader population or the ad as broadcast did not have the same appeal as it did to the focus group. For simplicity, in our first model we assume that RI/$ is proportional to NDR. Thus, we need only one parameter to convert from NDR to RI/$.

Finally, we must quantify the relationship between the number of drafts and the maximum (or best) NDR. We demonstrated with a thought experiment in the previous section that this relationship is probably increasing with diminishing returns, but we have no evidence yet on the degree of diminishing returns. A good starting point in most analyses is to assume proportionality or linearity, so we start there and test more complex relationships later. Another reason to start with a proportional model in this case is that if we find that purchasing more than a single draft is not optimal under linearity, then it is unlikely to be so under diminishing returns. So, in our first model, we assume MaxNDR increases in proportion to the number of drafts chosen. Later, in a second version of the model, we consider the possibility of diminishing returns in this relationship.

Figure 8.3 shows our first model in numerical form; Figure 8.4 displays the relationships. Four parameters drive this simple model: Budget (100); Cost per

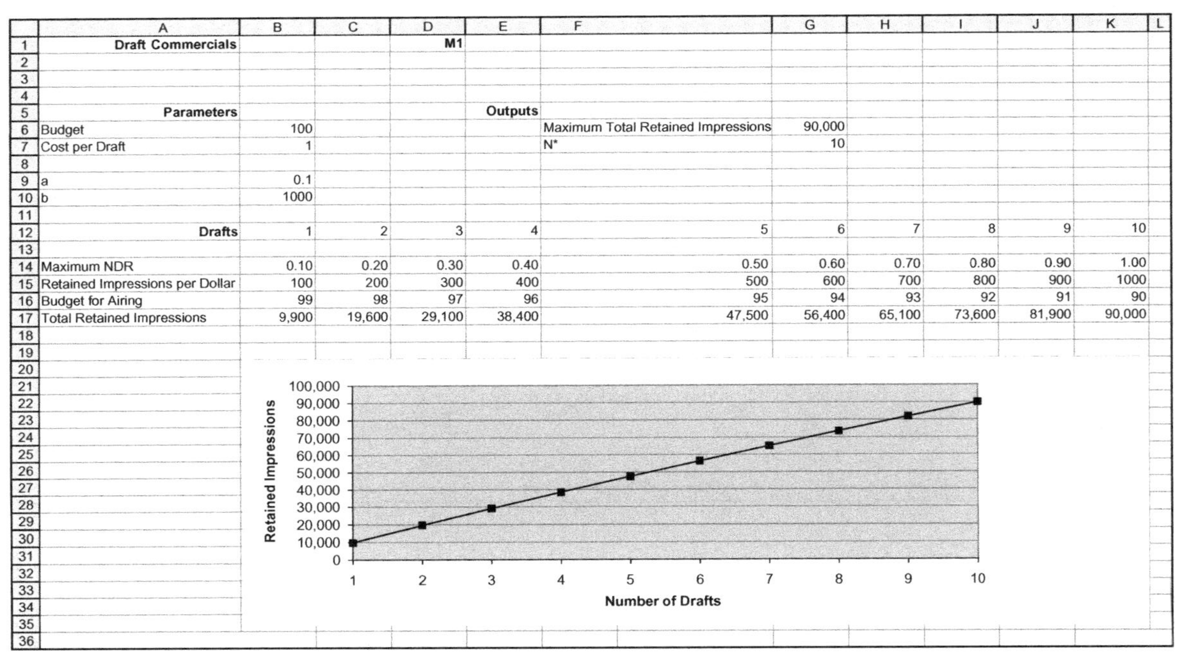

	A	B	C	D	E	F	G	H	I	J	K
1	**Draft Commercials**			**M1**							
2											
3											
4											
5	**Parameters**				**Outputs**						
6	Budget	100				Maximum Total Retained Impressions	90,000				
7	Cost per Draft	1				N*	10				
8											
9	a	0.1									
10	b	1000									
11											
12	**Drafts**	1	2	3	4	5	6	7	8	9	10
13											
14	Maximum NDR	0.10	0.20	0.30	0.40	0.50	0.60	0.70	0.80	0.90	1.00
15	Retained Impressions per Dollar	100	200	300	400	500	600	700	800	900	1000
16	Budget for Airing	99	98	97	96	95	94	93	92	91	90
17	Total Retained Impressions	9,900	19,600	29,100	38,400	47,500	56,400	65,100	73,600	81,900	90,000

Figure 8.3. M1—numerical view.

	A	B	C	D	E	F	G	H	I	J	K	L
1	**Draft Commercials**			**M1**								
2												
3												
4												
5	**Parameters**				**Outputs**							
6	Budget	100				Max RI	=MAX(B17:K17)					
7	Cost per Draft	1				N*	=HLOOKUP(G6,B17:K18					
8												
9	a	0.1										
10	b	1000										
11												
12	**Drafts**	1	2	3	4	5	6	7	8	9	10	
13												
14	Maximum NDR	=$B$9*B12	=$B$9*C12	=$B$9*D12	=$B$9*E12	=$B$9*F12	=$B$9*G12	=$B$9*H12	=$B$9*I12	=$B$9*J12	=$B$9*K12	
15	Retained Impressions per Dollar	=$B$10*B14	=$B$10*C14	=$B$10*D14	=$B$10*E14	=$B$10*F14	=$B$10*G14	=$B$10*H14	=$B$10*I14	=$B$10*J14	=$B$10*K14	
16	Budget for Airing	=$B$6-$B$7*B12	=$B$6-$B$7*C12	=$B$6-$B$7*D12	=$B$6-$B$7*E12	=$B$6-$B$7*F12	=$B$6-$B$7*G12	=$B$6-$B$7*H12	=$B$6-$B$7*I12	=$B$6-$B$7*J12	=$B$6-$B$7*K12	
17	Total Retained Impressions	=B15*($B$6-$B$7*B12)	=C15*($B$6-$B$7*C12)	=D15*($B$6-$B$7*D12)	=E15*($B$6-$B$7*E12)	=F15*($B$6-$B$7*F12)	=G15*($B$6-$B$7*G12)	=H15*($B$6-$B$7*H12)	=I15*($B$6-$B$7*I12)	=J15*($B$6-$B$7*J12)	=K15*($B$6-$B$7*K12)	
18												

Figure 8.4. M1—relationship view.

Draft (1); the constant *a,* which relates MaxNDR to the number of drafts (0.10); and the constant *b,* which relates the RI/$ to MaxNDR (1,000).

In row 14, we calculate the maximum NDR for one to ten ads. In row 15, we calculate the corresponding Retained Impressions per Dollar. Both relationships are proportional by assumption. In row 16, we calculate the budget that remains for airing ads after we have paid for the draft ads. Finally, in row 17, we calculate Total Retained Impressions by multiplying the Retained Impression per Dollar by the Budget for Airing. In cell G6, we calculate the Maximum Total Retained Impressions over all choices of the number of ads. Finally, in cell G7, we use the HLOOKUP function to find the optimal *number* of drafts N*. This stands at ten in our illustrative model, which immediately shows there are circumstances under which buying more than one draft ad can improve total retained impressions.

Insight 1: The optimal number of draft advertisements can be greater than one.

But surely ten drafts is not always the optimal number. If the cost per draft were higher, for example, we would expect fewer drafts to be optimal. We test this hypothesis directly with a sensitivity analysis on the Cost per Draft. The results are shown in Figure 8.5. When the Cost per Draft is between 1 and 5, the optimal number of drafts is ten. Above 5, the optimal number drops quickly, so that when the cost is 10, the optimal number of drafts is five. Thereafter, the optimal number of drafts drops slowly, so that more than one draft is optimal until the cost reaches 34. We generalize these observations as follows:

Insight 2: The optimal number of draft advertisements decreases at a decreasing rate with the cost per ad.

How sensitive are our results to the constants *a* and *b*? We might think that both the effectiveness of drafts in improving the maximum NDR (*a*) and the

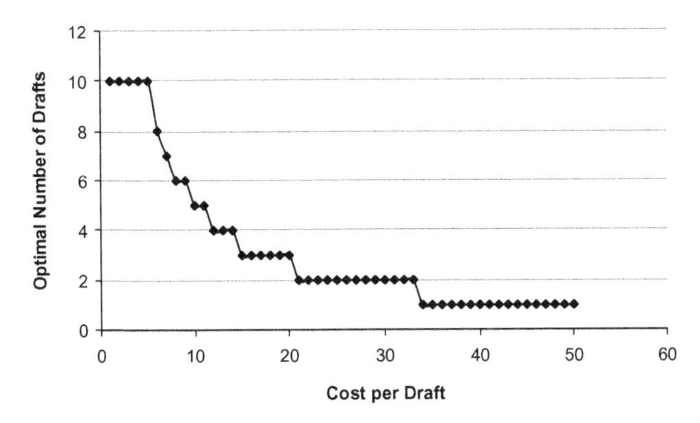

Figure 8.5. Sensitivity of Optimal Number of Drafts to Cost per Draft (M1).

effectiveness of NDR in generating retained impressions (b) would increase the optimal number of drafts, assuming all else is equal. But a little experimentation proves these expectations false. In fact, several Data Sensitivity runs show that changes to these parameters do not change the optimal number of drafts at all. What is the reason?

This model is simple enough to be expressed algebraically, as follows:

Retained Impressions
$= \text{RI/\$} \times \text{\$Spent}$
$= a \times b \times \text{Number of Drafts} \times (\text{Budget} - (\text{Cost per Draft} \times \text{Number of Drafts}))$
$= a \times b \times \text{Budget} \times \text{Number of Drafts} - a \times b \times \text{Cost per Draft} \times$
$\text{Number of Drafts}^2$

This is a simple quadratic equation in the Number of Drafts. A little calculus shows that the optimal number of drafts, N*, depends only on the Budget and the Cost per Draft:

$$N^* = \text{Budget}/(2 \times \text{Cost per Draft})$$

This analysis leads us to the following insight:

Insight 3: In a linear model, the optimal number of draft advertisements is *independent* of both the effectiveness of drafts in increasing NDR and the effectiveness of NDR in generating retained impressions.

Summary of M1

Our first model is extremely simple, and yet it has generated useful insights. The most important result of the model is that the optimal number of draft ads can exceed one. Beyond that, we suspect the insights we have derived are specific to this model and will not generalize when we elaborate the model. In particular, it seems unlikely that only the budget and the cost of each ad determine the optimal number of drafts in anything but the simplest model.

There are several directions in which we could go at this point to improve the model. We could introduce diminishing returns to the maximum NDR, we could introduce uncertainty in the quality of each draft ad, or we could introduce variability in the relationship between NDR and RI/$. All these complications were mentioned earlier; which one should we pursue first? We choose to introduce diminishing returns at this point primarily because in doing so we can continue to keep the model deterministic. Stochastic models are automatically more complex than deterministic ones, so it seems sensible to introduce diminishing returns first to see what its influence is without the added complication of randomness.

To the Reader: Before reading on, modify M1 to reflect diminishing returns in the relationship between the number of draft ads and the maximum NDR. Contrast the behavior of this version with M1.

8.5 M2 MODEL AND ANALYSIS

In some ways, it is remarkable that a model as simple as M1, with its strictly proportional relationships, generates the insights it does. But one of the relationships in this model seems particularly suspect to us, and that is the one between MaxNDR and the number of drafts. Whatever number of drafts we decide on, we are always going to adopt the *best* ad from among those drafts. As the number of drafts increases, it seems intuitive that the *next* draft will be less likely to be better than all the previous ones. This implies that the NDR of the best of N drafts increases with N but at a decreasing rate, hence the diminishing returns we alluded to before. Does this refinement change the basic implications of our model?

To test this new relationship in our model, we need a function that relates the number of drafts N to MaxNDR with diminishing returns. One such function is the power function:

$$\text{MaxNDR} = c \times \text{N}^d \quad d < 1$$

Three variants of this function are shown in Figure 8.6, along with the proportional relationship from M1. Each of these functions has $c = 0.1$, whereas d ranges from 0.1 to 0.9. This chart shows that the power function approaches linearity as d increases toward 1.0.

It is simple to replace the proportional relationship in row 14 in M1 with this power function to create M2. Does this new relationship change the behavior of the model? Figure 8.7 shows the sensitivity of the optimal number of drafts to the Cost per Draft. We see that the optimal number of drafts is ten when drafts are inexpensive and one when they are expensive. When compared with the results for M1, we see that the response of this model to changes in the Cost per Draft is essentially the same as in M1.

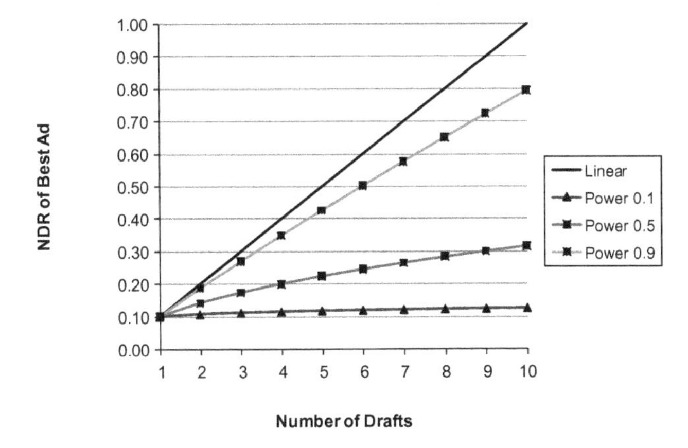

Figure 8.6. Diminishing returns to Number of Drafts.

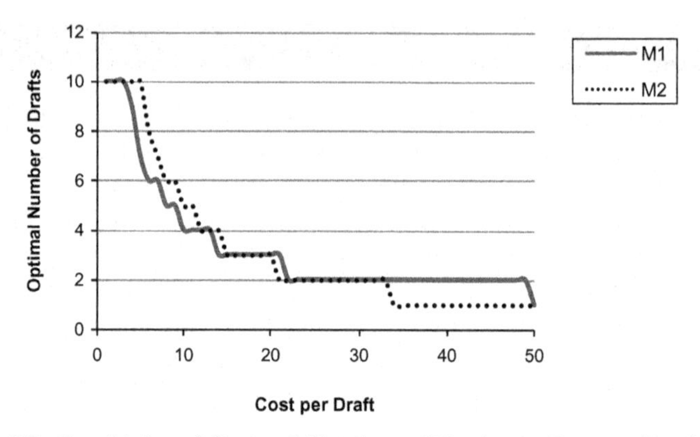

Figure 8.7. Sensitivity of Optimal Number of Drafts to Cost per Draft (M2).

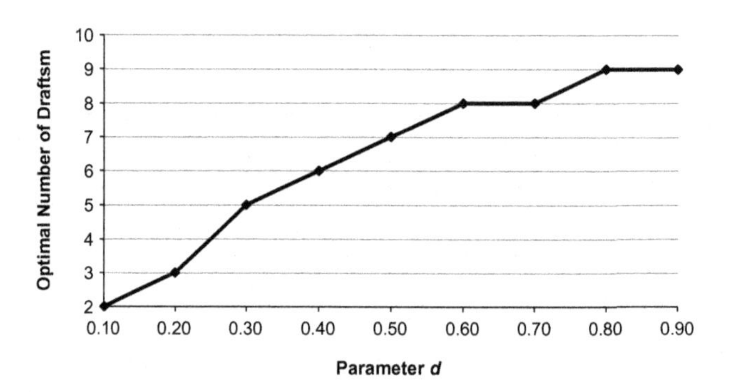

Figure 8.8. Sensitivity of Optimal Number of Drafts to parameter d (M2).

Insight 4: The optimal number of draft advertisements can exceed one in a model with diminishing returns to the number of draft ads.

If we optimize retained impressions by taking the derivative with respect to the number of drafts, as we did with M1, we see that the optimal number of drafts depends in this case on the Cost per Draft and on the exponent of the power curve, d. In fact, the optimal number of drafts is proportional to $d/(1 + d)$, which increases in d. The same effect can be observed in Figure 8.8, which shows the results of a sensitivity analysis on the parameter d between 0.1 and 0.9.

Insight 5: The greater the parameter d in the relationship between the number of drafts and the maximum NDR, the higher the optimal number of drafts.

Summary of M2

Modifying our first model to include diminishing returns has added some degree of realism, since we have a strong intuition that diminishing returns truly exist in the relationship between the number of drafts and the maximum NDR. However, it is still the case that the model can produce any outcome for the optimal number of ads, depending on the cost per ad.

We should point out that not every modeler would approach this problem in the way we have. Some modelers would have used a power function from the start, trusting their intuition about diminishing returns. Others would have introduced uncertainty in the quality of each ad from the start. Our approach, although it may take longer, does have the advantage that we know as soon as we study M1 that the optimal number of ads can exceed one *even in the simplest model.* Thus, it is a consequence of the problem's fundamental nature. Had we built a more complex model from the start, we would not know whether this property was fundamental or merely the result of an incidental assumption.

To the Reader: Before reading on, modify and improve M2 in whatever ways seem beneficial to you.

8.6 M3 MODEL AND ANALYSIS

The two models we have built to this point have generated several interesting insights into the factors that determine the optimal number of draft commercials. We have established the principle that purchasing multiple drafts may be optimal, depending on the model's parameters, if the goal is to maximize retained impressions. We have also determined how the cost of each draft and the effect of multiple drafts on NDR influence the choice of the number of drafts.

Two issues remain that we can explore more deeply. One issue involves the forces that drive the relationship between the number of drafts and the quality of the best ad. The other issue is uncertainty in the actual RI/$, given the NDR for the best ad chosen.

We initially chose a linear model for the relationship between number of drafts and best ad but later modified that model to allow for diminishing returns. Neither model, however, accounts for the *causes* of diminishing returns. Why is it that the best of N drafts increases at a decreasing rate? Our intuition supports this assumption, but can we model this process in a more transparent manner? If so, we may be able to shed additional light on the choice of the number of drafts.

The second issue is the uncertainty surrounding the actual quality of the ad we select. The NDR metric we use to determine an ad's quality does predict to some degree the number of retained impressions that ad will generate when it is aired. However, this relationship is subject to some uncertainty or error. Does this error change the conclusions we might otherwise draw about the selection of multiple drafts?

To derive as much insight as we can, we deal with these complications to the basic model one at a time. In M3, we model the quality of the best ad in more detail; then, in M4, we introduce error in the relationship between perceived quality and ultimate retained impressions.

Up to this point, we have not used the data provided on the retained impressions per dollar spent on airing (Figure 8.1). These data come from a wide variety of products, so they would not be directly applicable to any single product the company might be advertising. However, the data suggest there is a high degree of variability in RI/$. Moreover, the distribution is such that most ads have a modest RI/$ but a few have values many times the mean. In other words, the distribution is highly skewed.

Several factors could account for this pattern in the data. One factor is systematic differences among product types. It could be, for example, that cereal ads are generally more memorable than bath soap commercials. But this seems unlikely as a general rule, so we assume for modeling purposes that all ads are drawn from the same distribution.

We argued earlier that if we think of ad quality as being random, then we see the process of buying multiple drafts as akin to drawing random samples from a probability distribution. Presumably, the samples are independent, and under our assumption, they all come from the same distribution. If, hypothetically, the distribution has a small variance, it would seem likely that a single draft ad would be the best choice. That is because each draft is unlikely to differ much in quality from any other draft. But if the distribution has a high variance, sampling more than one ad makes more sense, since we are more likely to draw an especially good outcome on repeated samples. And what if the distribution is skewed, as is suggested by the data in Figure 8.1—does that change the results of this sampling process?

To incorporate uncertainty in our model, we need to invent a distribution for NDR, since that is our measure of ad quality. If our intuition is correct, it is the variance of that distribution that is important in determining the optimal number of drafts. So we want to use a distribution such as the normal or lognormal that can incorporate both high and low variances.

We begin by assuming the NDR for each draft ad is drawn randomly from a normal distribution, with a mean of 0.50 and a standard deviation of 0.25. Model M3 is displayed in numerical form in Figure 8.9 and in relationship form in Figure 8.10. In the ten columns from B to K, we simulate results for the ten different choices we have for the number of drafts. In cells B15 to K24, we take a sample from the NDR distribution for each draft. The first sample, in B15, is reused for the remaining nine cases by copying it across the row. Similarly, the second sample, in C16, is reused by copying it across the row. This approach reduces the number of random samples from 55 to 10 and speeds up the simulations.

	A	B	C	D	E	F	G	H	I	J	K	L
1	**Draft Commercials**			**M2**								
2												
3												
4												
5	**Parameters**				**Outputs**							
6	Budget	100				Max RE	103,436					
7	Ad Cost	5				N*	1					
8	RE Multiplier	1000										
9	Mean NDR	0.50										
10	SD NDR	0.25										
11												
12												
13	**Drafts**	**1**	**2**	**3**	**4**	**5**	**6**	**7**	**8**	**9**	**10**	
14												
15	**NDR**	1.09	1.09	1.09	1.09	1.09	1.09	1.09	1.09	1.09	1.09	
16			0.74	0.74	0.74	0.74	0.74	0.74	0.74	0.74	0.74	
17				0.41	0.41	0.41	0.41	0.41	0.41	0.41	0.41	
18					0.35	0.35	0.35	0.35	0.35	0.35	0.35	
19						0.09	0.09	0.09	0.09	0.09	0.09	
20							0.72	0.72	0.72	0.72	0.72	
21								0.36	0.36	0.36	0.36	
22									0.39	0.39	0.39	
23										0.56	0.56	
24											0.31	
25												
26	Maximum NDR	1.09	1.09	1.09	1.09	1.09	1.09	1.09	1.09	1.09	1.09	
27												
28	Retained Impressions/$	1,089	1,089	1,089	1,089	1,089	1,089	1,089	1,089	1,089	1,089	
29												
30	Total Retained Impressions	103,436	97,992	92,548	87,104	81,660	76,216	70,772	65,328	59,884	54,440	
31												
32	Mean RI	47,433	58,616	61,745	62,177	60,893	58,504	55,722	52,526	48,950	44,500	
33												

Figure 8.9. M3—numerical view.

In row 26, we determine Maximum NDR by taking the maximum of the ads available. Then, in row 28, we transform Maximum NDR into Retained Impressions per Dollar (RI/$) by multiplying by a constant. Finally, in row 30, we calculate Total Retained Impressions by multiplying RI/$ by the budget remaining after paying for the draft ads. We set row 30 to be Crystal Ball Forecast cells, and in row 32, we capture the mean of these cells by using the function CB.GETFORESTATFN.

The logic of this model is identical to that in M1 and M2 except in the determination of Maximum NDR. In the earlier models, this relationship was deterministic and somewhat abstract, since we simply assumed its shape, but in M3, the quality of each ad is random, as is the best outcome from a random sample. In general, we prefer models in which the basic mechanisms are transparent. Thus, we would claim that M3 is a better model than M2 or M1 because it more closely reflects the actual processes at work in the problem.

Figure 8.11 shows the relationship between the number of drafts and the maximum NDR that results from the random sampling process in our model. As we suspected, random sampling and choosing the best of the resulting samples lead to diminishing returns. This relationship has the same shape as Figure 8.6. The difference lies in the origins of the curve. In M2, we *assumed* diminishing returns; in M3, we *derive* it from more fundamental assumptions.

Figure 8.12 plots the mean values for Total Retained Impressions as the Number of Drafts varies from 1 to 10. This chart shows, as we would expect, that the optimal number of drafts can be larger than one in this new model.

	A	B	C	D	E	F	G	H	I	J	K	L
1	**Draft Commercials**			**M2**								
2												
3												
4												
5	**Parameters**				**Outputs**							
6	Budget	100				Max RE	=MAX(B30:K30)					
7	Ad Cost	5				N*	=HLOOKUP(G6,B30:K31,2,0)					
8	RE multiplier	1000										
9	Mean NDR	0.5										
10	SD NDR	0.25										
11												
12												
13	**Drafts**	**1**	**2**	**3**	**4**	**5**	**6**	**7**	**8**	**9**	**10**	
14												
15	**NDR**	=IF(CB.Normal($B$9,$B$10)>1, 1,IF(CB.Normal($B$9,$B$10)<0, 0,CB.Normal($B$9,$B$10)))	=B15	=C15	=D15	=E15	=F15	=G15	=H15	=I15	=J15	
16			=IF(CB.Normal($B$9,$B$10)>1, 1,IF(CB.Normal($B$9,$B$10)<0, 0,CB.Normal($B$9,$B$10)))	=C16	=D16	=E16	=F16	=G16	=H16	=I16	=J16	
17				=IF(CB.Normal($B$9,$B$10)>1, 1,IF(CB.Normal($B$9,$B$10)<0, 0,CB.Normal($B$9,$B$10)))	=D17	=E17	=F17	=G17	=H17	=I17	=J17	
18					=IF(CB.Normal($B$9,$B$10)>1, 1,IF(CB.Normal($B$9,$B$10)<0, 0,CB.Normal($B$9,$B$10)))	=E18	=F18	=G18	=H18	=I18	=J18	
19						=IF(CB.Normal($B$9,$B$10)>1, 1,IF(CB.Normal($B$9,$B$10)<0, 0,CB.Normal($B$9,$B$10)))	=F19	=G19	=H19	=I19	=J19	
20							=IF(CB.Normal($B$9,$B$10)>1, 1,IF(CB.Normal($B$9,$B$10)<0, 0,CB.Normal($B$9,$B$10)))	=G20	=H20	=I20	=J20	
21								=IF(CB.Normal($B$9,$B$10)>1, 1,IF(CB.Normal($B$9,$B$10)<0, 0,CB.Normal($B$9,$B$10)))	=H21	=I21	=J21	
22									=IF(CB.Normal($B$9,$B$10)>1, 1,IF(CB.Normal($B$9,$B$10)<0, 0,CB.Normal($B$9,$B$10)))	=I22	=J22	
23										=IF(CB.Normal($B$9,$B$10)>1, 1,IF(CB.Normal($B$9,$B$10)<0, 0,CB.Normal($B$9,$B$10)))	=J23	
24											=IF(CB.Normal($B$9,$B$10)>1, 1,IF(CB.Normal($B$9,$B$10)<0, 0,CB.Normal($B$9,$B$10)))	
25												
26	Maximum NDR	=MAX(B15:B23)	=MAX(C15:C23)	=MAX(D15:D23)	=MAX(E15:E23)	=MAX(F15:F23)	=MAX(G15:G23)	=MAX(H15:H23)	=MAX(I15:I23)	=MAX(J15:J23)	=MAX(K15:K23)	
27												
28	Retained Impressions/$	=$B$8*B26	=$B$8*C26	=$B$8*D26	=$B$8*E26	=$B$8*F26	=$B$8*G26	=$B$8*H26	=$B$8*I26	=$B$8*J26	=$B$8*K26	
29												
30	Total Retained Impressions	=B28*($B$6-$B$7*B13)	=C28*($B$6-$B$7*C13)	=D28*($B$6-$B$7*D13)	=E28*($B$6-$B$7*E13)	=F28*($B$6-$B$7*F13)	=G28*($B$6-$B$7*G13)	=H28*($B$6-$B$7*H13)	=I28*($B$6-$B$7*I13)	=J28*($B$6-$B$7*J13)	=K28*($B$6-$B$7*K13)	
31												
32	Mean RI	=CB.GetForeStatFN(B30,2)	=CB.GetForeStatFN(C30,2)	=CB.GetForeStatFN(D30,2)	=CB.GetForeStatFN(E30,2)	=CB.GetForeStatFN(F30,2)	=CB.GetForeStatFN(G30,2)	=CB.GetForeStatFN(H30,2)	=CB.GetForeStatFN(I30,2)	=CB.GetForeStatFN(J30,2)	=CB.GetForeStatFN(K30,2)	
33												

Figure 8.10. M3—relationship view.

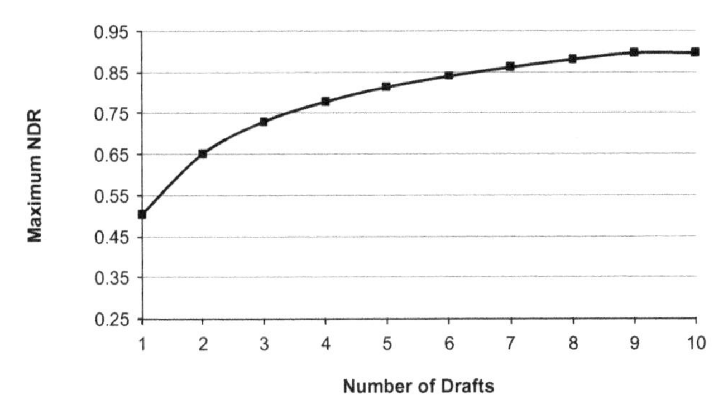

Figure 8.11. Sensitivity of MaxNDR to Number of Drafts (M3, $\sigma = 0.25$).

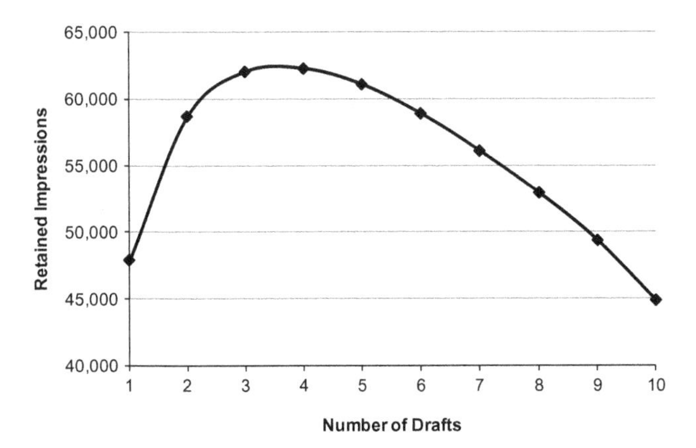

Figure 8.12. Sensitivity of Retained Impressions to Number of Drafts (M3, $\sigma = 0.25$).

Insight 6: The optimal number of draft advertisements can exceed one in a model in which uncertainty in the quality of ads leads to diminishing returns from the number of draft ads.

Of course, since the inputs to M3 are random, the outputs are random as well. For each possible choice of the number of drafts, we can create a probability distribution for Retained Impressions. Figure 8.13 shows one way to display these results. This Crystal Ball Trend Chart shows various certainty bands around the median outcome for each value of the number of drafts. The chart shows that the number of drafts that maximizes the *median* outcome is around four. It also shows that the *variance* in the distribution of Retained Impressions decreases with additional ads. This result is not immediately obvious. To understand it better, think of the extreme cases. When we take a single sample, the distribution of quality is the same as the distribution for ads themselves. But if we were to take a very large number of samples and select

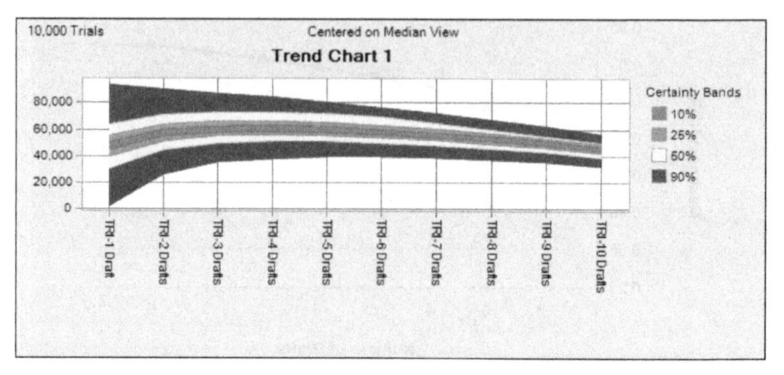

Figure 8.13. Distribution of Retained Impressions as a function of the Number of Drafts (M3, σ = 0.25).

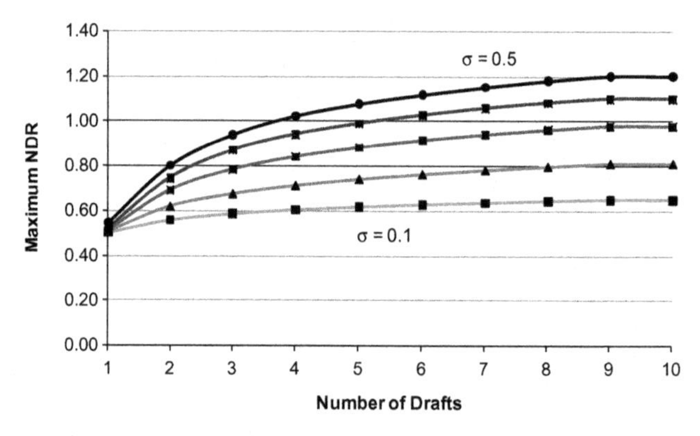

Figure 8.14. Effect of increasing variability of NDR on MaxNDR (M3).

the best, we would get the maximum possible outcome in every case. So the variability of the best of N samples declines with the number of samples. This is the phenomenon we observe in Figure 8.13.

Insight 7: The variability in retained impressions decreases with additional draft ads.

This model also allows us to answer a question neither of our previous models could: What is the effect of variability in ad quality on the optimal strategy? Figure 8.14 shows how the maximum NDR varies with the number of drafts for values of the standard deviation of the distribution of NDR between 0.10 and 0.50. In all five cases, the maximum NDR increases at a decreasing rate; but the higher the standard deviation of NDR, the higher the maximum NDR. This phenomenon is very important. In effect, variability in ad quality is a good thing in this case because it allows us to achieve a higher NDR by purchasing more draft ads.

Insight 8: Higher variability in the distribution of ad quality allows for achieving higher NDR through additional draft ads.

How does this effect carry over into retained impressions? Figure 8.15 suggests some answers. In this chart, we have again varied the standard deviation of the distribution for ad quality from 0.10 to 0.50. As the variability of quality increases, the entire curve for retained impressions shifts up. Thus, for any fixed number of drafts, retained impressions increases with the variability of quality. In addition, we can see in the chart that the *optimal* number of drafts increases with variability. At the lowest value for variability, the optimal number of drafts is two, but the optimal number increases to four as the variability increases. There is even a hint in these results that the optimal number of drafts increases at a decreasing rate, which means there is a level of uncertainty beyond which the optimal number may not change. We can summarize these conclusions as follows:

Insight 9: Higher variability in the distribution of ad quality increases retained impressions and increases the optimal number of drafts as well.

Finally, we return to our assumption that the distribution of ad quality is given by a normal distribution. The data in Figure 8.1 suggest, as we have noted, that the distribution of ad quality may be highly skewed, with some very high quality ads but a predominance of low-to-moderate quality ones. The normal distribution, of course, does not allow for this shape, but the lognormal distribution does. Figure 8.16 shows two distributions: one is the normal distribution we have used in M3, and the other is a lognormal distribution with the same mean (0.50) and standard deviation (0.25). Notice the high

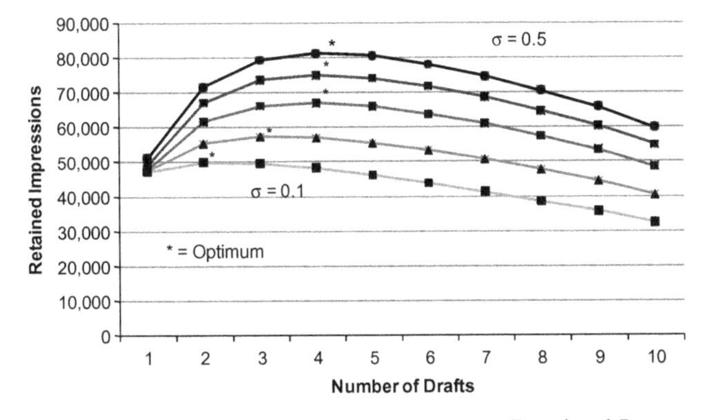

Figure 8.15. Effect of increasing variability of NDR on Retained Impressions (M3).

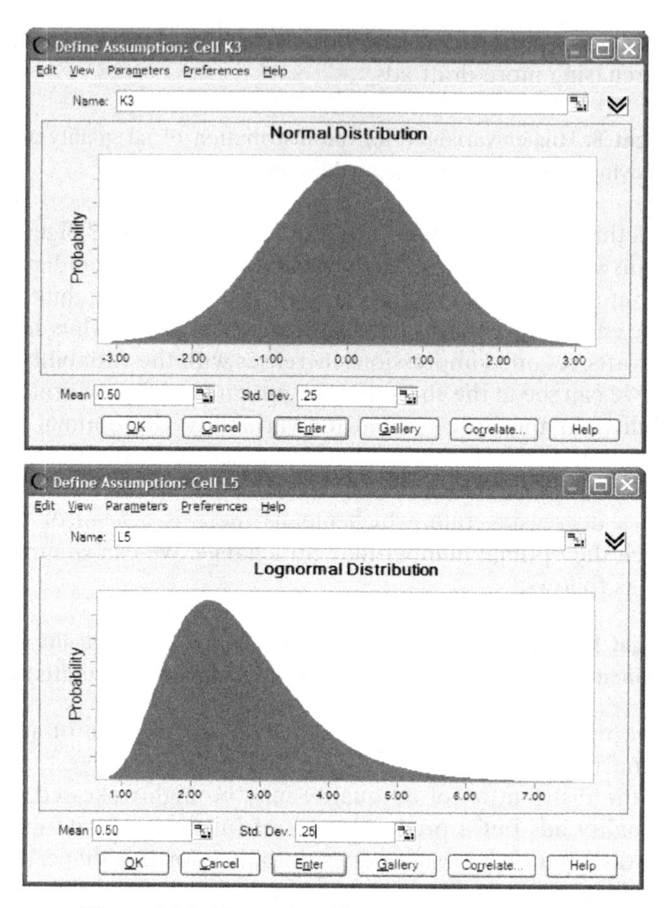

Figure 8.16. Normal and lognormal distributions.

skewness in the lognormal distribution and the similarity in shape to Figure 8.1. Does introducing the lognormal distribution change our results?

We change the normal distribution in M3 to a lognormal distribution and rerun the simulations. Figure 8.17 compares the RI/$ for the two model variants, with all other parameters the same. The lognormal distribution gives a higher RI/$, especially for a higher number of drafts. Skewness apparently works as we conjectured: the higher the chances of picking an extremely good ad, the higher the retained impressions. However, what is something of a surprise is that skewness seems to have little or no impact on the *optimal* number of drafts.

Insight 10: The shape of the distribution for ad quality may change the predicted level of retained impressions but not the optimal number of draft ads.

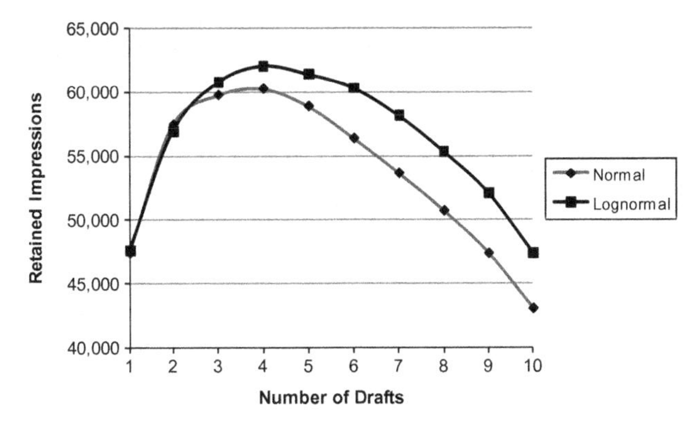

Figure 8.17. Retained Impressions for normal and lognormal models (M3).

Of course, this observation, just like any of our insights, may be true only for a limited range of parameters. In fact, when we increase the standard deviation of the distribution of NDR to 0.5, we find the lognormal distribution again gives a higher RI/$ than does the normal distribution. However, the optimal number of drafts increases from four to six. Thus, we can improve Insight 10 as follows:

Insight 10 (improved): Compared with a symmetric distribution, under a skewed distribution, the optimal number of draft ads is unchanged for low standard deviations of NDR but may increase for high standard deviations of NDR.

Summary of M3

M3 is a major improvement over previous models because it directly incorporates the factor that makes multiple drafts beneficial: randomness in ad quality. Whereas in the previous models we had assumed the relationship between the number of drafts and the best NDR, in this model, we derive that relationship from a simple random sampling process. It is always an advantage in modeling to build our models on more fundamental processes and assumptions.

M3 is also an improvement because it provides insights we could not have derived from the earlier models. For example, with M3, we can determine the impact that both variability and skewness have on the determination of the optimal number of drafts. We can also see how increasing the number of drafts changes the *distribution* of retained impressions per dollar, allowing us to consider the risks facing the decision makers, not just the average outcomes.

To the Reader: Before reading on, modify and improve M3 to reflect the possibility that NDR can only be measured with some error.

8.7 M4 MODEL AND ANALYSIS

One potentially important issue remains outside our model: the imprecision in our estimates of next day recall. The case indicates that NDR is a good, but not perfect, predictor of retained impressions. One way to interpret this is to assume the actual NDR of an ad *is* a perfect predictor, but that we can only estimate the true NDR through our consumer panels. In our models so far, of course, we have assumed we can determine NDR for each ad perfectly. We relax that assumption now and examine how this additional feature affects the choice of the optimal number of drafts.

A simple way to introduce the possibility of error in NDR is to take a sample from an error distribution and add it to the observed NDR. No data are available on the likely magnitude of these errors, so we need to be creative. If we assume observed NDR is correct on average, then the average error is zero. This leaves just the variance of the error to specify. We arbitrarily assume the error distribution itself is normal, since this is a simple distribution that is completely specified by its mean (zero, in this case) and its standard deviation.

Figures 8.18 and 8.19 show the numerical and relationship views of this latest version of our model (M4). In rows 15–24 we sample from the distribution of ad quality for each number of drafts, just as we did in M3. We interpret these data now as the actual (but unknown) NDR. Then, in rows 27–36, we take corresponding samples from the error distribution, which has a standard deviation of 0.1 in the base case. In rows 38–47, we add the errors to the matching NDR samples, which gives us the Predicted NDRs. These are the values we will observe when we test ads with a panel of consumers.

The next step is to determine the highest-quality draft for each possible number of drafts. For future use, we record both the Actual Best Draft (row 49) and the Predicted Best Draft (row 50). Both are calculated using a MATCH function that returns the number of the draft that has the highest relevant NDR. For the actual best, this is the actual NDR, whereas for the predicted best, this is the predicted NDR. The more drafts we have, the more likely it is that we will select a draft based on the predicted NDR that is not in fact the best.

From this point on in the model, we keep track of three sets of results: Best, Predicted, and Achieved. Best results are based on the actual NDRs (row 52). Predicted results are based on the predicted NDRs (row 53). Achieved results are based on the actual NDRs but using the draft selected based on the predicted NDRs (row 54). The idea here is simple. An omniscient observer would ignore the errors, base his or her decision on the actual NDRs, and obtain the best possible result from the actual NDR. In the real world, on the other hand, we cannot observe the errors but only the predicted NDRs (which are the actual NDRs plus the errors). Therefore, we make our decisions based on the erroneous predicted NDRs. But our actual outcomes will be based on the actual NDRs.

	A	B	C	D	E	F	G	H	I	J	K	L
1	Draft Commercials			M4								
2												
3												
4												
5	Parameters											
6	Budget	100										
7	Ad Cost	5										
8	RI multiplier	1000										
9	Mean NDR	0.50										
10	SD NDR	0.25										
11	SD NDR Error	0.10										
12												
13	Drafts	1	2	3	4	5	6	7	8	9	10	
14												
15	Actual NDR	0.57	0.57	0.57	0.57	0.57	0.57	0.57	0.57	0.57	0.57	
16			0.36	0.36	0.36	0.36	0.36	0.36	0.36	0.36	0.36	
17				0.12	0.12	0.12	0.12	0.12	0.12	0.12	0.12	
18					0.09	0.09	0.09	0.09	0.09	0.09	0.09	
19						0.25	0.25	0.25	0.25	0.25	0.25	
20							0.38	0.38	0.38	0.38	0.38	
21								0.40	0.40	0.40	0.40	
22									0.39	0.39	0.39	
23										0.37	0.37	
24											0.37	
25												
26												
27	NDR Error	-0.12	-0.12	-0.12	-0.12	-0.12	-0.12	-0.12	-0.12	-0.12	-0.12	
28			-0.12	-0.12	-0.12	-0.12	-0.12	-0.12	-0.12	-0.12	-0.12	
29				0.03	0.03	0.03	0.03	0.03	0.03	0.03	0.03	
30					0.09	0.09	0.09	0.09	0.09	0.09	0.09	
31						0.05	0.05	0.05	0.05	0.05	0.05	
32							0.24	0.24	0.24	0.24	0.24	
33								-0.09	-0.09	-0.09	-0.09	
34									-0.17	-0.17	-0.17	
35										-0.07	-0.07	
36											-0.02	
37												
38	Predicted NDR	0.45	0.45	0.45	0.45	0.45	0.45	0.45	0.45	0.45	0.45	
39			0.23	0.23	0.23	0.23	0.23	0.23	0.23	0.23	0.23	
40				0.15	0.15	0.15	0.15	0.15	0.15	0.15	0.15	
41					0.18	0.18	0.18	0.18	0.18	0.18	0.18	
42						0.31	0.31	0.31	0.31	0.31	0.31	
43							0.62	0.62	0.62	0.62	0.62	
44								0.31	0.31	0.31	0.31	
45									0.22	0.22	0.22	
46										0.30	0.30	
47											0.35	
48												
49	Actual Best Commercial	1	1	1	1	1	1	1	1	1	1	
50	Predicted Best Commercial	1	1	1	1	1	6	6	6	6	6	
51												
52	Maximum Best NDR	0.57	0.57	0.57	0.57	0.57	0.57	0.57	0.57	0.57	0.57	
53	Maximum Predicted NDR	0.45	0.45	0.45	0.45	0.45	0.62	0.62	0.62	0.62	0.62	
54	Maximum Achieved NDR	0.57	0.57	0.57	0.57	0.57	0.38	0.38	0.38	0.38	0.38	
55												
56	Best RI/$	568	568	568	568	568	568	568	568	568	568	
57	Predicted RI/$	447	447	447	447	447	621	621	621	621	621	
58	Achieved RI/$	568	568	568	568	568	383	383	383	383	383	
59												
60	Best Total RI	53,915	51,077	48,240	45,402	42,565	39,727	36,889	34,052	31,214	28,376	
61	Predicted Total RI	42,452	40,218	37,983	35,749	33,515	43,502	40,395	37,287	34,180	31,073	
62	Achieved Total RI	53,915	51,077	48,240	45,402	42,565	26,787	24,874	22,960	21,047	19,134	
63		1										
64												
65	Means											
66	Best RI	45,642	56,658	59,304	59,693	58,592	56,672	54,041	50,869	47,354	43,954	
67	Predicted RI	45,886	57,893	60,964	61,319	60,258	58,420	55,704	52,418	48,691	45,181	
68	Achieved RI	48,316	56,894	58,731	58,810	57,437	55,374	52,532	49,605	46,158	42,710	
69												
70	Difference	244	1,235	1,660	1,625	1,666	1,748	1,664	1,550	1,337	1,227	
71												

Figure 8.18. M4—numerical view.

Once we have calculated the three selected NDRs, it is straightforward to calculate the resulting total retained impressions. In rows 56–58, we calculate RI/$ based on the selected NDRs. Then in rows 60–62, we calculate the Total Retained Impressions based on the RI/$. Finally, in rows 66–68, we calculate the mean values for the three measures of Total Retained Impressions.

Before we analyze the behavior of this new model, we pause for a moment to ensure that our thinking is clear. It is crucial to distinguish two sets of decisions: the day-to-day decision as to which draft ad is the best, and the overall policy decision on how many drafts to purchase. Whatever the policy decision on the number of drafts, in a specific ad campaign, the company will choose the best draft based on the predicted NDRs. Sometimes, of course, the wrong draft will be chosen because the NDR of the actual best draft is disguised by error. However, once the draft is aired, the total retained impressions actually achieved will be based on its actual NDR, not on its predicted NDR. Thus,

	A	B	C	D	E	F	G	H	I	J	K
1	Draft Commercials			M4							
2											
3											
4											
5	Parameters										
6	Budget	100									
7	Cost per Draft	5									
8	RI Multiplier	1000									
9	Mean NDR	0.5									
10	SD NDR	0.25									
11	SD NDR Error	0.1									
12											
13	Drafts	1	2	3	4	5	6	7	8	9	
14											
15	Actual NDR	=CB.Normal($B$9,$B$10)	=B15	=C15	=D15	=E15	=F15	=G15	=H15	=I15	
16			=CB.Normal($B$9,$B$10)	=C16	=D16	=E16	=F16	=G16	=H16	=I16	
17				=CB.Normal($B$9,$B$10)	=D17	=E17	=F17	=G17	=H17	=I17	
18					=CB.Normal($B$9,$B$10)	=E18	=F18	=G18	=H18	=I18	
19						=CB.Normal($B$9,$B$10)	=F19	=G19	=H19	=I19	
20							=CB.Normal($B$9,$B$10)	=G20	=H20	=I20	
21								=CB.Normal($B$9,$B$10)	=H21	=I21	
22									=CB.Normal($B$9,$B$10)	=I22	
23										=CB.Normal($B$9,$B$10)	
24											
25											
26											
27	NDR Error	=CB.Normal(0,$B$11)	=B27	=C27	=D27	=E27	=F27	=G27	=H27	=I27	
28			=CB.Normal(0,$B$11)	=C28	=D28	=E28	=F28	=G28	=H28	=I28	
29				=CB.Normal(0,$B$11)	=D29	=E29	=F29	=G29	=H29	=I29	
30					=CB.Normal(0,$B$11)	=E30	=F30	=G30	=H30	=I30	
31						=CB.Normal(0,$B$11)	=F31	=G31	=H31	=I31	
32							=CB.Normal(0,$B$11)	=G32	=H32	=I32	
33								=CB.Normal(0,$B$11)	=H33	=I33	
34									=CB.Normal(0,$B$11)	=I34	
35										=CB.Normal(0,$B$11)	
36											
37											
38	Predicted NDR	=B15+B27	=C15+C27	=D15+D27	=E15+E27	=F15+F27	=G15+G27	=H15+H27	=I15+I27	=J15+J27	
39			=C16+C28	=D16+D28	=E16+E28	=F16+F28	=G16+G28	=H16+H28	=I16+I28	=J16+J28	
40				=D17+D29	=E17+E29	=F17+F29	=G17+G29	=H17+H29	=I17+I29	=J17+J29	
41					=E18+E30	=F18+F30	=G18+G30	=H18+H30	=I18+I30	=J18+J30	
42						=F19+F31	=G19+G31	=H19+H31	=I19+I31	=J19+J31	
43							=G20+G32	=H20+H32	=I20+I32	=J20+J32	
44								=H21+H33	=I21+I33	=J21+J33	
45									=I22+I34	=J22+J34	
46										=J23+J35	
47											
48											
49	Actual Best Draft	=MATCH(MAX(B15:B24),B15:B24,0)	=MATCH(MAX(C15:C24),C15:C24,0)	=MATCH(MAX(D15:D24),D15:D24,0)	=MATCH(MAX(E15:E24),E15:E24,0)	=MATCH(MAX(F15:F24),F15:F24,0)	=MATCH(MAX(G15:G24),G15:G24,0)	=MATCH(MAX(H15:H24),H15:H24,0)	=MATCH(MAX(I15:I24),I15:I24,0)	=MATCH(MAX(J15:J24),J15:J24,0)	
50	Predicted Best Draft	=MATCH(MAX(B38:B47),B38:B47,0)	=MATCH(MAX(C38:C47),C38:C47,0)	=MATCH(MAX(D38:D47),D38:D47,0)	=MATCH(MAX(E38:E47),E38:E47,0)	=MATCH(MAX(F38:F47),F38:F47,0)	=MATCH(MAX(G38:G47),G38:G47,0)	=MATCH(MAX(H38:H47),H38:H47,0)	=MATCH(MAX(I38:I47),I38:I47,0)	=MATCH(MAX(J38:J47),J38:J47,0)	
51											
52	Maximum Best NDR	=MAX(B15:B24)	=MAX(C15:C24)	=MAX(D15:D24)	=MAX(E15:E24)	=MAX(F15:F24)	=MAX(G15:G24)	=MAX(H15:H24)	=MAX(I15:I24)	=MAX(J15:J24)	
53	Maximum Predicted NDR	=MAX(B38:B47)	=MAX(C38:C47)	=MAX(D38:D47)	=MAX(E38:E47)	=MAX(F38:F47)	=MAX(G38:G47)	=MAX(H38:H47)	=MAX(I38:I47)	=MAX(J38:J47)	
54	Maximum Achieved NDR	=INDEX(B15:B24,B50)	=INDEX(C15:C24,C50)	=INDEX(D15:D24,D50)	=INDEX(E15:E24,E50)	=INDEX(F15:F24,F50)	=INDEX(G15:G24,G50)	=INDEX(H15:H24,H50)	=INDEX(I15:I24,I50)	=INDEX(J15:J24,J50)	
55											
56	Best RI/$	=$B$8*B52	=$B$8*C52	=$B$8*D52	=$B$8*E52	=$B$8*F52	=$B$8*G52	=$B$8*H52	=$B$8*I52	=$B$8*J52	
57	Predicted RI/$	=$B$8*B53	=$B$8*C53	=$B$8*D53	=$B$8*E53	=$B$8*F53	=$B$8*G53	=$B$8*H53	=$B$8*I53	=$B$8*J53	
58	Achieved RI/$	=$B$8*B54	=$B$8*C54	=$B$8*D54	=$B$8*E54	=$B$8*F54	=$B$8*G54	=$B$8*H54	=$B$8*I54	=$B$8*J54	
59											
60	Best Total RI	=B56*($B$6-$B$7*B13)	=C56*($B$6-$B$7*C13)	=D56*($B$6-$B$7*D13)	=E56*($B$6-$B$7*E13)	=F56*($B$6-$B$7*F13)	=G56*($B$6-$B$7*G13)	=H56*($B$6-$B$7*H13)	=I56*($B$6-$B$7*I13)	=J56*($B$6-$B$7*J13)	
61	Predicted Total RI	=B57*($B$6-$B$7*B13)	=C57*($B$6-$B$7*C13)	=D57*($B$6-$B$7*D13)	=E57*($B$6-$B$7*E13)	=F57*($B$6-$B$7*F13)	=G57*($B$6-$B$7*G13)	=H57*($B$6-$B$7*H13)	=I57*($B$6-$B$7*I13)	=J57*($B$6-$B$7*J13)	
62	Achieved Total RI	=B58*($B$6-$B$7*B13)	=C58*($B$8-$B$7*C13)	=D58*($B$6-$B$7*D13)	=E58*($B$6-$B$7*E13)	=F58*($B$6-$B$7*F13)	=G58*($B$6-$B$7*G13)	=H58*($B$6-$B$7*H13)	=I58*($B$6-$B$7*I13)	=J58*($B$6-$B$7*J13)	
63		=IF(B60=B62,1,0)									
64											
65	Means										
66	Best RI	=CB.GetForeStatFN(B60,2)	=CB.GetForeStatFN(C60,2)	=CB.GetForeStatFN(D60,2)	=CB.GetForeStatFN(E60,2)	=CB.GetForeStatFN(F60,2)	=CB.GetForeStatFN(G60,2)	=CB.GetForeStatFN(H60,2)	=CB.GetForeStatFN(I60,2)	=CB.GetForeStatFN(J60,2)	
67	Predicted RI	=CB.GetForeStatFN(B61,2)	=CB.GetForeStatFN(C61,2)	=CB.GetForeStatFN(D61,2)	=CB.GetForeStatFN(E61,2)	=CB.GetForeStatFN(F61,2)	=CB.GetForeStatFN(G61,2)	=CB.GetForeStatFN(H61,2)	=CB.GetForeStatFN(I61,2)	=CB.GetForeStatFN(J61,2)	
68	Achieved RI	=CB.GetForeStatFN(B62,2)	=CB.GetForeStatFN(C62,2)	=CB.GetForeStatFN(D62,2)	=CB.GetForeStatFN(E62,2)	=CB.GetForeStatFN(F62,2)	=CB.GetForeStatFN(G62,2)	=CB.GetForeStatFN(H62,2)	=CB.GetForeStatFN(I62,2)	=CB.GetForeStatFN(J62,2)	
69											
70	Difference	=B67-B66	=C67-C66	=D67-D66	=E67-E66	=F67-F66	=G67-G66	=H67-H66	=I67-I66	=J67-J66	
71											

Figure 8.19. M4—relationship view.

when the company comes to making the policy decision on how many drafts to purchase, it must optimize not over predicted total impressions but actual total impressions.

The first question we ask with this model is whether multiple drafts can be better than a single draft, as we have found with simpler models. Figure 8.20 shows that with our base case model, in which the standard deviation of the error in NDRs is 0.10, the answer is yes: The mean value for Achieved Retained Impressions for four drafts is around 60,000, whereas a single draft has a mean of only about 48,000.

Insight 11: The optimal number of draft ads can be larger than one, even with error in measuring next day recall.

We observed earlier that error in our observation of NDRs sometimes leads us to choose the wrong draft. When we choose the wrong draft, we give up the retained impressions we would have achieved had we been able to determine the truly optimal draft. How significant is this suboptimization effect? Figure 8.21 shows a comparison between the Best Retained Impressions and Achieved Retained Impressions. Recall that "best" reflects the outcomes for an omniscient observer, whereas "achieved" reflects the outcomes that can be achieved given the error in NDR. It is clear in this chart that the best outcomes are, as expected, somewhat higher than what we can achieve, and this effect grows as the number of drafts increases. Note, however, that the optimal number of drafts is four in both cases. In other words, an omniscient observer and a real one would both choose four draft ads as the optimal overall policy.

Insight 12: When NDR is subject to error, actual results will be worse than the ideal but the optimal number of drafts may not differ from the true optimum.

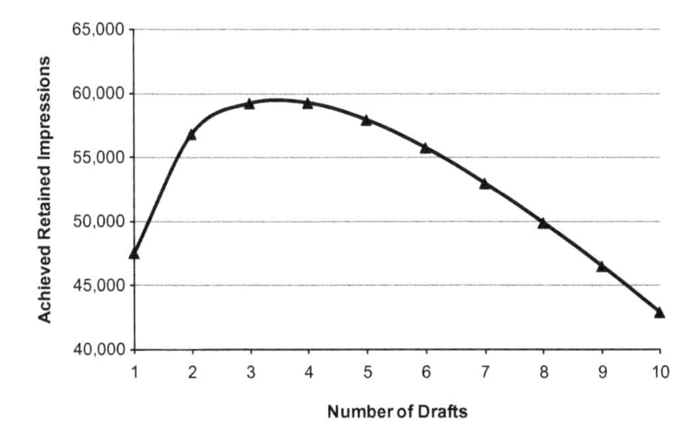

Figure 8.20. Achieved Retained Impressions (M4, σ error = 0.10).

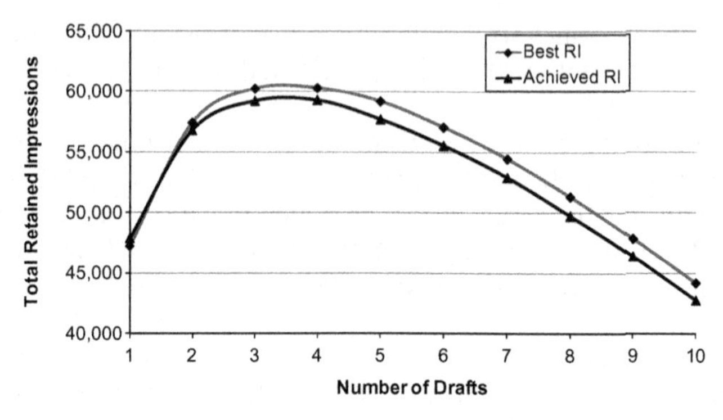

Figure 8.21. Best and Achieved Retained Impressions (M4, σ error = 0.10).

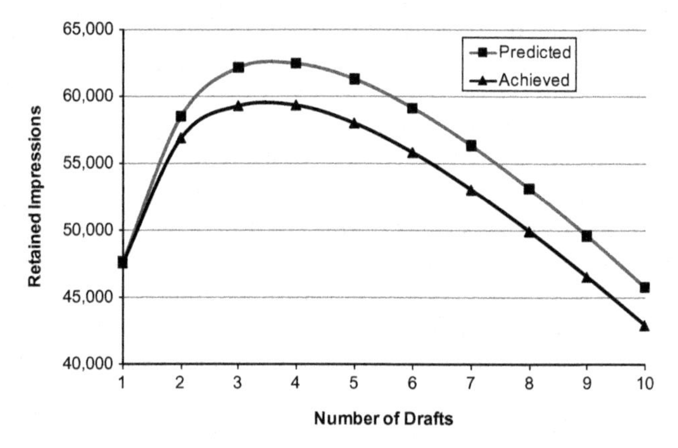

Figure 8.22. Predicted and Achieved Retained Impressions (M4, σ error = 0.10).

Although we choose the optimal number of drafts based on the achieved retained impressions, for day-to-day operations, we can only project results based on predicted retained impressions. Figure 8.22 shows the magnitude of the gap between these two results. Except when we use only one draft, predicted retained impressions exceeds actual retained impressions. The reasons for this difference are complex, but one driving force is a systematic bias created by the process of selecting the best from among predicted NDRs. In general, a particular draft having a high predicted NDR is likely to have both a high actual NDR and a positive error. It is the tendency for the ads we choose to have high positive errors that causes predicted retained impressions to exceed actual. This difference could have important implications for the implementation of a multiple-draft policy. If we systematically predict higher retained impressions than actually come about, we may also systematically predict higher than actual sales. This could lead to a loss of credibility as well as to excessive costs from over-production.

Insight 13: When NDR is subject to error, the predicted performance of an ad will be higher than the actual performance. This difference could cause problems in implementing a multiple-draft policy.

Our analysis up to this point has been based on the assumption that the error in NDR is fairly small. It remains to be seen whether these conclusions hold up when the error in estimating NDR is large. To test this, we raise the standard deviation of the error distribution to 0.50. This amount of error is large, given that the mean of the NDR itself is 0.50. We might anticipate in this case that the optimal number of drafts cannot be identified and that the entire idea of using multiple drafts is impractical.

Figure 8.23 shows the results for Best, Predicted, and Actual Retained Impressions for this case. The curve for Best Retained Impressions has not changed, as we would expect since it is not influenced by errors. However, increasing the variance of errors shifts the Predicted curve up and the Actual curve down. Not surprisingly, with high errors, the gap in retained impressions grows between what an omniscient observer and a real one can achieve. Likewise, the gap between predicted and actual grows, making the operational problems we discussed before more severe. Finally, we observe in this case that the optimal number of drafts is two, so even with substantial errors, the optimal policy may be to purchase multiple drafts.

One additional aspect of these results is worth pursuing here. Recall that the optimal number of drafts was four in the low error case. In the high error case, the optimal number is two. Apparently, the optimal number of drafts declines as the variance in errors increases, which suggests the possibility that the optimal number of drafts might actually be one if the errors are sufficiently large. We test this conjecture by raising the standard deviation of the error distribution to an extreme value. Sure enough, the optimal number of drafts is one for standard deviations somewhat above 1.0. So extreme degrees of uncertainty can eliminate the value of multiple drafts. We report this not because we think

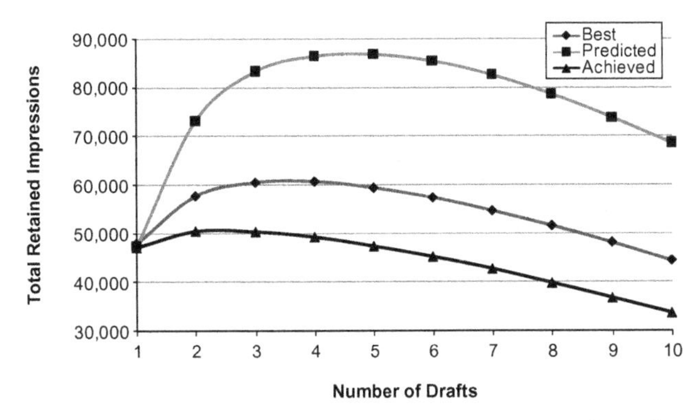

Figure 8.23. Best, Predicted, and Achieved Retained Impressions (M4, σ error = 0.50).

it is a realistic possibility, but to make the point that exploring the behavior of a model with extreme inputs often solidifies our understanding. In this case, the parameter values required to produce the result are so extreme that we feel more confident, not less, in recommending multiple drafts.

Insight 14: Only with extreme errors in evaluating NDRs is a single draft ad preferable to multiple drafts.

Summary of M4

Our latest model has taken us deeper into the problem than we might have anticipated at the start. Our first model, which was deterministic and assumed proportionality in all the key relationships, gave us the important result that multiple drafts could be an optimal strategy. We then fleshed out the analysis, first introducing diminishing returns by assumption (M2), then deriving it from more basic considerations (M3), and finally considering how errors in the process might affect outcomes (M4). This last addition to the model showed that multiple drafts can still be beneficial, even when we cannot measure NDR accurately. However, we will be frequently disappointed in the resulting retained impressions our ad creates because the process tends to overestimate the quality of the ads we air.

8.8 PRESENTATION OF RESULTS

Our client asked us to examine the feasibility of a new strategy for advertising. Since few data were available and the company had no experience with this policy, the approach we took was to build models of the process of ad creation and airing, and use those models to learn under what circumstances the new approach would be preferable. We built a succession of models that reveal several insights about why multiple ads may be advantageous. Now we want to present our results to our clients in a manner that helps them understand what we have done and that motivates them to act.

The first question to ask before we design our presentation is what can we reasonably assume about our audience? We can assume our clients are professionals with experience in advertising. They likely know how the current process works and are familiar with the terms "next day recall" and "retained impressions." However, we assume they do not have experience with the alternative process we have studied, and we also assume they are not experienced modelers. They come to the meeting wanting to know whether this new approach has promise. Our job is to help them understand when it has value, and why.

We normally take the direct approach in business presentations; we present our conclusions or recommendations at the start and then lay out the analysis that leads to those results. But in this case there is no single decision on which to focus our presentation. Rather, we have explored a generic problem and our results are more conceptual than quantitative. Our results do show that multiple ads can be optimal, but we want to focus our presentation on when this is the case and why. Thus, we mention this result at the start of the presentation, but devote the majority of it to discussing our approach and our insights.

What is the *story* we want to tell here? The most important discovery we have made is that, under certain circumstances, multiple drafts can deliver a greater number of total retained impressions than a single draft. This answers the basic question our client asked. But we have learned more than that. We have some insight into why multiple drafts can be better, and how various assumptions influence the choice of the number of drafts. Certain of our results, for example that the optimal number of drafts goes up with the uncertainty in ad quality, may be of particular interest because they are counterintuitive.

Our story divides naturally into three parts:

- What problem did we address?
- What approach did we take?
- What have we learned?

Refining this outline just a bit, we start by describing the problem (are multiple drafts attractive?), then describe our approach (modeling) and explain the major assumptions we made (for example, our objective is to maximize total retained impressions), and finally describe some of our most interesting insights.

This raises the question of which insights to present? In our analysis, we highlighted 14 distinct insights, which is many more than we can discuss in a short presentation. Our audience will remember 3 to 5 well-articulated points, but not 10 or 15. Moreover, some of our insights are essential and others are more narrow or technical.

We settle on five major points:

1. The quality of the best draft and total cost of drafts both increase with the number of drafts.
2. Multiple drafts can deliver more total retained impressions than a single draft (Insight 1).
3. The optimal number of drafts decreases with cost (Insight 2).
4. The optimal number of drafts increases with variability in quality (Insight 9).
5. Multiple drafts can be optimal even with measurement error (Insights 11–14).

We choose this set of results on the basis of several criteria. First, we want to start with insights that are easy to understand and conform to intuition, so our clients will gain confidence in our approach. Thus, the first point, although somewhat obvious, describes the fundamental trade-off that governs this decision. Second, we want to include some insights that answer the clients' questions directly, such as point 2. Finally, we want to include some surprises, or insights, that our clients could not have anticipated, such as points 4 and 5.

It is now time to storyboard the presentation (that is, to lay out the slides in more detail). One way to do this is to write down the titles of the slides, but not to include any content. Here is our list of 12 slide titles:

1. Overview
2. Results
3. Current and Proposed Policies
4. Modeling Assumptions
5. Structure of Model
6. Quality and Cost Increase with Number of Drafts
7. Multiple Drafts Can Be Optimal
8. Optimal Number of Drafts Decreases with Cost
9. Maximum NDR Increases with Variability in Quality
10. Optimal Number of Drafts Increases with Variability in Quality
11. Multiple Drafts Can Be Optimal even with Measurement Error
12. Summary

Our first slide, Figure 8.24, gives an overview of the presentation. It briefly describes the current and the proposed approaches and explains how we addressed the problem. The purpose of the second slide, Figure 8.25, is to give the audience a preview of our results so they have a sense of the where the presentation is headed. The third slide, Figure 8.26, contrasts the current and proposed approaches to advertising in sufficient depth for the audience to understand the problem. The next slide, Figure 8.27, lists the fundamental assumptions behind our analysis and gives a sense of how we abstracted the real problem into a model. The last of the introductory slides, Figure 8.28, shows the basic influence diagram for the problem. This allows us to get our audience thinking about the fundamental forces at play and to introduce the terms (such as NDR and RI/$) we will be using later. It also allows us to show how we use logic to decompose a seemingly intractable problem into pieces that we can model and recombine for insight.

We now get to the heart of the presentation. It is what we have learned that the audience is really interested in and that will convince them our approach is fruitful. We have selected five points from among our insights to convey what we have learned. We illustrate each of these points with a simple graph. When we present each slide, we will give more details about the model that gave rise to it and about the assumptions on which it is based. But we want our audience to leave with a clear picture in their heads not of the details of our models but of their power to lead to crisp and useful insights.

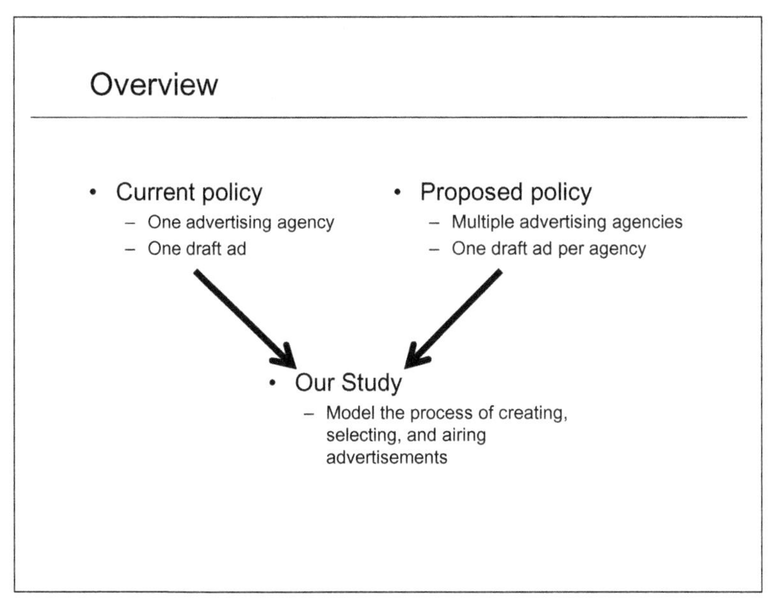

Figure 8.24. Slide 1.

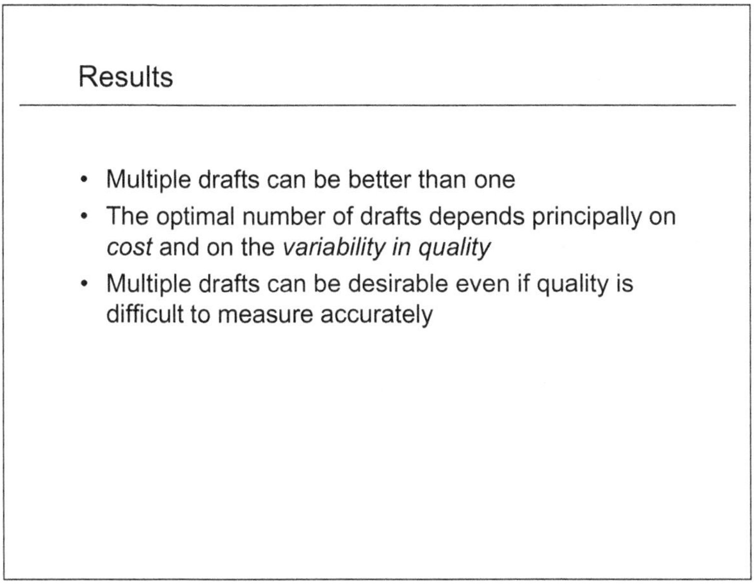

Results

- Multiple drafts can be better than one
- The optimal number of drafts depends principally on *cost* and on the *variability in quality*
- Multiple drafts can be desirable even if quality is difficult to measure accurately

Figure 8.25. Slide 2.

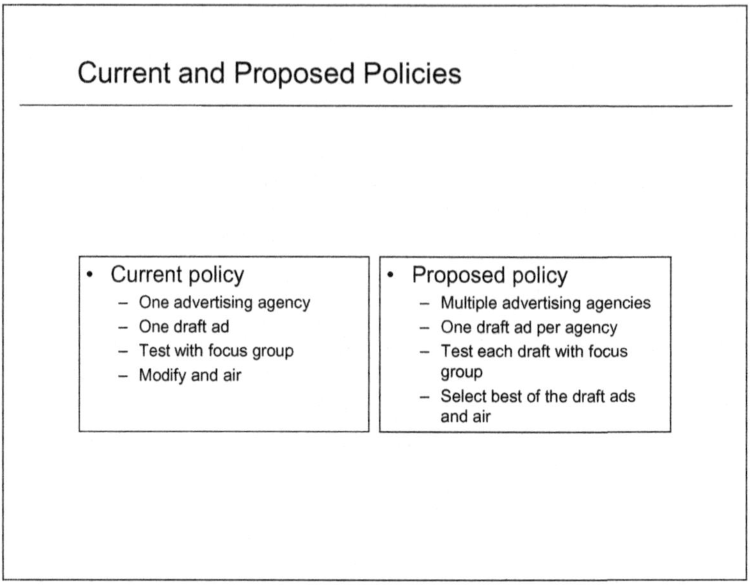

Current and Proposed Policies

- Current policy
  - One advertising agency
  - One draft ad
  - Test with focus group
  - Modify and air

- Proposed policy
  - Multiple advertising agencies
  - One draft ad per agency
  - Test each draft with focus group
  - Select best of the draft ads and air

Figure 8.26. Slide 3.

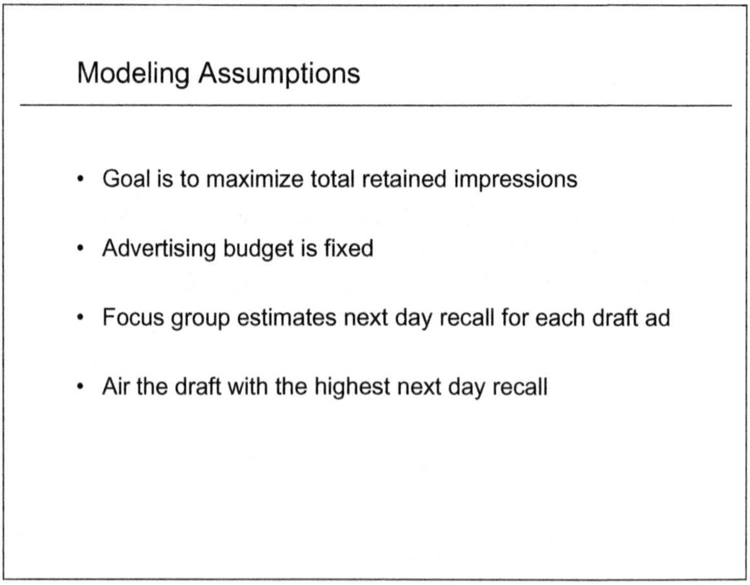

Modeling Assumptions

- Goal is to maximize total retained impressions
- Advertising budget is fixed
- Focus group estimates next day recall for each draft ad
- Air the draft with the highest next day recall

Figure 8.27. Slide 4.

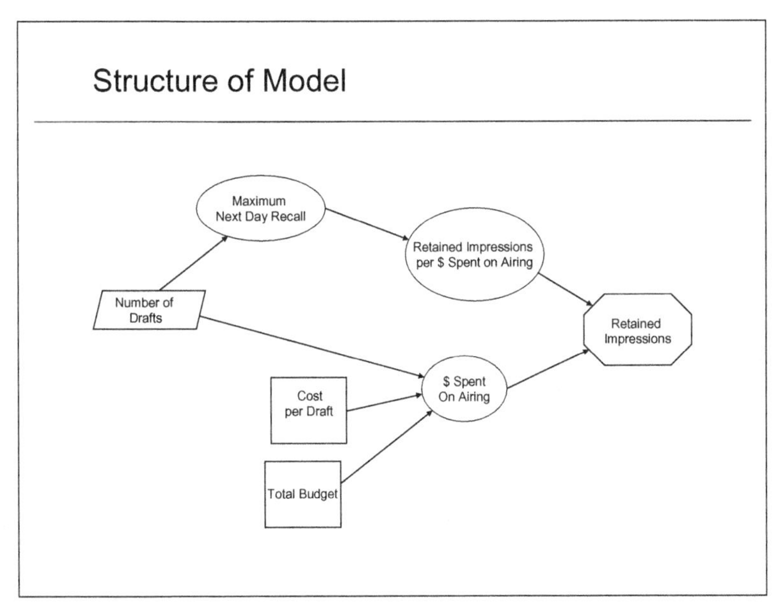

Figure 8.28. Slide 5.

The six slides that convey our insights (Figures 8.29– 8.34) tell our story in a natural progression. Figure 8.29 makes the fundamental point that as the number of draft ads purchased goes up, both the total cost of drafts and the quality of the best ad go up. The first point is obvious: The total cost is linear in the number of drafts. The second point is not obvious, but our models have shown how variability in ad quality causes the quality of the best ad to increase at a decreasing rate. So at this point we can discuss the intuition behind this all-important result.

The results shown in the next slide, Figure 8.30, follow naturally from the previous slide. As additional drafts are purchased, costs increase at a constant rate. Benefits, however, as measured by the quality of the best draft, show first increasing then diminishing returns. Since the quality of the best draft is directly related to total retained impressions, it follows that the optimal number of drafts will be greater than one and less than the maximum affordable number. This result answers our client's fundamental question.

At this point we turn to the results of sensitivity analyses to provide deeper insights into the choice of the number of drafts. Figure 8.31 shows that the optimal number of drafts declines (at a decreasing rate) with the cost of a single draft. This is obvious to a degree, but showing this result helps our client gain confidence in our approach. This slide also allows us to make the point that multiple ads are not *always* better than one, which is an important caveat.

The next two slides (Figures 8.32 and 8.33) bring the audience deeper into our analysis. At this point in the presentation, we remind the audience that an

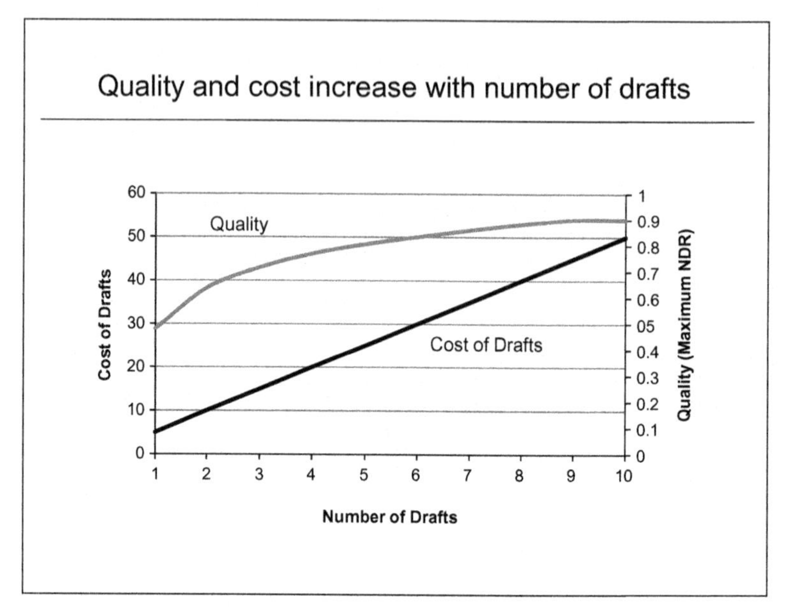

Figure 8.29. Slide 6.

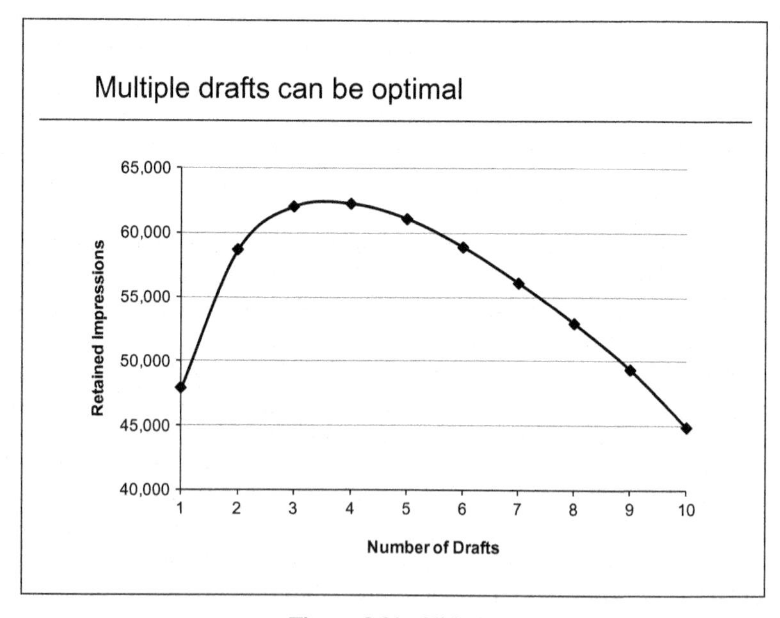

Figure 8.30. Slide 7.

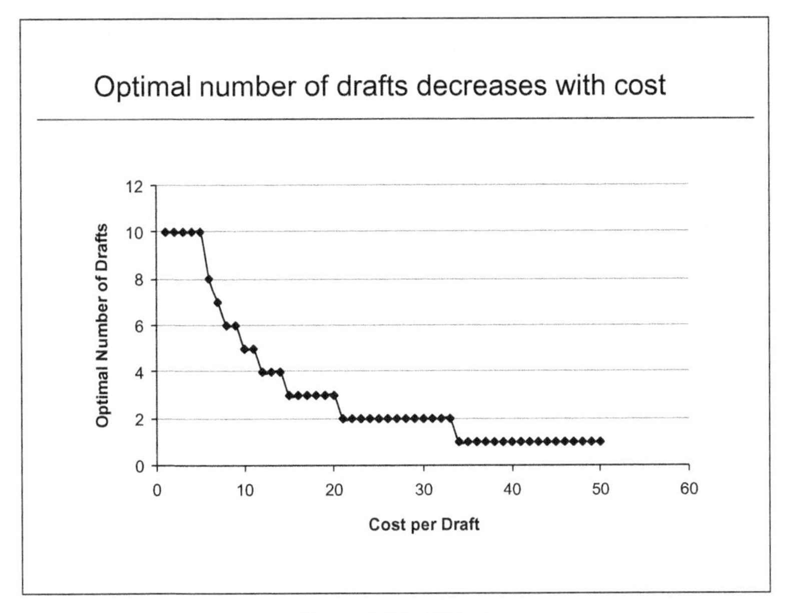

Figure 8.31. Slide 8.

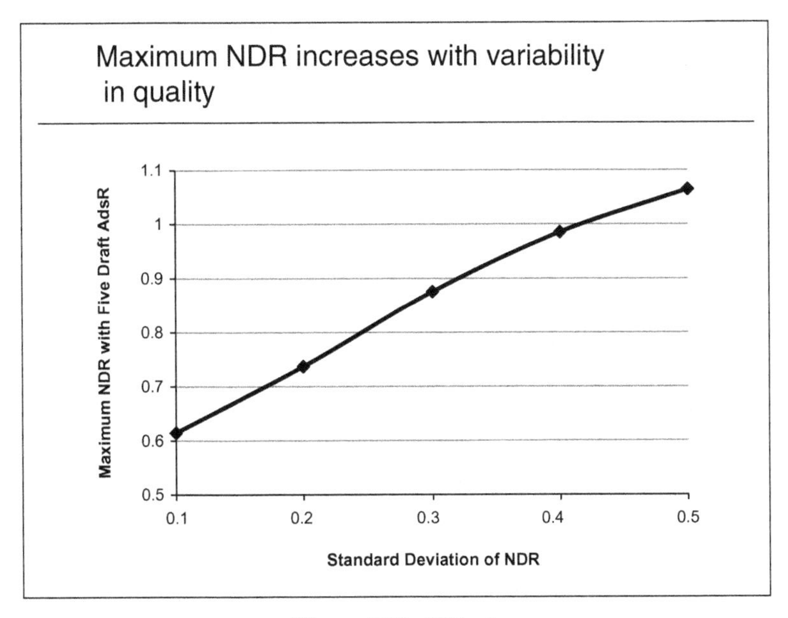

Figure 8.32. Slide 9.

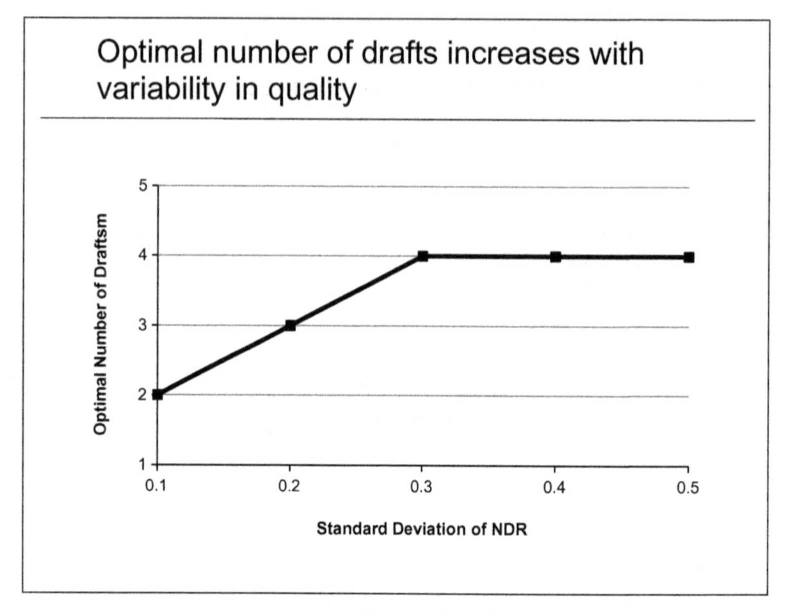

Figure 8.33. Slide 10.

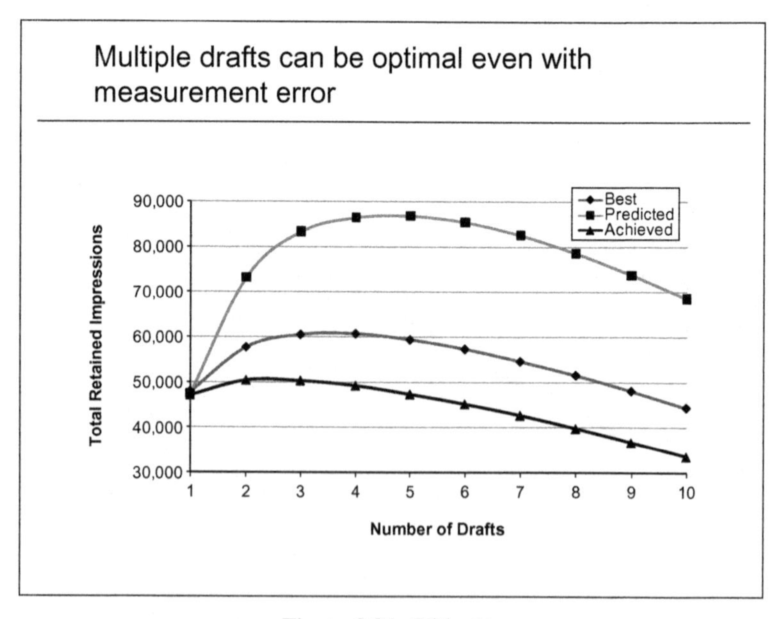

Figure 8.34. Slide 11.

individual ad's quality is unpredictable, and when we purchase multiple ads, we are in effect taking the best result from random samples. The unexpected implication is that we can actually benefit from this unpredictability. We make this point by first showing in Figure 8.32 that the maximum NDR of a fixed number of drafts (five in this case) increases with variability. It then follows (Figure 8.33) that the optimal number of ads increases with variability as well. It is also true, and we will mention it to the audience, that the ultimate number of total retained impressions also increases.

The last of the result slides (Figure 8.34) adds one more insight to the story. Here we introduce the possibility that draft quality cannot be observed precisely. Although, under this assumption, we cannot predict exactly how effective each ad will be, our results show that purchasing multiple ads can still be beneficial. We make this point by showing the actual number of retained impressions we would achieve, the best we could achieve without error, and what we would predict. The key points to take away from this graph are that multiple drafts are optimal, even though we do not achieve the best possible results and our predictions are systematically high.

The last slide (Figure 8.35) summarizes what has gone before. Here we make three points:

- Multiple drafts can be preferable to one.
- Costs and variability affect the choice of the number of drafts in a predictable way.

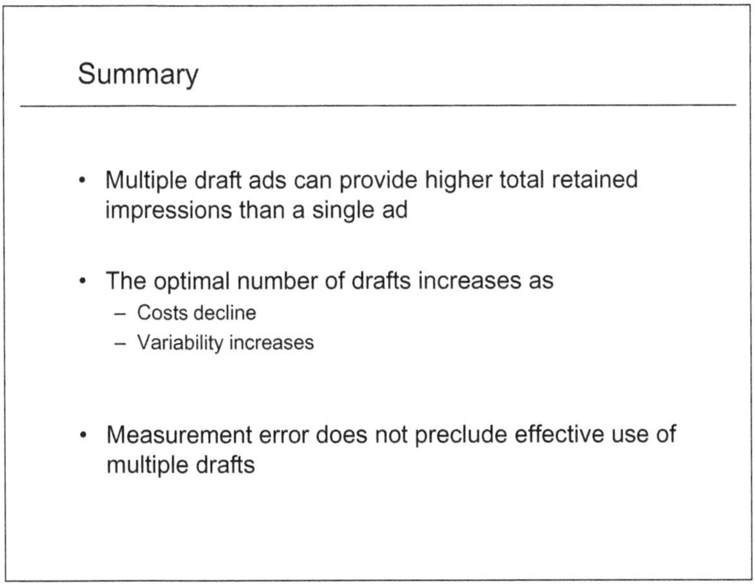

Figure 8.35. Slide 12.

- Multiple drafts can still be preferable even if we cannot measure quality perfectly.

Our goal with this summary is both to remind the audience of the main points of the presentation and to leave them with a sense that our analysis has been productive and has lead to insights they could not have gotten any other way.

The challenge in this presentation is to bring the audience along with us as we move from simple and obvious points (for example, whatever is not spent on draft ads can be spent on airing) to the more complex and subtle (the higher the uncertainty in ad quality, the better the results of purchasing multiple ads). If we do our job well, each result will seem insightful to the audience, and our listeners will leave the presentation with a clear sense that the proposed policy is worth considering seriously.

8.9 SUMMARY

This case challenges us to evaluate whether a standard business practice, purchasing a single draft advertisement, could be improved by purchasing multiple drafts. The question is not whether a particular advertising campaign should be conducted in this way, but whether there is a case for changing the entire business practice of advertising. This question is, in some ways, easier because it allows us to address the situation hypothetically and to introduce parameters where needed instead of tying our analysis to specific data on a specific case.

Our analysis proceeded from the simplest model that addressed just the basic questions to one that is quite complex but capable of shedding light on some subtle issues. The process of iterative modeling was essential, both in keeping the models simple and in establishing clear findings along the way. Each new model variant built on the previous one, adding in just one additional feature. In the same way, the insights we gained from each model built on those derived from a simpler model.

This case also illustrates the importance of making effective assumptions. All assumptions are not equal. Gross or arbitrary assumptions form a weaker foundation for modeling than assumptions that are clearly tied to first principles or fundamental behaviors. In this case, we first simply assumed diminishing returns but eventually derived it from more fundamental assumptions. The latter approach was stronger because it was more solidly grounded and because it gave deeper insights.

CHAPTER 9

New England College Skiway

9.0 INTRODUCTION

Many modeling and management science tools were first developed for use in operational settings, such as production lines or oil refinery operations. In recent years, these tools have become increasingly important in the service and financial sectors. The case we address in this chapter, which concerns the operation of a small ski area, has some elements from each of these domains.

Operational problems in the service sector share many common features. Whether it is an emergency room, a retail bank, or a ski area, service-sector operations typically serve human customers, whose arrival and consumption of resources are random. The quality of the service they receive is an important issue, sometimes more important than the profits the facility generates. Furthermore, *average* quality may not be an appropriate measure of customer satisfaction since every customer is important, and the customers who receive the worst service are probably going to complain the loudest.

These problems share another feature: a high degree of complexity. One source of complexity is randomness in arrival rates, service times, quality, and so on. Another source of complexity is the interaction between customers and resources. For example, a patient in an emergency room may draw on one or more of hundreds of hospital resources, often unpredictably, with little lead time and with serious consequences for any shortfalls in service quality. A ski area is not quite as complex, and the consequences of poor service may only be long lines and disgruntled skiers, but the operations of such facilities can also be challenging to model.

9.1 NEW ENGLAND COLLEGE SKIWAY CASE

New England College Skiway

Brian Shaw, manager of the New England College Skiway, has a problem. The lines of skiers waiting to use the chairlift at the Skiway are getting so long on weekends that customers are complaining of waits of 15 minutes or more.

Modeling for Insight: A Master Class for Business Analysts, by Stephen G. Powell & Robert J. Batt

The Skiway is served by a chairlift 3,540 feet long, with double chairs (each chair carries two people) spaced every 60 feet on the cable. The lift currently travels at a top speed of 450 feet/minute, depositing skier pairs at the top every eight seconds.

Once skiers arrive at the top of the lift, they dismount onto a relatively flat area of about an acre, a space that allows about 25 people to congregate comfortably before severe crowding occurs. All trails begin by funneling through a narrow chute, approximately 150 feet long, through a cliff band. The chute is only wide enough to accommodate two skiers side by side. Below it, three trails branch off and descend roughly parallel paths through the trees to the bottom. Each trail can accommodate at most three skiers side by side. The longest trail down the mountain is roughly one mile long, whereas the shortest is 4,000 feet, but the turns skiers take on these trails lengthen their actual paths by 25–50 percent. Skiers typically take four- to eight-minute trips down the mountain. The majority of the skiers at the Skiway are intermediate skiers who feel safe with a minimum distance of 75 feet between skiers (expert skiers can ski as close as 40 feet apart). At the foot of the mountain is a large flat area (partially occupied by ski schools and other facilities), across which skiers pass to rejoin the lift line.

The Skiway currently charges $40 for an all-day lift ticket. Brian is convinced the long lines on weekends are discouraging more people from using the Skiway, and he is considering three possible changes:

1. Speed up the chairlift to 600 feet/minute.
2. Replace the existing two-person lift with a quad lift (each chair holding four skiers).
3. Create a ski trail on the other side of the existing lift, separate from the current slope. This trail would be served by the existing lift but would not require skiers to pass through the chute.

Speeding up the existing lift would have minimal cost. Putting in a new quad lift would cost approximately $725,000. Creating another trail would cost $1.8 M.

Brian would appreciate your recommendations as to how he could improve the operations of his ski area.

To the Reader: Before reading on, frame this problem by setting boundaries for the analysis; making assumptions; listing questions; defining outputs, inputs, and decisions; and so on.

9.2 FRAME THE PROBLEM

As we begin work on this problem, we find that it is not the problem description that makes the case challenging, but our own personal experience that complicates things. Anyone who has been to a ski area recognizes how complex a task it is to keep track of the hundreds of people on the mountain. There are many sources of randomness: people's levels of expertise, when and where they stop, the configurations of their paths, and so on. Lifts often slow down temporarily and sometimes stop. There is a huge influx of skiers in the morning and an outflow in the afternoon, with lesser fluctuations in arrivals and departures throughout the day. How could we possibly capture all of this complex behavior in a model?

When confronted with this case, many people are inclined to model each individual skier as he or she goes through the system. Such an approach is possible and often used in problems of this type. The appropriate technique is known as Discrete Event Simulation (DES). Unfortunately, it is very difficult to execute in a spreadsheet. A spreadsheet model would require hundreds or thousands of rows in order to track the activities of each skier. DES is best done with specialized software that allows users to model the behavior of each element of the system (the lift, the wait line, the ski trails, etc.) and the arrivals to and departures from the system throughout the day. Although DES could be an appropriate way to attack this problem, the software is expensive, mastery of it takes many months, and clients might not trust software they do not know how to use. Spreadsheet software suffers from none of these drawbacks, so it will be far more practical to model this problem in a spreadsheet.

Another common reaction to this problem is that we cannot begin to model until we know much more about our customers. In particular, we have no information about the elasticity of demand relative to wait time or price. How many more people would come if the lift line was shorter? Although data on demand might be useful at some point, this information is not *necessary* to build a model and to gain insights about what affects wait time. It certainly would be a waste of time and money to collect data on our customers before we even know what changes we can offer them.

The essential features of the problem are quite simple. We have three options for improving the ski area, which we define here for future reference as follows:

1. *SpeedUp:* speed up the lift.
2. *Quad:* replace the double lift with a quad lift.
3. *Trail:* build another trail.

We could also recommend any combination of these three improvements. Each of these options could conceivably improve customer wait times and thereby improve customer satisfaction. Some of the options cost money, so unless the college wants to run the ski area at a loss, it will have to try to recoup the capital expenses by selling additional tickets. How the Skiway can do this and still keep its customers satisfied is the essential challenge before us.

9.2.1 Problem Kernel

The motivation behind identifying the problem kernel is that it helps us cut through the situation's complexity to see the real problem. Here is one possible statement of the kernel:

> Wait times at the ski area exceed 15 minutes during peak periods. Can any of the proposed changes to the system reduce the wait time and be financially feasible?

The kernel focuses our attention on two aspects of the system: customer wait times and financial results. Even at this early stage, we anticipate that much of our analysis will concern the trade-offs between these two outcome measures.

The kernel also draws attention to the fact that it is only periods of peak load that interest us. At this point, we are not concerned with the ramp up and ramp down at the beginning and end of the day. We want to focus on how the system behaves under its heaviest load, which we assume occurs during midday (roughly 11:00 a.m. to 1:00 p.m.).

Another important issue to consider as we frame the problem is whether to model the system in equilibrium or disequilibrium. In equilibrium, a system's *average* behavior is constant. In disequilibrium, the opposite is true. Equilibrium at the ski area means that we assume the number of skiers in the lift line, on the slope, and so on is constant. Thus, the total number of skiers in the system is also constant.

One of the advantages of modeling the system in equilibrium is that many of its aspects can be expressed in simple mathematical relationships. For example, the total number of skiers on the lift, in line, and on the mountain must equal the total number of skiers in the system. Likewise, the rate at which skiers enter any part of the system must equal the rate at which they exit that part. For example, every eight seconds, two skiers get on the lift at the bottom and two skiers get off the lift at the top.

9.2.2 Boundaries and Assumptions

We begin with some simplifying assumptions:

1. There are only three locations where a skier can be: waiting in line for the lift, riding the lift, or skiing down the mountain. We ignore all other areas of the Skiway.
2. The system is in equilibrium. The number of skiers in each location is constant, as is the wait time in the queue.
3. Capacity constraints are not binding. In particular, this means that for the moment, we ignore the details of how skiers behave when they get off the lift and ski through the chute to reach the slopes.
4. We assume the lift is always full and runs at a constant speed. Every seat is filled, and the lift never has to slow down or stop for inexperienced skiers or people who fall. We have to be particularly careful about this assumption because if our model ever predicts a queue length of zero, the lift obviously would not be full, and our calculations would break down.

For the mathematically inclined modeler, writing out equilibrium equations may be a fine first step in developing a model. In fact, that is how we started. However, we confess that we quickly found ourselves awash in a sea of symbols and equations, with little chance of gaining insight into the process. At this point, we remembered how useful an influence diagram is to sort out the system relationships.

To the Reader: Before reading on, sketch an influence diagram for this problem.

9.3 DIAGRAM THE PROBLEM

Figure 9.1 shows an initial influence diagram for the Skiway in equilibrium. Our goal with this diagram is to uncover the connections between the decisions we face (lift speed, number of seats, etc.) and an appropriate measure of system performance. In this first diagram, we take the time skiers wait in line, Queue Time, as our outcome measure. We realize that eventually we will need to calculate the financial outcomes from our investments, but for now, we concentrate on the system's physical performance.

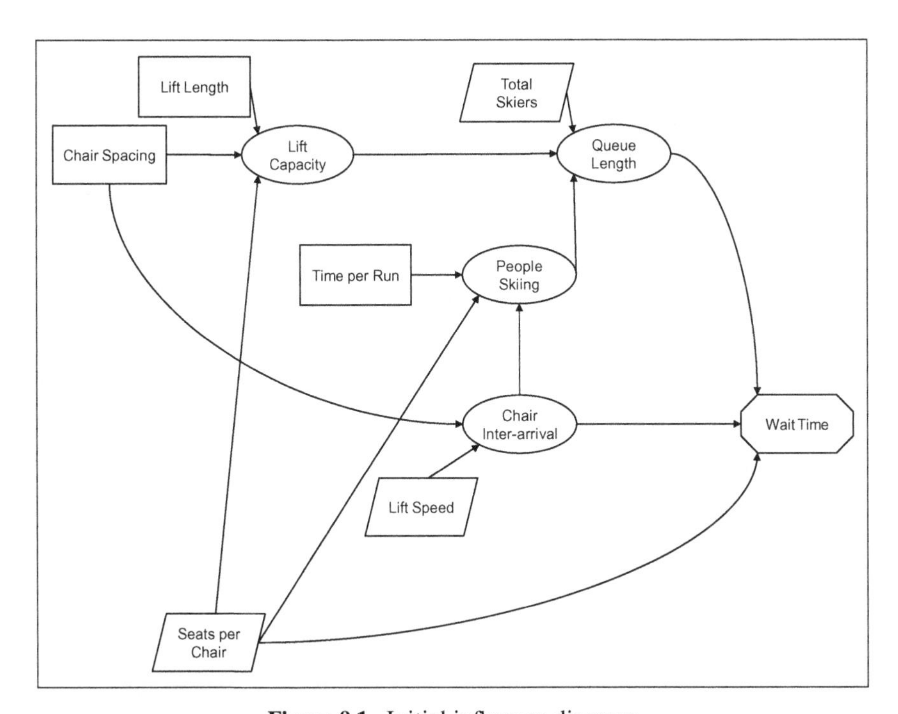

Figure 9.1. Initial influence diagram.

Queue Time is a function of three things: Queue Length, Chair Inter-arrival Time, and Seats per Chair. For example, if the lift has double chairs, the line has 60 people, and chairs arrive every minute, the time each pair of skiers waits in line is 30 minutes. This is because it takes 30 minutes for the lift to remove all 60 people from the line. In other words:

$$\text{Queue Time} = \text{Queue Length} \times \text{Chair Inter-arrival Time} \div \text{Seats per Chair}$$

Queue Length, which is the number of people in line, is simply the Total Skiers in the system less Lift Capacity (assuming the lift is always full) less People Skiing on the slopes.

$$\text{Queue Length} = \text{Total Skiers} - \text{Lift Capacity} - \text{People Skiing}$$

The Chair Inter-arrival Time is governed by the Lift Speed and the Chair Spacing. For example, if the lift travels at 100 feet/minute and the chairs are 50 feet apart, a chair arrives every 0.5 minute. Thus, in general:

$$\text{Chair Inter-arrival Time (min)} = \text{Chair Spacing (ft)}/\text{Lift Speed (ft/min)}$$

Lift Capacity, the number of people on the lift when it is full, is governed by three factors: Seats per Chair, Chair Spacing, and Lift Length. In general:

$$\text{Lift Capacity} = \text{Seats per Chair} \times \text{Lift Length} \div \text{Chair Spacing}$$

Finally, the influence diagram shows that People Skiing on the slopes is influenced by the Chair Inter-arrival Time, Seats per Chair, and Ski Time. Ski Time is the time from the moment a skier gets off the lift until the moment the skier joins the lift line again. Trail length, the distance between skiers, and skiing speed are not explicitly part of this calculation. These parameters are reflected in the Ski Time parameter. If two skiers arrive every 10 seconds and take 1,000 seconds to ski to the bottom, there must be 200 people on the slope. In general:

$$\text{People Skiing} = \text{Seats per Chair} \times \text{Ski Time} \div \text{Chair Inter-arrival Time.}$$

Note that Total Skiers is a parameter here, not a calculation. In fact, in later iterations of the model, it will be a decision variable. By making Total Skiers a parameter and Queue Time the output, we are effectively using the model to predict the time skiers wait in the queue under a given skier load.

To the Reader: Before reading on, sketch and build an initial model based on the influence diagram. Use the model to generate as many interesting insights into this situation as you can.

9.4 M1 MODEL AND ANALYSIS

M1 is designed to match the influence diagram closely. Figure 9.2 shows M1 in numerical format, and Figure 9.3 shows the relationship view.

There are six parameters in the model:

- Total Skiers
- Ski Time
- Lift Length
- Lift Speed
- Chair Spacing
- Seats per Chair

In cell C19, we calculate the number of chairs going up the mountain at any time by dividing the length of the lift by the chair spacing. We use the FLOOR

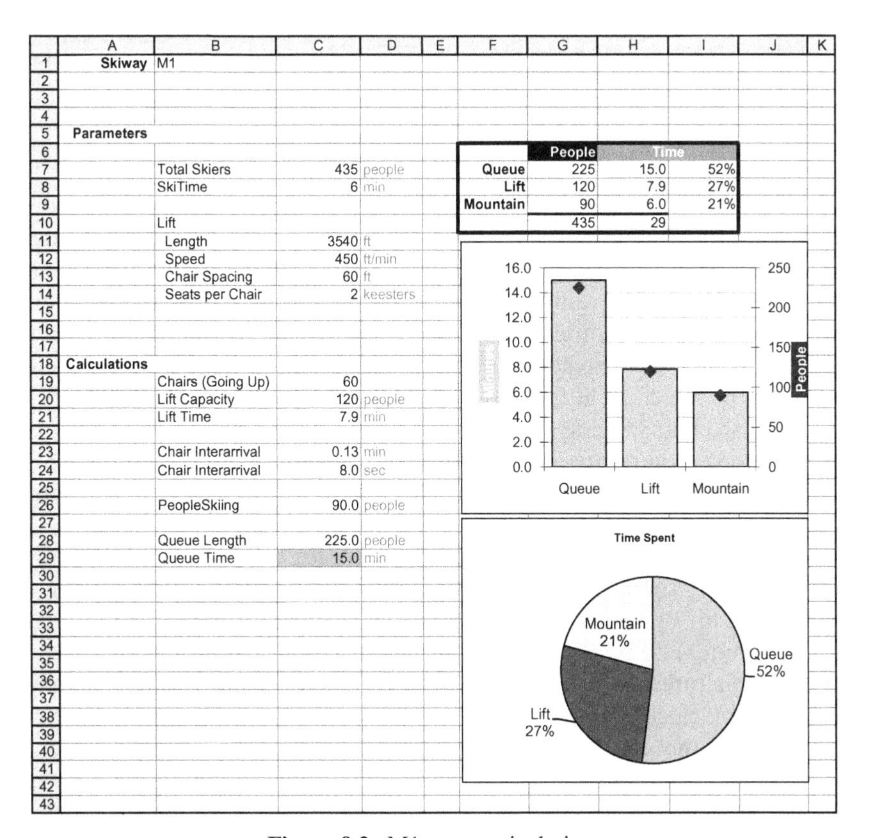

Figure 9.2. M1—numerical view.

	A	B	C	D	E
1	**Skiway**	M1			
2					
3					
4					
5	**Parameters**				
6					
7		Total Skiers	435	people	
8		SkiTime	6	min	
9					
10		Lift			
11		Length	3540	ft	
12		Speed	450	ft/min	
13		Chair Spacing	60	ft	
14		Seats per Chair	2	keesters	
15					
16					
17					
18	**Calculations**				
19		Chairs (Going Up)	=FLOOR(C11/C13,1)+1		
20		Lift Capacity	=C19*C14	people	
21		Lift Time	=C11/C12	min	
22					
23		Chair Interarrival	=C13/C12	min	
24		Chair Interarrival	=60*(C13/C12)	sec	
25					
26		PeopleSkiing	=C8*C14/C23	people	
27					
28		Queue Length	=C7-C20-C26	people	
29		Queue Time	=C28*C23/C14	min	
30					

Figure 9.3. M1—relationship view.

function here to round the calculation down to an integer number of chairs. This precaution is not strictly necessary with the given parameter values, since Chair Spacing divides evenly into Lift Length. However, if these parameters were to change at some point, the FLOOR function would maintain an integer number of chairs. We also add one chair to the result of the FLOOR function to account for the chair at the end of the lift. For example, if the lift were 100 feet long with chair spacing of 50 feet, dividing lift length by spacing gives two chairs. However, there are actually three chairs with people in them—one at the bottom picking up people, one halfway up, and one at the top dropping off people.

In cell C20, we calculate Lift Capacity (people) by multiplying Chairs by Seats per Chair. Then, in cell C21, we calculate the Lift Time by dividing the length of the lift by its speed. In cell C23, we calculate the Chair Inter-arrival time (in minutes) by dividing the Chair Spacing by the Speed. In cell C24, we convert Chair Inter-arrival Time to seconds simply for our own convenience. In cell C26, we calculate the number of people skiing by multiplying the Seats per Chair by the Ski Time, divided by the Chair Inter-arrival Time (in minutes).

Finally, in cell C28, we calculate the number of people in the lift queue as the total number of skiers in the system, less the number on the lift and the number skiing; and in cell C29, we calculate the Queue Time by multiplying the Queue Length by the Chair Inter-arrival Time and dividing by the Seats per Chair.

Figure 9.2 also shows the two graphical displays (time allocation in minutes and percentage) included in the model. Although Queue Time is our true output measure, we also gain knowledge of the system by looking at where people are distributed throughout the system and how much time they are spending in each segment. These graphical displays help us quickly get a sense of how the system's behavior changes as we change parameters. Note that the percentage of time spent in each activity and the percentage of people in each activity are practically the same.

Effective framing and boundary setting has allowed us to build a small, simple model of a very complex system. This model is easy to build and easy to debug, enabling us to move quickly to analysis and insights.

As mentioned, this model takes Total Skiers as a parameter and outputs Queue Time. The case tells us that the maximum queue time is around 15 minutes but does not give any information about the total number of skiers at the Skiway. Therefore, we must work backward to find the base-case value of Total Skiers. Goal Seek shows us that a total of 435 skiers in the system creates a 15-minute queue, given the other base-case parameters (Figure 9.4).

We begin the analysis of M1 by examining the base case (Figure 9.2). We know that skiers are waiting in line for 15 minutes, but without some context, it is hard to get a sense of how good or bad this is. From M1, we see that skiers are spending more than 50 percent of their time waiting in line. No wonder the skiers are upset! We also see that only 90 people out of a total of 435 are actually skiing down the mountain at any one time. A quick check reveals that 90 skiers spread out over three trails of roughly a mile in length makes for a practically empty mountain, since they are over 380 feet apart. This suggests that perhaps an effective solution will involve shifting people from the queue to the mountain.

Insight 1: Skiers spend more than 50 percent of their time waiting in line, while the mountain itself is very lightly utilized.

The primary purpose of M1 is to learn how the system works and what happens when we "turn the dials." This is a task for sensitivity analysis. Figures 9.5–9.8 show the results of single-variable data sensitivity analyses on four parameters:

Total Skiers (Figure 9.5): Queue Time increases linearly with Total Skiers. We calculate that the rate of increase is four seconds of queue time per additional skier. This makes sense since each additional pair of skiers in the queue requires a chair to pick them up, and a chair comes along every eight seconds.

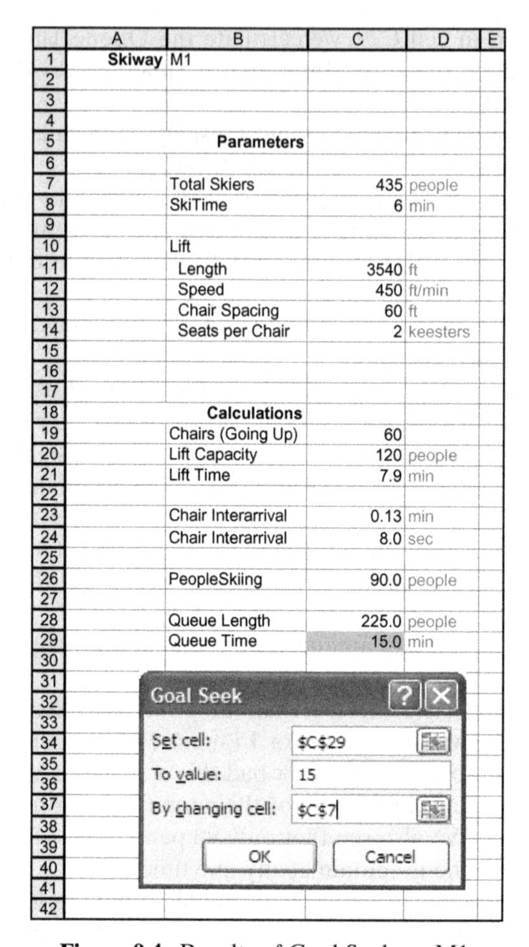

	A	B	C	D
1	Skiway	M1		
5		Parameters		
7		Total Skiers	435	people
8		SkiTime	6	min
10		Lift		
11		Length	3540	ft
12		Speed	450	ft/min
13		Chair Spacing	60	ft
14		Seats per Chair	2	keesters
18		Calculations		
19		Chairs (Going Up)	60	
20		Lift Capacity	120	people
21		Lift Time	7.9	min
23		Chair Interarrival	0.13	min
24		Chair Interarrival	8.0	sec
26		PeopleSkiing	90.0	people
28		Queue Length	225.0	people
29		Queue Time	15.0	min

Figure 9.4. Results of Goal Seek on M1.

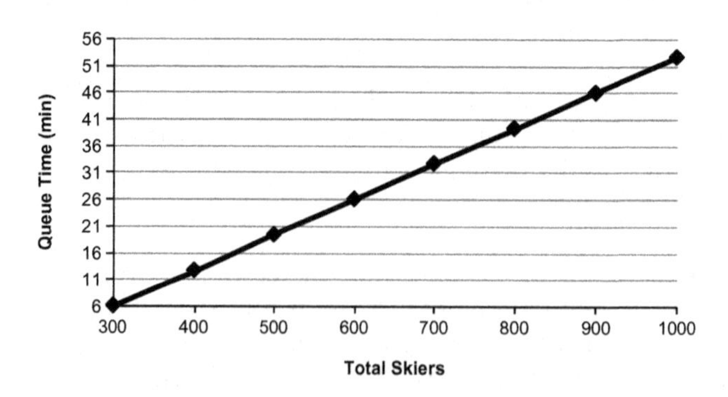

Figure 9.5. Sensitivity to Total Skiers (M1).

Ski Time (Figure 9.6): Queue Time decreases linearly with Ski Time. Interestingly, there is a one-for-one trade-off between Ski Time and Queue Time: One more minute spent skiing is one less minute spent in line.

Lift Speed (Figure 9.7): Queue Time decreases with Lift Speed; however, the effect is nonlinear, with Lift Speed having less effect on Queue Time at higher lift speeds. To understand this relationship, look again at the influence diagram in Figure 9.1. Lift Speed affects only Chair Interarrival Time, but Chair Inter-arrival Time affects Queue Time both directly and by way of People Skiing and Queue Length. Running the lift faster means chairs arrive at the top of the mountain more frequently. But Ski Time has not changed, so people are still skiing at the same speed. Faster arrivals with unchanged ski speed means the gap between groups of skiers decreases and the mountain becomes more crowded. Since Total Skiers is fixed and the number of skiers on the lift is fixed (the lift is always full), we know that if there are more people on the mountain, there must be fewer people in the queue. A shorter queue leads to a shorter Queue Time. Faster Lift Speed also reduces Queue

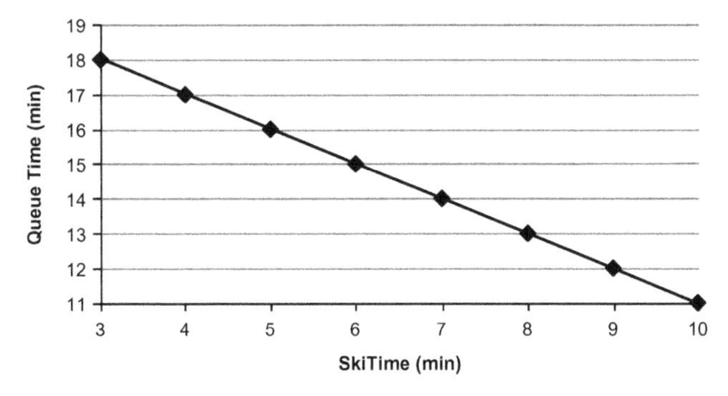

Figure 9.6. Sensitivity to Ski Time (M1).

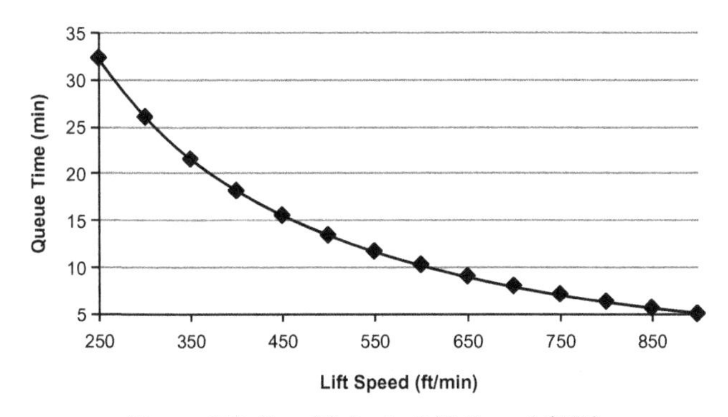

Figure 9.7. Sensitivity to Lift Speed (M1).

Time because chairs arrive more frequently at the bottom of the lift to whisk skiers away. In other words, when a skier gets in line, it takes less time to move to the front of the line because chairs are arriving more quickly to carry skiers away. However, as the queue length shortens, this effect diminishes.

Seats per Chair (Figure 9.8): Queue Time decreases dramatically with Seats per Chair. Like Lift Speed, the effect is nonlinear and decreases with higher numbers of seats. More seats per chair shifts skiers from the queue to both the mountain and the lift. The lift has a higher capacity, so it holds more people. The mountain holds more people because people are arriving at the top in larger groups. The shift of people out of the queue creates a shorter queue length and, thus, shorter queue time. Additionally, higher numbers of Seats per Chair mean more people are plucked out of line per unit of time, thus reducing the queue time. As with Lift Speed, this effect diminishes as the queue length gets shorter. Note that Queue Time in the model goes negative for Seats per Chair of five or more. As mentioned, the model calculations are not correct once the queue is empty.

The sensitivity analysis shows that we have four levers we can play with to affect wait time. Reducing total skiers is not a sensible choice since that obviously means fewer tickets sold and less revenue. In fact, we probably want to *increase* total skiers in order to offset the expense of any improvements we make. The other three levers, however, look promising for helping solve the Skiway's problem.

Insight 2: Increasing ski time, lift speed, or the number of seats per chair all reduce time spent waiting in the queue.

Although we do not explicitly consider capacity issues in M1, we are aware that there is a constraint and that, at the very least, increasing the number of

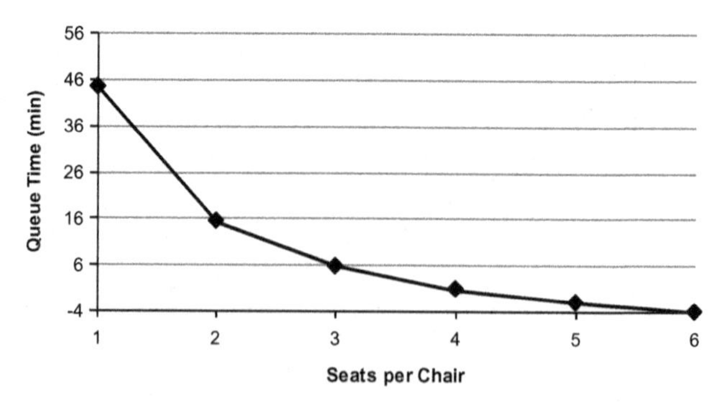

Figure 9.8. Sensitivity to Seats per Chair (M1).

skiers on the trails might degrade customer perception of the mountain. The number of people skiing is proportional to ski time and lift speed. Increasing ski time and lift speed by moderate amounts of 30 percent or so increases the load on the mountain by a corresponding 30 percent. With seats per chair, however, we must change in larger increments. (We cannot increase chair capacity by 10 percent.) Going from a double chair (two seats) to a quad (four seats) is a 100 percent increase and doubles the load on the mountain. This suggests that if any of our options is going to bump up against a capacity constraint, it is going to be the quad lift.

Insight 3: Increasing the number of seats per chair has a large impact on the number of people on the slopes.

Before moving on from M1, we consider the three scenarios presented by Brian, the Skiway manager, in light of what we have already learned. Figure 9.9 shows the results of our model under the base case, SpeedUp, Quad, and Trail scenarios. SpeedUp seems like an obvious choice since it reduces queue time, costs nothing, and does not dramatically change the load on the mountain. The Quad scenario would be highly effective at reducing queue times, but it costs a lot of money and it might overburden the mountain.

The Trail scenario is difficult to evaluate at this point because the benefit of a trail is increased capacity and we have not explicitly considered the mountain's capacity constraints. Thus, the Trail results in Figure 9.9 are identical to the base-case results. However, building a new trail might be helpful if it served to increase ski time. If the Skiway built a sufficiently long trail that took a long time to ski down, the average ski time would increase and queue time would decrease. However, since the trade-off between ski time and queue time is one-for-one, the Skiway would have to build a very long trail and increase ski time quite a bit in order to reduce queue time significantly. For example, increasing ski time by 50% from 6 to 9 minutes would only reduce the queue time by 3 minutes, from 15 to 12 minutes. Nevertheless, building a trail is something to consider.

	BaseCase	SpeedUp	Quad	Trail	
Lift Capacity	120	120	240	120	people
Lift Time	7.9	5.9	7.9	7.9	min
Chair Interarrival	0.13	0.10	0.13	0.13	min
Chair Interarrival	8	6	8	8	sec
People Skiing	90	120	180	90	people
Queue Length	225	195	15	225	people
Queue Time	15.0	9.8	0.5	15.0	min

Figure 9.9. Scenario results (M1).

Summary of M1

M1 has been an effective first model. We have a functioning basic model of the Skiway in equilibrium, and we have three insights in hand that point us toward possible solutions. What's next? There are three major items we have left out so far: mountain capacity, finances, and uncertainty. We explore mountain capacity next because it may impose limits on the effectiveness of one or more of our proposed changes.

To the Reader: Estimate the mountain capacity and determine when it will be a binding constraint.

9.5 ANALYZING MOUNTAIN CAPACITY

Before altering M1 to account for mountain capacity, we make some rough estimates to see whether capacity is a concern. There are two potential constraints on mountain capacity: the number of skiers the trails can hold, and the number of skiers who can pass through the narrow chute in the cliff band. We start with the capacity of the trails themselves.

We use a new worksheet, separate from the main model, to perform some "back-of-the-envelope" capacity calculations. The trails range from 4,000 feet to 5,280 feet (one mile), and each can support three skiers side by side. (The skiers do not have to be physically side by side, but the idea is that only three skiers can start down the trail at the same time.) Since skiers ski a series of turns down the mountain, the effective trail length is increased. We assume a 25 percent increase in trail length because of this factor. We also assume all skiers are intermediate skiers and prefer a gap of 75 feet between groups. These assumptions provide an estimate of the capacity for the mountain's three trails of just under 700 skiers (Figure 9.10). M1 showed there were only 90 out of 435 skiers on the mountain at any one time in the base case. Clearly, the trails' capacity is not a binding constraint on the system at present.

Insight 4: The existing trails are currently loaded at less than 15 percent of capacity.

		Trail 1	Trail 2	Trail 3	
A	Actual Length	4,000	4,640	5,280	
B	125% Length: (125 × A)	5,000	5,800	6,600	
C	Intermediate Spacing	75	75	75	
D	Groups @ Intermediate Spacing (B / C)	67	77	88	
E	Skiers per Group	3	3	3	TOTAL
F	Trail Capacity (D × E)	200	232	264	696

Figure 9.10. Trail capacity calculations.

The narrow chute at the top of the mountain could be a problem. To get through the chute, skiers must be in groups of no more than two and must maintain a minimum gap of 75 feet between groups (again assuming all skiers are intermediate skiers). Thus, we need to calculate the gap between skier groups created at the top of the mountain as skiers get off the lift. The gap is a function of how fast the skiers move away from the lift and the chair inter-arrival time. For example, if a pair of skiers dismounts the lift and skis away at 15 feet/second and the next chair arrives ten seconds later, the gap between groups is 150 feet. So we need to know skier speed to determine the gap between skiers. Also, we know that some of the improvement options change this simple gap calculation by adding more seats per chair or adding a trail that bypasses the chute. Thus, we need a more general method to calculate the gap.

An influence diagram helps us derive the generalized form of the gap calculation (Figure 9.11). As mentioned, the gap is determined by how far away from the lift one group can get before the next group arrives (assuming one group per chair). If there are more people on a single chair than can fit through the chute (Seats per Chair > Max Group Size), we must split the skiers up and reduce the gap spacing accordingly. For example, assume a quad chair delivers

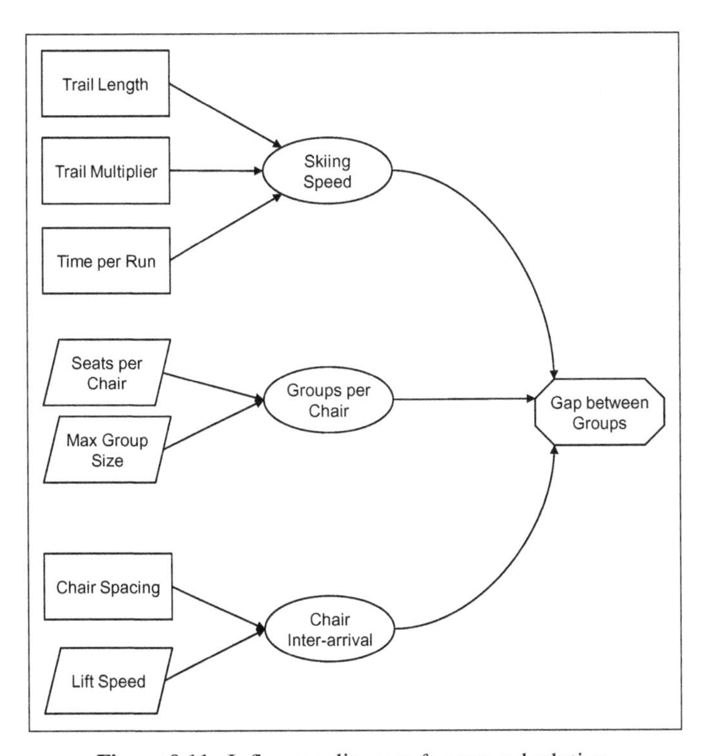

Figure 9.11. Influence diagram for gap calculation.

four people to the top of the mountain and they are able to ski 200 feet away before the next chair arrives. However, only two people can fit through the chute at a time. The four skiers must split into two groups. The first two skiers start down the chute immediately, and the other two skiers wait at the top. We assume the skiers share the gap equally, and thus, instead of a 200-foot gap between groups, the second pair of skiers will wait until the first pair is half a gap (100 feet) away before starting. Thus, the general equation for gap spacing is

$$\text{Gap Between Skiers} = \text{Skier Speed} \times \text{Chair Inter-arrival Time} \div \text{Groups per Chair}$$

Groups per Chair is determined by the number of seats in each chair and by the maximum group size that can start down the mountain. Any fractional group is rounded up to the next integer number of groups with the CEILING function. For example, a fully occupied triple chair creates two groups when the maximum group size is two, even though the second "group" is only one skier.

$$\text{Groups per Chair} = \text{CEILING}(\text{Seats per Chair}/\text{Max Group Size})$$

In the current state, the maximum group size is two because of the narrow chute. However, if the Skiway builds an additional trail that bypasses the chute and allowes groups of three, the total maximum group size would increase to five (two through the chute and three down the new trail).

Skiing speed is calculated based on effective trail length and the time required to ski down the trail. Remember, we are ignoring the details of each individual run and are assuming constant travel from getting off the lift to entering the queue at the bottom.

$$\text{Skiing Speed} = \text{Trail Length} \times \text{Trail Multiplier} \div \text{Time per Run}$$

Chair Inter-arrival Time is already calculated in the model as described in Section 9.3.

We perform the gap spacing calculations on a separate sheet of the workbook (Figure 9.12). Our calculations show that in the base-case scenario, the gap is more than 75 feet for all three trails and thus is not a concern. The gap is different for the three trails because the different trail lengths lead to a different result for skier speed. This happens because we assume the same ski time for each trail. In reality, it probably takes less time to get down the shorter trail and more time for the longer trail. Therefore, for all future calculations, we use the middle trail length of 4,640 feet. Keep in mind the purpose of our modeling is to inform major operational decisions, not to predict precisely the gap spacing or skier speed.

Figure 9.13 shows the gap calculation for three scenarios: Base case, SpeedUp, and Quad. For the first two scenarios, the gap is well above the

		Trail 1	Trail 2	Trail 3	
A	Actual Length	4,000	4,640	5,280	ft
B	125% Length (1.25 × A)	5,000	5,800	6,600	ft
C	Avg. Ski Time	6.0	6.0	6.0	min
D	Implied Speed (B / C)	833	967	1,100	ft/min
E	Implied Speed (D / 60)	13.89	16.11	18.33	ft/sec
F	Seats per Chair	2	2	2	
G	Max Group Size	2	2	2	
H	Group per Chair (Ceiling (F / G))	1	1	1	
I	Chair Spacing	60.00	60.00	60.00	ft
J	Lift Speed	450.00	450.00	450.00	ft/min
K	Chair Interarrival Time (60 × I / J)	8	8	8	sec
L	Group Spacing After Lift (E × K / H)	111	129	147	ft

Figure 9.12. Gap calculations for three trails.

		Base	Speed	Quad	
A	Actual Length	4,640	4,640	4,640	ft
B	125% Length (1.25 × A)	5,800	5,800	5,800	ft
C	Avg. Ski Time	6.0	6.0	6.0	min
D	Implied Speed (B / C)	967	967	967	ft/min
E	Implied Speed (D / 60)	16.11	16.11	16.11	ft/sec
F	Seats per Chair	2	2	4	
G	Max Group Size	2	2	2	
H	Group per Chair (Ceiling (F / G))	1	1	2	
I	Chair Spacing	60.00	60.00	60.00	ft
J	Lift Speed	450.00	600.00	450.00	ft/min
K	Chair Interarrival Time (60 × I / J)	8	6	8	sec
L	Group Spacing After Lift (E × K / H)	129	97	64	ft

Figure 9.13. Gap calculations for scenarios.

constraint of 75 feet. However, in the Quad scenario, the gap spacing drops below that constraint to only 64 feet. The violation occurs because with a maximum group size of two, the quad chair is depositing two groups of skiers at the top of the mountain every eight seconds. This cuts the gap size in half since two of the skiers must wait for the other two skiers to get at least 75 feet away before starting down the mountain.

Insight 5: Under some scenarios, such as building a quad lift, the gap constraint is binding.

This constraint violation is a major problem and must be addressed before our analysis can progress. The influence diagram shows several ways to alleviate the gap violation. Reducing the number of seats per chair or increasing the maximum group size reduces the number of groups per chair; however, these changes require either building a new trail or rejecting the possibility of a quad lift. Increasing the skiing speed would help alleviate the gap problem, but we assume skiing speed is essentially outside the control of Skiway management. Increasing the chair inter-arrival time would also help alleviate the gap problem; although this could be accomplished by spacing the chairs farther apart on the cable, it would require a lot of effort and reduce the lift capacity. Slowing down the lift can also alleviate the gap problem; it has the advantages that it is free and can be adjusted as frequently as necessary. Modifying the model to automatically adjust for the capacity constraint in this way will be the focus of M2.

To the Reader: Before reading on, build M2 to take into account the capacity constraint. Use the model to generate as many interesting insights into this situation as you can.

9.6 M2 MODEL AND ANALYSIS

We have added four new parameters to M2 to allow for calculation of skier speed and the gap between skiers. The calculations are now divided into two sections: Max Speed Calculations and Actual Speed Calculations (Figures 9.14 and 9.15). The Max Speed section calculates Queue Time and Queue Length based on the Lift Max Speed parameter. These calculations are identical to the calculations in M1. The model also calculates the Gap Between Skiers if the lift runs at the maximum speed. These calculations are identical to those described in Section 9.5 and check whether the capacity constraint is binding.

The Actual Speed Calculations section uses an IF statement to determine whether the gap calculated under the Max Speed section violates the Minimum Gap constraint. If not, the Actual Lift Speed is set equal to the Lift Max Speed parameter. If a gap violation does occur at maximum speed, the model calculates the Lift Speed necessary to make the skier gap equal to the constraint value (75 feet). The Chair Inter-arrival Time, People Skiing, and Queue Time values are all then calculated based on this reduced lift speed. Thus, the model automatically slows down the lift to ensure the minimum gap constraint is not violated.

Since the only change between M1 and M2 is the addition of the capacity constraint, the results of M2 should be identical to M1 except when the capacity constraint is violated. We perform the same four sensitivity studies as we did in M1. As expected, sensitivity to Total Skiers (Figure 9.16) is the same as M1 because the number of skiers in the system does not affect the gap between groups. Sensitivity to Ski Time (Figure 9.17) is the same for M1 and M2 over the range of values tested. However, Ski Time does affect the Gap Spacing, as

	A	B	C	D	E
1	Skiway	M2	Add Capacity Constraint		
2					
3					
4					
5	Parameters				
6					
7		Total Skiers	435	people	
8		SkiTime	6	min	
9					
10		Lift			
11		Length	3540	ft	
12		Max Speed	450	ft/min	
13		Chair Spacing	60	ft	
14		Seats per Chair	2	keesters	
15					
16		Average Trail Length	4640	ft	
17		Trail Length Multiplier	125%		
18		Max Group Size	2	people	
19		Minimum Gap	75	ft	
20					
21	Calculations				
22			Max Speed Calculations		
23		Chairs	60		
24		Lift Capacity	120	people	
25					
26		Lift Time	7.87	min	
27		Chair Interarrival	0.13	min	
28		Chair Interarrival	8	sec	
29					
30		PeopleSkiing	90	people	
31					
32		Queue Length	225.0	people	
33		Queue Time	15.0	min	
34					
35		Skiing Speed	966.7	ft/min	
36		Gap Between Skiers	128.9	ft	
37					
38			Actual Speed Calculations		
39		Gap Too Small?	0		
40					
41		Lift Speed	450		
42		Lift Time	7.87		
43					
44		Chair Interarrival	0.13	min	
45		Chair Interarrival	8.00	sec	
46					
47		People Skiing	90.0	people	
48					
49		Queue Length	225.0	people	
50		Queue Time	15.0	min	
51					
52		Gap Between Skiers	128.9	ft	
53					

Figure 9.14. M2—numerical view.

shown on the lower graph of Figure 9.17. Had we tested a larger range of values, we would have seen the capacity constraint become binding.

Increasing Max Lift Speed (Figure 9.18) reduces both Queue Time and the gap between groups. Once the maximum lift speed exceeds about 770 feet/minute, however, the capacity constraint becomes binding. At this point, the model will not let the lift go any faster, regardless of what maximum speed is entered. Since the actual lift speed does not change, the queue time and gap remain constant for Max Lift Speed values above 770 feet/minute.

	A	B	C	D	E
1	Skiway	M2	Add Capacity Constraint		
2					
3					
4					
5	Parameters				
6					
7		Total Skiers	435	people	
8		SkiTime	6	min	
9					
10		Lift			
11		Length	3540	ft	
12		Max Speed	450	ft/min	
13		Chair Spacing	60	ft	
14		Seats per Chair	2	keesters	
15					
16		Average Trail Length	4640	ft	
17		Trail Length Multiplier	1.25		
18		Max Group Size	2	people	
19		Minimum Gap	75	ft	
20					
21	Calculations				
22				Max Speed Calculation	
23		Chairs	=FLOOR(C11/C13,1)+1		
24		Lift Capacity	=C23*C14	people	
25					
26		Lift Time	=C11/C12	min	
27		Chair Interarrival	=C13/C12	min	
28		Chair Interarrival	=60*(C13/C12)	sec	
29					
30		PeopleSkiing	=$C$8*$C$14/C27	people	
31					
32		Queue Length	=$C$7-$C$24-C30	people	
33		Queue Time	=C32*C27/$C$14	min	
34					
35		Skiing Speed	=C16*C17/C8	ft/min	
36		Gap Between Skiers	=(C28*C35/60)/CEILING(C14/C18,1)	ft	
37					
38				Actual Speed Calculat	
39		Gap Too Small?	=IF(C36<C19,1,0)		
40					
41		Lift Speed	=IF(C39,(C13*C35)/(C19*CEILING(C14/C18,1)),C12)		
42		Lift Time	=C11/C41		
43					
44		Chair Interarrival	=C13/C41	min	
45		Chair Interarrival	=C44*60	sec	
46					
47		People Skiing	=$C$8*$C$14/C44	people	
48					
49		Queue Length	=$C$7-$C$24-C47	people	
50		Queue Time	=C49*C44/$C$14	min	
51					
52		Gap Between Skiers	=(C44*C35)/CEILING(C14/C18,1)	ft	
53					

Figure 9.15. M2—relationship view.

Figure 9.19 shows that the results of M1 and M2 differ slightly for values of Seats per Chair greater than two. This is because the maximum allowable group size at the top of the mountain is two, and therefore, any chair larger than a double requires skiers to split into multiple groups and cuts the gap size by half or more. So although increasing Seats per Chair still has a large effect on Queue Time, the effect is somewhat attenuated by the capacity constraint.

Insight 5 noted that building a quad lift leads to a violation of the capacity constraint. Figure 9.20 shows M2 under the Quad scenario. The Max Speed Calculations section shows results if the lift were to run at the set maximum speed of 450 feet/minute. These results are identical to the M1 results for this scenario. The Actual Speed Calculations section shows that the lift must be slowed down to 387 feet/minute to maintain the minimum gap spacing. This

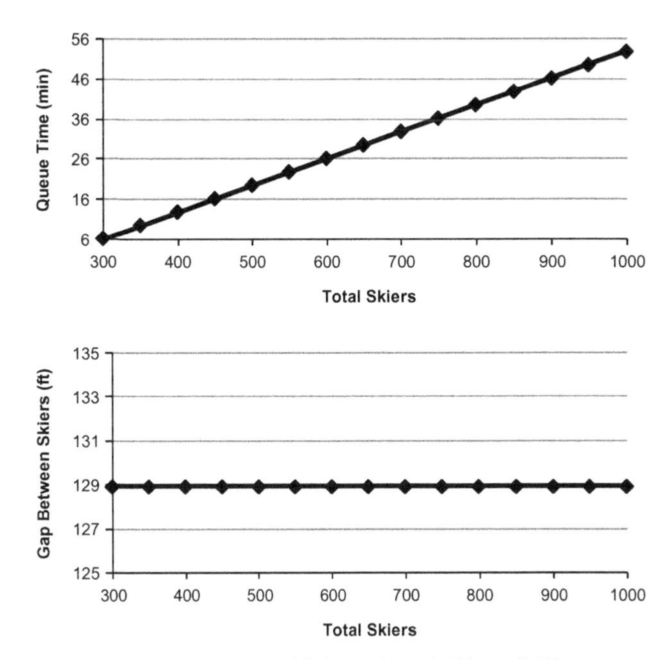

Figure 9.16. Sensitivity to Total Skiers (M2).

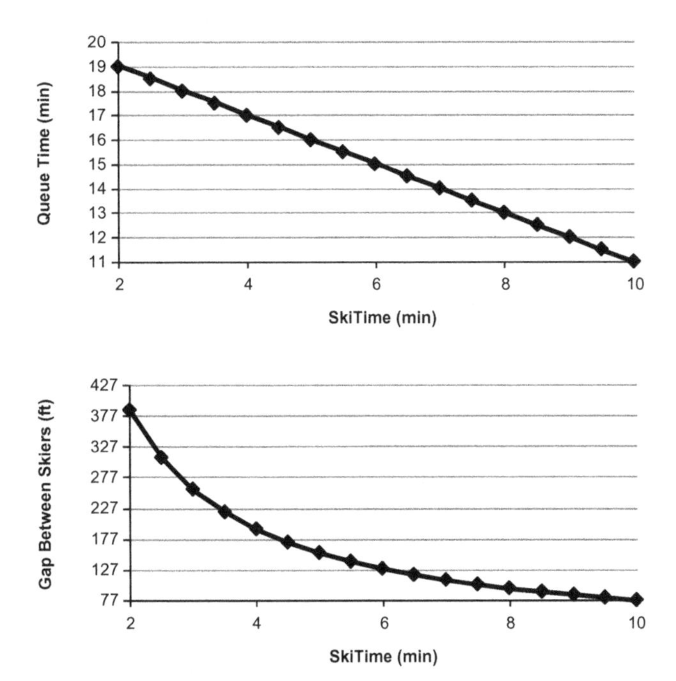

Figure 9.17. Sensitivity to Ski Time (M2).

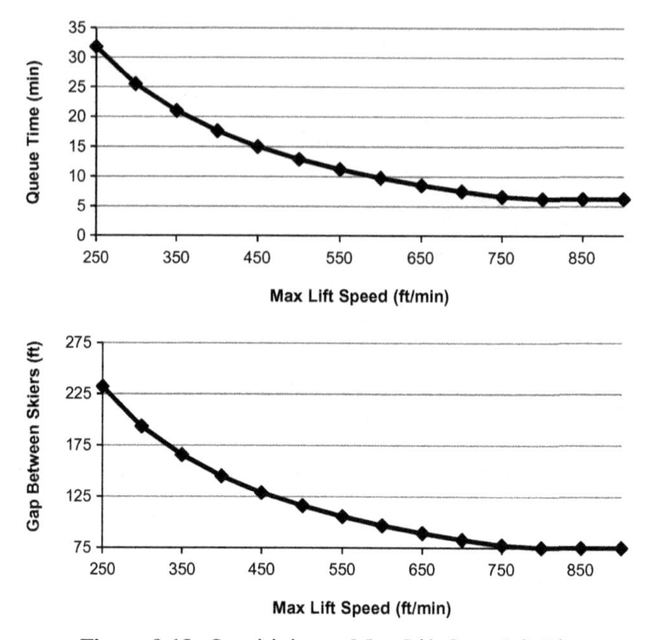

Figure 9.18. Sensitivity to Max Lift Speed (M2).

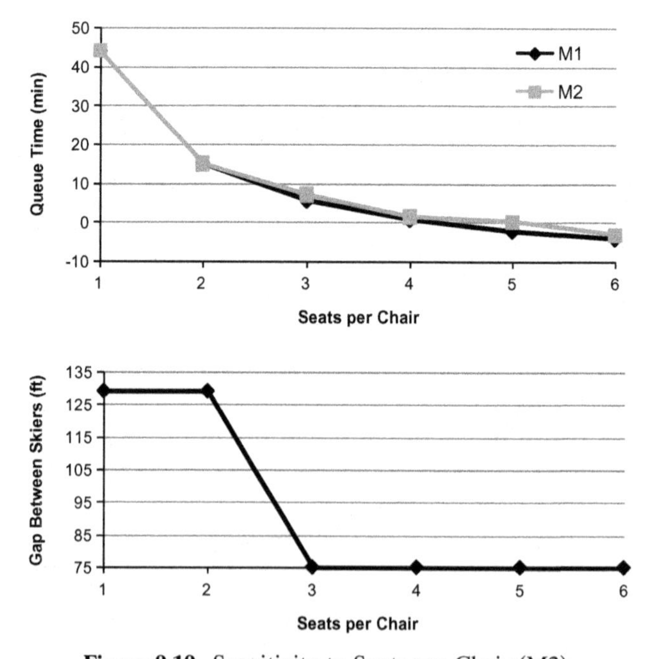

Figure 9.19. Sensitivity to Seats per Chair (M2).

	A	B	C	D	E
1	**Skiway**	M2	Add Capacity Constraint		
2					
3					
4					
5	**Parameters**				
6					
7		Total Skiers	435	people	
8		SkiTime	6	min	
9					
10		Lift			
11		Length	3540	ft	
12		Max Speed	450	ft/min	
13		Chair Spacing	60	ft	
14		Seats per Chair	4	keesters	
15					
16		Average Trail Length	4640	ft	
17		Trail Length Multiplier	125%		
18		Max Group Size	2	people	
19		Minimum Gap	75	ft	
20					
21	**Calculations**				
22			**Max Speed Calculations**		
23		Chairs	60		
24		Lift Capacity	240	people	
25					
26		Lift Time	7.87	min	
27		Chair Interarrival	0.13	min	
28		Chair Interarrival	8	sec	
29					
30		PeopleSkiing	180	people	
31					
32		Queue Length	15.0	people	
33		Queue Time	0.5	min	
34					
35		Skiing Speed	966.7	ft/min	
36		Gap Between Skiers	64.4	ft	
37					
38			**Actual Speed Calculations**		
39		Gap Too Small?	1		
40					
41		Lift Speed	387		
42		Lift Time	9.16		
43					
44		Chair Interarrival	0.16	min	
45		Chair Interarrival	9.31	sec	
46					
47		People Skiing	154.7	people	
48					
49		Queue Length	40.3	people	
50		Queue Time	1.6	min	
51					
52		Gap Between Skiers	75.0	ft	
53					

Figure 9.20. M2 Results for Quad Scenario.

results in fewer people on the mountain and in a Queue Time of 1.6 minutes rather than 0.5 minutes.

Insight 6: The capacity constraint at the top of the mountain sometimes forces the lift to be run at a reduced speed and, therefore, increases the Queue Time.

With capacity now accounted for in our model, we can add Building a New Trail to our list of scenarios (Figure 9.21). The benefit of building a trail is an

	Base Case	Speed Up	Quad	Quad & Trail	Quad & Speed up & New Trail	
Gap Too Small?	0	0	1	0	0	
Lift Speed	450	600	387	450	600	ft/min
Lift Time	7.9	5.9	9.2	7.9	5.9	min
Chair Interarrival	0.13	0.10	0.16	0.13	0.10	min
Chair Interarrival	8.0	6.0	9.3	8.0	6.0	sec
People Skiing	90	120	155	180	240	people
Queue Length	225	195	40	15	–45	people
Queue Time	15.0	9.8	1.6	0.5	–1.1	min
Gap Between Skiers	129	97	75	129	97	ft

Figure 9.21. Scenario results (M2).

increase in the maximum group size at the top of the mountain from two to five. Thus, adding a trail is only beneficial when the mountain is capacity constrained. This occurs if the Skiway installs a quad chair. Note that when we install a quad, the actual lift speed drops to 387 feet/minute, but if we also add a trail, the actual lift speed can return to the maximum lift speed of 450 feet/minute.

Insight 7: Adding a trail is only beneficial if a quad lift is also installed.

Summary of M2

Although we only generated two new insights from M2, we now have a more realistic model that automatically adjusts lift speed to compensate for capacity constraints. We have also examined four scenarios that lead to queue times shorter than the base case. M2 suggests that the Skiway should install a new quad lift, speed it up, and build a new trail. This is the obvious result of considering only benefits and ignoring costs. Clearly, it is time to bring costs and revenues into the picture to help us reach a more realistic conclusion.

To the Reader: Before reading on, build M3 to take into account the financial costs and benefits.

9.7 M3 MODEL AND ANALYSIS

Brian, our client, believes the 15-minute wait time that occurs at peak periods is hurting the attractiveness of the Skiway and is therefore limiting ticket sales. The implication is that if queue time could be reduced, the Skiway could sell more tickets and increase revenues. (Perhaps there is also a social benefit of serving more customers, regardless of revenue.) In economic terms, we assume there is some elasticity of demand relative to queue time such that demand goes up as queue time goes down. M1 and M2 showed that queue times can be reduced through various improvement projects. However, these improvements are only economically beneficial if they result in revenues that are higher than costs. So, as we consider revenues and costs in this model, we must consider not only the immediate cost of the improvements but also the potential increased revenue from ticket sales over the life of the improvement. The costs are easy to deal with because they are given. The increased revenue takes a little more work.

Recall that in M1, we used Goal Seek to determine the total number of skiers in the system that led to a 15-minute wait (435 people). Phrased differently, the Skiway must sell no more than 435 tickets at present if it wants to limit queue times to 15 minutes. Similarly, we can use Goal Seek to find the maximum number of tickets that can be sold to attain any given queue time. Doing this for each of the improvement scenarios allows us to estimate the increase in ticket sales for any given queue time and improvement scenario. For example, if building a quad allows the Skiway to have 650 people in the system while maintaining a ten-minute queue time, the Skiway will realize an increase of 215 tickets per day. Actual demand with a ten-minute queue time, however, could be a limiting factor, but we will deal with that later. For now, we create a schedule of maximum allowable ticket sales and queue times for each scenario.

What is the value of being able to sell one more ticket? We make several assumptions to help estimate this value. We assume that increased ticket sales only occur on weekends, when the mountain faces peak demand. (During weekdays, demand is low and queue times are already short.) We also make an assumption about the number of weekends in the ski season. From Thanksgiving to the beginning of April is about 18 weekends, giving us 36 days of peak demand per season. At $40 per ticket, that means that an increase of one skier each of those 36 days brings in an extra $1,440 ($40 × 36 days) per season. Of course, a major improvement project like a new lift or trail has a life much longer than a single season, so we include the present value of the increased revenue over several years. This requires selecting a time horizon and a discount rate.

M2 requires little modification to create M3 (Figures 9.22 and 9.23). A set of new parameters is added that cover lift ticket price, days in the season, and so on. The new calculations determine the change in total skiers from the base case and the resulting change in revenue. We use the PV function to calculate the present value of the incremental revenue over the life of the improvement. (We assume a ten-year life and a ten percent discount rate.) Note that costs are not included in the model. At this point we just want to find the change in revenue regardless of the costs of the improvements.

Another assumption we make for now is that demand is unlimited. This allows us to use M3 to explore what happens to Queue Time and NPV under

	A	B	C	D	E
1	**Skiway**	M3	Add Money		
2					
3					
4					
5	**Parameters**				
6					
7		Total Skiers	618	people	
8		SkiTime	6	min	
9					
10		Lift			
11		Length	3540	ft	
12		Max Speed	450	ft/min	
13		Chair Spacing	60	ft	
14		Seats per Chair	4	keesters	
15					
16		Average Trail Length	4640	ft	
17		Trail Length Multiplier	125%		
18		Max Group Size	2	people	
19		Minimum Gap	75	ft	
20					
21		Base Case Skiers	435		
22		Weekends per Season	18		
23		Days per Week at Capacity	2		
24		Discount Rate	10%		
25		Project Life	10	yrs	
26		Lift Ticket Price	$40		
27					
28		Target Queue Time	14		
29					
30	**Calculations**				
31			Max Speed Calculations		
32		Chairs	60		
33		Lift Capacity	240	people	
34		Lift Time	7.87	min	
35					
36		Chair Interarrival	0.13	min	
37		Chair Interarrival	8	sec	
38					
39		PeopleSkiing	180	people	
40					
41		Queue Length	198.0	people	
42		Queue Time	6.6	min	
43					
44					
45		Skiing Speed	966.7	ft/min	
46		Gap Between Skiers	64.4	ft	
47					
48			Actual Calculations		
49		Gap Too Small?	1		
50					
51		Lift Speed	387	ft/min	
52		Lift Time	9.16	min	
53					
54		Chair Interarrival	0.16	min	
55		Chair Interarrival	9.31	sec	
56					
57		People Skiing	154.7	people	
58					
59		Queue Length	223.3	people	
60		Queue Time	8.7	min	
61					
62		Gap Between Skiers	75.0	ft	
63					
64			Financial Calculations		
65		Additional Skiers	183	people/day	
66		Additional Revenue	$7,320	$/day	
67		Additional Rev for Season	$263,520	$/year	
68					
69		PV of Incremental Revenue	$1,619,216		
70					

Figure 9.22. M3—numerical view.

	A	B	C	D	E
1	Skiway	M3	Add Money		
2					
3					
4					
5	Parameters				
6					
7		Total Skiers	618	people	
8		SkiTime	6	min	
9					
10		Lift			
11		Length	3540	ft	
12		Max Speed	450	ft/min	
13		Chair Spacing	60	ft	
14		Seats per Chair	4	keesters	
15					
16		Average Trail Length	4640	ft	
17		Trail Length Multiplier	1.25		
18		Max Group Size	2	people	
19		Minimum Gap	75	ft	
20					
21		Base Case Skiers	435		
22		Weekends per Season	18		
23		Days per Week at Capacity	2		
24		Discount Rate	0.1		
25		Project Life	10	yrs	
26		Lift Ticket Price	40		
27					
28		Target Queue Time	14		
29					
30	Calculations				
31				Max Speed Calculations	
32		Chairs	=FLOOR(C11/C13,1)+1		
33		Lift Capacity	=C32*C14	people	
34		Lift Time	=C11/C12	min	
35					
36		Chair Interarrival	=C13/C12	min	
37		Chair Interarrival	=60*(C13/C12)	sec	
38					
39		PeopleSkiing	=$C$8*$C$14/C36	people	
40					
41		Queue Length	=$C$7-$C$33-C39	people	
42		Queue Time	=C41*C36/$C$14	min	
43					
44					
45		Skiing Speed	=C16*C17/C8	ft/min	
46		Gap Between Skiers	=(C37*C45/60)/CEILING(C14/C18,1)	ft	
47					
48				Actual Calculations	
49		Gap Too Small?	=IF(C46<C19,1,0)		
50					
51		Lift Speed	=IF(C49,(C13*C45)/(C19*CEILING(C14/C18,1)),C12)	ft/min	
52		Lift Time	=C11/C51	min	
53					
54		Chair Interarrival	=C13/C51	min	
55		Chair Interarrival	=C54*60	sec	
56					
57		People Skiing	=$C$8*$C$14/C54	people	
58					
59		Queue Length	=$C$7-$C$33-C57	people	
60		Queue Time	=C59*C54/$C$14	min	
61					
62		Gap Between Skiers	=(C54*C45)/CEILING(C14/C18,1)	ft	
63					
64				Financial Calculations	
65		Additional Skiers	=C7-C21	people/day	
66		Additional Revenue	=C65*C26	$/day	
67		Additional Rev for Season	=C66*C22*C23	$/year	
68					
69		PV of Incremental Revenue	=-PV(C24,C25,C67)		
70					

Figure 9.23. M3—relationship view.

the various improvement scenarios. Later on we consider how limited demand might alter our insights and decisions.

To the Reader: Before reading on, use M3 to generate as many interesting insights into the situation of unlimited demand as you can.

9.7.1 Unlimited Demand

We pointed out above that we could use Goal Seek to create a schedule of maximum tickets sold for given maximum queue times. For example, with M3 set for the base-case scenario, we run Goal Seek to set Queue Time to 14 minutes by changing Total Skiers. Goal Seek finds that 420 total skiers lead to a queue time of 14 minutes. The financial calculations section shows that Total Skiers are reduced by 15 from the base case, leading to a daily revenue loss of $600 and a present value loss of $132,723. This answer makes intuitive sense, since without any improvements to the system, the only way to reduce the queue time is to reduce the number of skiers and this results in a loss of revenue.

We could use Goal Seek to calculate the maximum number of skiers for queue times from 1 to 15 minutes for the base case and the four improvement scenarios. However, this would be quite laborious, involving manually running Goal Seek 75 times. A faster way is to use Solver to perform the goal-seeking task and to use Solver Sensitivity to repeat the process for different target queue times. (For more information on Solver Sensitivity, see Appendix C.)

We first set up Solver to perform the goal-seeking function. With Goal Seek, Excel is looking for a single input value (Total Skiers) that yields the target output (Queue Time). A different, perhaps more realistic way to think about the problem is to maximize revenue subject to a maximum allowable queue time. In Solver lingo, revenue is the objective function that we want to maximize subject to a constraint on Queue Time. Total Skiers is the decision variable that Solver can use to maximize the objective. We add a parameter called Target Queue Time to the model. This parameter serves only as the constraint value for Solver and is not used in any calculations. Because revenue increases continuously as Total Skiers increases, we know there is exactly one solution to the optimization problem, and Solver should have no problem finding it. Running Solver on the base-case scenario with a target queue time of 14 minutes gives the same result obtained using Goal Seek.

We use Solver Sensitivity to perform the optimization 15 times for target queue times between 1 and 15 minutes. We set up Solver Sensitivity to report both Total Skiers and PV of Revenue. We run Solver Sensitivity for the base case and four improvement scenarios and combine the results into one table, which is presented in Figure 9.24. (Had we thought of it earlier, we could have built the model with a scenario selection variable, as we did in Chapter 7, that plugged in the appropriate parameter values for each scenario. Then we would have used a two-way Solver Sensitivity to automate completely the analysis described above for all target queue times and improvement scenarios.)

Figure 9.25 shows the number of skiers (or tickets sold) that leads to a given queue time for each of the five scenarios under consideration. Using terminology borrowed from economics, these lines are *supply curves*. They represent the number of tickets the Skiway would supply if customers accept a given queue time. For example, if customers accept a ten-minute queue time, the

Target	Base Case		Speed Up		Quad		Quad & Trail		Quad, Trail & Speed	
Wait Time	Total Skiers	PV of 10 Years	Total Skiers	PV of 10 Years	Total Skiers	PV of 10 Years	Total Skiers	PV of 10 Years	Total Skiers	PV of 10 Years
1	225	($1,858,117)	260	($1,548,431)	420	($128,790)	450	$132,723	520	$752,095
2	240	($1,725,394)	280	($1,371,467)	446	$99,296	480	$398,168	560	$1,106,022
3	255	($1,592,672)	300	($1,194,504)	472	$327,383	510	$663,613	600	$1,459,949
4	270	($1,459,949)	320	($1,017,540)	498	$555,469	540	$929,059	640	$1,813,876
5	285	($1,327,226)	340	($840,577)	524	$783,555	570	$1,194,504	680	$2,167,803
6	300	($1,194,504)	360	($663,613)	549	$1,011,642	600	$1,459,949	720	$2,521,730
7	315	($1,061,781)	380	($486,650)	575	$1,239,728	630	$1,725,394	760	$2,875,657
8	330	($929,059)	400	($309,686)	601	$1,467,814	660	$1,990,840	800	$3,229,584
9	345	($796,336)	420	($132,723)	627	$1,695,901	690	$2,256,285	840	$3,583,512
10	360	($663,613)	440	$44,241	652	$1,923,987	720	$2,521,730	880	$3,937,439
11	375	($530,891)	460	$221,204	678	$2,152,073	750	$2,787,176	920	$4,291,366
12	390	($398,168)	480	$398,168	704	$2,380,160	780	$3,052,621	960	$4,645,293
13	405	($265,445)	500	$575,131	730	$2,608,246	810	$3,318,066	1000	$4,999,220
14	420	($132,723)	520	$752,095	756	$2,836,332	840	$3,583,512	1040	$5,353,147
15	435	$0	540	$929,059	781	$3,064,419	870	$3,848,957	1080	$5,707,074

Figure 9.24. Scenario results (M3).

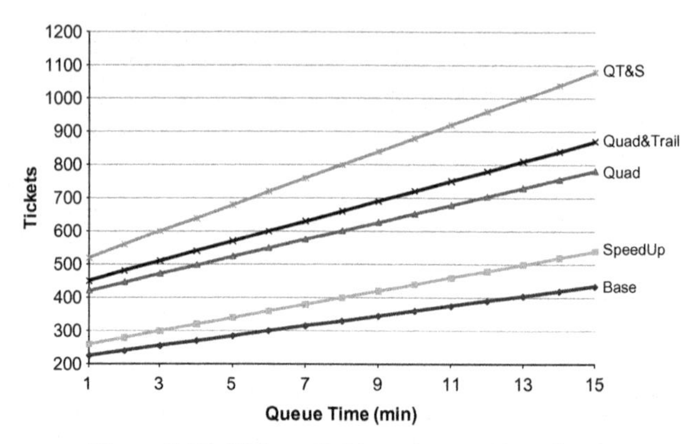

Figure 9.25. Tickets Sold vs. Queue Time (M3).

Skiway will supply only 360 tickets in the base case. If it sells one more ticket, the queue time will exceed ten minutes. In comparison, under the Quad scenario, the Skiway can sell 652 tickets and maintain a ten-minute queue time. The results show that each improvement increases the number of tickets that can be sold at any given queue time and, likewise, increases the revenues.

Insight 8: Improvements to the mountain increase the maximum number of tickets that can be sold at each target queue time.

Subtracting the costs of the various improvements from the results in Figure 9.24 gives us the net present value (NPV) of each queue time and scenario, as shown in Figure 9.26. Subtracting the costs now, instead of in the model itself, allows us more flexibility to deal with changes in costs. Had we included costs in the main model, we would have had to rerun the entire Solver Sensitivity tables if one of the cost parameters changed. Subtracting costs later, as we do here, allows us to change the cost parameter and instantly see the effect on NPV without rerunning Solver.

In Figure 9.26, we see that, as discussed above, any reduction in Queue Time in the base case results in negative NPV because the Skiway must sell fewer tickets. The SpeedUp scenario, however, does not cost anything and offers some improvement. Speeding up the lift allows the system to hold 540 skiers, rather than 435, with a 15-minute Queue Time. SpeedUp yields positive NPVs for Target Queue Times down to ten minutes. At a Target Queue Time of ten minutes, the system can hold 440 skiers, five more than the base case, resulting in a positive NPV of $44,000. Furthermore, if a Queue Time of 13 minutes is acceptable, then the Skiway can realize an NPV of more than a half-million dollars by just speeding up the lift and selling more tickets.

Insight 9: The SpeedUp scenario has a positive NPV down to a ten-minute queue time.

Target	Base Case		Speed Up		Quad		Quad & Trail		Quad, Trail & Speed			
Wait Time	Total Skiers	NPV	Total Skiers	NPV	Total Skiers	NPV	Total Skiers	NPV	Total Skiers	NPV	DEMAND	
1	225	($1,858,117)	260	($1,548,431)	420	($853,790)	450	($2,392,277)	520	($1,772,905)	841	1
2	240	($1,725,394)	280	($1,371,467)	446	($625,704)	480	($2,126,832)	560	($1,418,978)	812	2
3	255	($1,592,672)	300	($1,194,504)	472	($397,617)	510	($1,861,387)	600	($1,065,051)	783	3
4	270	($1,459,949)	320	($1,017,540)	498	($169,531)	540	($1,595,941)	640	($711,124)	754	4
5	285	($1,327,226)	340	($840,577)	524	$58,555	570	($1,330,496)	680	($357,197)	725	5
6	300	($1,194,504)	360	($663,613)	549	$286,642	600	($1,065,051)	720	($3,270)	696	6
7	315	($1,061,781)	380	($486,650)	575	$514,728	630	($799,606)	760	$350,657	667	7
8	330	($929,059)	400	($309,686)	601	$742,814	660	($534,160)	800	$704,584	638	8
9	345	($796,336)	420	($132,723)	627	$970,901	690	($268,715)	840	$1,058,512	609	9
10	360	($663,613)	440	$44,241	652	$1,198,987	720	($3,270)	880	$1,412,439	580	10
11	375	($530,891)	460	$221,204	678	$1,427,073	750	$262,176	920	$1,766,366	551	11
12	390	($398,168)	480	$398,168	704	$1,655,160	780	$527,621	960	$2,120,293	522	12
13	405	($265,445)	500	$575,131	730	$1,883,246	810	$793,066	1000	$2,474,220	493	13
14	420	($132,723)	520	$752,095	756	$2,111,332	840	$1,058,512	1040	$2,828,147	464	14
15	435	$0	540	$929,059	781	$2,339,419	870	$1,323,957	1080	$3,182,074	435	15

Figure 9.26. NPV of improvements (M3).

Looking at the results of adding a quad lift, we see that the maximum number of tickets for each queue time goes up dramatically. Adding a quad lift is profitable down to a target queue time of five minutes, with 524 tickets sold.

Insight 10: The Quad scenario has a positive NPV down to a five-minute queue time.

Note that we do not consider the scenario of building a quad and speeding up the maximum lift speed because we already know the quad must be run at less than 450 feet/minute because of the capacity constraint. Increasing the maximum speed to 600 feet/minute would have no effect.

Adding a quad lift and building a new trail is never practical financially. Adding both a quad and a trail does allow the lift to run at full speed and increases the maximum number of skiers but not enough to offset the high cost of adding the trail. At every Target Queue Time, the Skiway earns a higher NPV by building just the quad. Thus, Quad + Trail is completely dominated by the Quad scenario. This is helpful because it means we can take the Quad + Trail option off the table.

Insight 11: The Quad + Trail scenario has a lower NPV than the Quad scenario for every target queue time.

If the Skiway goes all out and builds a quad lift, speeds it up to 600 feet/minute, and adds a new trail, it can sell a lot of tickets and still keep queue times under control. This scenario remains profitable with Queue Times down to almost seven minutes. In fact, it looks as if the Skiway can achieve an NPV of $350,657 with just a seven-minute queue. However, note that the Quad scenario has a larger NPV of $514,728 at seven minutes. Similarly, at a Target Queue Time of eight minutes, the Quad scenario still has a higher NPV than does the Quad + SpeedUp + Trail scenario. This is a major insight!

Insight 12: There are situations where building a quad alone is better than implementing all three improvements (Quad + SpeedUp + Trail).

Furthermore, the Quad scenario involves less risk since fewer skiers are needed to make the project profitable. For example, at seven minutes, the Quad + SpeedUp + Trail scenario requires 760 skiers, whereas the Quad scenario only requires 575. Lastly, building the quad alone requires less capital. The Skiway must invest $2.525M to build a quad and trail. That's a lot of money, even for a well-established, well-endowed college.

The NPV results from Figure 9.26 are shown in graphical form in Figure 9.27. This figure shows the NPV of each scenario for each Queue Time. The Skiway managers should select the scenario with the highest NPV for any desired Queue Time. For Target Queue Times between one and eight minutes, building a quad yields the highest NPV. For Target Queue Times above eight minutes, the Quad + SpeedUp + Trail scenario is best.

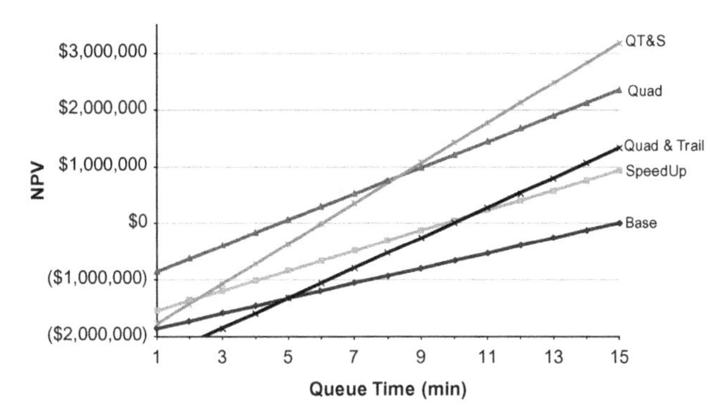

Figure 9.27. NPV of improvement scenarios (M3).

Insight 13: With unlimited demand, the Skiway should either choose the Quad or Quad + SpeedUp + Trail scenario, depending on the allowable queue time.

Remember, however, that these conclusions assume demand is unlimited. But, is this realistic for the Skiway? What happens when demand is limited?

To the Reader: Before reading on, use M3 to generate as many interesting insights into the situation of limited demand as you can.

9.7.2 Limited Demand

In the previous section we created supply lines representing the number of tickets the Skiway will sell for various target queue times. But what about the *demand* for tickets? How many people *want* to ski if the queue time is ten minutes? We have mentioned demand in some of the previous analysis, but we have not yet tried to create a demand curve. The Skiway could conduct a market survey to estimate the demand curve, but this would cost time and money. Furthermore, it is unclear how accurate the results would be. It is doubtful potential customers can say accurately how likely they are to go skiing if the queue time were six minutes and how their behavior would change if it were seven minutes. Perhaps we can make progress with a hypothetical demand curve.

We already know one point on the demand curve: the base case. Demand for lift tickets is 435 when the queue time is 15 minutes. If, for simplicity, we assume the demand curve is linear, then all we need is one other point to estimate the line. For example, if we knew what demand is when there is no queue, we could connect this point and the base-case point to estimate a demand curve. The Skiway manager can probably make a pretty good guess at what maximum demand would be if there was never a line at the lift. He might not know it exactly, but he probably has a sense of whether it is 500 or 5,000 skiers. For now, we assume demand would double to 870 if there were

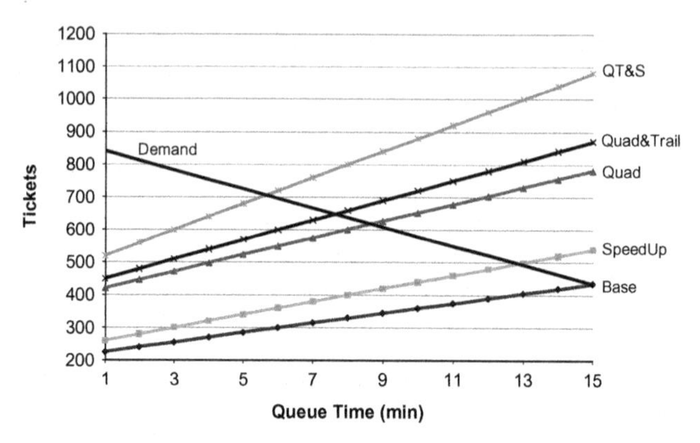

Figure 9.28. Supply and demand curves (M3).

no queue, and we draw the resulting demand line on the graph with the supply curves (Figure 9.28).

The effect of limiting demand is that many outcomes on the supply curves become infeasible. For example, the supply curve for the Quad + SpeedUp + Trail scenario shows that selling 1,080 tickets results in a queue time of 15 minutes. (If fewer tickets are sold, the queue time will be less than 15 minutes.) But demand is only 435 at a 15-minute wait. The Skiway cannot sell 1,080 tickets at the same time it offers a 15-minute wait. Thus, it is *impossible* to create a 15-minute wait under the Quad + SpeedUp + Trail scenario. The 15-minute data point of that supply curve must be removed from consideration. Likewise, at a 14-minute queue time, 1,040 tickets must be sold, but demand is only 464. Again, the outcome is infeasible and the data point gets removed. We progress to the left along the supply curve, eliminating infeasible outcomes, until we reach the intersection of the supply curve with the demand curve. In more general terms, any outcome above and to the right of the demand curve is infeasible. Outcomes below and to the left of the demand curve are feasible because there is more demand than the Skiway would supply. At a wait of five minutes, demand for lift tickets is 725, but the Skiway will only sell 680 tickets in the Quad + SpeedUp + Trail scenario. If it sells any more, the queue time will go up.

Insight 14: Outcomes above and to the right of the demand curve are infeasible.

There are several ways to find the intersection points of the supply and demand curves. The simplest but least automated way is to read the point off the graph. However, we can also use Goal Seek if we make a slight modification to M3 (Figure 9.29). M3 is already set up with Total Skiers as a parameter and Queue Time as a calculation. The Total Skiers parameter is the number of tickets the Skiway will supply at the calculated queue time. We then use

	A	B	C	D	E
1	**Skiway**	M3	Add Money		
2					
3					
4					
5	**Parameters**				
6					
7		Total Skiers	618	people	
8		SkiTime	6	min	
9					
10		Lift			
11		Length	3540	ft	
12		Max Speed	450	ft/min	
13		Chair Spacing	60	ft	
14		Seats per Chair	4	keesters	
15					
16		Average Trail Length	4640	ft	
17		Trail Length Multiplier	125%		
18		Max Group Size	2	people	
19		Minimum Gap	75	ft	
20					
21		Base Case Skiers	435		
22		Weekends per Season	18		
23		Days per Week at Capacity	2		
24		Discount Rate	10%		
25		Project Life	10	yrs	
26		Lift Ticket Price	$40		
27					
28		Target Queue Time	14		
29					
30	**Calculations**				
31			**Max Speed Calculations**		
32		Chairs	60		
33		Lift Capacity	240	people	
34		Lift Time	7.87	min	
35					
36		Chair Interarrival	0.13	min	
37		Chair Interarrival	8	sec	
38					
39		PeopleSkiing	180	people	
40					
41		Queue Length	198.0	people	
42		Queue Time	6.6	min	
43					
44					
45		Skiing Speed	966.7	ft/min	
46		Gap Between Skiers	64.4	ft	
47					
48			**Actual Calculations**		
49		Gap Too Small?	1		
50					
51		Lift Speed	387	ft/min	
52		Lift Time	9.16	min	
53					
54		Chair Interarrival	0.16	min	
55		Chair Interarrival	9.31	sec	
56					
57		People Skiing	154.7	people	
58					
59		Queue Length	223.3	people	
60		Queue Time	8.7	min	
61					
62		Gap Between Skiers	75.0	ft	
63					
64			**Financial Calculations**		
65		Additional Skiers	183	people/day	
66		Additional Revenue	$7,320	$/day	
67		Additional Rev for Season	$263,520	$/year	
68					
69		PV of Incremental Revenue	$1,619,216		
70					
71					
72			**Demand Calculations**		
73		Slope of Demand	-29		
74		y-Intercept	870		
75					
76		Actual Demand @ Given QTime	619		
77					
78		Difference: Allow-Actual	-0.75		
79					

Figure 9.29. M3 Modified to find supply–demand intersection.

linear interpolation of the demand line to calculate demand for the above-calculated Queue Time (C76). For example, in the Quad + SpeedUp + Trail scenario, 1,080 total skiers create a Queue Time of 15 minutes. The model calculates demand at 15 minutes to be 435 skiers. We add a formula that calculates the difference between supply and demand (1,080 – 435 = 645). We use Goal Seek to set this difference to zero by changing the value of Total Skiers. This forces the supply and demand to be the same and gives us the intersection point of the two curves. In this scenario (Quad + SpeedUp + Trail), supply equals demand when Total Skiers is set to 706.

Because we are dealing with linear supply and demand curves, we can also solve for the intersection point algebraically. We can solve for the intersection of some nonlinear curves, as long as we know the algebraic expressions involved. Otherwise, using Goal Seek is our only option.

Since supply and demand are linear, both follow the standard $y = mx + b$ form, where m is the slope and b is the y-intercept. Because supply and demand are functions of queue time, x is in units of minutes and y is in units of tickets or skiers. We define the two lines as

$$\text{Demand: } y_d = m_d x + b_d$$

$$\text{Supply: } y_s = m_s x + b_s$$

We estimate values for m and b for both supply and demand based on the data points we used to draw the graphs. In Figure 9.30a, we use the Excel SLOPE and INTERCEPT functions to calculate the m and b values of the supply lines, based on the data in Figure 9.25. Likewise, in Figure 9.30b, we use the SLOPE and INTERCEPT functions to calculate the m and b values of the demand line based on the assumptions about demand.

To find the intersection of the supply and demand lines, we set them equal to each other and solve for x.

$$\text{Set } y_d = y_s$$

$$m_d x + b_d = m_s x + b_s$$

$$x = (b_s\text{-}b_d) \div (m_d\text{-}m_s)$$

This x value is the queue time at the intersection of supply and demand. It is the maximum queue time that can be achieved in the given scenario. We can plug x back into either the supply or the demand equation to find the number of tickets sold at the intersection point. This is the maximum number of tickets that can be sold in the given scenario. We cannot sell one more ticket because demand does not support it.

We still do not know which scenario is best. It is NPV, not Total Skiers or Queue Time, that determines which choice is optimal. Previously, we developed Figure 9.27 to show the NPV for each Queue Time in each scenario. In Figure 9.27, NPVs for all Queue Times, from 1 to 15, in all scenarios are shown

	A	B	C
59		Base Case	
60	Target Wait Time	Total Skiers	
61	1	225	
62	2	240	
63	3	255	
64	4	270	
65	5	285	
66	6	300	
67	7	315	
68	8	330	
69	9	345	
70	10	360	
71	11	375	
72	12	390	
73	13	405	
74	14	420	
75	15	435	
76			
77	Slope	=Slope(B61:B62,A61:A62)	
78	y-Intercept	=Intercept(B61:B62,A61:A62)	
79			

Figure 9.30. (a) Calculating supply line.

	A	B	C	D
90			Demand Line	
91		BaseCase	Estimate	
92	Time	15	0	
93	Demand	435	870	
94				
95			Demand Calculations	
96	Slope of Demand	=Slope(B93:C93,B92:C92)		
97	y-Intercept	=Intercept(B93:C93,B92:C92)		
98				

Figure 9.30. (b) Calculating demand line.

because we were not considering demand limitations. But now we know that any Queue Time beyond the intersection point of supply and demand is infeasible. For example, the Quad + SpeedUp + Trail scenario intersection queue time is 5.7 minutes, and thus, all data points to the right of this are infeasible. On Figure 9.27, all NPVs from points to the right of 5.7 minutes are infeasible for the Quad + SpeedUp + Trail line, and thus, we can remove these points from the graph.

As with the intersection point, we can find the maximum NPV visually on the graph or we can solve for it algebraically. Since the NPV curve appears to be linear, we use the SLOPE and INTERCEPT functions again to find the m and b values for the NPV lines of Figure 9.27. We plug the intersection x value into the NPV equation to find the NPV at the intersection of the supply and demand curves. This NPV is the maximum NPV attainable from the given scenario and is the new endpoint of the NPV line.

Figure 9.31 shows the result of removing all the infeasible data points from the Figure 9.27 NPV graph. We see that the base case is the only scenario that is feasible for all values of Queue Time from 1 to 15 and that its NPV is never positive, as we have noted before. We also see that the Quad + Trail and Quad

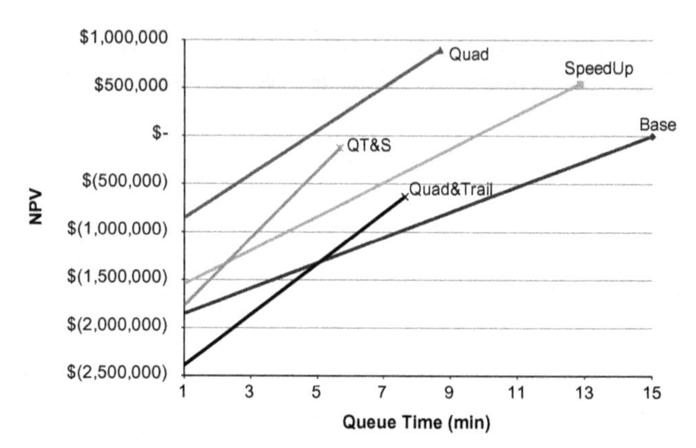

Figure 9.31. NPV under limited demand (M3).

+ SpeedUp + Trail lines never reach positive NPV territory! Apparently, the Skiway cannot get enough demand to justify the costs of these improvements. This is quite helpful to discover because these options can be removed from consideration. (This confirms our earlier decision to remove the Quad + Trail scenario from consideration.)

Insight 15: Because of limited demand, building an additional trail never pays off.

That leaves building a quad and speeding up the lift as the two viable alternatives. Remember that the lines in Figure 9.31 show only feasible solutions and that the Skiway can choose to operate anywhere along these lines by controlling how many tickets are sold. Would the Skiway ever choose to operate somewhere along one of these lines other than at the right end-point? If the Skiway builds a quad, it could choose to sell 524 tickets, have a queue time of five minutes, and realize an NPV of $58,555. However, it could sell one more ticket (the demand is there) and increase its NPV. The queue time would go up slightly, but demand would still exceed supply. It only makes sense to increase ticket sales (and queue time) until the maximum possible NPV is reached. Therefore, the right end points of each line are the only necessary information on Figure 9.31.

Insight 16: The NPV for each scenario is maximized when the Skiway operates at the intersection of the supply and demand curves.

It follows that the Skiway should choose the scenario with the end point that gives the highest NPV.

Insight 17: The Skiway maximizes NPV by choosing the scenario with the highest NPV at the supply–demand intersection.

Given our assumptions, the Skiway should build a quad, sell 618 tickets, have a queue time of 8.7 minutes, and realize an NPV of just under $900,000.

Let us stop and review the analysis so far. We created supply curves for each scenario by calculating the number of tickets that must be sold to achieve various queue times. We then estimated a demand curve from two points: the known data point of the base case and the assumed level of demand when there is no queue. When we combine the demand curve with the supply curves, all points above and to the right of the demand curve are *infeasible*. All points below and to the left of the demand curve are *suboptimal* because NPV can be increased by selling more tickets. Thus, the Skiway should only consider operating at the intersection of the demand and supply curves and should operate under the scenario that has the highest feasible NPV.

Our analysis suggests that building a quad is the indisputable optimal decision for the Skiway. However, this conclusion is based on an estimate of maximum demand. What if the estimate of maximum demand were higher or lower? Would that change the decision? Also, how accurate does the estimate need to be to determine the optimal decision? We return to sensitivity analysis to answer these questions.

As the estimate of maximum demand changes, the slope of the demand curve in Figure 9.28 changes. This, in turn, changes the intersection points with the supply curves, thus changing the maximum NPV of each scenario. With our current maximum demand estimate of 870 skiers, the Quad scenario has the highest maximum NPV. We want to know whether a different estimate of maximum demand causes a different scenario to have the highest maximum NPV. In other words, is the optimal *decision* sensitive to the estimate of maximum demand?

Because we have built a flexible model, we can easily run a sensitivity analysis on the maximum demand estimate and record the results of the maximum NPV of each scenario (Figure 9.32). Notice that the scenario with

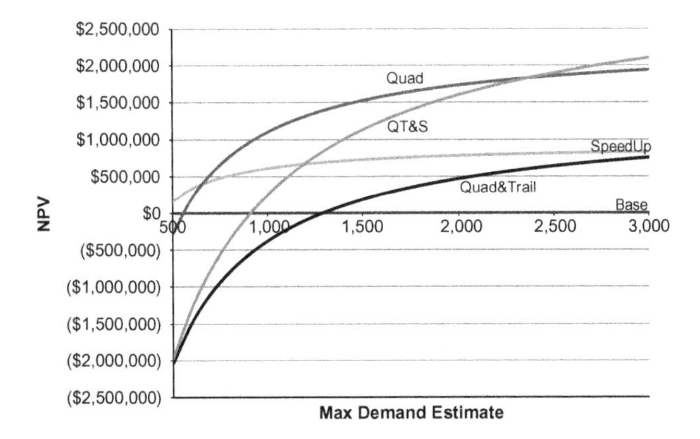

Figure 9.32. Sensitivity of NPV to Max Demand Assumption (M3).

the maximum NPV switches only twice: from SpeedUp to Quad at about 650 skiers, and from Quad to Quad + SpeedUp + Trail at about 2,400 skiers. It is hard to believe that demand would go to 2,400 if the queue length went to zero, so we ignore this crossover point. But the 650 crossover point is worth considering. After all our work, the problem boils down to whether Brian thinks maximum demand (with no queue) is at least 650 skiers. This is a pretty startling realization!

Insight 18: If maximum demand is greater than 650, the Skiway should build a quad. Otherwise, it should just speed up the double lift.

It may be that the Skiway manager is not sure what the answer is, but at least he now knows the right question. This makes doing a customer survey much easier should he choose to conduct one. The survey simply needs to ask whether a person would go to the Skiway if there was never a queue at the lift. This is much simpler than trying to construct an entire demand curve.

Of course, the survey still might not be necessary. If Brian's estimate of maximum demand is well above the 650 skier crossover point, then a little bit of uncertainty in the estimate does not affect the decision. Better yet, the Skiway can learn more about customer demand by running an experiment. The Skiway can speed up the lift and observe how demand changes. Queue time will certainly decrease, and we expect demand will increase. This new data point will help define the demand line, refine the estimate of maximum demand, and clarify the optimal decision.

Figure 9.32 illustrates the trade-offs between the two best scenarios. If the Skiway chooses to just speed up the double lift, it does not risk a negative NPV, and the only downside is it will not realize the larger NPV from building a quad if maximum demand actually turns out to be greater than 650. On the other hand, the Skiway might decide the risk of having a negative NPV with the quad is so low that it is offset by the potential gains of high demand. Also, building the quad allows the Skiway to serve more people, which may well be important in a small town.

Summary of M3

What a long way we have come! We certainly would not have guessed as we started work on this problem that it would all boil down to such a concise question (although we are pleased it did). We have generated some amazing insights so far, but they have all come from deterministic models. We next consider uncertainty to see what additional insights we can gain.

To the Reader: Before reading on, consider which aspects of uncertainty should be modeled.

9.8 CONSIDERING UNCERTAINTY

As we mentioned at the beginning of this chapter, there is a lot of randomness at a ski area. Each skier takes a different path at a different speed with various stops and starts. Skiers come and go throughout the day. Sometimes people fall getting off the lift and the lift stops. Sometimes the lift breaks down. If we were using Discrete Event Simulation to model the Skiway, we would be able to model all the details of an individual skier's experience. However, in our equilibrium model, we are not modeling individuals but system averages, or what can be thought of as the average for a given day.

Consider the Ski Time parameter. In our models so far, this is deterministically set at six minutes. One interpretation is that every skier takes exactly six minutes to get from the top to the bottom of the mountain. If we allow uncertainty, we recognize that skiers take different times to get down the mountain. But the Ski Time parameter really represents the average time down the mountain, not the time for each individual. On any given day, that average ski time is different, and this is the randomness we can simulate in our model.

What would the distribution of that randomness look like? In the base case, there are more than 5,000 individual runs down the mountain in a single day (435 skiers skiing for six hours, at roughly 30 minutes per cycle), and we are told that these individual runs vary from four to eight minutes. Our Ski Time parameter is the *average* of these individual runs. If the ski times are symmetrically distributed, then the mean ski time is six minutes (which is why we picked six minutes for the deterministic base case). When we consider randomness in Ski Time, we are really considering randomness in the mean of individual ski times. However, there is very little variability in the estimate of the mean because we are taking the average of 5,000 individual runs.

We prove this to ourselves by creating a model with 5,000 independent samples from a uniform distribution between four and eight. These samples represent individual ski runs down the mountain. We calculate the mean of the 5,000 individuals; it is six. This is the equivalent of calculating the average Ski Time for a single day at the Skiway. We then run several trials of this experiment and see that the answer is always quite close to six. In fact, the distribution of means is normally distributed around six, with a standard deviation of about 0.02. (We have just demonstrated the central limit theorem.) With so little variability in the Ski Time parameter, there is really no value in bothering to simulate it in our model.

What about Lift Time? Lift Time is a function of the length of the lift and the Lift Speed parameter. In our deterministic models, Lift Time is constant because Lift Speed is constant. (Obviously, the length of the lift is also constant.) When uncertainty is allowed, we recognize that sometimes the lift slows or stops during the day. Any given skier's ride up the lift might be delayed by a few seconds or even a few minutes. Once again, however, we are not modeling individual rides up the lift, but the system average. There are approximately 2,500 rides up the lift in the base case (half as many rides up as runs down since it is a double chair). The Lift Time is the average of these 2,500 rides. Regardless of the distribution of individual delays (it is probably exponential), the distribution of the average lift time will be normal, with a very small standard deviation. So, although the mean delay is greater than zero, there is almost no variability in its value. Thus, we could add a delay parameter into the model, but there is no need to simulate uncertainty in its value. The delay parameter would have the same effect as a constant reduction in Lift Speed, and we saw in M1 that slowing the lift increases the queue time.

It seems that simulating uncertainty adds no value to this problem. This is a startling conclusion for a system that we know has so much uncertainty in it!

Insight 19: Simulating uncertainty adds little or no value to our analysis.

In an equilibrium model like ours, all the parameters are either constants (Lift Length, Chair Spacing) or averages across thousands of independent events (Ski Time, Lift Time). Does this mean we built the wrong kind of model? We think not. Recall what question we are answering. We are trying to determine what policy or capital expenditure best addresses the long queue times that occur only at peak system loading. Our model addresses this problem quite well. However, if we were trying to answer a question about staffing throughout the day or about individual skier experiences, we would need to consider using a discrete event model. The question at hand truly dictates the form of the model. This is why we start with the problem kernel. It focuses us on the key question and removes elements that might distract us and lead us down the wrong path.

Truth be told, we did not recognize the futility of simulating uncertainty in this case when we should have. We plunged ahead and built M4 before stopping to think about what we were really doing. We put in distributions that were logical for individuals but not for means. It was not until we had done a fair bit of analysis that we realized our error. We mention this to clear our consciences and to highlight how easy it is to get lulled into overconfidence in modeling. To prevent wasting time and effort and, more importantly, to avoid erroneous analysis, it is crucial to think through how each element of the model translates to the real world. Perhaps our last insight of this chapter is that it is easy to get carried away with the tools of modeling and not think about what we are really trying to accomplish.

9.9 PRESENTATION OF RESULTS

It is now time to pull together our analysis into a presentation for our client. We assume the audience for our presentation will include the Skiway manager, Brian, a few key members of his staff, a couple of administrators from the college, and perhaps a few potential donors who might foot the bill for the improvements we recommend.

Normally we prefer a direct presentation style with the recommendation stated at the beginning of the talk. However, for this presentation, the indirect method is the better choice. We are presenting a startling result: The optimal decision is directly linked to the Skiway's estimate of maximum demand. The audience will be much more receptive to this conclusion after we have established our credibility and gained their trust in our model.

What story should we tell to connect with our audience and help them understand our work? To build credibility we show base case results and sensitivities that conform to their intuition of how the system works. Our story then moves on to explain how the improvement scenarios reduce queue times and can increase both tickets sold and profits. By this point, our audience will be hooked. We can then dive into the more complex issues of estimating maximum demand and how limited demand limits the improvement options. We show that the optimal decision is different if the estimate of maximum demand is above or below 650 skiers, but it is otherwise insensitive to the estimate. Finally, we show that building a quad lift leads to less waiting, more skiers, a more balanced system, and higher profits.

Here is an outline of the presentation:

- The Skiway Problem
- Approach and Model (Influence Diagram)
- Base Case
- Sensitivities
    - Lift Speed
    - Seats per Chair
- The Value of Improvements
    - Tickets
    - NPV
- Demand Limits Options
- Sensitivity to Estimate of Demand
- Summary

Our PowerPoint slides are shown in Figures 9.33–9.45. The first slide introduces the problem, the three improvement options, and our "peak load" approach to the problem (Figure 9.33). In slide 2, we use the influence diagram to show the logic behind the model. This is the first step toward building credibility with the audience (Figure 9.34). We point out that their ideas of speeding up the lift and/or building a quad fit logically into the model structure.

Slide 3 shows a graphical representation of the base-case scenario under peak load (Figure 9.35). We point out that over 50 percent of a customer's day is spent standing in line, which is the least enjoyable part of skiing. We also

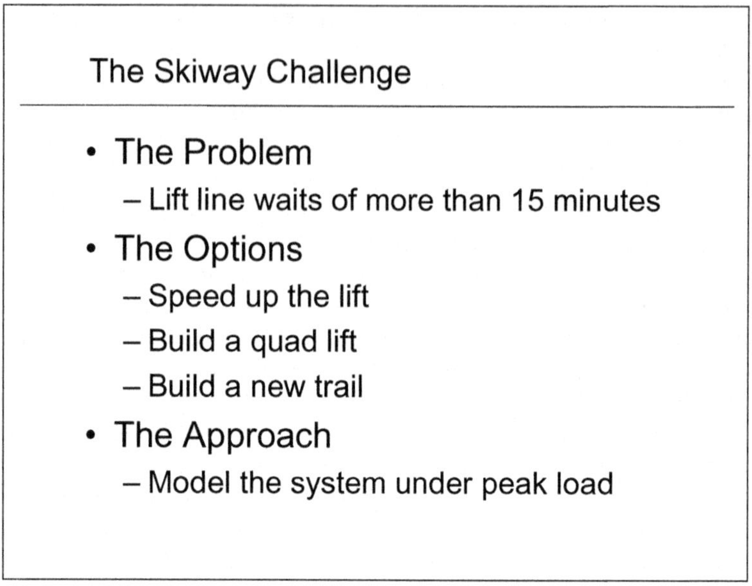

The Skiway Challenge

- The Problem
  - Lift line waits of more than 15 minutes
- The Options
  - Speed up the lift
  - Build a quad lift
  - Build a new trail
- The Approach
  - Model the system under peak load

Figure 9.33. Slide 1.

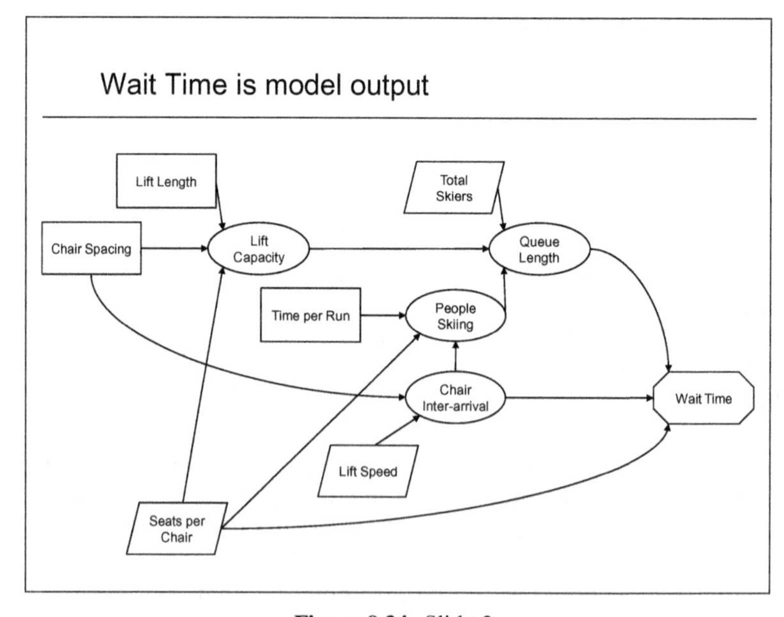

Figure 9.34. Slide 2.

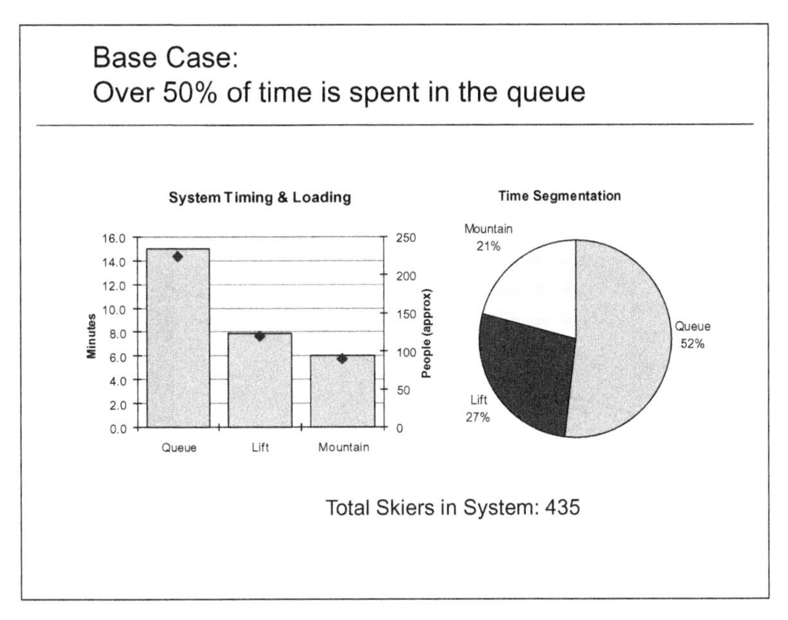

Figure 9.35. Slide 3.

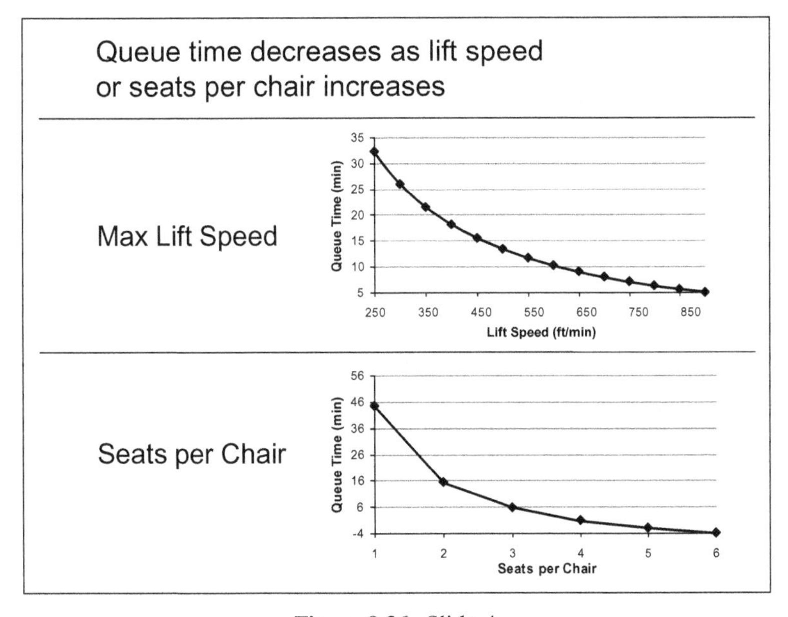

Figure 9.36. Slide 4.

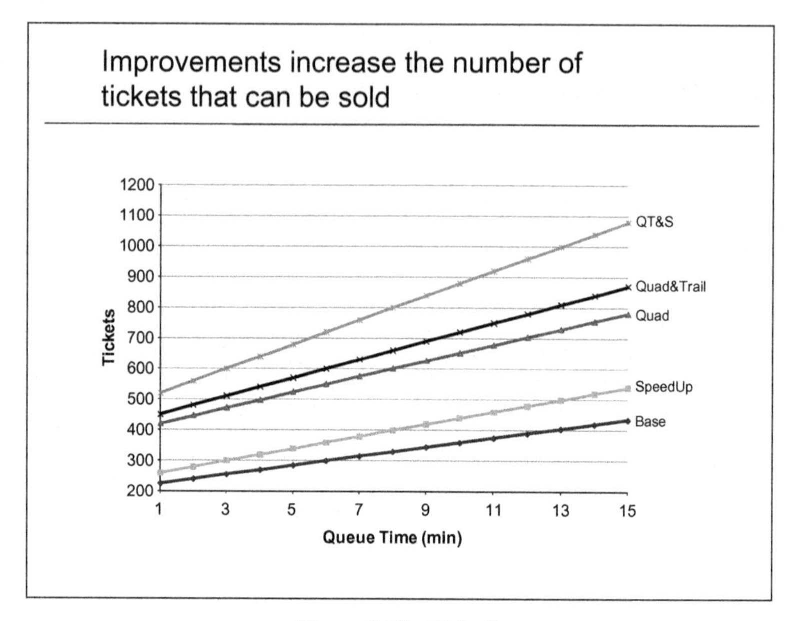

Figure 9.37. Slide 5.

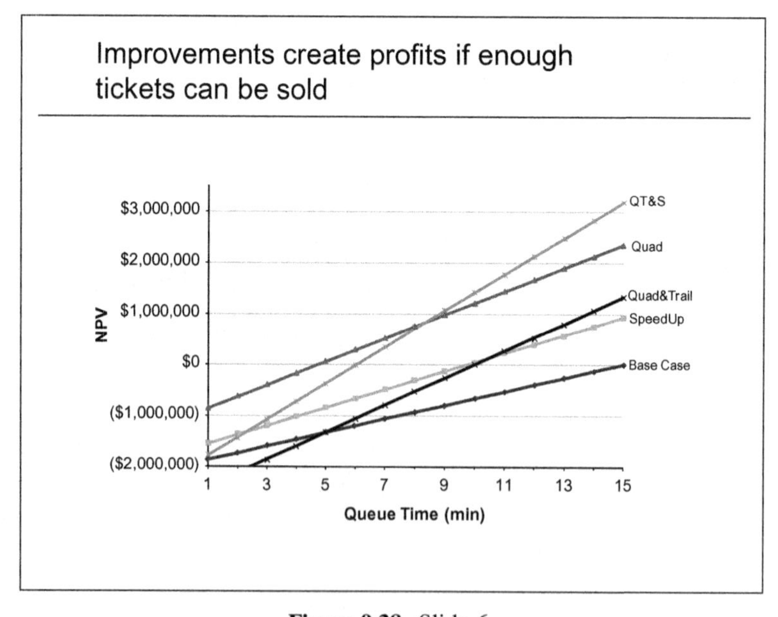

Figure 9.38. Slide 6.

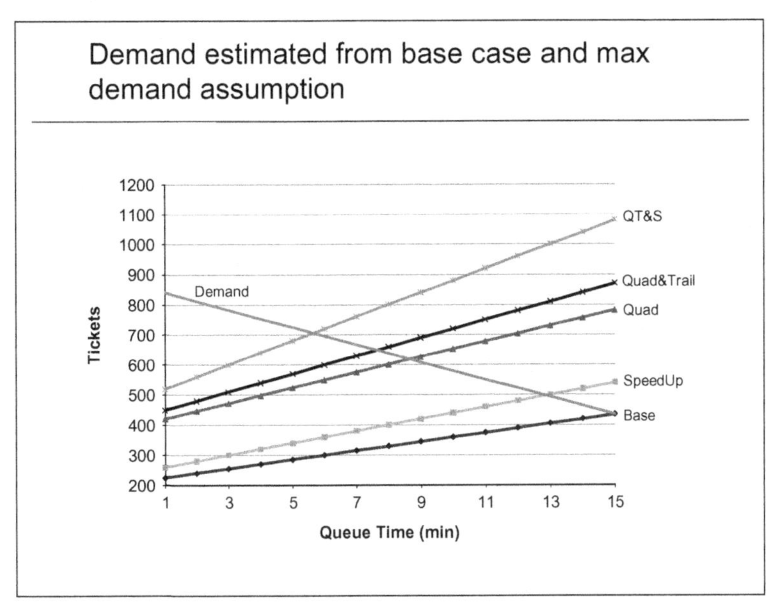

Figure 9.39. Slide 7.

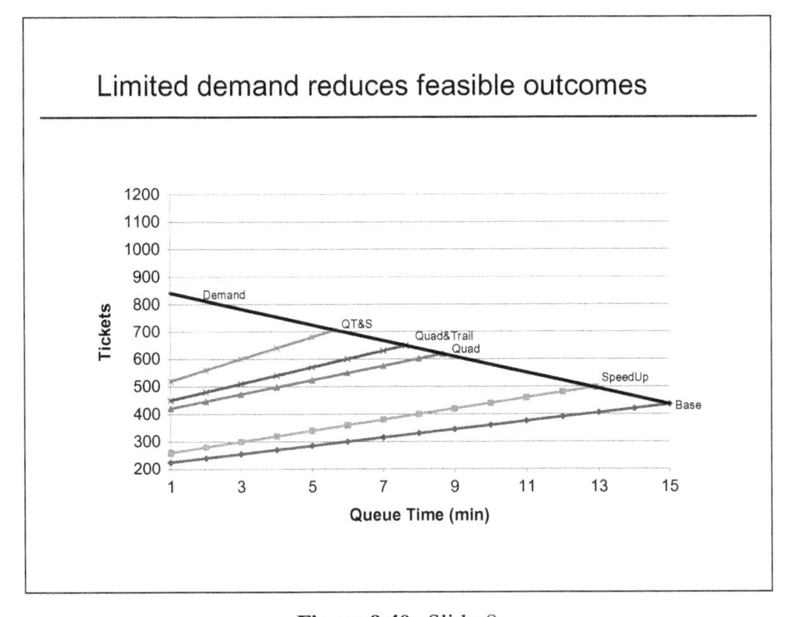

Figure 9.40. Slide 8.

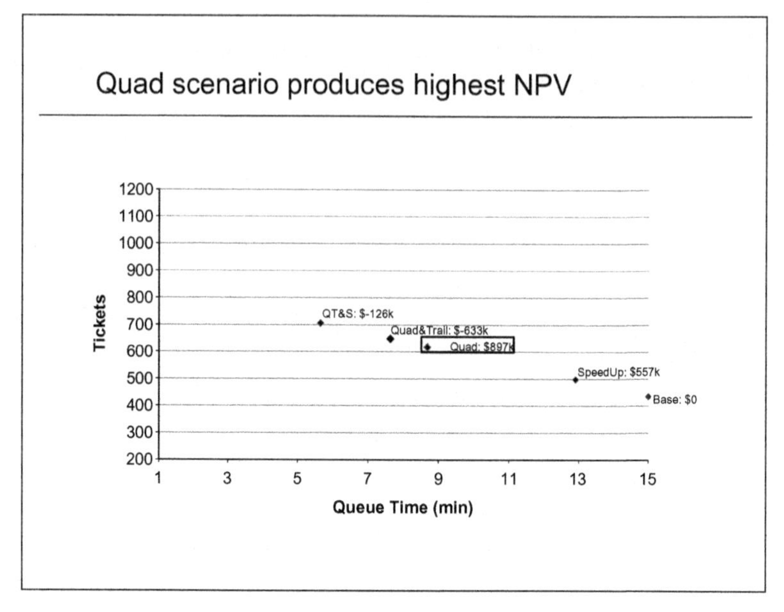

Figure 9.41. Slide 9.

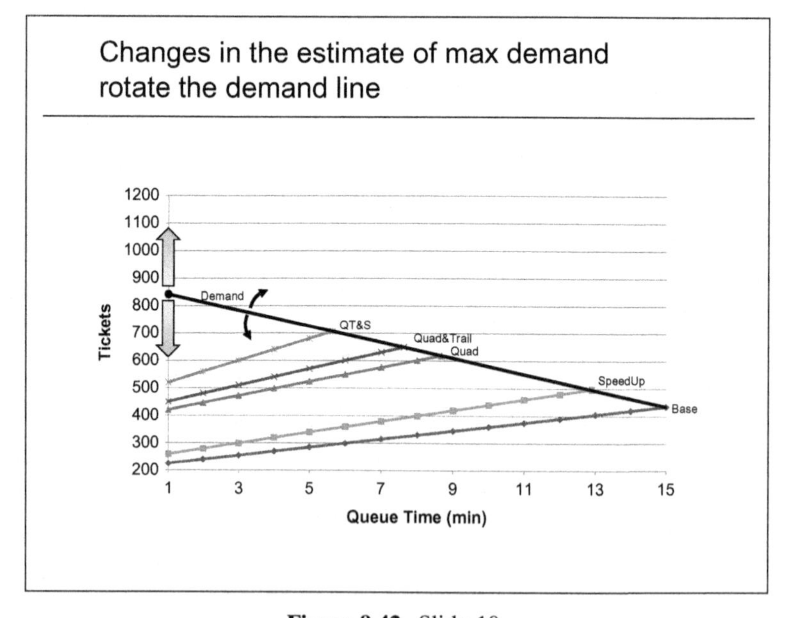

Figure 9.42. Slide 10.

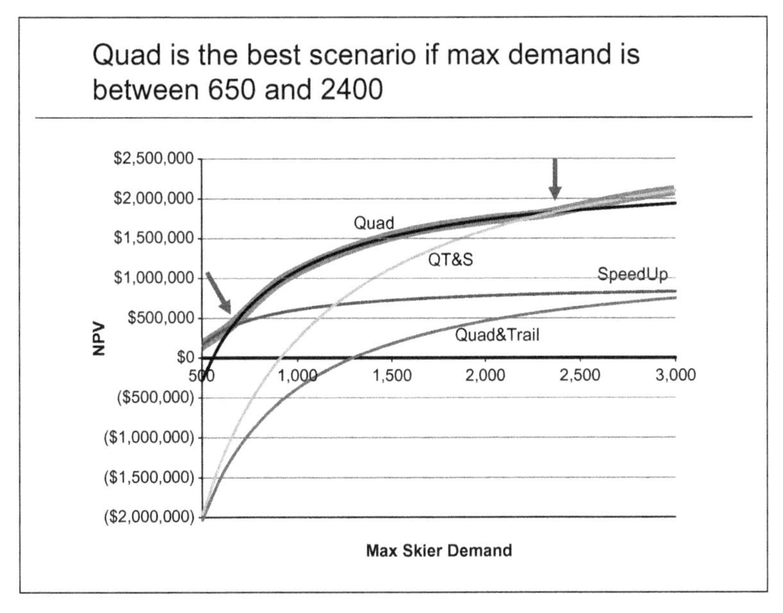

Figure 9.43. Slide 11.

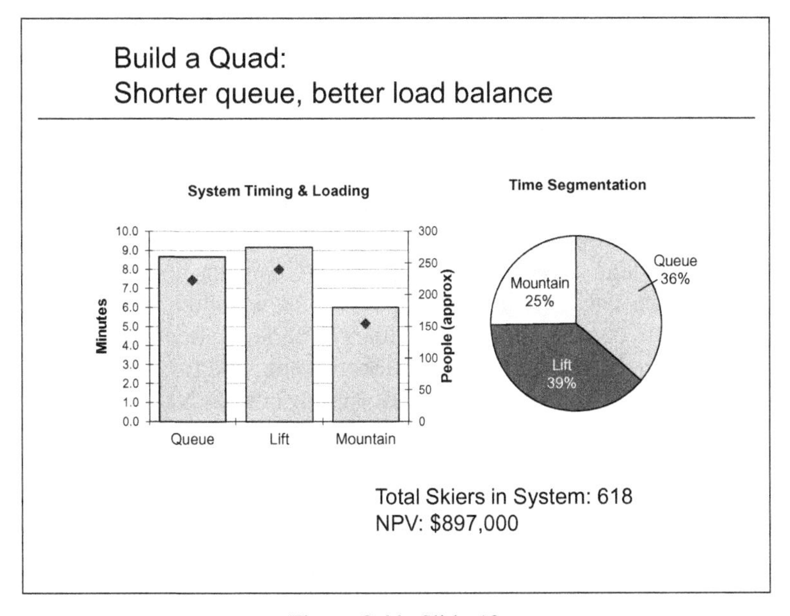

Figure 9.44. Slide 12.

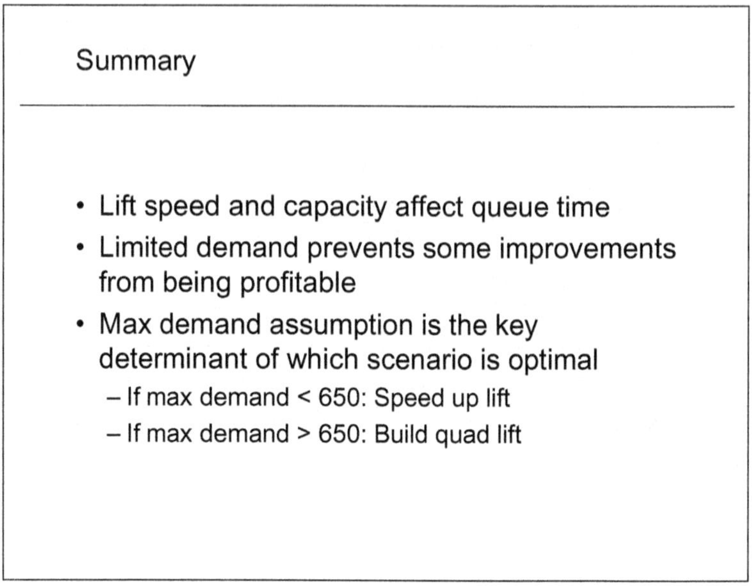

Summary

- Lift speed and capacity affect queue time
- Limited demand prevents some improvements from being profitable
- Max demand assumption is the key determinant of which scenario is optimal
  - If max demand < 650: Speed up lift
  - If max demand > 650: Build quad lift

Figure 9.45. Slide 13.

point out that the segmentation of time is similar to the segmentation of people. In other words, 50 percent of a skier's time is spent in queue, and approximately 50 percent of the people in the system are in the queue at any given point. This slide should resonate with the Skiway employees as they see that our model projects an outcome (system loading) similar to what they see in reality at the Skiway.

We repeat the point that their ideas of speeding up the lift and/or building a quad are good ideas and illustrate this with slide 4 (Figure 9.36). The sensitivity graphs of Lift Speed and Seats per Chair show that increasing either of these reduces Queue Time and that Seats per Chair has a stronger effect.

The discussion becomes more complicated as we show the supply curves in Slide 5 and the corresponding NPV lines in Slide 6. Slide 5 shows that each scenario allows a different number of tickets to be sold for each target queue time (Figure 9.37). These are the scenario supply curves. Next, we explain that each possible outcome leads to an NPV that is a function of the number of tickets sold and the cost of the improvements (Figure 9.38).

The next three slides work together to show how we determine the optimal decision. Slide 7 (Figure 9.39) adds the demand line to the graph of supply lines show in Slide 5. We explain that we estimated the demand curve from the base-case data point and an estimate of demand when there is no queue (in the graph shown, demand doubles to 870). The managers might suggest that the demand curve is not linear in reality, and that is fine. We can adapt the model to include whatever demand curve shape they prefer. The form of the curve will probably not dramatically change the results. We explain that all points above and to the right of the demand curve are infeasible outcomes and, thus, are eliminated from consideration. This is shown in Slide 8 (Figure 9.40). Note that we use fundamental concepts like supply and demand with which they are likely to be familiar to explain our model and to build their confidence in us.

Finally, we describe how all the points below and to the left of the demand line are suboptimal and are eliminated. This leads to Slide 9 in which only the intersection points of supply and demand remain (Figure 9.41). We also show on this graph the NPV of each point and that the Quad scenario produces the highest NPV. We would not mention that the lift has to be slowed down with the quad chair. This would likely serve to confuse our listeners and to distract them from the stunning results coming next. Under the current assumptions, the Skiway should build a quad, sell 618 tickets, have a queue time of 8.7 minutes, and realize an NPV of just under $900,000. These numbers should grab the attention of everyone in the room. They probably have no idea that investing in a quad can improve the queue time problem, allow the Skiway to serve more customers, balance the load on the mountain, *and* be highly profitable!

Although most people will, hopefully, be wowed by these results, there may well be one person who demands to know "what if your assumptions are wrong?" First, we gently point out that we are using data and estimates provided to us by the Skiway staff, then we show them Slide 10 (Figure 9.42). This slide shows that changing the max demand assumption has the effect of rotating the demand line up and down, which changes the intersections of the supply and demand lines. This, in turn, changes the maximum achievable NPV from each scenario. Slide 11 shows how these maximum NPVs change as we change the estimate of maximum demand (Figure 9.43). It is only around the crossover points, marked on the slide with arrows, where the decision is sensitive to the estimate. If management is confident that maximum demand is greater than 650 skiers, then building the quad is the right choice. The forecast of NPV will be incorrect if the estimate of demand is incorrect, but the decision will be correct. We also use this slide to show that the costs of making the wrong decision are low.

Our bottom line recommendation is that the Skiway should build a quad lift unless the managers are confident that maximum demand is less than 650 skiers. Slides 12 shows the system loading under this recommended scenario (Figure 9.44). Queue Time is under nine minutes and the system is much more evenly loaded. We close with a summary slide of our analysis and recommendation (Figure 9.45).

9.10 SUMMARY

This case demonstrated the use of an equilibrium model to gain insight into how best to reduce queue times at a small ski area. Using an equilibrium model allowed us to maintain a high-level view of how the system as a whole functions without having to worry about the details of each skier and ski run. If we had been concerned with moment-by-moment decisions, we would probably have needed to create a Discrete Event Simulation rather than an equilibrium model.

There were two items of modeling methodology of particular note in this case. We used Solver Sensitivity for the first time. Solver Sensitivity automated the process of finding the maximum number of tickets the Skiway could sell given a constraint on queue time that varied from 1 to 15 minutes. The other methodological note was that simulation of uncertainty was not helpful. Because an equilibrium model is based on mean values, not on individual events, simulation could not add any insight to this case.

We believe modeling adds value to a situation like the Skiway decision because it allows managers to explore what drives operations. Understanding operations then allows managers to ask the right questions. In this case, the right question—what is maximum demand?—was not at all obvious. Thus, modeling proved useful by highlighting the correct question for managers to focus on in order to make good operational decisions.

CHAPTER 10

National Leasing, Inc.

10.0 INTRODUCTION

Leasing has become an increasingly popular way to own an automobile, especially among affluent consumers. Leasing has several advantages over ownership, the most important of which is the option to drive a new car every three or four years without making large expenditures at any one time.

The auto-leasing business has at times been very profitable for credit companies. The essential economics of an auto lease involve the leasing company borrowing money at a low interest rate to purchase the car, and then leasing it to the consumer at a higher rate. The leasing business is substantially complicated by the purchase option that typical leases offer consumers. In most leases, the consumer can buy the car from the leasing company for a prespecified price at the end of the lease. Consumers who choose not to buy can simply return the car when the lease ends. The option to purchase is part of what makes leasing attractive to many consumers, but it has, at times, led to large financial losses for the leasing companies.

This case focuses on the issue of how leasing companies should set the purchase price of cars at the end of the lease. As you will read in the case, some managers believe the purchase price in the contract should be set equal to the best available *forecast* of the used car's value at the end of the lease. From this point of view, the real challenge is to develop better forecasts of used car prices. Others believe the purchase price should not be set equal to the forecast of the used car's value. You will be asked to shed light on this issue through modeling.

A complete and operational model for setting lease terms would take many months of labor to develop. However, the case asks for a *prototype* model—a stripped-down, inexpensive version of the complete model that can show that modeling *works*. Here we will develop a prototype to show decision makers that a modeling approach can shed light on the lease decisions they face and can be expanded later to include the data and details we deliberately leave out. Building effective prototypes is an essential modeling skill because, in many situations, the approval and funding for a full-scale modeling effort will only be secured after a successful proof of concept is made using a prototype.

10.1 NATIONAL LEASING CASE

National Leasing

In the 1990s, new-vehicle leasing grew to 40 percent of new car sales. Consumers who might otherwise purchase a new car every few years are attracted by monthly lease payments that are lower than those for financing a new-car purchase. Thus, leasing allows consumers to drive a nicer vehicle than they could afford to buy or finance. The most popular leases are for expensive or mid-range vehicles and carry a term of twenty-four or thirty-six months.

Most leases are sold via so-called captive leasing companies, run by a vehicle manufacturer. About 40 percent of leases are sold by independent leasing companies, primarily banks and other financial firms. Among the independents, there are six major national players, which compete against a host of smaller regional companies.

Increasing competition among leasing companies and online pricing information has made vehicle leasing nearly a commodity. Consumers care most about getting the lowest monthly payment, other factors being equal. Online information sources at dealers readily report the lowest lease payments for a given vehicle.

Demand for any one lease is highly unpredictable. However, it is generally accepted that it is sensitive to the gap between a leasing company's monthly payments and the going market rate for that car model, which is usually set by the lease with the lowest monthly payments. Other factors, such as the purchase price at lease-end, appear to be secondary in the consumer's mind.

The most common form of leasing is the closed-end lease. In this arrangement, the monthly payment is computed based on three factors:

1. Capitalized cost: purchase price for the car, net of trade-ins, fees, discounts, and dealer-installed options
2. Contract residual value: value of the vehicle at the end of the lease, specified by the leasing company in the lease contract; the consumer has the right to purchase the vehicle at this price upon termination of the lease
3. Money factor: interest rate the leasing company charges in the monthly payments

A typical leasing company acts much like a bank, borrowing money at a low interest rate and leasing cars to consumers at a higher rate. Thus, the leasing company essentially is making a monthly payment to its bank on the full price of the vehicle, while receiving a monthly lease payment from its customer based on the difference between the full price and the contract residual value.

For a given vehicle, a lower contract residual value implies a greater drop in value over the term of the lease, prompting higher monthly payments. Therefore, a leasing company that offers the highest residual value usually has the lowest, and most competitive, monthly payment. Such a high residual value, relative to competitors, is likely to sell a lot of leases. However, this can be a hidden time bomb if the leasing company expects to receive payment of the contract residual value at the end of the lease. If the *actual* end-of-lease market value is lower than the contract residual value, the consumer is likely to return the car to the leasing company. The leasing company then typically sells the vehicle, usually at auction, and realizes what is called in the industry a "residual loss." Note that a residual loss is not a negative cash flow but simply a lower cash flow than anticipated.

If a leasing company sets a low contract-residual value, then the corresponding monthly payments are higher. This reduces the number of leases sold in this competitive market. And, at the end of these leases, if the actual market value is greater than the contract residual, the consumer is more likely to purchase the vehicle at the contract residual value. By then selling the vehicle for the prevailing market value, the consumer in essence receives a rebate for the higher monthly payments during the term of the lease. (Of course, the consumer may also decide to keep the car.) When consumers exercise their purchase option, the leasing company loses the opportunity to realize "residual gains."

The economically rational thing for the lease owner to do at the end of the lease is to purchase the car when the actual market value exceeds the contract residual (and resell at the higher price) and to return the car to the leasing company when the actual market value falls below the contract residual value. However, not all consumers behave in this fashion when their leases end. Some will buy the vehicle regardless of the actual market value, presumably because they have become attached to the vehicle or because the transactions costs of acquiring a new vehicle are too high. Others will not purchase even when the actual market value is well above the contract residual value. Then there are those who will purchase even when the actual market value is well below the contract residual.

The primary challenge for companies offering a closed-end lease is to select intelligently the contract residual value of the vehicle 24, 36, 48, or even 60 months into the future. Intelligent selection means that the leasing company must offer competitive monthly payments on the front end, while not ignoring the risk of being left with residual losses on the back end two or more years in the future. To cushion financial performance against this risk, leasing companies set aside a reserve against residual losses. This practice is similar to insurance companies' reserves against future claims.

During the period 1990–95, used car prices rose faster than inflation (5–6 percent per year on average in nominal terms). This price rise was driven by the higher quality of used cars (itself a result of higher manufacturing quality standards in new cars), high new-car prices relative to used-car prices, and a greater social acceptance of used vehicle ownership.

In this environment, leasing companies rarely suffered residual losses, since they generally were conservative in residual setting, forecasting low used-vehicle prices. They missed some opportunity to capture more volume, but this trend caught all players off guard and, therefore, no single leasing company was able to take advantage of it by offering lower monthly payments.

In 1996–97, used vehicle prices first leveled off then dropped dramatically. This shift was driven largely by the oversupply of nearly new used vehicles returned at the end of their leases. The oversupply and attendant price drops were particularly evident for sport-utility vehicles and light trucks. Suddenly, leasing companies found themselves with mounting residual losses, as much as $2,500 per vehicle in some cases.

Company Profile

National Leasing, Inc. is a major independent provider of auto leases, with $10 billion in vehicle assets on its books. National sold just over 100,000 leases in 1997. Buoyed by the general used-vehicle price strength described above, the company experienced very fast growth and excellent profitability from 1990 to 1994. Competition drove down share dramatically shortly thereafter, slowing growth and reducing profitability.

From 1995 to 1997, fewer than 20 vehicle types comprised the bulk of National Leasing's portfolio. These were the models in which the company offered a competitive (high residual value) monthly payment. Six models accounted for about half the total volume. One sport utility vehicle in particular accounted for nearly 20 percent of total vehicles in the portfolio. These concentrations were the result of adverse selection: National sold large volumes of leases for which it set the highest residual compared with the competition. Such competitive rates were the key to success in a period of generally rising used-car prices.

But in 1997, when used car prices dropped 8 percent in the first six months of the year, National was left with significant residual losses. Consequently, the company reported a loss of net income of $400M in fiscal year 1997, prompting an internal audit. The audit revealed that many of the losses were related to operational errors and inefficiencies, including improper data entry, inadequate information systems, and faulty reporting procedures.

The audit also revealed the following flaws in the process used to determine contract residual values:

- No explicit consideration was given to the risks of setting residual values too high or too low.
- Estimates were made by a small group of internal analysts, ignoring external market information and expertise.
- Setting future contract residual values relied excessively on current market residual values.

Current Situation

During the first half of 1998, National Leasing revamped its operations, thereby correcting most of the data entry and information technology-related problems. At the same time, the internal residual-value forecasting group adopted a very conservative posture, setting residuals for new leases at low levels across the board. Rumors suggest that a new manager will be brought in to run residual setting and a new process will be developed.

Senior management is divided on the question of what new residual forecasting method to use. Some believe National should simply use values from the Auto Lease Guide (ALG), a standard industry-reference publication. Others strongly disagree with this approach on the grounds that using ALG eliminates National's competitive advantage. This faction supports their opinion with analysis showing that using the ALG would not have avoided the 1997 losses. Management has also considered hiring a consulting company to improve the used-car-price forecasting process.

Despite a general consensus among industry insiders that most other major leasing companies experienced similar net income losses in 1997, National's major competitors did not follow its lead in setting lower residuals. The higher monthly payments associated with National's low residual values resulted in a 50 percent drop in sales volume during the first six months of 1998. Used car prices have continued to decline since then, apparently driven by flat (or falling) new car prices. National Leasing's senior management, fearing the industry is entering a period of sustained used-car-price deflation, is reluctant to raise residuals to competitive levels. They think the competition is crazy.

In recent meetings, senior managers have discussed a number of possible solutions to these problems, including improving residual-setting techniques, acquiring competitors, entering the downstream used-car business, and even getting out of the new-vehicle leasing business. An influential senior manager at National believes a modeling approach might assist National in making better decisions on lease terms on individual vehicles. In order to give you a forum in which to promulgate your ideas, she has arranged time for you to make a short presentation to the board next week. In this presentation, your goal will be to convince board members that your modeling approach will significantly improve their management of the lease business.

To the Reader: Before reading on, frame this problem by setting boundaries for the analysis; making assumptions; listing questions; defining outputs, inputs, and decisions; and so on.

10.2 FRAME THE PROBLEM

This case raises a host of issues that could easily overwhelm an undisciplined modeler. Fortunately, the client does not expect us to develop a full-scale model but merely to demonstrate that a modeling approach might help National make better leasing decisions. This is an ideal setting for a prototyping approach.

Our first task is to simplify National's problem to the point where we can address it successfully in a week. With limited data and time available to us, we cannot expect to show our client that we can uncover the "right" answer to the problem. Rather, we will attempt to identify the key trade-offs that define the leasing decision and show how modeling can shed light on these trade-offs.

In reality, decisions about lease terms are made periodically (perhaps quarterly or monthly) on dozens of car models. In principle, the decision on a single lease type should take into account the lease terms offered on other car types as well as the existing book of leasing business, which may total hundreds of car models and lease lengths. This is the portfolio problem to which the client alluded. A key simplification is to ignore the portfolio problem and to concentrate on a single car lease. In other words, we set out to determine the essential economic forces that influence the choice of lease parameters, including the contract residual value and the money factor. We presume a model that captures the trade-offs involved in the choice of terms for a single lease would be useful in modeling the larger portfolio problem.

To make the problem more concrete, we focus on a specific class of lease, say, a four-year lease on a Honda Accord, with a purchase price of $25,000. We further assume National can borrow money at 5 percent. The company's decisions are the remaining terms of the lease: the money factor (used to calculate the monthly payments) and the contract residual value, which is the price at which the consumer can purchase the car at lease end. Of course, the monthly payments are related to the difference between the initial sale price of the car and the contract residual value.

Some of the trade-offs in this problem are self-evident. A higher money factor raises the monthly payment and increases revenue (all else equal) but, presumably, decreases demand. A higher contract residual value decreases the monthly payment and decreases the revenue from a single lease (all else equal) but raises demand; it may also reduce the probability that the consumer will purchase the car at lease end. Although these effects are readily understood individually, the combined effects of all the forces in this problem are complex.

There are two areas in this problem where modeling is especially challenging. One area involves the demand for leases. Once we set the money factor and the contract residual value, the monthly payment on the lease is set. We have to find a way to estimate the *number* of leases we will sell at that price in order to estimate the overall profitability of leases on this class of vehicle. The other area that requires some creativity involves consumer decision making at the end of the lease. The lease contract allows the consumer to buy the vehicle at the contract residual value (assuming the other terms of the contract have been met, such as not exceeding a mileage limit). How do consumers actually behave at lease end? Are they economically rational, in which case, they will buy the vehicle if its value on the used car market exceeds the contract residual value by more than the cost of selling the vehicle? Do consumers prefer to buy their vehicle because it is familiar, regardless of the going rate for used cars? Do they purchase their leased vehicle even when a cheaper equivalent vehicle is available on the used market? These are all questions of fact, which, in principle, we could research. But we have limited time to develop a model and common sense suggests that consumers engage in all possible behaviors in different proportions.

Would collecting data help us to build the type of model we need or merely confuse the issue?

As always, we find it useful to reduce the problem to its essentials before proceeding too far with modeling. Here is our statement of the problem kernel for National Leasing:

> Automobile leasing companies compete by setting the money factor and contract residual value on their leases. (All other terms of consumer leases, such as the leases' length, are standard across the industry.) These two decisions determine the lease's profitability through their influence on the monthly payment, the number of leases sold, and the cash flows at the end of lease. How should the client best determine the money factor and contract residual value on a particular vehicle?

To the Reader: Before reading on, sketch an influence diagram for this problem.

10.3 DIAGRAM THE PROBLEM

Drawing an influence diagram helps us better understand the modeling challenge we face. Recall our earlier description of a leasing company as a kind of bank, which borrows at a low rate and sells at a high rate. What makes the business complicated is the contract residual value. Not only does the lease contract give the consumer the right to buy the car at lease end for the contract residual value, but also the contract residual value (along with the purchase price) determines the monthly payment.

We start with the assumption that our client's goal is to maximize cash flows from leases. Each category of leases generates three distinct cash flows:

1. Monthly lease payments from consumers to the leasing company (inflow)
2. The cash flow that occurs at the end of the lease when the consumer either buys the vehicle at the contract residual value or returns it to the leasing company, which then sells it on the used car market (inflow)
3. Payments from the leasing company to its lender to cover the cost of borrowing funds to purchase the car (outflow)

Since actual leases cover two, three, or four years and a significant portion of the total cash flow comes in at the end of the lease, it is necessary to account for the time value of money. Thus, we take as our objective the Net Present Value of Total Cash Flows (Figure 10.1).

The first level of decomposition of the objective shows the three cash flow components: Total Cost of Borrowing, Total Lease Payments, and Total

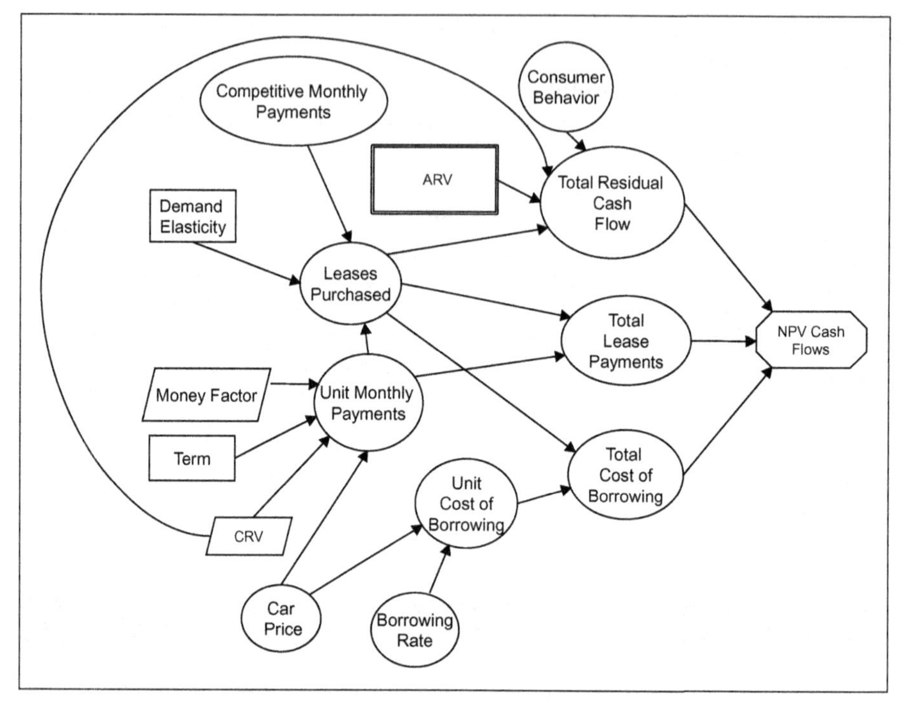

Figure 10.1. Influence diagram for National Leasing.

Residual Cash Flow. The Total Cost of Borrowing depends on the Unit Cost of Borrowing (per vehicle) and on the Leases Purchased. The Unit Cost of Borrowing, in turn, depends on the Car Price and on the Borrowing Rate (the rate at which the leasing company can obtain funds).

The Total Lease Payments depend on the Unit Monthly Payments and on the Leases Purchased. The Unit Monthly Payment depends on the Term, Contract Residual Value (CRV), and Money Factor (MF). (The monthly payment is calculated as if the consumer were borrowing the difference between the price of the car and the CRV at the MF.)

The Total Residual Cash Flow represents whatever cash flow the leasing company receives at the end of the lease. We simplify somewhat and assume only two things can happen at the end of the lease: either the consumer buys the vehicle at the CRV or the consumer returns the vehicle and the leasing company sells the vehicle at the going used car price. We refer to the used car price as the Actual Residual Value or ARV. We include an influence called Consumer Behavior in Figure 10.1 to represent the as-yet unspecified rules we will use in our model to represent the way consumers make this choice.

The final aspect of the model that needs explication is how the demand for leases is determined. We assume the unit monthly payment influences demand. We also assume the menu of competing leases available in the market for the same vehicle influences demand. Exactly how this influence is felt we leave for elaboration later. In this first influence diagram, we merely add a variable called Demand Elasticity to suggest that consumers are sensitive to the cost of leasing from our client's company, relative to the cost of alternative choices.

Our initial influence diagram leaves us with two modeling challenges: how to model demand and how to model the end-of-lease decision. Common sense suggests that the demand for our leases depends on the relationship between our monthly payment and that of competitors. If leases were really a pure commodity, then all demand would go to the lowest-priced competitor. But we suspect this does not happen. Leases probably differ from one another, at least in consumers' eyes, because they are offered by companies with different reputations or name recognition. If we assume our competitors offer leases at a series of different prices, then as we increase our monthly payment, we assume more consumers choose leases from our competitors. This basic demand effect can be captured with a simple relationship between our monthly payment and the number of leases purchased. In later versions of the model, we elaborate on this simple mechanism.

The consumer's decision at the end of the lease is another potentially critical aspect of the model. Actual consumer behavior is undoubtedly complex. Some proportion of consumers make economically rational decisions, whereas others do not. As always, our goal is to start with the simplest credible model we can construct, one that can be expanded easily later. In this case, the simplest approach seems to be to assume all consumers are economically rational; in other words, the consumer's decision depends only on the relationship between the CRV and the actual used-car price (ARV). If the CRV is lower than the ARV, consumers purchase their leased vehicle (and, presumably, sell it to gain the difference between the CRV and the ARV), whereas if the CRV is higher than the ARV, consumers return the vehicle to the leasing company, which must sell it at the ARV. Among other things, we ignore the costs of selling a used car and acquiring a new car in this initial model.

To the Reader: Before reading on, design and build a first model based on the influence diagram.

10.4 M1 MODEL AND ANALYSIS

Our first model implements most of the ideas expressed in the influence diagram in Figure 10.1. We use a simple demand curve, in which demand in number of leases sold is governed only by the price of National's lease. The constant-elasticity demand curve we use (see Figure 10.2) has the following form:

$$\text{Demand} = a \times \text{Price}^{-b}$$

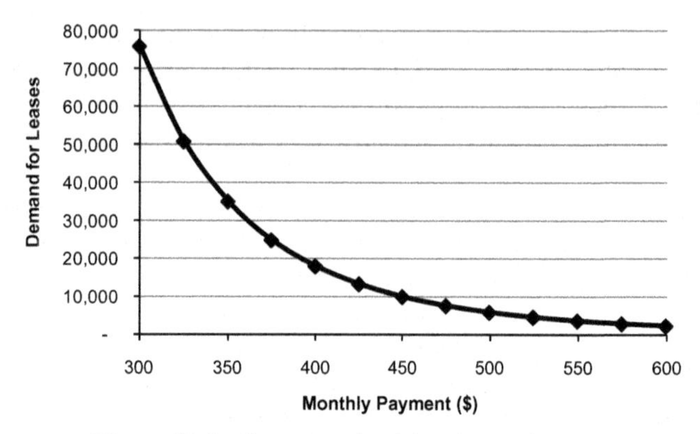

Figure 10.2. Constant elasticity demand curve.

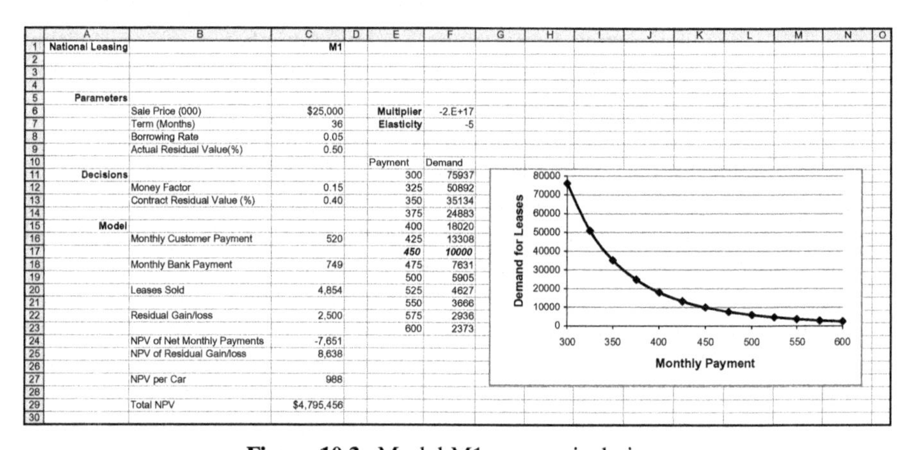

	A	B	C	D	E	F
1	**National Leasing**		**M1**			
2						
3						
4						
5	**Parameters**					
6		Sale Price (000)	$25,000		**Multiplier**	-2.E+17
7		Term (Months)	36		**Elasticity**	-5
8		Borrowing Rate	0.05			
9		Actual Residual Value(%)	0.50			
10					Payment	Demand
11	**Decisions**				300	75937
12		Money Factor	0.15		325	50892
13		Contract Residual Value (%)	0.40		350	35134
14					375	24883
15	**Model**				400	18020
16		Monthly Customer Payment	520		425	13308
17					***450***	***10000***
18		Monthly Bank Payment	749		475	7631
19					500	5905
20		Leases Sold	4,854		525	4627
21					550	3666
22		Residual Gain/loss	2,500		575	2936
23					600	2373
24		NPV of Net Monthly Payments	-7,651			
25		NPV of Residual Gain/loss	8,638			
26						
27		NPV per Car	988			
28						
29		Total NPV	$4,795,456			
30						

Figure 10.3. Model M1—numerical view.

This functional form is known as a constant-elasticity demand function because the percentage change in demand that results from a given percentage change in price is given by the parameter b. With $b = 5$, a 10 percent change in the price from any price level results in a 50 percent decline in demand.

We also assume in this model that consumers behave rationally at the end of the lease and purchase their leased vehicles only when it is in their economic interest.

This first model is shown in Figures 10.3 and 10.4. The parameters in the problem are Sale Price, Term, Borrowing Rate, Actual Residual Value (the used car price as a percent of Sale Price), Demand Multiplier, and Elasticity. The decisions are the Money Factor and the Contract Residual Value (the contract residual as a percent of the Sale Price). Note that we express both the CRV and the ARV as a percentage of the sale price, rather than in absolute dollars, to keep the model as general as possible.

	A	B	C	E	F	G
1	**National Leasing**		**M1**			
2						
3						
4						
5	**Parameters**					
6		Sale Price (000)	25000	**Multiplier**	-184528124998844000	
7		Term (Months)	36	**Elasticity**	-5	
8		Borrowing Rate	0.05			
9		Actual Residual Value (%)	0.5			
10				Payment	Demand	
11	**Decisions**			300	=-$F$6*(E11^$F$7)	
12		Money Factor	0.15	325	=-$F$6*(E12^$F$7)	
13		Contract Residual Value (%)	0.6	350	=-$F$6*(E13^$F$7)	
14				375	=-$F$6*(E14^$F$7)	
15	**Model**			400	=-$F$6*(E15^$F$7)	
16		Monthly Customer Payment	=-PMT($C$12/12,$C$7,$C$6-C13*C6)	425	=-$F$6*(E16^$F$7)	
17				450	=-$F$6*(E17^$F$7)	
18		Monthly Bank Payment	=-PMT($C$8/12,$C$7,$C$6)	475	=-$F$6*(E18^$F$7)	
19				500	=-$F$6*(E19^$F$7)	
20		Leases Sold	=-$F$6*C16^$F$7	525	=-$F$6*(E20^$F$7)	
21				550	=-$F$6*(E21^$F$7)	
22		Residual Gain/loss	=C6*(C9-C13)	575	=-$F$6*(E22^$F$7)	
23				600	=-$F$6*(E23^$F$7)	
24		NPV of Net Monthly Payments	=-PV($C$8/12,$C$7,C16-C18)			
25		NPV of Residual Gain/loss	=IF(C22>0,C13*C6,C9*C6)*(1/(1+$C$8)^(C7/12))			
26						
27		NPV Per Car	=C24+C25			
28						
29		Total NPV	=C27*C20			
30						

Figure 10.4. Model M1—relationship view.

We input the Demand Multiplier and Elasticity, which are the parameters *a* and *b* in the demand relationship, in cells F6 and F7. Then, in E11:F23, we create a table that shows the demand for monthly payments between \$300 and \$600. The demand curve is plotted next to the table. Plotting the demand curve provides a convenient way to visualize the relationship and to experiment with alternative parameters.

The Monthly Customer Payment (cell C16) is calculated using the PMT function, which determines the monthly payments on a loan of the Sale Price less the CRV, at the Money Factor for the Term of the lease. A similar calculation is used to determine National's payments for the loan used to finance the Sale Price (cell C18). Note that the interest rate in this calculation is the Borrowing Rate, not the Money Factor. In cell C20, we calculate the number of leases sold, using the demand curve described above and the Monthly Bank Payment as the price. In cell C22, we calculate the Residual Gain/Loss, which we define as the ARV less the CRV. In cells C24 and C25, we calculate the net present values of the two key cash flows: the difference between the monthly payments to the bank and from the consumer (cell C24), and the residual cash flow (cell C25). For the latter, we use an IF statement that states that if the Residual Gain/Loss in cell C20 is positive (so the ARV exceeds the CRV), the cash flow to the leasing company is the CRV; otherwise, it is the ARV. This is a mathematical statement of our assumption that the consumers always act rationally at lease end. Finally, in cell C27, we calculate the NPV per vehicle, and in cell 29, we calculate the Total NPV by multiplying the NPV per vehicle by the number of leases sold.

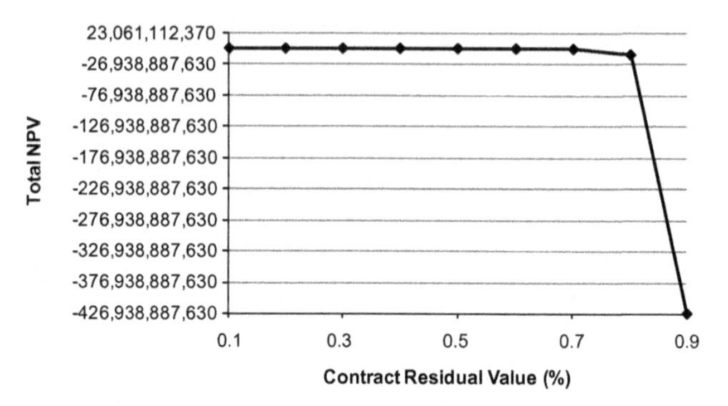

Figure 10.5. Sensitivity to CRV (M1).

The result shows that for the arbitrary but plausible parameter values we have chosen, the NPV for this block of leases is $3.1M. We also see that the leasing company pays far more to the bank to finance the vehicle than it gets in return in monthly payments, but it more than makes up for this loss when the vehicle is purchased at lease end. Of course, these observations apply only to the one scenario captured in our model, in which the contract residual and the actual residual are both 50 percent.

To the Reader: Before reading on, use M1 to generate as many insights into the situation as you can.

Since we built this model primarily to see whether we could shed light on the choice of CRV, the first sensitivity test we perform is on this decision variable. Figure 10.5 shows how the NPV changes as we vary the CRV percentage from 10 to 90 percent. Over this range of CRVs, the NPV seems to peak around a CRV of 0.4 and then drops off precipitously. High CRVs lead to huge losses. This is because a high CRV leads to very small monthly payments and a high demand for leases, whereas the vehicle's value at lease end is fixed at the actual residual (which is set to 50 percent in our base case).

Figure 10.6 shows the sensitivity of the NPV over a narrower range of values for the CRV. Given the model's other parameters, the optimal CRV is 0.42, which gives an NPV for this block of leases of $4.8M.

These first results suggest several interesting hypotheses. First, we see that the optimal CRV is not necessarily equal to the ARV. Second, at least for this numerical illustration, the optimal CRV is *below* the ARV. Finally, the NPV we earn at the optimal CRV is substantially higher than the NPV when the CRV is set equal to the ARV—36 percent higher, in fact. Now, not all of these generalizations apply to all cases. For example, it is possible that the optimal CRV may be above the ARV in some circumstances. So we should note these hypotheses but not assume they are universally true. We record our most significant and solid finding as follows:

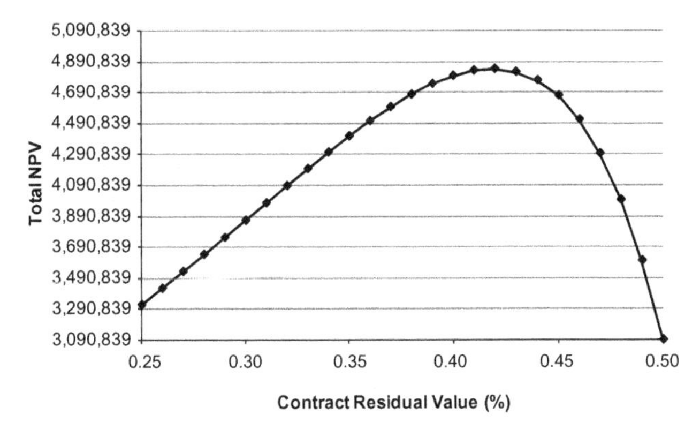

Figure 10.6. Refined sensitivity to CRV (M1).

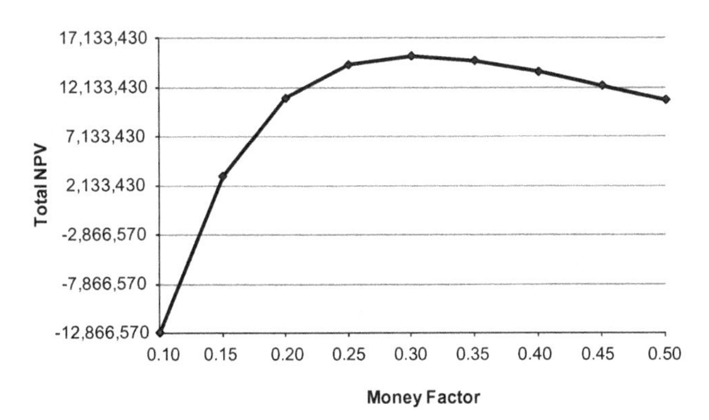

Figure 10.7. Sensitivity to MF (M1).

Insight 1: The optimal contract residual value can differ from the actual residual value.

This first insight is significant because this possibility is not obvious to everyone at National Leasing, as the problem description makes clear. Some managers at the company feel the challenge they face is essentially one of improving the forecasts of used car prices for their leased cars. This belief has led them to consider hiring a consulting firm to develop better forecasts on the assumption that the CRV will be set to the ARVs produced by this improved process. But our first result shows that whatever the ARV, the optimal CRV may be different. Better forecasting procedures *may* have value, but that remains to be seen. We address this issue later on.

The Money Factor, or interest rate the leasing company charges its consumers, is also a decision variable. The sensitivity analysis in Figure 10.7 shows that, all else equal, the NPV-maximizing MF is around 30 percent, which is much higher than the nominal 15 percent we assumed in our base case. At least for this version of the model, as we raise the MF above 15 percent, the monthly lease payments increase faster than demand decreases (and the residual cash flow does not change), so overall NPV increases. Again, we note any hypotheses suggested by our early numerical results, while being careful not to overgeneralize. However, the following conclusion seems warranted:

Insight 2: The optimal money factor can be much higher than the borrowing rate.

These sensitivity results suggest another question: What *combination* of CRV and MF maximize NPV? If the optimal CRV is 42 percent with the MF fixed at 15 percent, and the optimal MF is about 30 percent with the CRV fixed at 50 percent, are these still the optimal decisions when both parameters are free to vary? We can answer this question by creating a two-way data sensitivity table or by optimizing using Solver. Although our model is nonlinear, the smooth, concave shape of the two graphs of NPV against the decision variables (Figures 10.6 and 10.7) suggests that the nonlinear search algorithm in Solver is likely to find a global optimum. The Solver specification is as follows:

Set cell:	C29 (max)
By changing variable cells:	C12:C13
Constraints:	none
Solver:	Standard GRG Nonlinear

We run Solver and find that the optimal NPV is \$15.4M, with a MF of 61 percent and a CRV of 65 percent. This NPV is much higher than we found when we optimized the CRV alone, but just about the same as when we optimized the MF (with the CRV at 50 percent). To see in more detail what is going on here, we create the two-way data sensitivity graph shown in Figure 10.8. This graph shows that there are many combinations of CRV and MF that give about the same near-optimal NPV. When the CRV is between 55 and 65 percent, and the MF is between 40 and 65 percent, the NPV is above \$15M.

Insight 3: The optimal combination of the contract residual value and money factor in the base case is roughly 60 percent and 65 percent, respectively, although there is a range of values that give essentially the same NPV.

A Money Factor of 60 percent is, of course, unrealistic. It is far above comparable interest rates and may well violate laws regulating lending rates. However, there is value in these results. Most importantly, they demonstrate

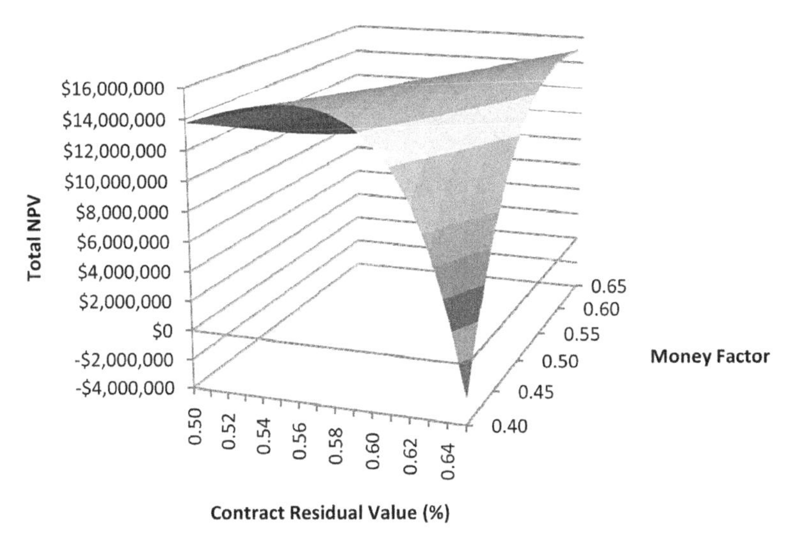

Figure 10.8. Sensitivity to MF and CRV (M1).

that there is an interaction between the choice of MF and the choice of CRV: We cannot set one without taking the other into consideration. For example, if we set the MF at 15 percent, the optimal CRV is 42 percent; if we increase the MF to 20 percent, the optimal CRV becomes 50 percent.

One shortcoming of our analysis so far is that we have treated the ARV as if it were known, when its uncertainty is at the core of the challenge leasing companies face. We explicitly introduce uncertainty later on, but for now, we examine the impact the ARV has on the results by repeating the optimization process for various values of the ARV. The Solver Sensitivity tool is perfectly suited for this operation; we change the ARV and then use Solver to find the corresponding optimal values for the CRV and MF.

We first examine the relationship between the optimal CRV and the ARV with the MF fixed at 15 percent. The results are shown in Figure 10.9. When the ARV is low, the optimal CRV is equal to the ARV. In other words, if we knew that used car prices at lease end would be low, we would set the CRV to the used car price. Although this would result in high monthly payments, it limits residual losses. That is to say, if we set the CRV higher, we would find at lease end that consumers returned their vehicles and we would have to sell them at the going used-car price. The graph also shows that at some point, the CRV should be set below the ARV. This occurs in our model at about 42 percent of the sale price. When the ARV rises above this point, the optimal CRV does not increase but remains fixed at 42 percent. Why set the CRV below the ARV? In this scenario, consumers realize a residual gain at lease end, when they purchase the vehicle for the CRV and sell it at the higher used car price. By raising the CRV, we could increase the cash flow to the leasing

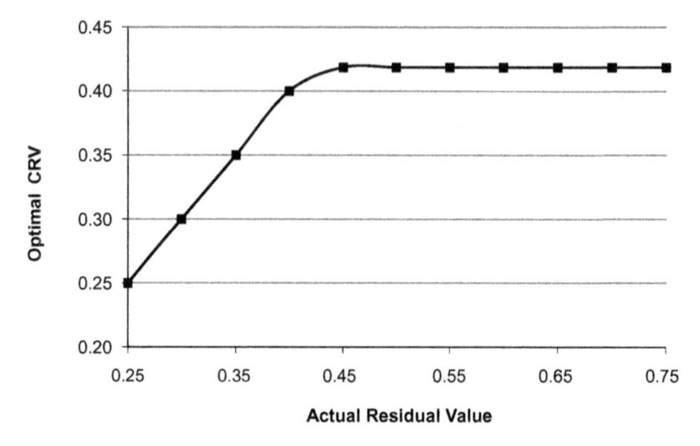

Figure 10.9. Sensitivity of optimal CRV to ARV with MF fixed (M1).

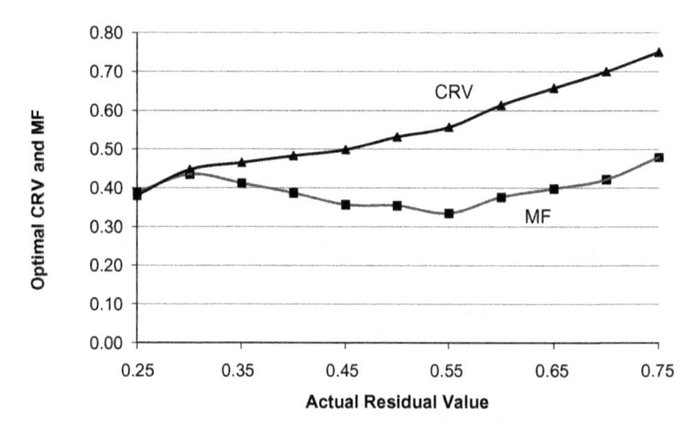

Figure 10.10. Sensitivity of optimal CRV and MF to ARV (M1).

company at lease end. However, we would simultaneously decrease the monthly payments and increase the demand. Apparently the net effect on revenue is such that the optimal CRV never exceeds 42 percent.

Insight 4: For a fixed money factor, the optimal contract residual value increases with the actual residual value up to a point and then remains constant.

Figure 10.10 shows the results of a related experiment, in which we allow Solver to choose both the CRV and the MF as we vary the ARV. In this case, we see that the optimal CRV increases almost in lockstep with the ARV, whereas the optimal MF varies within a narrow range around 40 percent until the ARV gets very high. To understand these results, it is important to recognize that the MF influences only the monthly payment, whereas the CRV influences both the monthly payment and the residual cash flow. What we see in this graph is a pattern in which the MF is relatively insensitive to the ARV, whereas the CRV tracks it closely. The optimal response when both decisions can vary is to use the CRV to mimic the ARV so as to gain as much as possible at lease end, and then to use the MF to balance the trade-off between the monthly payments and demand to generate favorable cash flows during the lease.

Insight 5: When both the money factor and the contract residual value can be chosen optimally, the money factor is relatively insensitive to the actual residual value. The optimal contract residual value CRV, on the other hand, matches the actual residual value ARV.

How sensitive is the optimal solution to demand elasticity? We answer this question by changing the elasticity and running Solver. However, we must be careful to ensure the demand curves that result when we change the elasticity are comparable. We do this by changing the constant term in the demand curve so that, regardless of the elasticity, the demand with a monthly payment of $450 is 10,000 leases, as it is in the base-case model. The results are summarized in Figure 10.11.

The higher the absolute value of the elasticity, the more sensitive consumers are to the monthly lease cost and the lower the NPV of the lease will be, even with optimization of the CRV and MF. There is no obvious pattern in the optimal values of the decision variables, although in all three cases, the MF is very high compared with normal interest rates and the CRV is above 50 percent. A somewhat clearer pattern emerges when we fix the MF at a more normal value, such as 15 percent, and find what the optimal CRV is in that case. As we increase the absolute value of the elasticity from –3 to –7, the optimal CRV increases from 30 percent to 42 percent and then to 46 percent. Thus, we conclude that with a fixed MF, the optimal CRV increases as the demand elasticity increases in absolute value.

Insight 6: The optimal contract residual value increases with the absolute value of demand elasticity (for fixed values of the money factor).

The ARV, which under our assumptions is the same as the price of the used car at the end of the lease, is another parameter whose impact we should examine. Figure 10.12 shows how the NPV in the base case varies as we change

Elasticity	Optimal CRV	Optimal MF	NPV
–3	51%	47%	$24.9M
–5	63	55	$15.5M
–7	53	29	$11.9M

Figure 10.11. Sensitivity of optimal CRV and MF to Elasticity of Demand.

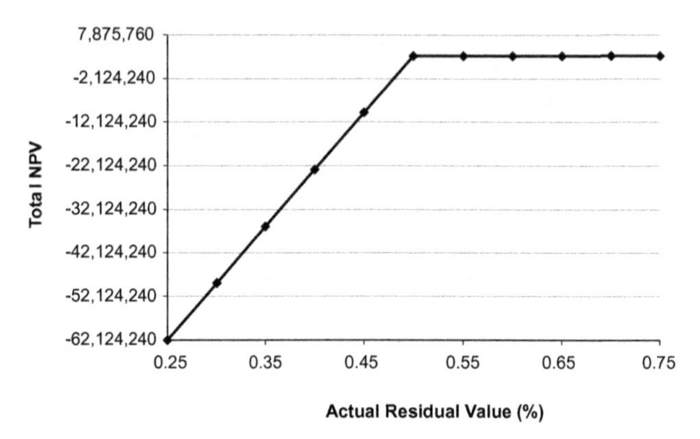

Figure 10.12. Sensitivity to ARV (M1).

the ARV. Recall that in the base case, we assume the CRV is 50 percent. The chart shows that when the ARV is significantly below the CRV, the block of leases has a negative NPV. This is because when the ARV is low, consumers return their vehicles to the leasing company to buy the inexpensive used cars. The leasing company, in turn, must sell its returned vehicles for the (low) used car price. In the opposite case, when the ARV is higher than the CRV, the NPV does not increase. This follows from the fact that consumers (in our model) buy their vehicles at the CRV as soon as the ARV is above the CRV. Thus, the cash flow to the leasing company is fixed at the CRV.

Insight 7: Low actual residual values relative to the contract residual value lead to large losses, whereas high actual residual values do not lead to correspondingly large gains.

Finally, a tornado chart shows us which of our inputs has the biggest impact on the NPV. Figure 10.13 shows that the most significant factor by far is the CRV. More specifically, if we increase the CRV we lose a large amount, whereas if we decrease it, we gain relatively little. The second most important factor is the ARV, which has no impact if it decreases but a large impact if it increases. The remaining parameters—Money Factor, Borrowing Rate, Term, and Sale Price—have relatively modest effects.

Insight 8: The contract residual value and the realized value of the uncertain actual residual value at the end of the lease are the most significant factors in determining the profitability of a block of leases.

Before we leave the issue of which parameters have the biggest impact on the NPV, we recall the limitations of tornado charts. Remember that in a tornado chart, we vary each parameter around its base-case value. So it is possible that starting with different base-case values would give different

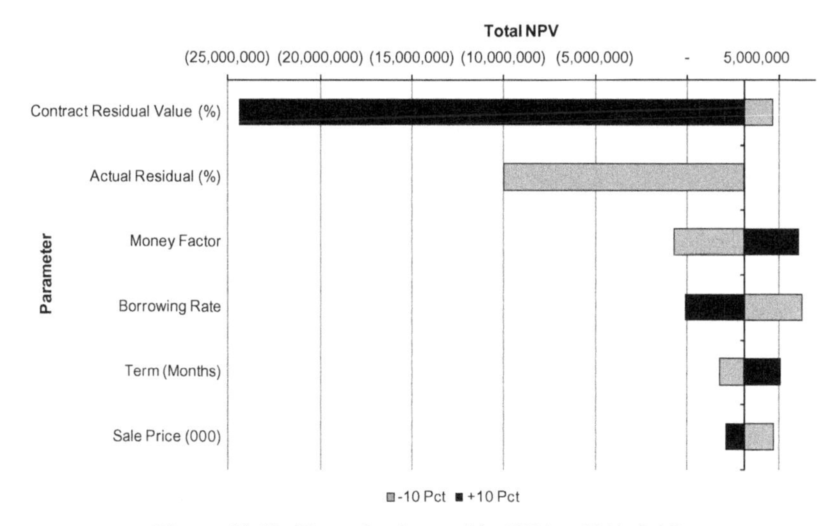

Figure 10.13. Tornado chart with CRV = 50% (M1).

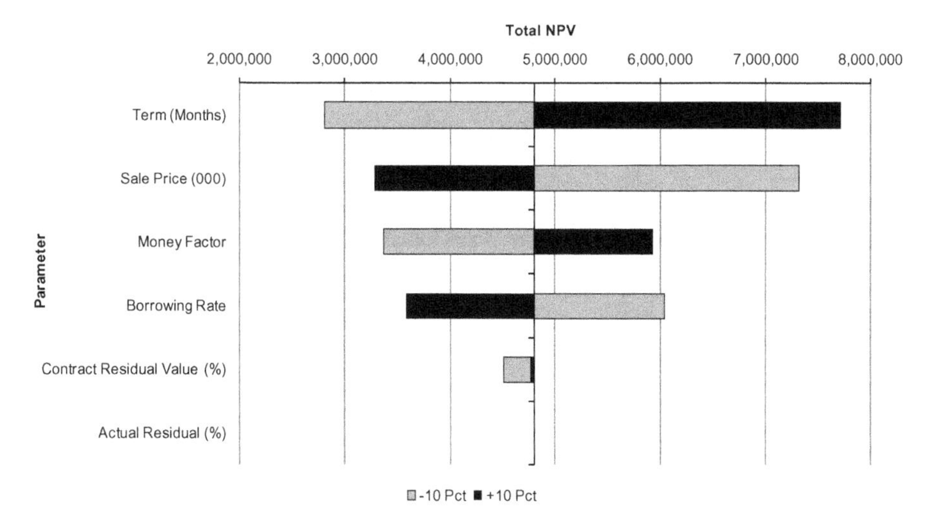

Figure 10.14. Tornado chart with CRV = 40% (M1).

results. In our base case, both the CRV and the ARV are set at 50 percent. Since we have seen that the optimal CRV is closer to 40 percent, it makes sense to create a tornado chart around that value and see whether the results differ significantly.

Figure 10.14 shows the results, and they are different from those in Figure 10.13. Here, the significant parameters are Term, Sale Price, Money Factor, and Borrowing Rate. Both CRV and ARV have only minor impacts on the NPV.

Why the difference? When the CRV is well below the ARV, consumer behavior at lease end is insensitive to changes in parameters (that is, consumers buy the car at the low CRV). Changing the ARV by ±10 percent has no impact on the lease-end cash flow, and a similar percentage change in the CRV has only a minor impact. So we modify the conclusions we came to earlier about which parameters are most important.

Insight 9: When the contract residual value is chosen optimally, the most significant factors in determining the NPV of a block of leases are Term, Sale Price, Money Factor, and Borrowing Rate.

Summary of M1

Our first modeling efforts captured some of the essential economic forces that determine the profitability of a lease and suggested several insights that bear additional exploration. This model uses a very simple demand relationship that completely ignores the cost of competing leases. We want to elaborate on this relationship in future models. The model also assumes consumers behave rationally at lease end. We also want to explore alternative models that allow for more realistic behavior in this area.

To the Reader: Before reading on, create a list of changes you would like to make to M1 and decide on the order in which to make them.

10.5 M2 MODEL AND ANALYSIS

We can improve our first model in three ways. One option is to improve our demand module. Another option is to improve our model of the consumer's lease-end decision. The third option is to consider the impact of uncertainty in the ARV, or used car price, at lease end. In M1 we took the ARV to be a fixed parameter, but in reality, it is uncertain to the leasing company when it must decide on the CRV and MF for a given block of leases. Since introducing uncertainty into a single parameter represents a modest change to the model, we pursue this issue first. (The downside of introducing simulation at this stage is that simulation outputs are more complex to analyze.)

M1 is easily modified to incorporate uncertainty in the ARV. Essentially, all that is required is to replace the fixed parameter for the ARV with a probability distribution. But what distribution should we use?

The uncertainty in the ARV the leasing company faces at the time a lease is offered reflects its ability to predict used car prices three or four years out. Many factors influence used car prices, including economic conditions, new car prices and quality, used car quality, and the supply of used cars. It is unlikely anyone can predict used car prices with perfect accuracy years in advance, so some degree of error is unavoidable. For the purposes of building a prototype, we must ask two questions: Are the forecasts systematically biased, and how do they vary around the true values? If the forecasts are biased, then they are higher or lower than the true values *on average*. If they are unbiased, then they predict the true values accurately on average, but they vary around the true value. In the absence of any information about the actual forecasting process National Leasing uses, we make some simplifying assumptions. If, in fact, the forecasts are biased (and the bias is known), it is a simple matter to correct the bias by adding or subtracting it from the actual forecast. Thus, from this point on, we assume the forecasts are unbiased. That implies that the mean of the distribution is the same as the true mean. Second, we assume the ARV is given by a normal distribution, so we can reflect the forecast error in the standard deviation. (We could have used any distribution, but the normal is convenient here because its variability is captured in a single parameter.) Specifically, we assume the distribution has a mean of 50 percent of the sale price and we arbitrarily pick a standard deviation of 6 percent. The corresponding normal distribution is shown in Figure 10.15.

Our new model is shown in Figure 10.16. The mean and standard deviation of the ARV distribution are in cells C10 and C11, respectively. In cell C9, we calculate the ARV by taking a sample from a normal distribution with these parameters using the Crystal Ball function

CB.NORMAL(C10, C11)

The only other change required in this model is to define the NPV of cash flows, cell C32, as a Crystal Ball Forecast cell. We need to do a few test runs to determine an appropriate run length. At 1,000 trials, the mean standard error is about 10 percent of the NPV, which seems high even for a prototype. At 10,000 trials, the mean standard error is around 3 percent, which is probably sufficient for our purposes.

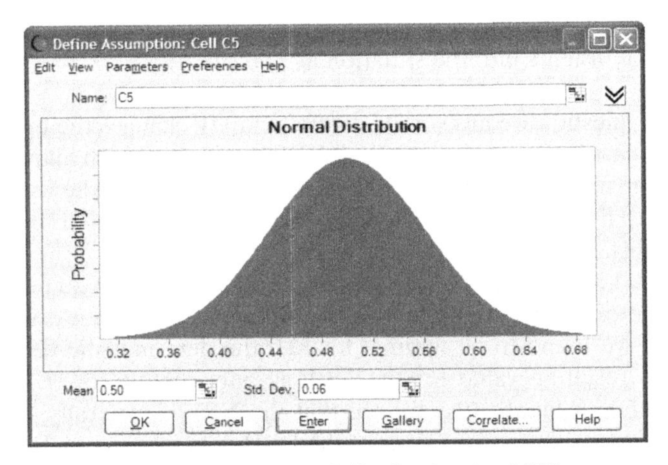

Figure 10.15. Normal distribution for ARV.

	A	B	C	D	E	F	G
1		**Auto Leasing Prototype**	**M2**				
2							
3							
4							
5	**Parameters**						
6		**Sale Price (000)**	$25,000		Multiplier	-2.E+17	
7		**Term (Months)**	36		Elasticity	-5	
8		**Borrowing Rate**	0.05				
9		**Actual Residual (%)**	0.44				
10		**Mean**	0.50				
11		**St. Dev**	0.06				
12							
13					Payment	Demand	
14	**Decisions**				300	75938	
15		**Money Factor**	0.15		325	50892	
16		**Contract Residual Value (%)**	0.40		350	35134	
17					375	24883	
18	**Model**				400	18020	
19		Monthly Customer Payment	520		425	13308	
20					***450***	***10000***	
21		Monthly Bank Payment	749		475	7631	
22					500	5905	
23		Leases Sold	4,854		525	4627	
24					550	3666	
25		Residual Gain/Loss	932		575	2936	
26					600	2373	
27		NPV of Net Monthly Payments	-7,651				
28		NPV of Residual Gain/Loss	8,638				
29							
30		NPV Per Car	988				
31							
32		Total NPV	$4,795,456				
33							
34			0				
35							

Figure 10.16. Model M2—numerical view.

To the Reader: Before reading on, use M2 to generate as many interesting insights into the situation as you can.

The first question we answer is how uncertainty changes the optimal CRV and MF. The tool to use here is Crystal Ball Sensitivity, which allows us to run a simulation model over a range of input parameters. (When you do this, remember to first fix the random number seed in Crystal Ball to reduce spurious variability between runs.)

Figure 10.17 shows the results of varying the CRV from 20 percent to 50 percent. The optimal value of the CRV is 40 percent, which is essentially the same value we found to be optimal for M1, the deterministic version of the model. The resulting value for the NPV is only slightly lower. Comparing Figure 10.17 with Figure 10.6, we see that the NPV is generally lower under uncertainty, especially so for values of the CRV near 50 percent. The following conclusion seems justified at this point:

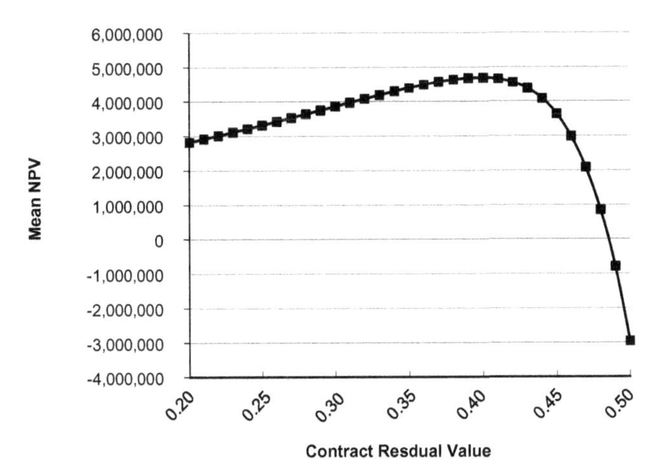

Figure 10.17. Sensitivity to CRV (M2, σ = 0.06).

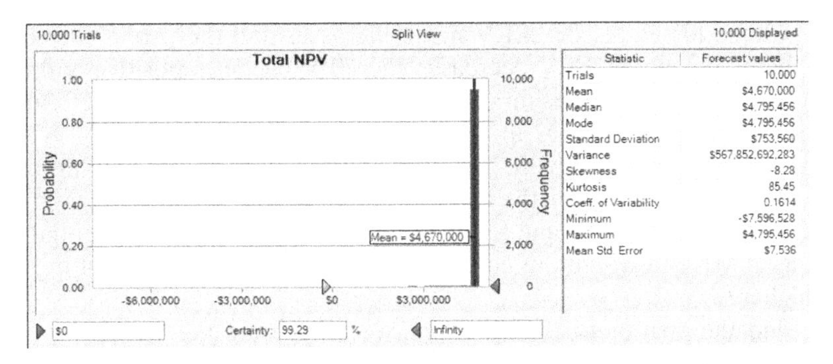

Statistic	Forecast values
Trials	10,000
Mean	$4,670,000
Median	$4,795,456
Mode	$4,795,456
Standard Deviation	$753,560
Variance	$567,852,692,283
Skewness	-8.28
Kurtosis	85.45
Coeff. of Variability	0.1614
Minimum	-$7,596,528
Maximum	$4,795,456
Mean Std. Error	$7,536

Figure 10.18. Distribution of the NPV with optimal CRV (M2).

Insight 10: The optimal contract residual value under uncertainty in the actual residual value is essentially the same as under certainty.

Figure 10.18 shows the distribution of NPV for the case where the CRV is set to its optimal value of 40 percent. A careful analysis of the results shows that although the mean is $4.7M and 99 percent of the outcomes exceed zero, the *minimum* outcome is negative $7.6M, while the maximum is only $4.8M. So the financial outcome facing the leasing company is extremely risky in the sense that the worst outcome is truly catastrophic.

Insight 11: The distribution of NPV under uncertainty in the actual residual value is extremely skewed, with a small probability of a huge loss and little chance of a large gain relative to the mean.

Why is there a large spike in the distribution of NPV in Figure 10.18? This spike indicates that in about 95 percent of the simulation trials, the outcome is identical: $4.795M. Recall that the ARV has a mean of 50 percent and a standard deviation of 6 percent, whereas the CRV is set to 40 percent. Whenever the ARV exceeds the CRV, the cash flow at lease end is identical since, in these cases, the consumer buys the car from National at the CRV. And this occurs roughly 95 percent of the time since the CRV value of 40 percent is 1.67 standard deviations below the mean of the normal distribution for the ARV.

It is always a good idea to ask periodically whether the results we are getting are general or are due to the specific numerical assumptions we have made. The results in Figure 10.18 remind us that the effect of uncertainty in the ARV depends on the relationship between the ARV and the CRV. When the CRV is chosen to be considerably smaller than the mean of the ARV, the actual ARV will only rarely be below the CRV. Thus, most of the time consumers buy at the CRV.

But what if the mean of the ARV shifts? Figure 10.19 shows the results of varying the mean of the ARV distribution from 40 percent to 85 percent while the CRV is held constant at 50 percent. We see that low ARV mean values lead to huge losses. As the ARV mean increases and the probability of the ARV being below the CRV decreases, the NPV becomes positive and levels off. Again, it is the relationship of the ARV to the CRV that seems to matter, just as we observed with M1 in Insight 7.

The probability of the ARV being below the CRV increases with the variance of the distribution for the ARV. In the base case for M2, we assumed a standard deviation of 0.06 and found that the optimal CRV was essentially the same as in M1. But would our conclusion change if the variability of the ARV was higher?

To find the answer to this question, we exaggerate the variability in the ARV beyond what we might expect to see in practice, just to establish whether a conclusion such as Insight 10 is sound. Accordingly, we change the standard deviation of the ARV from 0.06 to 0.20 and repeat the sensitivity analysis on

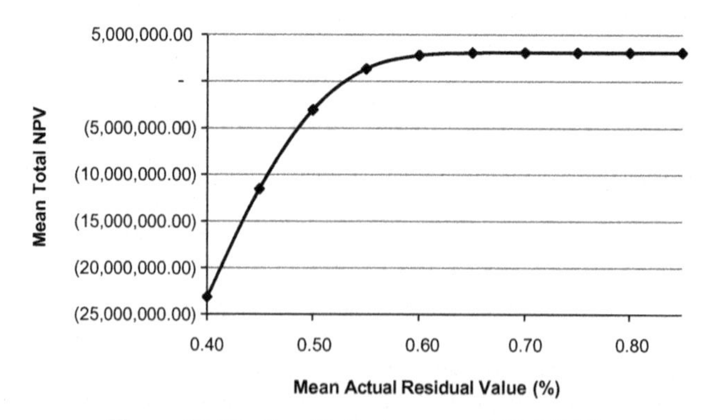

Figure 10.19. Sensitivity to mean ARV (M2).

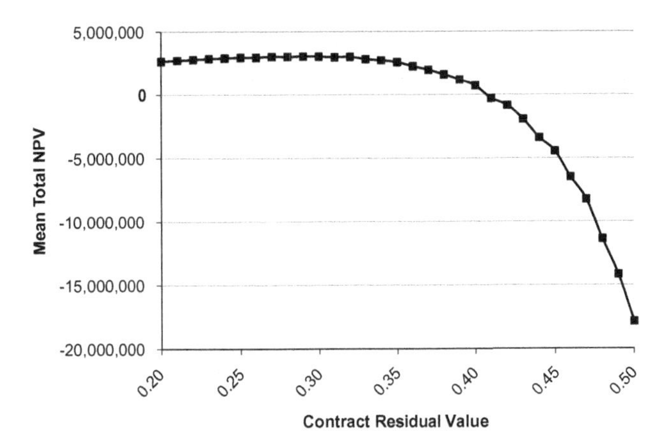

Figure 10.20. Sensitivity to CRV (M2, $\sigma = 0.20$).

the CRV. Note that this distribution for ARV implies there is a 95 percent chance the actual ARV will lie between 0.10 and 0.90, which surely is more variability than even the poorest forecasting procedure would show. The results are shown in Figure 10.20. It turns out our hunch is right: Variability *can* change the optimal CRV. In this case, the optimal CRV is 29 percent, considerably below the base-case value of 40 percent. It is still an open question whether this result is of practical importance, since we suspect the true uncertainty in the ARV is much less than assumed here. We do know, however, that if the uncertainty is small, the optimal CRV is the same as in the deterministic case, and if it is huge, the optimal CRV is much lower. We capture this improved insight in a form that modifies our original Insight 10:

Insight 10 (improved): Under uncertainty in the actual residual value, the optimal contract residual value declines with increasing variability in the distribution of actual residual values and can be substantially below the optimal contract residual value of the deterministic case.

The extremely skewed shape of the distribution of NPVs shown in Figure 10.18 suggests we take a look at the risks of very low NPVs as we vary the CRV. This is another question for which the Crystal Ball Sensitivity tool is well suited. However, that tool only captures certain aspects of a Forecast cell, in particular, the mean, standard deviation, minimum, and maximum. We could capture the minimum outcome as we vary the CRV from low to high values, but this would tend to overstate the risks. A better measure of risk would be the probability that the NPV falls below a certain value, say, \$2.5M. We capture this measure of the distribution of NPVs by adding to the model an IF statement that takes the value 1 when the NPV falls below \$2.5M and 0 otherwise. Since this cell will only take the values 1 and 0, the mean of the forecast for

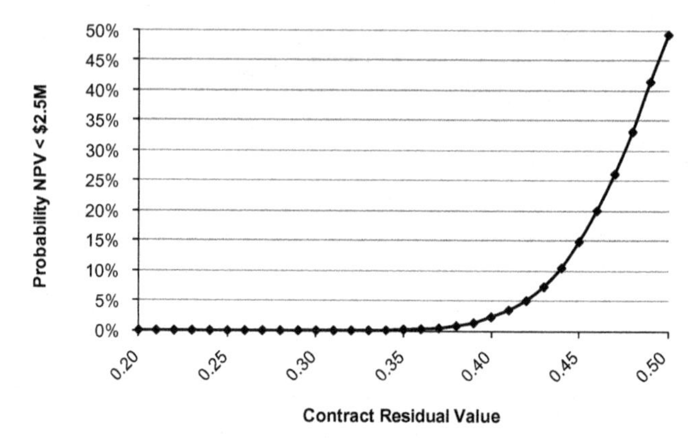

Figure 10.21. Sensitivity to CRV (M2, $\sigma = 0.06$).

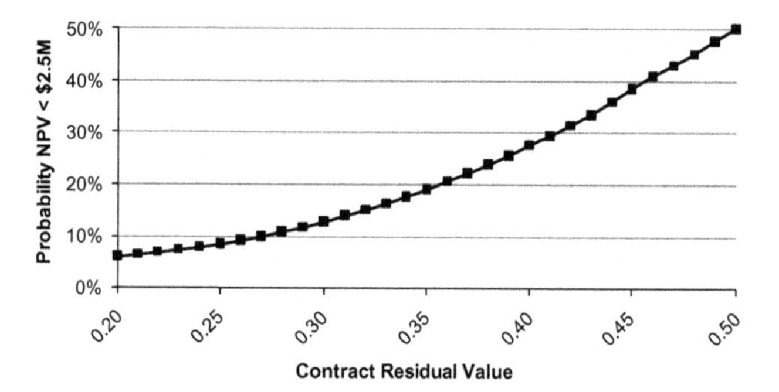

Figure 10.22. Sensitivity to CRV (M2, $\sigma = 0.20$).

this cell will be the percentage of ones, or the percentage of trials on which the NPV falls below $2.5M.

Figure 10.21 shows this measure of risk as we vary the CRV from 20 to 50 percent, with the standard deviation of the ARV distribution set at 0.06. As we saw earlier, the CRV that maximizes the mean NPV is 40 percent. At this level, the risk of the NPV falling below $2.5M is only 2.3 percent, which is probably acceptable to most decision makers. However, the picture changes when we increase the standard deviation of the ARV distribution to 0.20. Figure 10.22 shows the results for this case. Under this assumption, the optimal CRV is 29 percent, but the probability of falling below $2.5M is 11.7 percent. Therefore, the risk increases with uncertainty. If this is too high a risk, the leasing company might choose a lower value of the CRV and sacrifice some expected profit to decrease the probability of an extremely low return. The figure shows, for example, that to reduce the risk of falling below $2.5M to less than 5 percent, we should set the CRV just below 20 percent.

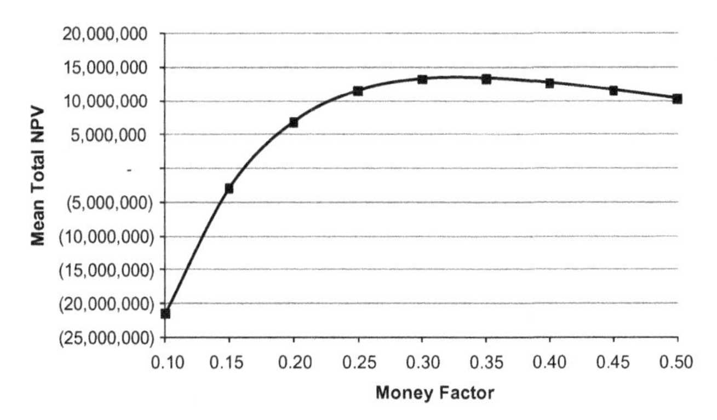

Figure 10.23. Sensitivity to MF (M2).

Insight 11: The risk of a low NPV increases with the contract residual value.

We next perform a similar sensitivity analysis for the optimal MF. The results are shown in Figure 10.23, which we compare with Figure 10.7 for the deterministic model. The optimal MF moves from 30 percent to 35 percent, and the NPV values are generally lower under uncertainty, but overall the picture has not changed much.

Insight 12: The optimal money factor under uncertainty in the actual residual value is essentially the same as under certainty.

Of course, we should also check that this generalization holds for more variable distributions for the ARV. When we repeat the sensitivity analysis for a standard deviation of 0.20, we find that the optimal MF increases again, to 40 percent. But this is an extreme degree of variability, so the conclusion we drew in Insight 12 still seems appropriate.

The question of how the range of uncertainty in the ARV influences the optimal CRV and MF is extremely important to National Leasing, since it is considering hiring a consulting company to develop better forecasts of the ARV. Unless the current forecasting process is biased, "better" forecasts are those with a *lower variance* around the true value. We have already seen that changing the variability in the ARV influences the optimal CRV and MF. Therefore, a more careful analysis is warranted.

The results of a two-way Crystal Ball sensitivity run varying the standard deviation of the ARV distribution and the CRV are shown in Figure 10.24. Here, we vary the standard deviation of the ARV distribution from 0.03 to 0.12, which seems to be a range of plausible values. We see that the optimal CRV decreases by a modest amount, from 42 percent to 35 percent, as the standard deviation increases from 0.03 to 0.12. Increased variance also reduces

	A	B	C	D	E	F
1	**Total NPV: Mean**					
2		Standard Deviation				
3	Contract Residual Value (%)	0.03	0.06	0.09	0.12	
4	0.30	$3,862,679	$3,862,276	$3,847,614	$3,763,038	
5	0.31	$3,974,153	$3,973,553	$3,951,116	$3,841,741	
6	0.32	$4,085,052	$4,084,070	$4,050,362	$3,909,797	
7	0.33	$4,194,491	$4,192,903	$4,143,285	$3,964,581	
8	0.34	$4,301,399	$4,298,722	$4,226,674	$4,001,647	
9	0.35	$4,404,479	$4,399,201	$4,296,230	**$4,016,011**	
10	0.36	$4,502,172	$4,491,574	$4,347,009	$4,001,174	
11	0.37	$4,592,600	$4,572,229	**$4,372,705**	$3,950,308	
12	0.38	$4,673,482	$4,636,810	$4,366,197	$3,853,157	
13	0.39	$4,742,017	$4,678,682	$4,317,422	$3,694,780	
14	0.40	**$4,795,020**	**$4,687,773**	$4,214,163	$3,463,038	
15	0.41	$4,828,381	$4,651,172	$4,040,952	$3,141,577	
16	0.42	$4,836,788	$4,553,378	$3,773,209	$2,709,881	
17	0.43	$4,810,333	$4,372,764	$3,389,515	$2,145,174	
18	0.44	$4,733,539	$4,081,743	$2,860,664	$1,416,430	
19	0.45	$4,581,008	$3,634,869	$2,153,806	$486,090	
20	0.46	$4,306,216	$2,988,616	$1,221,526	-$687,219	
21	0.47	$3,842,500	$2,087,493	$11,505	-$2,154,213	
22	0.48	$3,074,869	$855,251	-$1,538,251	-$3,976,451	
23	0.49	$1,870,960	-$791,211	-$3,503,639	-$6,228,909	
24	0.50	$67,328	-$2,956,184	-$5,979,695	-$9,003,206	
25						

Figure 10.24. Sensitivity to CRV and ARV standard deviation (M2).

the optimal NPV by about 16 percent over a wide range of variability. We translate these model results into an insight for National Leasing as follows:

Insight 13: Reducing the variability in forecasts of the actual residual value leads to a slight change in the optimal contract residual value and to a higher NPV.

A similar analysis for the optimal MF is reported in Figure 10.25. As the standard deviation of the ARV forecast decreases from 0.12 to 0.03, the optimal MF decreases from 35 percent to 30 percent and the NPV increases by about 18 percent. Again, reducing the variance in the forecasts allows National Leasing to modify the MF slightly and achieve a significantly higher NPV on this block of leases.

Insight 14: Reducing the variability in forecasts of the actual residual value leads to a slightly lower optimal money factor and a higher NPV.

Of course, these results do not prove that better forecasts are worth paying for. That depends on how much improved forecasting reduces the magnitude of forecast errors.

What is the *jointly* optimal combination of CRV and MF? We could use OptQuest for this purpose, but with only two decision variables, it is more straightforward to use Crystal Ball Sensitivity to enumerate a wide range of

	A	B	C	D	E	F
1	**Total NPV: Mean**					
2		Standard Deviation				
3	Money Factor	0.03	0.06	0.09	0.12	
4	0.10	-$17,193,528	-$21,520,486	-$25,847,443	-$30,174,401	
5	0.15	$67,328	-$2,956,184	-$5,979,695	-$9,003,206	
6	0.20	$8,934,879	$6,799,867	$4,664,855	$2,529,843	
7	0.25	$12,964,015	$11,440,784	$9,917,553	$8,394,322	
8	0.30	**$14,266,803**	$13,169,014	$12,071,225	$10,973,435	
9	0.35	$14,091,357	**$13,292,323**	**$12,493,289**	**$11,694,255**	
10	0.40	$13,168,179	$12,580,954	$11,993,729	$11,406,504	
11	0.45	$11,918,043	$11,482,398	$11,046,753	$10,611,107	
12	0.50	$10,577,046	$10,250,876	$9,924,706	$9,598,537	
13						

Figure 10.25. Sensitivity to MF and ARV standard deviation (M2).

combinations and choose the best one. (For more information on these tools, refer to Appendix C.) The results confirm what we have seen up to this point: If the objective is to maximize the mean NPV, the optimal combination is a CRV of 65 percent and a MF of 60 percent, about the same as in the deterministic model. If this value for the MF is unrealistically high for reasons not captured in the model, the optimal value of the CRV is also lower, as we saw previously.

Summary of M2

In our second model, we introduced the crucial element of uncertainty in the Actual Residual Value, which represents the price of an equivalent used car at the end of the lease. This new model allowed us to examine the impact uncertainty has on the optimal choices for CRV and MF and to estimate the value of reducing variability in the ARV. Reducing variability is what we would expect an improved forecasting process to accomplish. Our results showed that reducing variability in the ARV can help National achieve higher NPVs. However, the policy decision of how to set the CRV and MF is relatively insensitive to variability in the ARV, at least, within what we think are plausible ranges. In short, better forecasts have value in that they raise profits, but they do not materially affect policy decisions.

To the Reader: Before reading on, modify and improve the demand module in M2.

10.6 M3 MODEL AND ANALYSIS

The next challenge we take on is to improve the demand module in our model. Our current model subsumes all influences on demand into a single elasticity parameter. This demand model shows the same response to (percentage) increases in demand as it does to decreases. It also does not include in any explicit way the prices other leasing companies charge. Yet, common sense suggests that the prices competitors charge strongly influence demand for what is largely a commodity product. Moreover, it is possible that price increases may have different effects on demand than price decreases. In this section, we modify the demand model from M1 to take these effects into account.

For most vehicles, dozens of leases are available to the consumer. These leases offer a range of monthly payments and CRVs. Demand for any one lease depends on the monthly payment, perhaps the CRV itself, the terms of competing leases, the reputation of the leasing company, and possibly, the reputation of the car dealer. Including all these factors in a model at this point would not be productive. But reflecting the relationship between the monthly payment one company charges and that charged by its competition may offer fresh insights.

We begin our model building with an important simplification. Although, in fact, dozens of leases are available at different prices, we assume just *one* alternative lease is offered at a given monthly payment. We also assume that the monthly payment is the only factor that influences demand. This single alternative price could be interpreted as the price of the major competing company, the minimum price of similar leases, or the price of the competitor with the largest market share. Remember, our task in this case is to show how model building and analysis can shed light on the leasing decisions, not to develop a final model for daily use. Thus, our analysis has a somewhat hypothetical flavor, as is evident in previous sections. We are more interested in establishing interesting insights about how the CRV and Money Factor should be set than in perfecting an operational model.

The idea behind our new demand module is to anchor demand at the monthly payment set by the competition and then to model independently how many *more* leases we would sell if we lowered our monthly payment below this reference level and how many *fewer* we would sell if we raised our monthly payment above the reference level. To make this model as similar as possible to our earlier one, we set the reference monthly payment at \$433 and assume we sell 10,000 leases at that monthly payment, just as in the M1 base case.

Our model uses six parameters: Reference Price, Base, Multiplier1, Multiplier2, α, and β. The reference price, again, represents the monthly payment on competing leases. The α and β parameters are used to represent the change in demand for monthly payments above and below the reference price, respectively. The two multiplier parameters scale the results appropriately. In pseudocode, the demand function has the following form:

```
MAX(0,
        IF(Monthly Payment > Reference Price)
          Base − Multiplier1 × (Monthly Payment − Reference Price)^α
        ELSE
          Base + Multiplier2 × (Reference Price − Monthly Payment)^β)
```

This formula first ensures that demand will not be negative. Then it determines whether the Monthly Payment is above or below the Reference Price. If it is above, it uses a power function with coefficient α to determine demand; if below, it uses an equivalent power function with coefficient β.

Figure 10.26 shows an example of this new demand relationship. Here we graph both the original constant-elasticity demand model from M1 and the new model with $\alpha = \beta = 0.9$. Note that demand with this new model drops off more slowly than in the original model as we increase our monthly payment above the reference price, and increases more slowly as we decrease the monthly payment below the reference price. Figure 10.27 shows two alternative versions of this new demand model, one with α reduced to 0.5 and another with β reduced to 0.5. As we would expect, decreasing α flattens the demand curve for monthly payments above the reference price, whereas decreasing β flattens demand for monthly payments below the reference price.

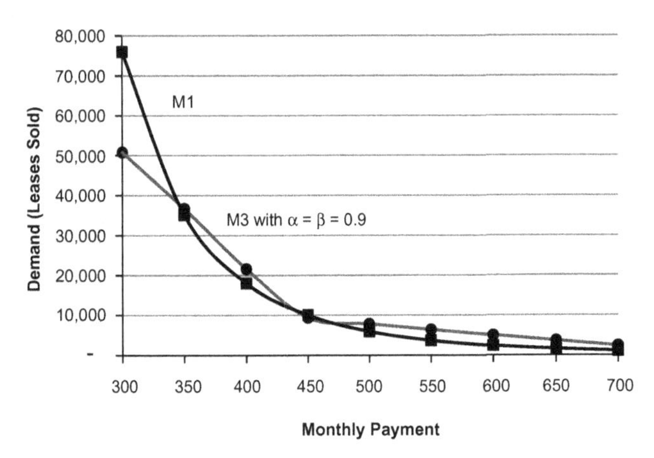

Figure 10.26. Demand curves for M1 and M3.

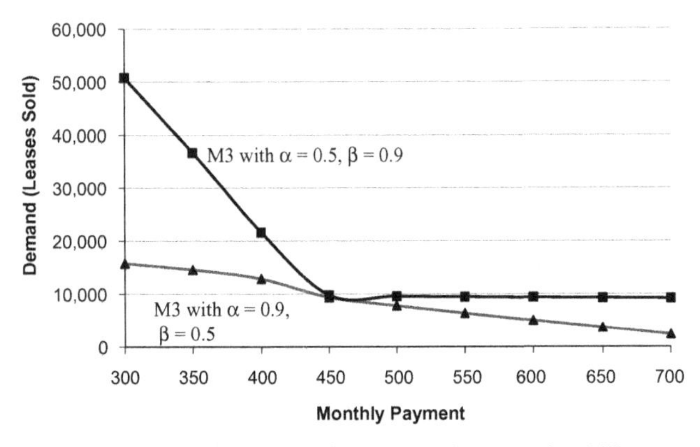

Figure 10.27. Alternative demand curves for M3.

Although we benchmark this demand curve so that it generates a demand of 10,000 leases at a Monthly Payment of $433, as did the simpler demand curve in M1, we cannot make this demand curve identical to the earlier one (nor do we wish to do so). Thus, we should not directly compare the numerical results from this version of the model with those developed earlier. What is of interest is which of the insights we found earlier continue to hold and which change under this new model.

To the Reader: Before reading on, use M3 to generate as many interesting insights into the situation as you can.

Our new model, M3, is identical to M2 except that the demand curve now has the more complex form described above. The first question we pose for the model is, what are the optimal values of the CRV and MF?

Figures 10.28 and 10.29 show Crystal Ball sensitivity runs for CRV and MF. No dramatic changes from our earlier results are evident. The optimal CRV

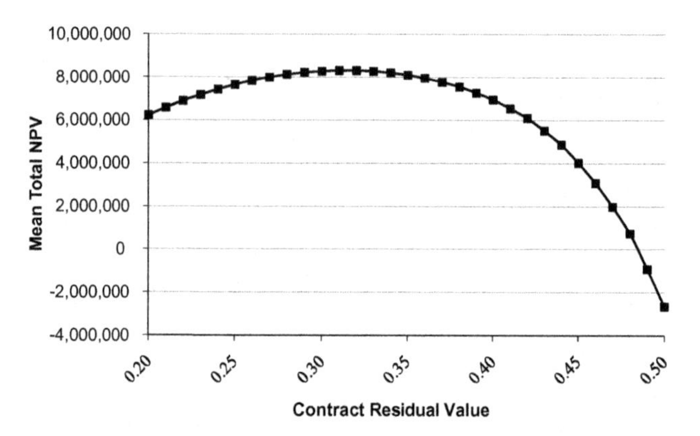

Figure 10.28. Optimal CRV (M3).

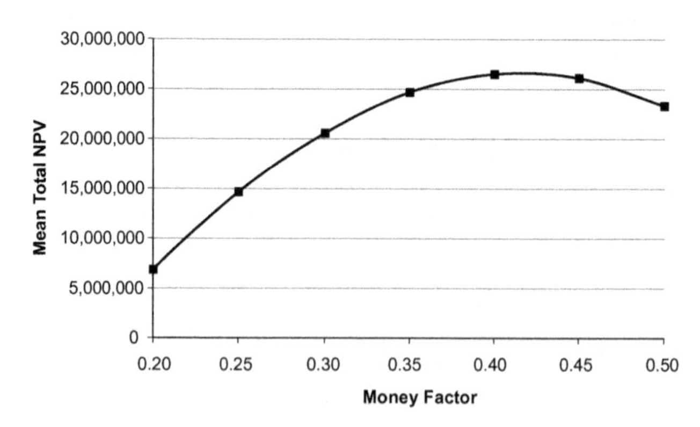

Figure 10.29. Optimal MF (M3).

for this model is 31 percent (versus 40 percent for M2 in the base case) and 40 percent for the MF (versus 35 percent for M2 in the base case). These results are not too surprising since this model is identical to M2 except in how demand is parameterized.

More interesting is the sensitivity of the optimal CRV to the reference price itself. Since the market for leases is highly competitive, we expect leasing companies to change their leases in response to the prices set by others in the market. Without attempting to model this competition in detail, we examine how the optimal CRV changes as the reference price changes. Figure 10.30 shows the results of a two-way Crystal Ball sensitivity with reference prices between $300 and $600. The patterns in these results are complex and not easily explained. When the reference price is low, our monthly payment is above the reference price for all values of the CRV between 20 percent and 50 percent. This leads to low demand and low NPVs. The optimal CRV is 38 percent. As we increase the reference price, all else remaining the same, we raise demand for our leases at all levels of the monthly payment. The number of leases sold is higher, as is the NPV. The optimal CRV declines from 38 percent, at a reference price of $300, to 28 percent, at $500. Then the pattern reverses: The NPV at all levels of CRV continues to go up, but the optimal CRV jumps back up to the 40 percent range. What can explain this reversal in the pattern?

	A	B	C	D	E	F	G	H	I
1	**Total NPV: Mean**								
2		Reference price							
3	Contract Residual Value (%)	$300	$350	$400	$450	$500	$550	$600	
4	0.20	$0	$1,045,002	$4,145,565	$7,299,715	$10,519,851	$13,825,394	$17,250,988	
5	0.21	$0	$1,531,843	$4,548,175	$7,618,255	$10,755,000	$13,979,003	$17,328,171	
6	0.22	$0	$1,988,103	$4,919,944	$7,905,703	$10,958,852	$14,101,302	$17,374,823	
7	0.23	$0	$2,413,695	$5,260,777	$8,161,952	$11,131,301	$14,192,214	$17,391,109	
8	0.24	$91,072	$2,808,528	$5,570,546	$8,386,895	$11,272,236	$14,251,668	$17,377,287	
9	0.25	$540,246	$3,172,510	$5,849,230	$8,580,419	$11,381,543	$14,279,599	$17,333,489	
10	0.26	$958,770	$3,505,544	$6,096,641	$8,742,406	$11,459,106	$14,275,722	$17,261,212	
11	0.27	$1,346,507	$3,807,532	$6,312,683	$8,872,735	$11,504,746	$14,240,709	$17,160,709	
12	0.28	$1,703,466	$4,078,275	$6,497,093	$8,971,190	**$11,518,512**	$14,173,848	$17,033,855	
13	0.29	$2,029,360	$4,317,492	$6,650,162	$9,037,353	$11,500,092	$14,075,024	$16,883,741	
14	0.30	$2,324,276	$4,525,635	$6,771,453	$9,071,600	$11,448,422	$13,945,035	$16,725,685	
15	0.31	$2,588,011	$4,701,654	$6,860,445	**$9,072,609**	$11,365,072	$13,781,929	$18,015,443	
16	0.32	$2,819,621	$4,845,519	$6,915,614	$9,042,446	$11,248,328	$13,585,729	$22,360,156	
17	0.33	$3,020,676	$4,958,986	**$6,939,848**	$8,973,486	$11,097,724	$13,364,475	$25,783,645	
18	0.34	$3,189,462	$5,035,428	$6,928,501	$8,878,799	$10,911,586	$13,103,313	$28,563,773	
19	0.35	$3,322,814	$5,084,057	$6,879,345	$8,744,637	$10,690,274	$12,825,838	$30,797,341	
20	0.36	$3,426,971	**$5,091,923**	$6,801,625	$8,572,392	$10,433,830	$12,506,389	$32,532,087	
21	0.37	$3,490,988	$5,066,486	$6,680,620	$8,346,152	$10,132,175	$14,107,094	$33,799,566	
22	0.38	**$3,517,459**	$4,992,052	$6,520,966	$8,087,303	$9,793,602	$16,785,372	$34,474,384	
23	0.39	$3,513,083	$4,886,500	$6,312,697	$7,765,022	$9,372,078	$18,639,961	**$34,775,021**	
24	0.40	$3,444,070	$4,736,459	$6,040,150	$7,437,840	$8,905,826	$19,940,055	$34,485,526	
25	0.41	$3,343,649	$4,517,699	$5,714,572	$6,988,862	$8,359,097	$20,551,805	$33,681,691	
26	0.42	$3,173,110	$4,241,082	$5,330,195	$6,439,891	$7,794,359	**$20,552,486**	$32,059,354	
27	0.43	$2,955,837	$3,909,216	$4,834,783	$5,865,315	$8,617,966	$19,821,902	$29,711,534	
28	0.44	$2,634,094	$3,439,894	$4,252,181	$5,130,224	$8,999,558	$18,504,863	$26,931,124	
29	0.45	$2,262,067	$2,838,920	$3,564,611	$4,304,660	$8,723,075	$16,155,148	$22,678,521	
30	0.46	$1,742,750	$2,220,843	$2,770,599	$3,248,416	$7,433,847	$13,059,115	$18,048,009	
31	0.47	$1,121,285	$1,377,488	$1,679,541	$2,107,736	$5,133,320	$8,866,634	$11,260,501	
32	0.48	$375,588	$462,904	$622,325	$678,975	$1,896,991	$2,693,092	$3,767,786	
33	0.49	-$531,014	-$596,793	-$689,279	-$1,143,111	-$2,644,694	-$3,389,303	-$4,875,772	
34	0.50	-$1,510,706	-$1,913,931	-$2,359,651	-$4,159,755	-$8,387,077	-$12,258,092	-$15,419,715	
35									

Figure 10.30. Sensitivity to CRV and Reference Price (M3).

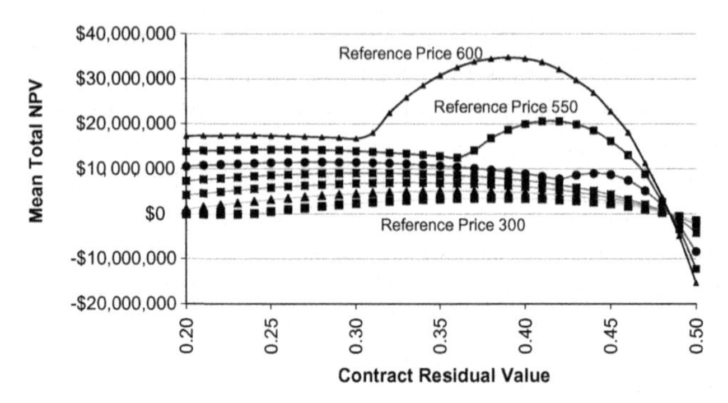

Figure 10.31. Sensitivity to CRV and Reference Price (M3).

As we change the reference price, we are also changing the relationship between the monthly payments that results from a particular value we set for the CRV and the reference price. When the reference price is low, say, $300, all our monthly payments are above the reference price. Thus, as we change the CRV, we are only seeing the effect of one side of the demand curve—the side governed by the parameter α. However, when the reference price is high, say, $600, at some point as we increase the CRV, our monthly payments go from being above the reference price to below it and the effective elasticity of demand switches from α to β. We see the effect of this switch in Figure 10.31, in which we graph the data from Figure 10.30. The top curve, where the reference price is $600, shows how the sensitivity of the NPV to the CRV changes at a CRV of 31 percent, which is precisely where the monthly payment crosses over from being above to being below the reference price. We also observe the optimal CRVs for each reference price in this figure and see why they decline for low reference prices and then eventually increase. This is a good illustration of a graph being more powerful than a table. In the tables, we did not see the patterns that are evident in the graphs, and it was only when we graphed the data that we began to develop an understanding of our results. We capture this analysis in the following insight:

Insight 15: Starting with low reference prices, the optimal contract residual value decreases as the reference price in the market increases, but eventually this pattern is interrupted and the optimal contract residual value jumps back up to a high value. This is because changing the reference price shifts the effective elasticity of demand at different monthly payments and, thus, the sensitivity of the NPV to the contract residual value.

Summary of M3

M3, unlike M2, involved quite a bit of additional work and time. The most effective way to parameterize a demand curve that would include a competitor's price and allow for different elasticities above and below that price did not occur to us immediately. In fact, we had many false starts before we felt we got it right. We have spared you a full description of all the mistakes we made and all the dead-end paths we went down. But looking back, we cannot help but wonder whether the effort was worth it.

Did we gain important additional insights from M3 that we did not have in M2? Not really. We do have a more realistic model. And we can evaluate the impact of changes to more factors, such as the reference price or the elasticity above the reference price (α). Both may be beneficial, especially in our client's eyes. So, to that extent, the results may justify the effort. But it is the nature of modeling that not all efforts generate brilliant new insights. We simply cannot always know in advance which refinements to a model will be most productive. In retrospect, it was a wise decision to add uncertainty to M2, from which we did learn a great deal, and to delay treatment of the demand model until M3.

To the Reader: Before reading on, modify and improve the consumer repurchase module in M2.

10.7 M4 MODEL AND ANALYSIS

The final elaboration we make to this model is to examine the impacts of a more realistic treatment of consumer behavior at lease end. We know that all consumers do not behave rationally in the strict economic sense when their leases end. However, in all three of our previous models, we assumed consumers purchase the leased vehicle at the CRV if the used car price (ARV) is above the CRV, and not otherwise.

The rational model for consumer behavior rests on the assumption that consumers compare the CRV on the vehicle they have been leasing with the actual used car price for similar vehicles in the market. (Recall that we assume throughout this analysis that consumers at lease end will either buy their leased vehicle or buy an equivalent used car, but not lease another vehicle or buy a different brand of used car. These complications would require significant changes to the model.) If the used car price is *above* the CRV, the least expensive way for the consumer to obtain a vehicle is to buy the leased vehicle at its CRV. If the used car price is *below* the CRV, the consumer can save money by returning the leased vehicle and purchasing the equivalent used vehicle at the lower price.

Actual behavior is more complex. Some consumers choose to purchase their leased vehicle when the CRV is above the ARV, despite the financial gain they could enjoy by returning the vehicle and buying the cheaper used car. Other consumers may decide not to purchase their leased vehicle even when the CRV is below the ARV. In this case, they will have to buy a used car at the higher ARV.

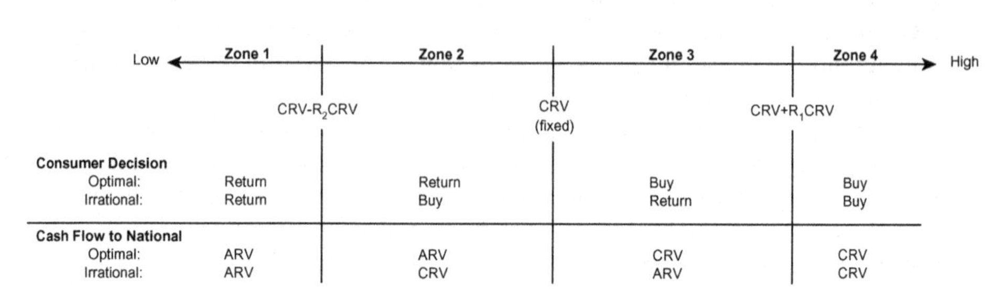

Figure 10.32. Consumer behavior at lease end. (M4).

We model these possibilities by introducing two new parameters, *R1* and *R2*. *R1* defines the region in which consumers will return the leased vehicle (and buy a used vehicle at the ARV) when the ARV exceeds the CRV. In this region, the cash flow at lease end to National is the ARV, since it will sell the returned vehicle at the used car price. *R2* defines the region in which consumers will buy the leased vehicle at the CRV, even though the ARV is less than the CRV. In this region, the cash flow at lease end to National is the CRV.

Figure 10.32 shows how this new model differs from the previous one that was based on rational lease-end behavior. Four zones are defined by the relationship between the ARV and the CRV. (Remember that the CRV is fixed but the ARV is random.) The key point to notice here is that when the ARV exceeds the CRV by less than R_1CRV (Zone 3), the cash flow to National changes from the CRV to the ARV. Since the ARV is above the CRV in this range, this represents an increase in cash flow to National. Likewise, when the ARV is less than the CRV by less than R_2CRV (Zone 2), the cash flow to National changes from the ARV to the CRV, which also represents an increase in the cash flow. Thus, all else equal, irrational consumer behavior at lease end improves the NPV on a set of leases.

To create this new model only requires a change in the formula in M3 for the NPV of the residual cash flow. The new formula for the cash flow at lease end has this structure:

$$
\begin{aligned}
&\text{IF}(\text{ARV} \le \text{CRV} \times (1 - R2))\\
&\qquad \text{ARV},\\
&\text{ELSE IF}(\text{ARV} > \text{CRV} \times (1 - R2) \text{ AND ARV} \le \text{CRV})\\
&\qquad \text{CRV},\\
&\text{ELSE IF}(\text{ARV} > \text{CRV AND ARV} \le \text{CRV} \times (1 + R1))\\
&\qquad \text{ARV},\\
&\text{ELSE IF}(\text{ARV} > \text{CRV} \times (1 + R1))\\
&\qquad \text{CRV}.
\end{aligned}
$$

This formula is too complex to enter safely into a single cell. A better approach is to calculate the cash flow in each of the four zones separately and combine the results. We do this in M4 in cells G29:J29, in each of which we calculate the cash flow to National if ARV and CRV fall in that zone. When ARV and CRV fall in another zone, each of these cells is zero. That approach allows us to add all four of these formulas to calculate the overall cash flow in cell C29, since only one of the four will be non-zero on a given simulation trial.

To the Reader: Before reading on, use M4 to generate as many interesting insights into the situation as you can.

Since this new model is identical to M3 when $R1 = R2 = 0$, we would expect that the optimal CRV in that base case would be around 31 percent, as we saw in Figure 10.28. To determine the impact of irrational behavior on the optimal CRV, we first test the impact of $R1$ with $R2 = 0$, then the impact of $R2$ with $R1 = 0$, and finally the combined impact of both parameters. Figure 10.33 shows how the NPV varies with the CRV in this case, using 0.25 for the two parameters.

As we expected, in the base case with $R1 = R2 = 0$, the NPV reaches a maximum of about \$8.3M at a CRV of 32 percent. When $R1 = 0.25$ (and $R2 = 0$), the optimal CRV increases to 42 percent and the NPV increases to 11.7M, which is an increase of about 41 percent. When $R2 = 0.25$ (and $R1 = 0$), the optimal CRV and the resulting NPV remain unchanged from the base case. Finally, when $R1 = R2 = 0.25$, the optimal CRV is 43 percent and the NPV increases 47 percent over the base case.

Why the striking difference in the effects of irrationality when the ARV is above the CRV versus below it? When the ARV is above the CRV and

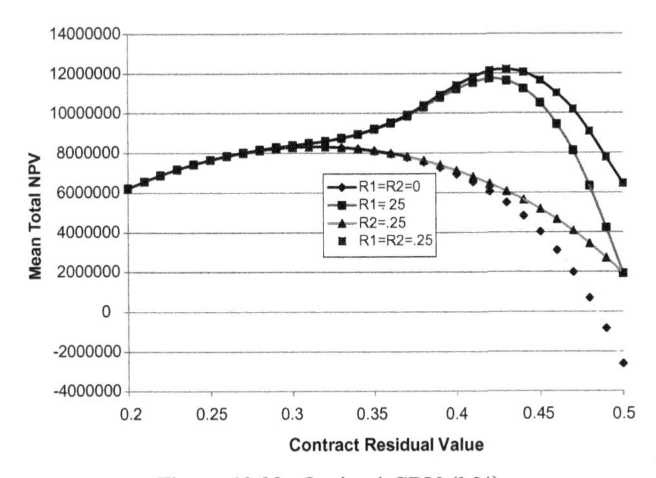

Figure 10.33. Optimal CRV (M4).

consumers return their leased vehicles, National sells them for the ARV. Since the ARV is random in our model, this leads to a random cash flow. To capture this increased cash flow (when compared with the rational case), the optimal response is to increase the CRV. In effect, the leasing company can raise the CRV, reduce monthly payments, and increase leases sold because the lease-end cash flow is higher.

On the other hand, when the ARV is below the CRV, irrational consumers buy their leased vehicles at the CRV. In this case the cash flow to National is always the fixed value of the CRV. In this case, the optimal CRV remains unchanged from the base case. At a CRV around 30 percent, the effects of this form of irrationality are muted because the random ARV is hardly ever below the CRV. If we were to raise the CRV from this level, we would have more consumers buying their vehicles for the CRV, but the gain from this effect is more than offset by the negative effects on other cash flows.

Insight 16: If consumers return their leased vehicles when the actual residual value exceeds the contract residual value, the optimal contract residual value and the resulting NPV are both higher. If consumers buy their leased vehicles when the actual residual value is less than the contract residual value, the optimal contract residual value and the resulting NPV are unchanged.

Up to this point, all our conclusions have been based on one set of values for the parameters *R1* and *R2*. We have found that *R2* has little impact on the results, so we next examine the relationship between the optimal CRV and *R1*. Figure 10.34 shows the results of a two-way Crystal Ball sensitivity analysis in which we vary *R1* from 0 percent to 25 percent. Over this range of parameters, the optimal CRV varies between 31 and 42 percent. *R1* has little influence on the optimal CRV until *R1* reaches a fairly high value, around 20 percent. Below that value, irrational lease-end behavior raises the NPV but has little impact on the choice of CRV.

Insight 17: The optimal contract residual value is insensitive to irrational behavior unless consumers return leased vehicles when the actual residual value significantly exceeds the contract residual value.

Again, we should be careful not to draw overly general conclusions from a limited number of model runs. All of our previous results were based on a model in which the variability in the ARV was modest (standard deviation of 0.06 relative to a mean of 0.50). To check whether variability has a substantial effect on our results we repeat the analysis behind Figure 10.33, this time with a standard deviation for the distribution of ARV of 0.20. The results are shown in Figure 10.35.

This figure suggests that high variability in the ARV mutes the impact of consumer irrationality at lease end. In the base case ($R1 = R2 = 0$), the optimal CRV is 26 percent. When $R1 = R2 = 0.25$, the optimal CRV increases to 29

	A	B	C	D	E	F	G	H
1	**Total NPV: Mean**							
2		R1						
3	Contract Residual Value (%)	0	0.05	0.1	0.15	0.2	0.25	
4	0.20	$6,220,626	$6,220,626	$6,220,626	$6,220,626	$6,220,626	$6,220,626	
5	0.21	$6,567,708	$6,567,708	$6,567,708	$6,567,708	$6,567,708	$6,567,708	
6	0.22	$6,883,780	$6,883,780	$6,883,780	$6,883,780	$6,883,780	$6,883,780	
7	0.23	$7,168,740	$7,168,740	$7,168,740	$7,168,740	$7,168,740	$7,169,458	
8	0.24	$7,422,483	$7,422,483	$7,422,483	$7,422,483	$7,423,103	$7,426,020	
9	0.25	$7,644,899	$7,644,899	$7,644,899	$7,645,401	$7,647,957	$7,651,843	
10	0.26	$7,835,873	$7,835,873	$7,836,239	$7,838,032	$7,841,736	$7,846,025	
11	0.27	$7,995,287	$7,995,355	$7,996,613	$7,999,526	$8,003,633	$8,013,535	
12	0.28	$8,122,995	$8,123,290	$8,125,617	$8,128,894	$8,135,927	$8,156,144	
13	0.29	$8,218,771	$8,219,778	$8,222,246	$8,226,691	$8,244,059	$8,286,193	
14	0.30	$8,282,240	$8,283,405	$8,285,915	$8,297,039	$8,327,373	$8,381,648	
15	0.31	**$8,312,556**	**$8,313,391**	$8,320,042	$8,341,395	$8,388,871	$8,491,189	
16	0.32	$8,309,674	$8,311,372	**$8,322,853**	**$8,358,628**	$8,433,873	$8,608,600	
17	0.33	$8,273,306	$8,276,626	$8,299,445	$8,345,353	$8,469,514	$8,767,994	
18	0.34	$8,202,552	$8,209,258	$8,236,612	$8,322,406	$8,538,065	$8,950,004	
19	0.35	$8,095,917	$8,105,356	$8,151,010	$8,298,859	$8,612,485	$9,207,155	
20	0.36	$7,951,124	$7,963,176	$8,034,263	$8,256,888	$8,715,013	$9,502,702	
21	0.37	$7,765,320	$7,788,424	$7,905,867	$8,208,985	$8,834,661	$9,892,328	
22	0.38	$7,537,029	$7,572,109	$7,746,777	$8,178,515	$8,989,968	$10,306,213	
23	0.39	$7,259,642	$7,311,668	$7,556,699	$8,115,685	$9,136,210	$10,784,645	
24	0.40	$6,926,934	$7,001,518	$7,321,055	$8,070,830	$9,315,050	$11,196,187	
25	0.41	$6,527,442	$6,630,551	$7,045,936	$7,946,416	$9,472,788	$11,523,639	
26	0.42	$6,050,881	$6,185,344	$6,718,542	$7,831,715	**$9,552,795**	**$11,723,884**	
27	0.43	$5,483,095	$5,654,616	$6,293,273	$7,598,705	$9,451,054	$11,630,784	
28	0.44	$4,807,705	$5,021,872	$5,824,548	$7,249,526	$9,201,986	$11,203,678	
29	0.45	$4,007,676	$4,260,758	$5,173,311	$6,710,584	$8,623,963	$10,477,431	
30	0.46	$3,060,091	$3,374,749	$4,377,975	$5,998,181	$7,763,640	$9,385,165	
31	0.47	$1,946,956	$2,298,512	$3,372,538	$4,954,194	$6,607,045	$8,060,133	
32	0.48	$645,105	$1,036,861	$2,163,900	$3,634,481	$5,093,117	$6,292,101	
33	0.49	-$878,041	-$466,658	$651,522	$2,032,031	$3,333,038	$4,167,883	
34	0.50	-$2,650,090	-$2,206,643	-$1,156,474	$63,719	$1,163,460	$1,854,060	
35								

Figure 10.34. Sensitivity of CRV and R1 (M4).

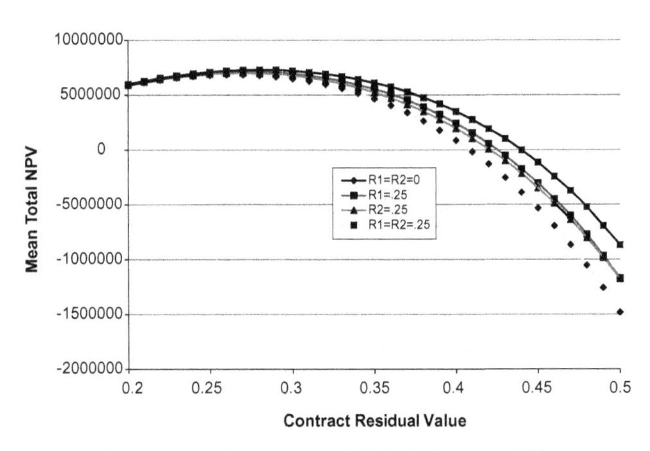

Figure 10.35. Optimal CRV (M4, σ = 0.20).

percent and the resulting NPV increases just 6 percent. So consumer irrationality is still relevant when ARVs are highly variable, but the adjustment required in setting the CRV is negligible.

Insight 18: High variability in the actual residual value reduces the impact of consumer irrationality on the optimal contract residual value to a negligible level.

Summary of M4

M4 has led us to some valuable insights. Given that we do not know much about consumer behavior at lease end, it may prove useful when we present our results to management to be able to say how much it matters. We have found that one form of irrational behavior leads to a higher optimal CRV and NPV but that another form has little impact. This specific insight may, of course, depend on the numerical values we have chosen and even on the way we have formulated the model, but at the very least, it serves to show how we can formulate alternative models of consumer behavior and determine their impacts on the choices open to management.

10.8 PRESENTATION OF RESULTS

National Leasing offered us an opportunity to show that modeling can help them think through the process of setting the terms on auto leases. We have responded by building a series of models that capture many of the problem's essential elements. Although our results are based on arbitrary data and, therefore, cannot be taken as definitive, they nonetheless demonstrate the power of modeling. They also suggest several provocative hypotheses about the leasing decision. Our final task is to prepare a presentation to management that will convince them to adopt our approach.

We assume our audience consists of knowledgeable senior managers at National Leasing. Like most managers, they are likely to have more knowledge of their own business (auto-leasing) than of modeling. So we focus our presentation on showing how our modeling approach can shed light on issues of concern to them. The problem description gives us important clues as to what these issues are. For example, how should CRVs be set, can residual losses be avoided, and can improved forecasts of ARVs help?

When there is a single decision to provide the focus for a presentation, we take a direct approach and offer our recommendations at the beginning. But the task in this case is to show that modeling could provide insight. We have several insights to share, but no single result to provide the focus. Accordingly, we will take a somewhat indirect approach and devote most of our presentation to developing the logic behind our approach and to discussing the intuition behind our results. We will, however, summarize early in our presentation our key insights in the form of hypotheses.

The strategy we will use in the presentation is to convince the audience that our approach allows them to think more deeply about the decisions they face than they have been able to in the past. This is how we use the insights we have generated in our analysis—not to prove a point, but to raise issues that challenge the managers' intuition. After reviewing our four models and 18 insights, we settle on five points that address important questions and showcase our approach:

- The optimal CRV is not necessarily the same as the ARV.
- The optimal CRV increases with the elasticity of demand.
- Uncertainty in the ARV reduces the optimal CRV and leads to a highly skewed distribution of NPVs.
- Reducing uncertainty in the ARV leads to a higher optimal CRV and a higher NPV.
- The CRV is insensitive to irrational consumer behavior at lease end.

We design our presentation around the following outline:

- Introduction
- Limitations of the pilot study
- Hypotheses
- Structure of the model
- Insights
- Summary

In the first slide (Figure 10.36), we set the stage by describing the major challenges National Leasing is facing and why we were called in to help. Next, we explain the limitations under which we worked, so the audience will have appropriate expectations (Figure 10.37). It is important to strike the proper balance at this point between overselling and underselling our work. If we

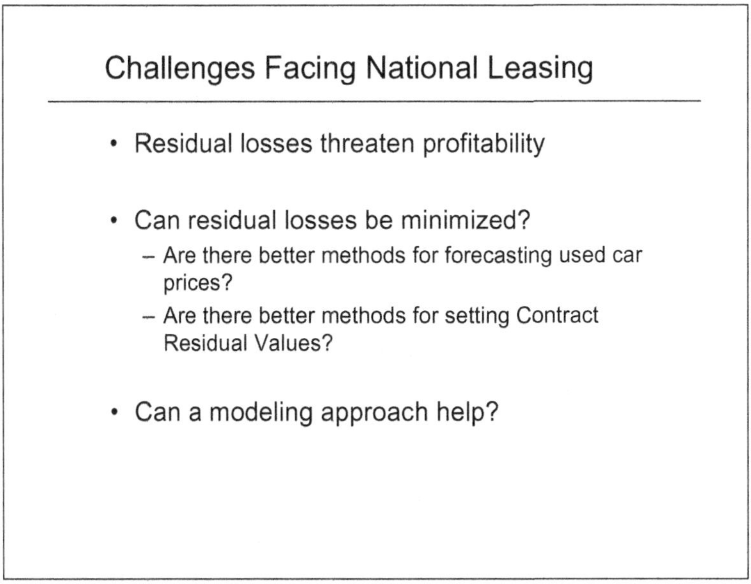

Figure 10.36. Slide 1.

Limitations of Pilot Study

- Time and data were limited.
- We focused on one block of leases.
- Data used are reasonable, not precise.
- Relationships used are plausible, not derived from actual data.
- Our results are illustrative, not definitive.

Figure 10.37. Slide 2.

stress the limitations too much, the audience will not think our work is credible; if we do not stress them enough, they may take our results as the last word. We want to avoid both outcomes.

The third slide (Figure 10.38) offers five hypotheses to whet the audience's appetite. We will not pause at this point in the presentation to explain or justify these points. We will simply mention each one and point out that we will provide more detail and explanation later on. If we have chosen these points well, the audience will begin to believe that our approach answers some questions they struggle with, such as how to set CRVs relative to ARVs, and some important questions they have not thought to ask, such as how uncertainty in the ARV affects their decisions.

Having set the stage, we then walk the audience through the influence diagram (Figures 10.39– 10.43) to show them that their problem is not simple but that we have captured its essential features. This also allows us to describe the key trade-offs that make determining the optimal CRV difficult. We present the influence diagram in a sequence of stages, starting with just the outcome measure, the decisions, and the key uncertainty. Then we add details step by step. In Figure 10.39, we show how the Net Cash Flow is broken down into three components. Then in Figures 10.40 and 10.41, we show how Total Residual Cash Flow is determined. Finally, in Figures 10.42 and 10.43, we show how Total Lease Payments and Total Cost of Borrowing are determined.

We now turn to the hypotheses that our work has led us to (Figures 10.44–10.48). The goal here is to challenge the audience to think about the problem

Five Hypotheses

- The optimal CRV does not equal the ARV.
- The optimal CRV increases with the elasticity of demand.
- Uncertainty about the ARV reduces the optimal CRV.
- Reducing uncertainty in the ARV leads to a higher optimal CRV and a higher NPV.
- The optimal CRV is insensitive to irrational lease-end behavior.

Figure 10.38. Slide 3.

Structure of Model (1)

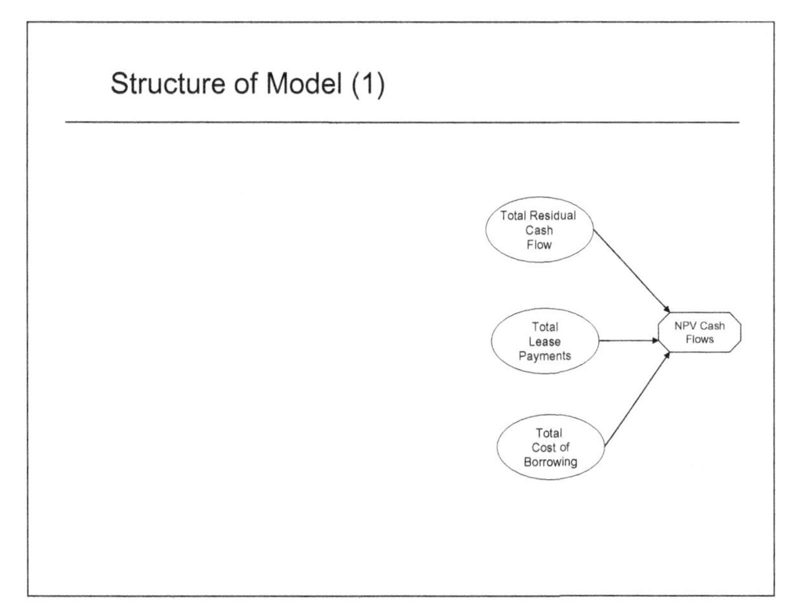

Figure 10.39. Slide 4.

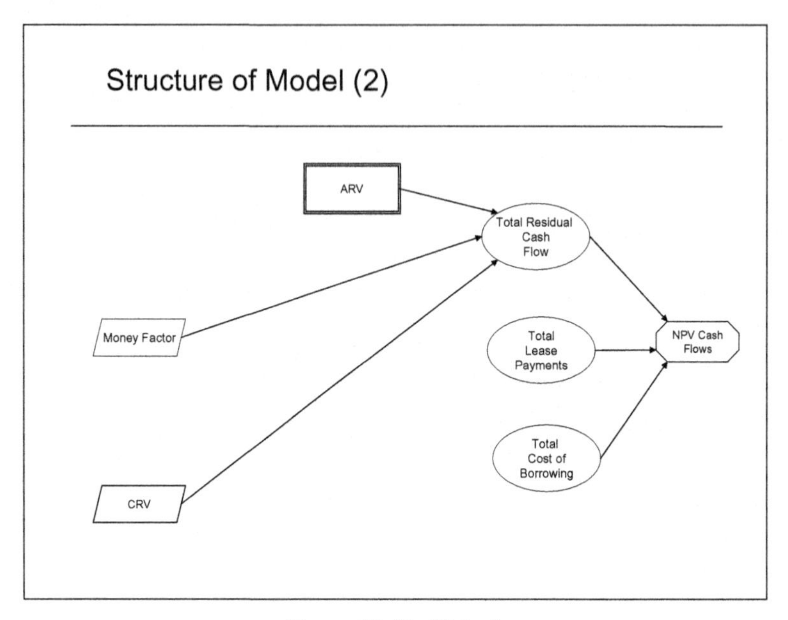

Figure 10.40. Slide 5.

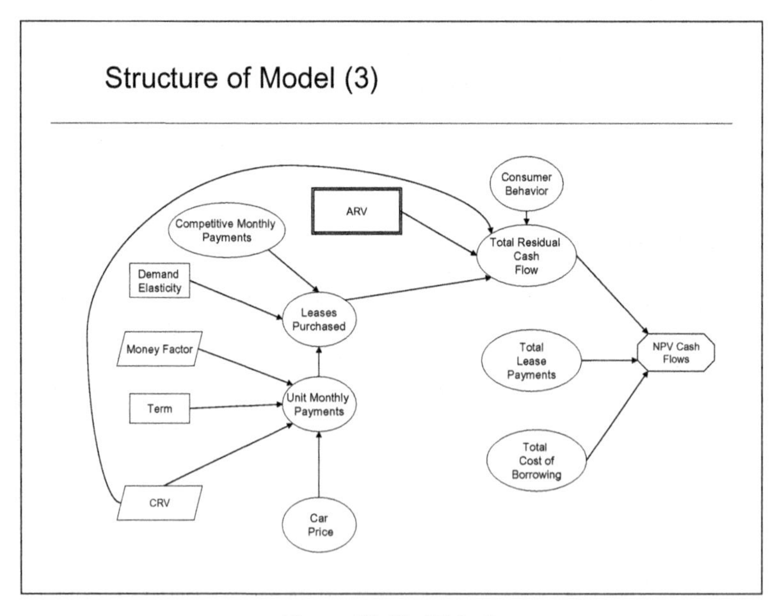

Figure 10.41. Slide 6.

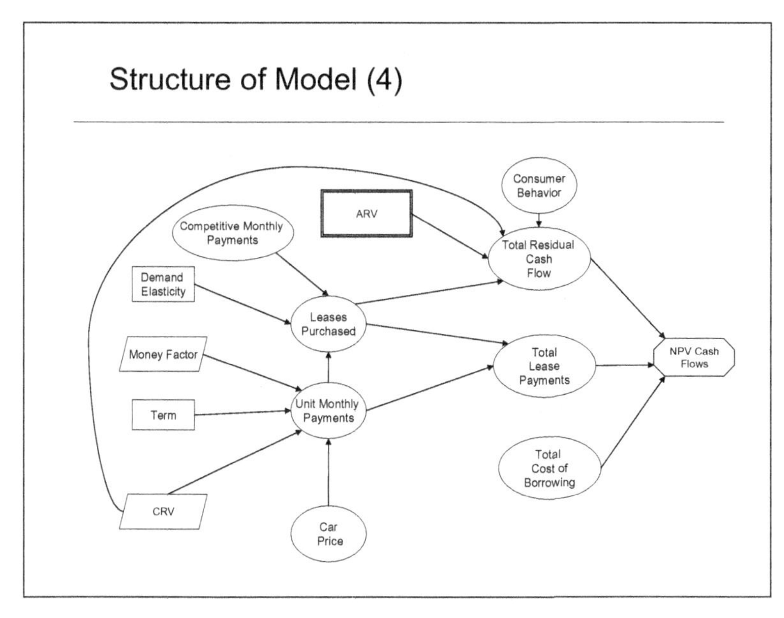

Figure 10.42. Slide 7.

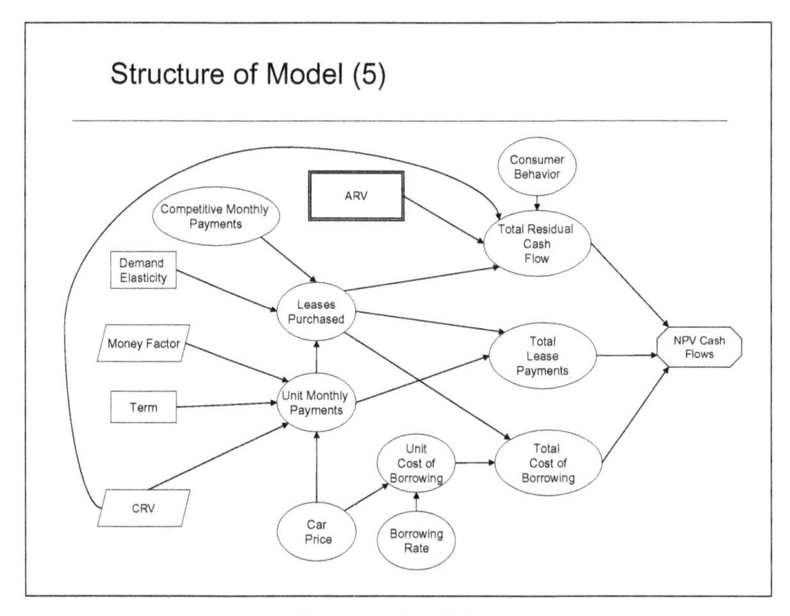

Figure 10.43. Slide 8.

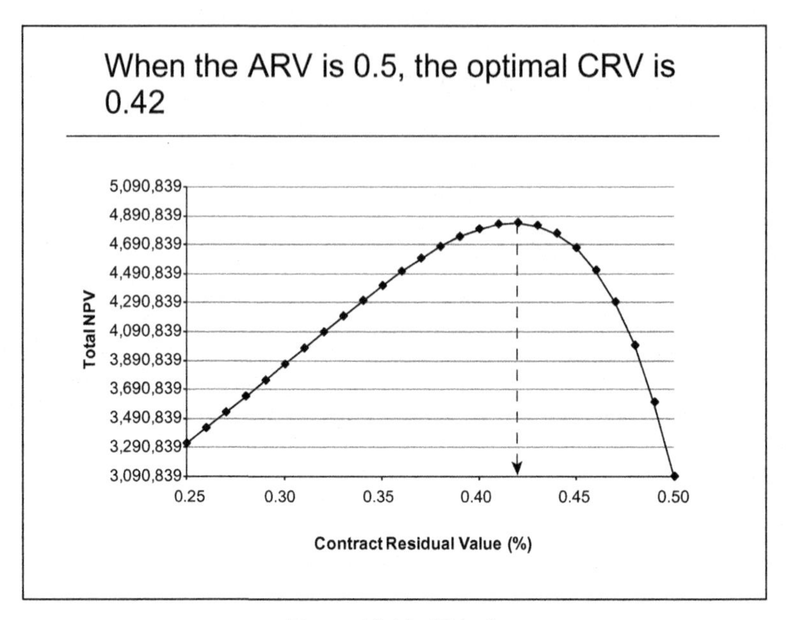

Figure 10.44. Slide 9.

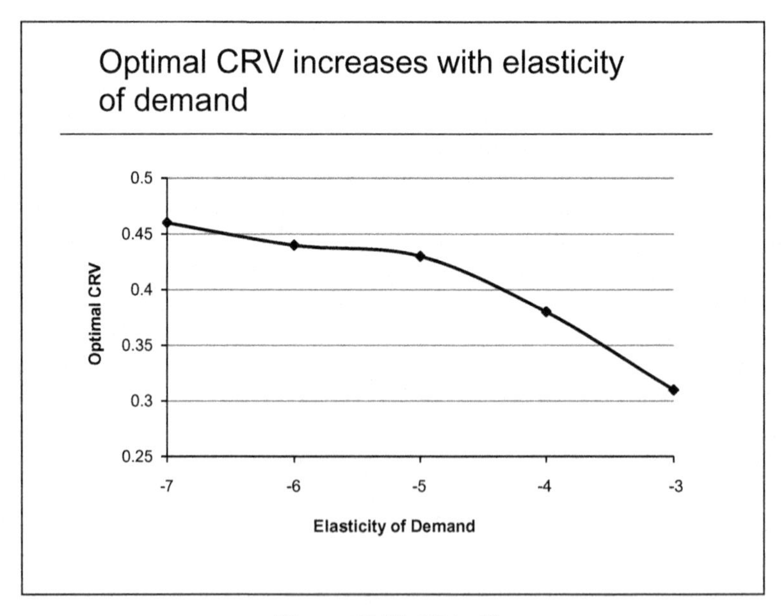

Figure 10.45. Slide 10.

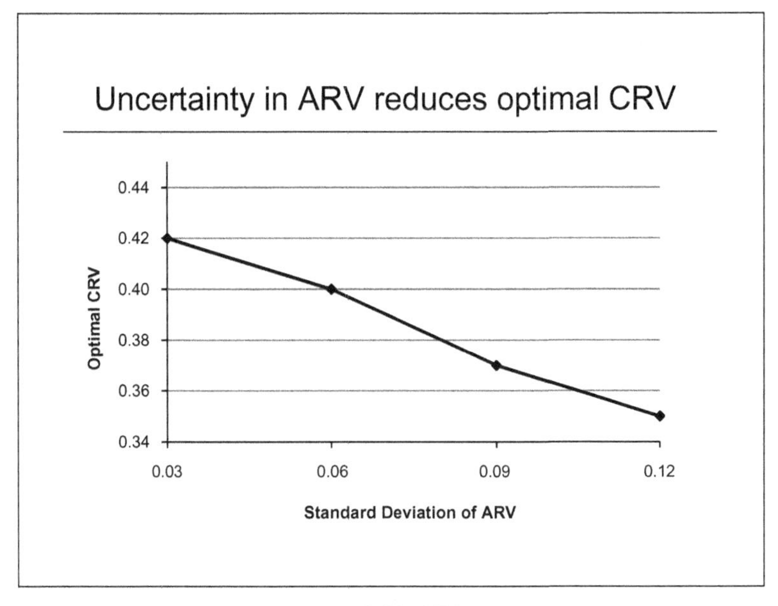

Figure 10.46. Slide 11.

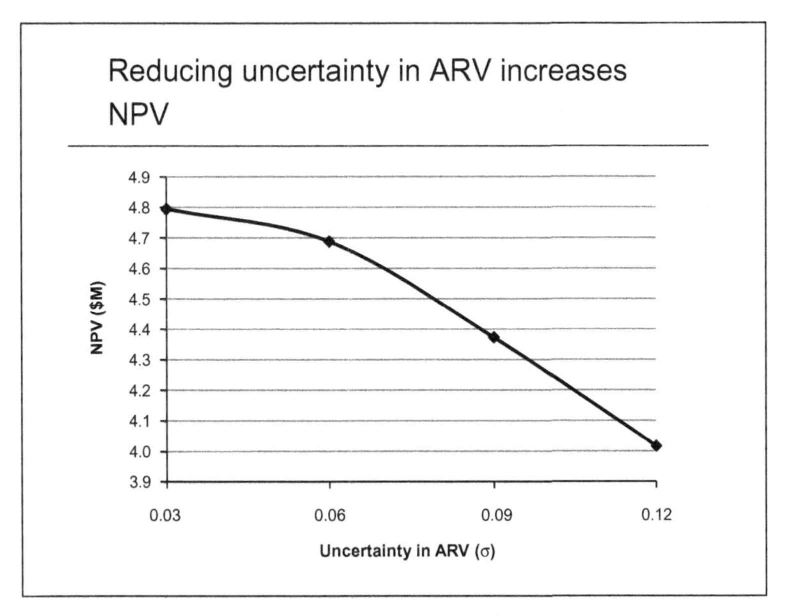

Figure 10.47. Slide 12.

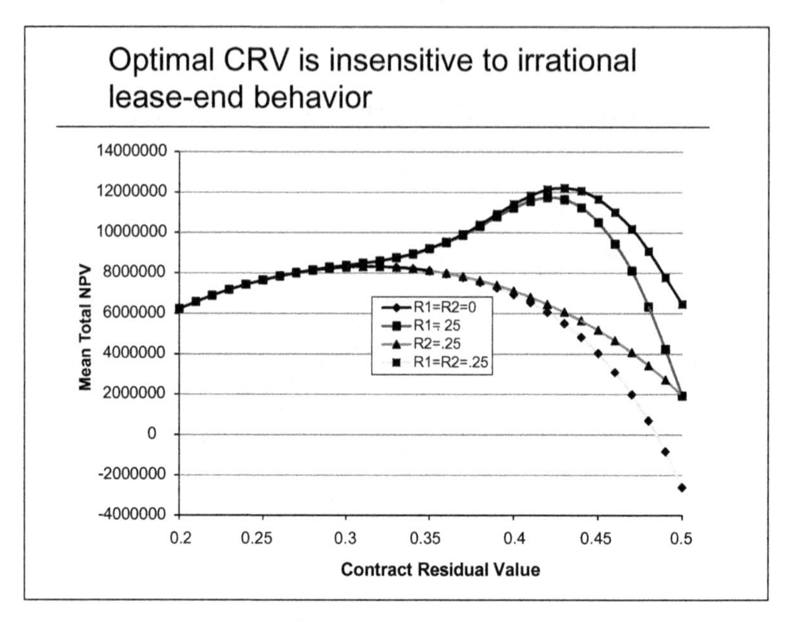

Figure 10.48. Slide 13.

Summary

- Purpose of study: explore factors affecting residual losses using modeling
- Limitations: time, data
- Tentative results
  - Optimal CRV < ARV
  - Use CRV to control effect of uncertainty in ARV on residual losses
  - Reducing uncertainty in ARV is potentially valuable
  - Irrational lease-end behavior can affect optimal CRV

Figure 10.49. Slide 14.

and to test whether they can think through the implications of our hypotheses in a logical way. If some audience members think one or another of our hypotheses are wrong, we will engage them to explain their reasoning and try to compare their mental model with our spreadsheet model to identify the source of the difference.

We end our presentation with a summary slide (Figure 10.49) that restates what we were asked to do and the limitations under which we worked, and reiterates some of the highpoints of our results.

10.9 SUMMARY

Automobile leasing has become a highly competitive and risky business, even for the large financial firms that dominate the industry. Leasing companies compete to attract the attention of consumers with attractive upfront terms on their leases, but they must constantly beware of the risk of residual losses, which can occur when actual used car prices deviate from forecasts.

National Leasing is considering changing its approach to setting the contract residual value on its leases. The company has given us the opportunity to show how modeling could help to inform these decisions. This is a situation in which a prototype model is necessary, since the goal is to demonstrate what we can do with models, not to develop a fully operational approach. This task is ideal for our modeling approach, which is tailored to situations in which both time and data are limited.

We built a sequence of models, starting with a simple model and adding complexity in carefully considered stages. At each stage we used our model to develop insights into the questions facing National. Given the tentative nature of our models, we consider these insights more as hypotheses that help to illuminate the economic trade-offs decision makers face in setting lease terms, rather than as firm conclusions. These insights also are a natural vehicle for conveying the potential power of modeling.

CHAPTER 11

Pharma X and Pharma Y

11.0 INTRODUCTION

The pharmaceutical industry, more than most industries, is driven by research and development (R&D). Firms typically invest a high proportion of their annual profits in R&D and consider management of the R&D process a strategic concern. Pharmaceutical companies often engage in intense legal competition over the intellectual property they and their competitors develop. They also frequently undertake buyouts or joint ventures, sometimes to reduce the risks of their large and uncertain investments in drugs and other times to share their unique capabilities in science or marketing.

In this chapter, we consider two pharmaceutical firms that are developing drugs for a new class of treatments. One of the firms is very large, and the other one is quite small. The small firm has a dominant patent position, which means that the threat of a lawsuit from this firm is credible. If it were to engage in such a suit and win, it could completely block the large firm from marketing a drug it had spent millions to develop. The large firm, however, has a much stronger marketing presence in the large Japanese and European markets. Our client is the large firm.

The case we analyze here takes the large firm's perspective, but a critical aspect of the analysis is to try to anticipate the small firm's negotiating position. We consider several different deals the large firm could offer to its small competitor to avoid a devastating patent suit and to ensure both firms benefit from their work on this drug. The case asks us to structure the best possible deal for our client, but any deal we propose must also be acceptable to the small firm.

11.1 THE PHARMA X AND PHARMA Y CASE

Pharma X and Pharma Y

Pharma X is a large pharmaceutical company with year 2010 sales of $5B (all figures are in constant 2010 dollars). Pharma X is developing a drug, code-named XCardia, for a particular variety of cardiovascular disease. XCardia is expected

Modeling for Insight: A Master Class for Business Analysts, by Stephen G. Powell & Robert J. Batt

to reach the market in 2013. Recently, Pharma X learned that a competitor, Pharma Y, is developing a similar chemical compound, called YCardia, which is also expected to reach the market around the year 2013. Pharma Y is a small pharmaceutical company with sales of $500M per year. These two compounds are instances of a new class of treatments, and no other companies are thought to be developing competitive products. Patent experts at Pharma X believe that Pharma Y is likely to get a patent on the new compound, so Pharma Y stands a good chance of preventing Pharma X from marketing its own compound. (Note: Pharma Y can only receive a patent and sue Pharma X for patent infringement if Pharma Y is successful in developing a marketable product.)

Pharma X executives are considering making a deal with Pharma Y that will allow both companies to profit from the sale of this new drug without getting tied down in litigation or other costly competitive actions.

A typical pharmaceutical product goes through a standard series of development phases, as follows:

- Preclinical phase: animal trials focusing on safety (typically 13 weeks)
- Phase 1: safety studies conducted on 50 to 100 normal healthy male volunteers (typically three to six months)
- Phase 2: efficacy and safety trials on a diseased target population under controlled medical conditions (typically six to nine months)
- Phase 3: efficacy and safety trials on the target population under actual conditions of use (typically six to nine months)
- FDA submission: preparation of a new drug application (NDA), involving extensive statistical analysis and report writing (typically six months)
- FDA review: The FDA evaluates the NDA based on the preclinical and clinical data (typically 17 to 24 months)

Both XCardia and YCardia are currently about to enter Phase 2.

Pharma X believes its compound has a 50 percent chance of success in Phase 2 and an 80 percent chance of success in Phase 3. The likelihood health authorities in the major markets (the United States, Europe, and Japan) will reject it is negligible given successful Phase 3 results. Phase 2 studies will cost $10M, and Phase 3 studies will cost $40M. Regulatory review support will cost $2M.

According to Pharma X's marketing staff, sales of the new therapeutic class represented by XCardia and YCardia are expected to peak at $500M worldwide two years after launch. Sales should stay near the peak until the patent expires, which, for both compounds, will occur in 2022. After that, sales will taper off over the next eight years. The contribution margin over the product's lifetime is expected to be 75 percent of sales revenues. ("Contribution" measures revenues net of variable costs, such as manufacturing and marketing expenses.) Pharma X believes that its market share of the new therapeutic class worldwide will be 50 percent if YCardia is also in the market but 100 percent otherwise. Since the products are almost identical, they will very likely succeed or fail together; clinicians estimate that if one product is brought to market successfully, there is a 90 percent chance the other will succeed.

Patent infringement litigation typically begins when a drug is first marketed. Pharma X believes Pharma Y will almost certainly sue if it can and has a 50 percent chance of winning the suit, thereby entirely preventing Pharma X from marketing its product.

At Pharma X, decisions to develop drugs are based on a Net Productivity Index (NPI), which is the ratio of the contribution to the development costs. The reason management does not simply evaluate projects on the basis of Net Present Value (NPV) is that development funds are limited, so there is an opportunity cost (not reflected in the NPV) associated with spending money on one project because it cannot be spent on another. The NPI allows management to compare the net returns from various projects with the net development costs each project incurs.

The NPI is calculated by determining the flow of contribution that can be expected from the product after launch, discounting this to the present at an appropriate discount rate, and then taking an expected value over different possible scenarios. Similarly, development costs are discounted over time and an expected value is taken over all future scenarios. Generally speaking, Pharma X management would like to see the NPI exceed five, but there is no specific hurdle rate.

Figure 11.1 provides a generic illustration of an NPI calculation. The upper panel shows two random events: R&D Success and Market Size. Four outcomes are possible, and the expected contribution over all four is 180. The lower panel shows three random events that influence costs: Phase 2, Phase 3, and FDA Review. Again, there are four possible outcomes. The expected costs are 19.9 for an NPI of 9.05 (9.05 = 180 ÷ 19.9).

Analysts at Pharma X believe the same success probabilities and cost estimates apply to both companies because their drugs are so similar. The contribution Pharma Y receives differs slightly from Pharma X's case in that the smaller firm does not have the marketing strength to sell its product in Japan or Europe. It will need to find a partner in those markets and likely will receive a 10 percent royalty on sales as opposed to a 75 percent margin. Pharma X anticipates 50 percent of sales will come from the United States, 30 percent from Europe, and 20 percent from Japan.

Several months ago, Pharma X decided to offer Pharma Y $50M for the rights to YCardia. Pharma Y declined the offer. Pharma X executives believe the reason was that Pharma Y expects this new class of drugs to compete with traditional treatments. One Pharma Y executive suggested peak sales for the new treatment market would likely be $900M (rather than the $500M Pharma X's marketing department estimated).

Pharma X executives have asked us to work on this problem and report back in a week when the executive team meets to prepare for the next round of negotiations with Pharma Y. The following issues particularly interest them:

1. Is there a dollar amount Pharma X should be willing to pay for the rights to YCardia that Pharma Y would be likely to accept? (If Pharma X buys the rights to the drug, Pharma Y will agree not to develop it and not to sue.)

Expected Contribution

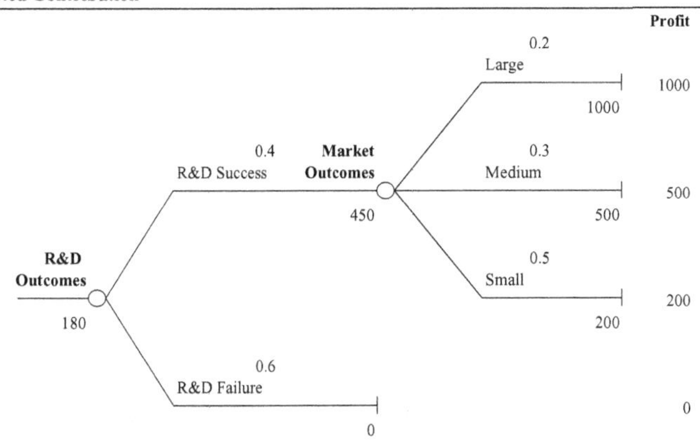

Expected Value of Contribution = 0.4 x (0.2 x 1000 + 0.3 x 500 + 0.5 x 200) + 0.6 x 0 = 180

Expected Costs

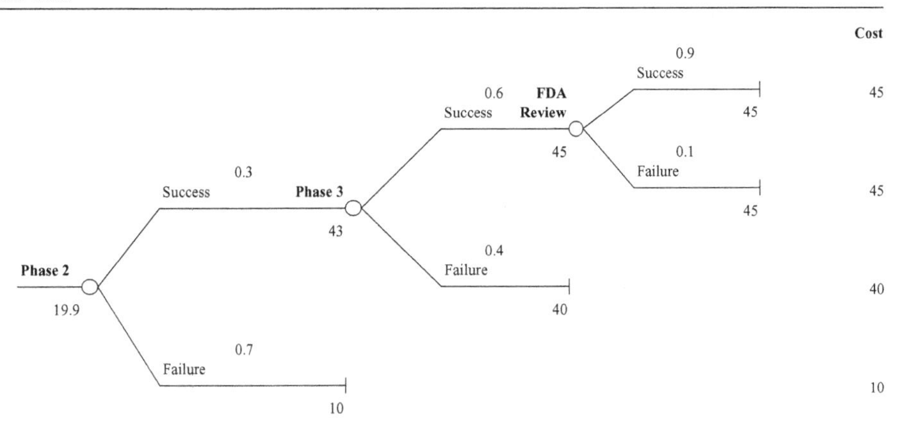

Expected Value of Costs = 0.3 x 0.6 x 0.9 x 45 + 0.3 x 0.6 x 0.1 x 45 + 0.3 x. 0.4 x 40 + 0.7 x 10 = 19.9

NPI

NPI = EV Contribution/EV Cost = 180 ÷ 19.9 = 9.05

Figure 11.1. Sample calculation of NPI.

2. Rather than buying outright the rights to the compound, is there a possible deal in which Pharma X could purchase a license from Pharma Y to avoid a patent battle? (Under a license agreement, Pharma X would pay Pharma Y a percentage of its revenues if Pharma Y is successful in the market. Pharma Y could still develop its own drug but could not sue.) This would allow both companies to market their drugs separately. If this option is viable, how much should Pharma X be willing to pay and Pharma Y be willing to accept?
3. Pharma X's CEO is interested in whether a codevelopment deal could be struck—a joint effort in which Pharma X and Pharma Y would share

development costs and commercial returns. Under such an agreement, both development labs could continue to operate in parallel or could be combined as a single team.

Our task is to develop recommendations for the negotiation team as to which of the various alternatives might be best for Pharma X and acceptable to Pharma Y.

To the Reader: Frame the problem facing Pharma X, and describe the problem kernel.

11.2 FRAME THE PROBLEM

The case statement gives us quite a bit of information about the two firms and the drug development process. We anticipate that both firms will analyze the choices open to them in the light of the scientific and market uncertainties they face. For example, Pharma X must recognize that Pharma Y has a strong patent position. Pharma Y, on the other hand, must recognize that Pharma X has a marketing advantage abroad.

Since Pharma X uses the NPI criterion, we assume Pharma Y uses it as well. We also reserve the possibility of looking behind the NPI to its components, the present value (PV) of contribution and PV of costs, since these details may reveal deeper insights. A related issue has to do with how we might incorporate risk into our evaluation. Risk may play an important role in this analysis, and the risks Pharma X faces may differ from those faced by Pharma Y. For example, Pharma X has much larger revenues and may be better able to take on financial risks. The NPI is an expected value criterion, so it does not directly measure risk. Therefore, we may want to broaden our outcome measures to include a risk measure.

Another assumption is needed to help us analyze deals between the two rivals. The question we face is how does each party decide whether to accept an offer? A simple and plausible assumption is that any firm will potentially accept an offer that leaves it better off than it would be without the deal. This requires us to estimate the NPI that each firm could achieve on its own, without a deal. These estimates provide the baselines against which the companies will measure proposed deals. Presumably, they also want to get the best deal they can, so they will bargain for the highest NPI they can get.

Some other assumptions are implicit in the case, but as a general rule, we try to make assumptions explicit in the problem-framing process. For example, it seems both firms face the same economics, but their estimates of total market size differ (between $500M and $900M).

We summarize the essential features of the problem in the problem kernel:

Two firms are simultaneously developing compounds that could compete in the same market. Each company has strengths the other lacks: X is big and has marketing strength abroad; Y is small but has a strong patent position. Each firm must evaluate its options and decide whether a deal with the other firm is preferable to going it alone.

To the Reader: Before reading on, draw an influence diagram for Pharma X.

11.3 DIAGRAM THE PROBLEM

Figure 11.2 shows an influence diagram for this problem. The outcome measure is NPI for Pharma X. (We eventually need to model the NPI for Pharma Y, but the structure of those calculations will be similar.) The main components of NPI are the Expected Present Value (PV) of Contribution and the Expected PV of Costs.

The expected costs to develop a drug are simple to determine if we know the stages the compound must go through, the costs of each stage, and the probabilities of success at each stage. Figure 11.1 gives an example of the calculations. In the influence diagram, we record the components of total cost: Probability of Success by Stage and Cost by Stage.

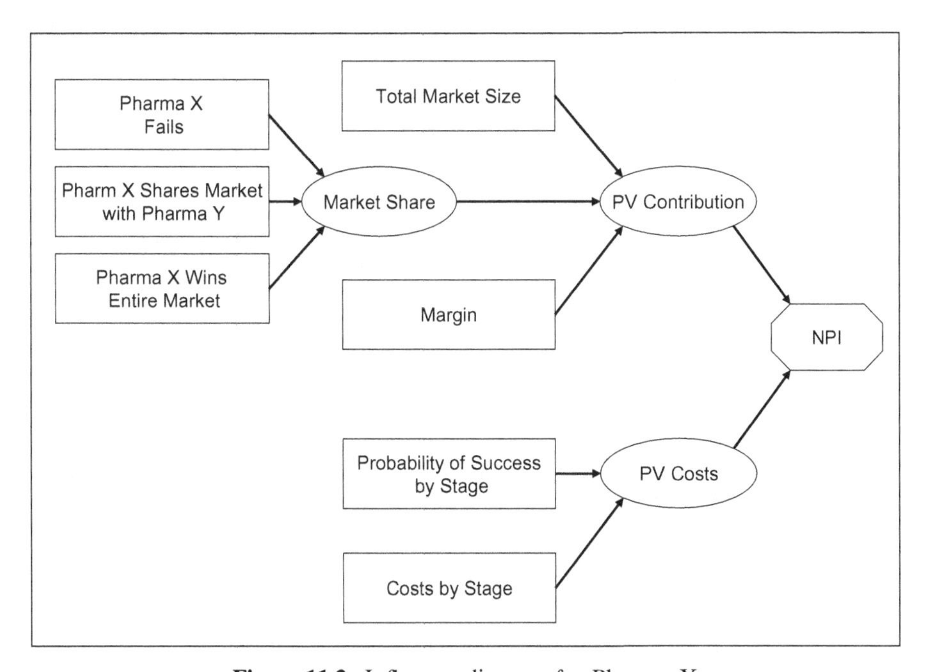

Figure 11.2. Influence diagram for Pharma X.

We decompose Contribution into three elements: Total Market Size, Market Share, and Margin. The total size of the market seems to be independent of the actions of both firms. The case suggests the market will gradually rise to a peak of around $500M and then decline to nothing after the patent expires. We model this behavior in a spreadsheet using a parameterized function, but for the purposes of the influence diagram, these details are not essential.

The second component of Contribution is Margin, which in the case of Pharma X, is 75 percent. (The situation is more complex for Pharma Y since it can only receive a 10 percent royalty in Japan and Europe.)

Market Share is more complicated. The actual share Pharma X can achieve is the outcome of its own development efforts, those of Pharma Y, and the possible patent suit. There are three possible outcomes for Pharma X's share:

- 0 percent: Pharma X fails to develop the drug or Pharma X and Pharma Y both succeed and Pharma Y sues and wins.
- 100 percent: Pharma X succeeds and Pharma Y fails
- 50 percent: Pharma X and Pharma Y both succeed and Pharma Y either does not sue, or sues and fails.

We list these three cases as components of Market Share and again leave the details to the spreadsheet model.

11.4 EXPECTED VALUE OR SIMULATION?

The influence diagram for this problem is quite simple and suggests that our models should be simple as well. However, there is one issue that we have not dealt with that becomes critical as soon as we begin to translate the influence diagram into a spreadsheet model. Pharma X and Pharma Y face substantial uncertainties, both in the development of their compounds and in the outcomes of a possible suit. How should we model uncertainty in this problem?

The most comprehensive approach to modeling uncertainty is simulation. In this case, a simulation would create random outcomes for the success of Pharma X and Pharma Y in Phases 2 and 3, as well as Pharma Y's choice to sue and the outcome of the suit. Each of these six random outcomes is a Bernoulli trial: Yes/No or Success/Failure. Other parameters in the problem, such as market size, may also be uncertain. In a simulation model, we would create a random outcome for each event and then calculate the costs and revenues for Pharma X based on those outcomes. For example, in a single simulation trial, we might get these random outcomes:

- Success for Pharma X in Phase 2
- Success for Pharma X in Phase 3
- Success for Pharma Y in Phase 2

- Success for Pharma Y in Phase 3
- Pharma Y sues
- Pharma Y loses the lawsuit

With this combination of outcomes, Pharma X's market share would be 50 percent, and its contribution, costs, and NPI could be calculated accordingly. By repeating this process hundreds or thousands of times, we would create a probability distribution of the NPI for Pharma X.

An alternative to simulation is to base a model on *expected* outcomes—in this case, expected market shares. In this approach, we draw a probability tree for Pharma X that shows all possible combinations of outcomes of the random events with their probabilities and resulting market shares. We then calculate expected contributions based on expected market shares. This approach gives us a single estimate of the expected NPI, not a distribution of NPIs. Thus, it does not tell us much about the risk facing each firm.

Which approach is better? As we have said elsewhere, this is not a question that can be answered in the abstract without considering the importance of the problem and the resources available to work on it. The expected value approach is simpler and quicker but gives a less complete answer. If our time was extremely limited or this was a relatively routine and unimportant decision, we might build an expected value model and not expand it into a simulation model. But if the problem is important, as this one seems to be to both companies, the more complete analysis may be justified.

In keeping with our philosophy of iterative modeling, we begin our analysis with an expected value approach and see how much we can learn from it before building a simulation model. This means analyzing all three possible types of deals twice—once under an expected value approach and again under a simulation approach. However, this approach also allows us to avoid building a simulation model if we feel it is not warranted.

To the Reader: Before reading on, build an initial model for Pharma X based on your problem frame and influence diagram.

11.5 M1 MODEL AND ANALYSIS

During the problem-framing process, we realized that, ultimately, we need to understand contributions and costs for both companies. However, to avoid unnecessary work, we model outcomes for Pharma X first and add Pharma Y later. It is possible, after all, that some alternatives are unworkable for Pharma X; if this is the case, we certainly do not need to evaluate them for Pharma Y.

We start with a probability tree to lay out the decision and random events that govern the market share for Pharma X. Then, we develop a spreadsheet model for calculating the NPI.

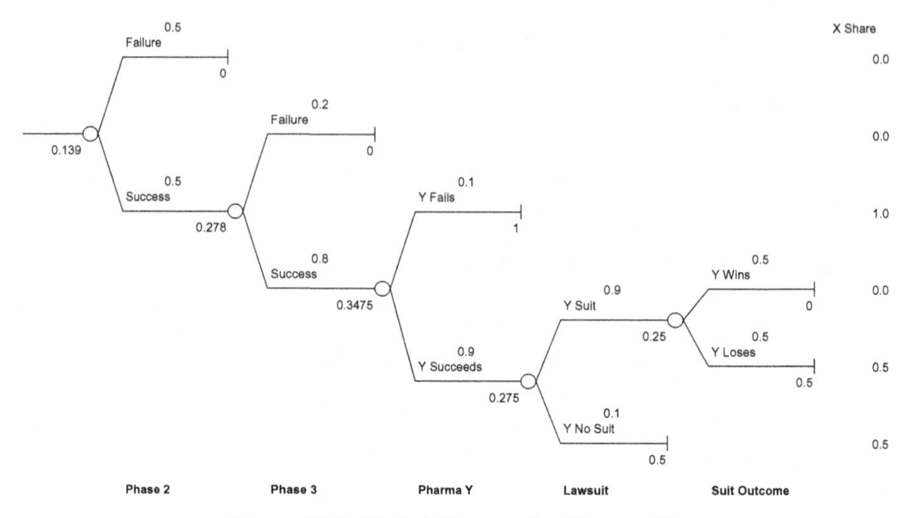

Figure 11.3. Probability tree for Pharma X.

11.5.1 Probability Tree

Figure 11.3 shows the outcomes of the development and lawsuit process from the viewpoint of Pharma X. The random events shown in the diagram are the outcomes of Phases 2 and 3 and the combined outcome of these phases for Pharma Y, the Pharma suit, and the outcome of the suit. (Note that we have to include the outcome of research for Pharma Y here because that determines whether it can sue.) The probabilities of success for Pharma X at Phases 2 and 3 are given to us in the case description. However, we are not told explicitly how likely it is Pharma Y will succeed at each stage, only that there is a 90 percent chance it will succeed if Pharma X succeeds. We interpret this to mean there is a 90 percent probability that Pharma Y succeeds at both phases, providing Pharma X also succeeds at both stages. Similarly, we assume the probability that Pharma Y succeeds if Pharma X fails is 10 percent. Finally, we assume that Pharma Y will sue with 90 percent probability if it is successful at development.

This tree leads to six outcomes. Pharma X's market share is zero on three paths: when its own development fails (two paths) and when the suit succeeds. It gains 100 percent share when it succeeds and Pharma Y fails to develop the drug. And it shares the market equally with Pharma Y in two situations: when both succeed in developing the drug and the suit fails and when Pharma Y does not sue. The overall expected share for Pharma X is 13.9 percent.

11.5.2 Calculation of NPI

The probability tree describes the possible market shares for Pharma X, but we need a spreadsheet model to calculate the NPI. Figures 11.4 and 11.5 show M1 in numerical view and relationship view, respectively. We list 18 parameters

	A	B	C	D	E
1	**Pharma X-Y**		**M1**		
2					
3					
4	**Parameters**				
5		**Discount Rate**		10%	
6					
7		**Costs**			
8			Phase II	10	
9			Phase III	40	
10			Regulatory Approval	2	
11					
12		**Outcome Probabilities**			
13			**Pharma X Succeeds**		
14			Phase II	0.5	
15			Phase III	0.8	
16					
17			**Pharma Y**		
18			Succeeds if X Succeeds	0.9	
19			Succeeds if X Fails	0.1	
20			Pharma Y Sues?	0.9	
21			Pharma Y Wins	0.5	
22					
23		**Market Size**			
24			Start	2012	
25			Peak	2015	
26			Patent Expires	2022	
27			End	2030	
28			Max	500	
29			Upslope	166.7	
30			Downslope	-62.5	
31					
32		**Margins**			
33			US Margin	0.75	
34	**Calculations**				
35		**Pharma X share**			Share
36			Pharma X Fails	0.76	0.0
37			Pharma X Shares	0.20	0.5
38			Pharma X Alone	0.04	1.0
39				1.000	0.139
40					
41				**2012**	**2013**
42		**Total Market Revenue**		0	167
43					
44		**Pharma X Contribution**		0	17
45					
46		**PV Pharma X Contribution**	$298		
47					
48		**Pharma X EV Costs**	30.8		
49					
50		**PV Pharma X NPI**	9.7		
51					

	F	G	H	I	J	K	L	M	N	O	P	Q	R	S	T	U	V
41	**2014**	**2015**	**2016**	**2017**	**2018**	**2019**	**2020**	**2021**	**2022**	**2023**	**2024**	**2025**	**2026**	**2027**	**2028**	**2029**	**2030**
42	333	500	500	500	500	500	500	500	500	438	375	313	250	188	125	63	0
44	35	52	52	52	52	52	52	52	52	46	39	33	26	20	13	7	0

Figure 11.4. M1—numerical view.

in the range D5:D33. Then, we calculate the expected share for Pharma X in E39.

In row 42, we calculate the Total Market Revenue using a flexible family of functions that allow revenue to start in a given year, grow linearly to a maximum level in a peak year, remain level until the patent expires, and then decline linearly to a final year. This function uses five parameters: four dates (Start, Peak, Patent Expires, and End) and a fifth parameter (Max) for the maximum revenue. The slopes during the increase and decrease periods are calculated in cells D29 (Upslope) and D30 (Downslope). The function itself has the following form:

IF(Current Date < Start Date)

 0,

ELSE IF(Current Date < Peak)

 Upslope × Years since Start Date,

ELSE IF(Current Date < Patent Expires)

 Max,

ELSE IF(Current Date < End),

 Max – (Downslope × Years since Patent Expires),

ELSE

 0.

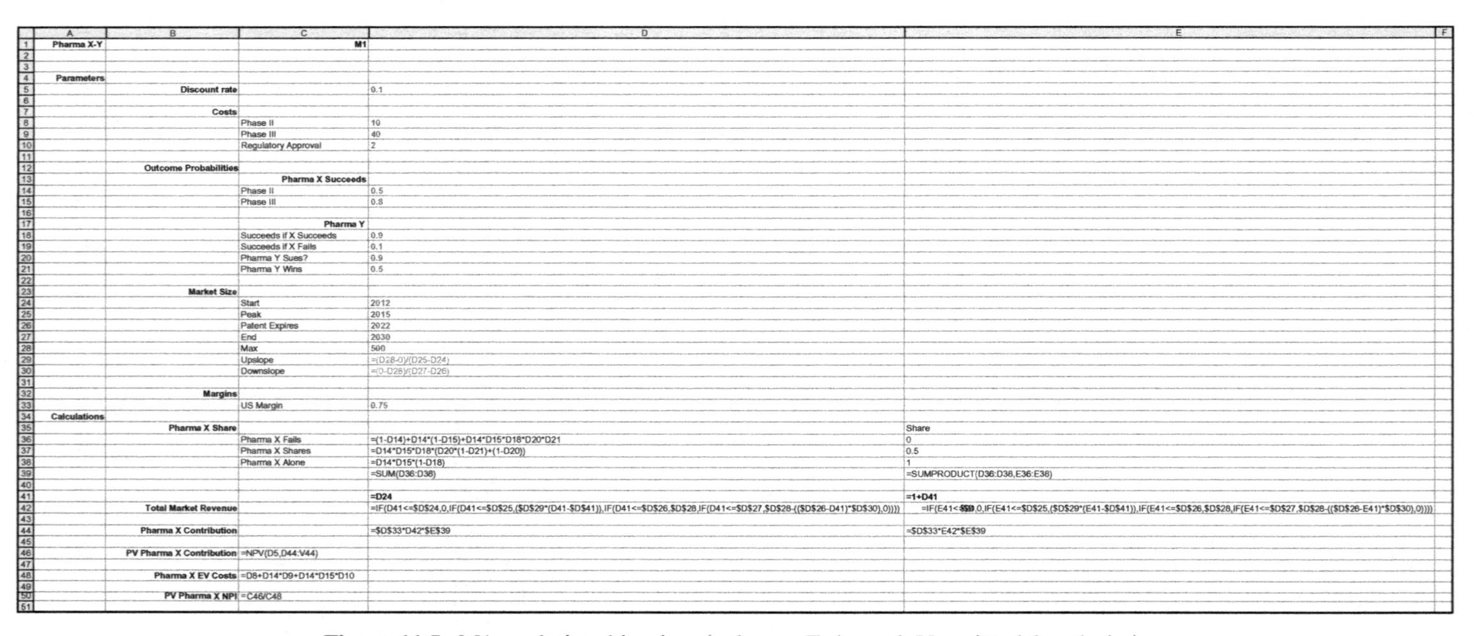

	A	B	C	D	E	F
1	**Pharma X-Y**		**M1**			
2						
3						
4	**Parameters**					
5		**Discount rate**		0.1		
6						
7		**Costs**				
8			Phase II	10		
9			Phase III	40		
10			Regulatory Approval	2		
11						
12		**Outcome Probabilities**				
13			**Pharma X Succeeds**			
14			Phase II	0.5		
15			Phase III	0.8		
16						
17			**Pharma Y**			
18			Succeeds if X Succeeds	0.9		
19			Succeeds if X Fails	0.1		
20			Pharma Y Sues?	0.9		
21			Pharma Y Wins	0.5		
22						
23		**Market Size**				
24			Start	2012		
25			Peak	2015		
26			Patent Expires	2022		
27			End	2030		
28			Max	500		
29			Upslope	=(D28-0)/(D25-D24)		
30			Downslope	=(0-D28)/(D27-D26)		
31						
32		**Margins**				
33			US Margin	0.75		
34	**Calculations**					
35		**Pharma X Share**			Share	
36			Pharma X Fails	=(1-D14)+D14*(1-D15)+D14*D15*D18*D20*D21	0	
37			Pharma X Shares	=D14*D15*D18*(D20*(1-D21)+(1-D20))	0.5	
38			Pharma X Alone	=D14*D15*(1-D18)	1	
39				=SUM(D36:D38)	=SUMPRODUCT(D36:D38,E36:E38)	
40						
41				**=D24**	**=1+D41**	
42		**Total Market Revenue**		=IF(D41<=$D$24,0,IF(D41<=$D$25,($D$29*(D41-$D$41)),IF(D41<=$D$26,$D$28,IF(D41<=$D$27,$D$28-(($D$26-D41)*$D$30),0))))	=IF(E41<[illegible],0,IF(E41<=$D$25,($D$29*(E41-$D$41)),IF(E41<=$D$26,$D$28,IF(E41<=$D$27,$D$28-(($D$26-E41)*$D$30),0))))	
43						
44		**Pharma X Contribution**		=$D$33*D42*$E$39	=$D$33*E42*$E$39	
45						
46		**PV Pharma X Contribution**	=NPV(D5,D44:V44)			
47						
48		**Pharma X EV Costs**	=D8+D14*D9+D14*D15*D10			
49						
50		**PV Pharma X NPI**	=C46/C48			
51						

Figure 11.5. M1—relationship view (columns F through V omitted for clarity).

Figure 11.6 shows a plot of this function for the parameter values in M1. Note that in our model, revenues do not begin until one year *after* the entered Start Date. Thus, we set Start Date to 2012, one year before the products are expected to enter the market. This is purely a result of how we wrote the IF statements and has no effect on the results.

The contribution for Pharma X is calculated in row 44 by multiplying total market revenue by the expected share for Pharma X and by the 75 percent margin.

We now have the ingredients we need to calculate the NPI for Pharma X. In cell C46, we calculate the present value of Pharma X's Contribution ($298M). In cell C48, we calculate the Expected Costs without discounting, under the assumption that both phases take little time relative to revenues. Pharma X pays $10M for Phase 2. If Phase 2 is successful (probability 0.5), Pharma X pays $40M for Phase 3. If Phase 3 is successful (probability $0.5 \times 0.8 = 0.4$), Pharma X pays $2M to support the FDA review. The Expected Cost is, therefore, $30.8M. Finally, in cell C50, we calculate the NPI for Pharma X by dividing the PV of Contribution by Expected Costs. Our first results show that Pharma X achieves an NPI of 9.7 ($298 ÷ $30.8 = 9.7). This is our first insight.

Insight 1: Based on an expected value analysis, Pharma X will achieve almost twice the minimum NPI of five if it pursues the new drug independent of Pharma Y.

11.5.3 Sensitivity Analysis

Before we begin our analysis of the three types of deals possible between Pharma X and Pharma Y, it is important to explore our basic model using sensitivity analysis. In the base case, Pharma X stands to achieve an attractive NPI of 9.7. Sensitivity analysis gives us some idea of how dependable that conclusion really is.

We have already been warned by the experience Pharma X had in its first round of negotiations with Pharma Y that the two companies may have radically different views of the ultimate size of the market. Pharma Y may believe the total revenues will be closer to $900M than $500M. So it is natural to start our sensitivity analysis with this parameter.

The base case establishes that a market size of $500M leads to an attractive NPI for Pharma X. Certainly, a higher market size will increase the NPI. In fact, the NPI for Pharma X is 17.4 when the total market is $900M. Could the market be so small that Pharma X fails to meet its NPI target? A Goal Seek exercise establishes that Pharma X will just meet its NPI target of five with a maximum market size of $258M.

Insight 2: Pharma X can expect to exceed the minimum NPI of five as long as the total market exceeds about $250M. Its NPI will be 17.4 if the market is actually $900M.

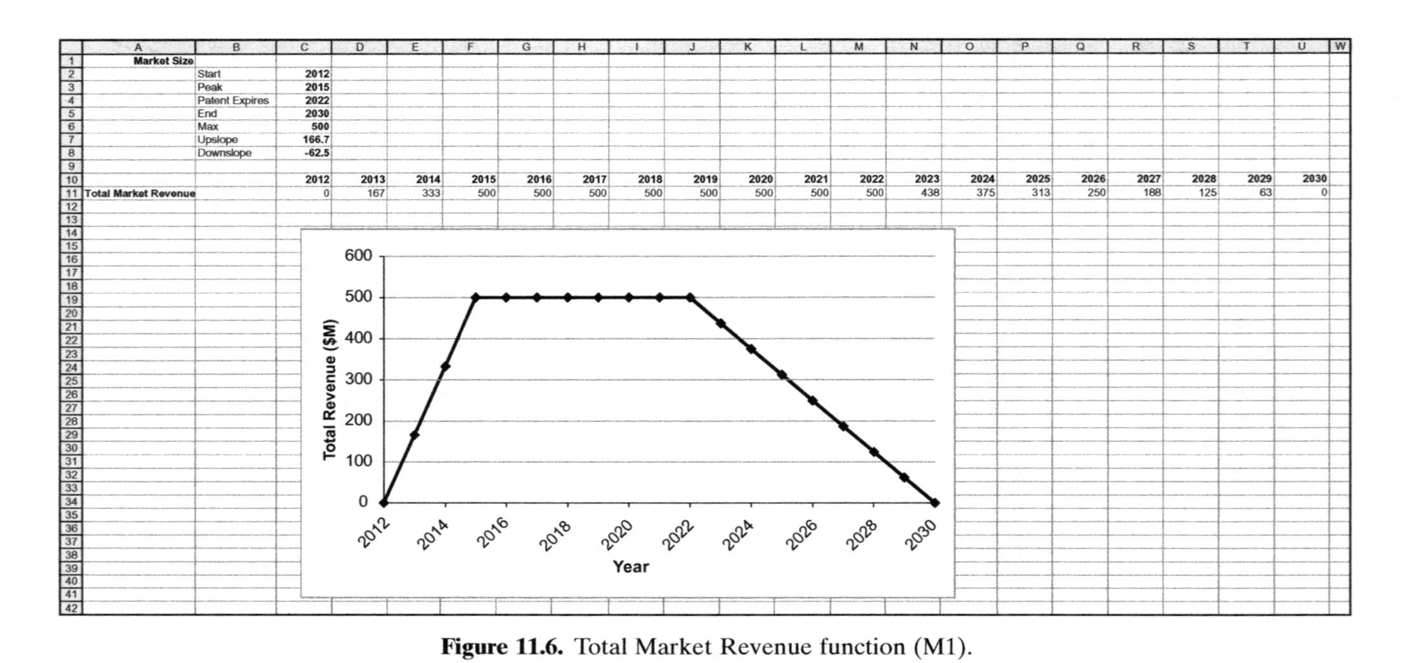

	A	B	C	D	E	F	G	H	I	J	K	L	M	N	O	P	Q	R	S	T	U
1	**Market Size**																				
2		Start	**2012**																		
3		Peak	**2015**																		
4		Patent Expires	**2022**																		
5		End	**2030**																		
6		Max	**500**																		
7		Upslope	**166.7**																		
8		Downslope	**-62.5**																		
9																					
10			**2012**	**2013**	**2014**	**2015**	**2016**	**2017**	**2018**	**2019**	**2020**	**2021**	**2022**	**2023**	**2024**	**2025**	**2026**	**2027**	**2028**	**2029**	**2030**
11	**Total Market Revenue**		0	167	333	500	500	500	500	500	500	500	500	438	375	313	250	188	125	63	0

Figure 11.6. Total Market Revenue function (M1).

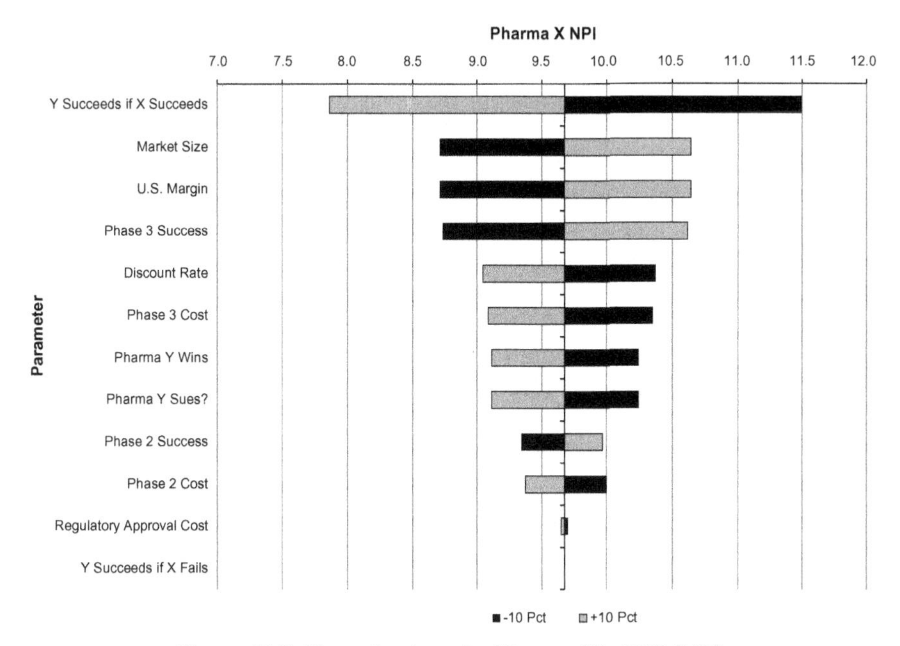

Figure 11.7. Tornado chart for Pharma X's NPI (M1).

Figure 11.7 shows a tornado chart for Pharma X's NPI. We see that just a few parameters are particularly critical for this outcome measure. The most critical is the (conditional) probability that Pharma Y succeeds in developing the drug, which is set at 90 percent in the base case. The second important parameter is the maximum Market Size. The margin on Pharma X's sales and its probability of success in Phase 3 are also important. The remaining parameters have relatively unimportant effects, at least within the limits of ±10 percent changes in all parameters. Also note that none of the changes in parameters reflected in the chart drive the NPI below five.

Insight 3: The maximum size of the market, the U.S. margin, and the probability Pharma Y succeeds are all critical parameters for Pharma X's NPI.

Note that the tornado chart excludes the four time parameters (Start, Peak, Patent Expires, and End) in the total market revenue function. We exclude these parameters because a ±10 percent change in a date makes no sense in this context (a ±10 percent variation around a start date of 2012 gives us a range from 1811 to 2213). To examine the sensitivity of the NPIs to these four parameters, we can either use the percentiles option in a tornado chart, which requires us to specify explicitly the lower and upper values for each parameter, or we can change the dates one at a time. Both approaches show that adding or subtracting a single year from any one of these parameters changes the NPI at most from 9.1 to 10.3, so the results are not very sensitive to these parameters.

One last question occurs to us in the context of sensitivity analysis. How much would Pharma X have to pay Pharma Y to both abandon the product and not sue? To investigate this question, we eliminate YCardia and the lawsuit from the model by setting Pharma Y's probability of success and of a suit to zero. As expected, Pharma X's expected market share increases (from 14 to 40 percent), its contribution goes up (by $560M, from $298M to $858M), and its NPI increases (from 9.7 to 27.8). However, we are not yet in a position to determine whether buying out Pharma Y will increase *total* contribution for the two companies, since we have not yet calculated results for Pharma Y.

Insight 4: If Pharma X can convince Pharma Y to abandon the project and not to sue, Pharma X's contribution increases by $560M (188 percent).

11.5.4 Deal Analysis

Now that we have built a base-case model and have explored it using sensitivity analysis, we are ready to consider the deals that Pharma X could offer Pharma Y: a buyout, a license arrangement, or a codevelopment agreement.

11.5.4.1 Buyout Option We begin with the buyout option. The first question we consider is *when* would Pharma X offer to buy out Pharma Y? Suppose Pharma X were to make an immediate offer, before either firm begins Phase 2 development. Pharma X has an expected contribution of $298M and an expected NPI of 9.7. We saw in Insight 4 that Pharma X could offer up to $560M to buy out Pharma Y ($858M – $298M) and still achieve an NPI of 9.7.

Would Pharma Y accept an offer of $560M to stop development of this new drug? We know that it previously rejected an offer of $50M, but $560M is more than ten times higher. A lump-sum offer of $560M would technically result in an infinite NPI for Pharma Y, since no development costs would be incurred. But we do not know whether the offer is better than Pharma Y could do on its own. Furthermore, $560M is a lot of money for Pharma X to pay upfront for a project that still has a 60 percent chance of failure. So, a successful buyout at the present time is questionable. We need to revisit this question later once we have added Pharma Y to our model.

Another option is for Pharma X to sign a conditional buyout contract with Pharma Y. The terms of the contract would state that if both companies are successful in developing the drug, Pharma X will pay Pharma Y the contract amount (β) to abandon the drug and to not sue. In terms of *contribution* to Pharma X, this buyout model is the same as the previous one in that Pharma X still has a 40 percent chance of claiming the entire market (regardless of whether Pharma Y is successful). Thus, the expected contribution is still \$858M. This expected contribution is the probability weighted average of two outcomes:

- 40 percent chance of successful development leading to 100 percent market share and contribution of \$2.1B
- 60 percent chance of failed development leading to 0 share and \$0 contribution

The difference between a conditional buyout and an immediate one is that there is a 36 percent chance that Pharma Y will also be successful and that Pharma X will have to pay the contract amount (β). The expected profit (contribution – β) with a conditional buyout of β is therefore given by the formula:

$$E[\text{Profit}] = 0.40 \times \$2.1\text{B} + 0.60 \times \$0 - 0.36 \times \beta$$

How large a contract payment can Pharma X offer and still be better off than without a deal? Pharma X will not offer a contract with an expected profit less than \$298M since it can get that with no deal. So, we set the above equation equal to \$298M and find that β equals \$1.6B.

Would Pharma Y accept a conditional lump sum of \$1.6B? That depends on what it expects to make without a deal. Again, we need to revisit this question later once we have added Pharma Y to our model.

Insight 5: Pharma X could offer up to \$560M in an immediate buyout agreement or up to \$1.6B in a conditional buyout agreement.

11.5.4.2 License Option A license agreement is actually easier to analyze than a buyout. Under a license agreement, Pharma X would agree to pay Pharma Y some percentage of the contribution (revenue) it earns. In return, Pharma Y would agree not to sue. M1 is easily modified to reflect this agreement. We first set the probabilities of a lawsuit (and of winning the suit) to zero. Next, we modify the calculation of Contribution for Pharma X to deduct the license payments to Pharma Y. Last, we vary the License Fee percentage and see how high a license agreement Pharma X can afford to offer.

Figure 11.8 shows the results of a Data Sensitivity analysis in which we vary the License Fee from 0 percent to 100 percent in steps of 5 percent. Recall that Pharma X can achieve an NPI of 9.7 without a license arrangement. The sensitivity analysis shows that Pharma X can offer to pay up to about 35 percent while maintaining an NPI above 9.7.

Insight 6: Pharma X could pay Pharma Y a license fee of up to 35 percent of its revenues.

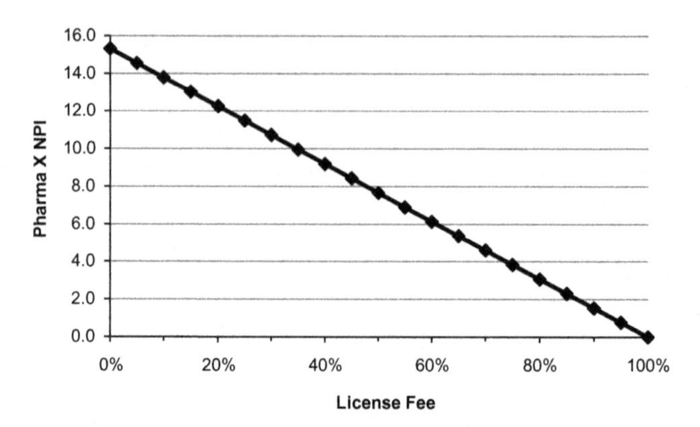

Figure 11.8. Sensitivity to License Fee (M1).

11.5.4.3 Codevelopment Option In a codevelopment agreement, two (or more) companies pool their resources to develop a new product and share the resulting costs and revenues in some agreed-upon manner. There are an infinite variety of terms for codevelopment deals, limited only by the creativity of the legal departments, and we cannot hope to model all these details here. Rather, we attempt to determine whether a simple codevelopment deal might be feasible. Our analysis is most useful in setting terms for the initial negotiations between Pharma X and Pharma Y, not in determining the final contract terms.

The underlying premise of a codevelopment agreement between Pharma X and Pharma Y is that Pharma Y will give up its right to sue in return for a share of the eventual contribution. The benefit of a codevelopment deal for Pharma X is that it avoids being sued and potentially locked out of the market; for Pharma Y, the benefit is access to increased revenue from Japan and Europe. Both companies may benefit from sharing development costs.

Since the two firms both currently have development teams working on the new drug, two options exist for codevelopment: The companies could pick one team to develop the drug, or they could allow both teams to develop it. The latter option is not necessarily wasteful since there is some chance Pharma Y would be successful even if Pharma X were not, and vice versa. Of course, this option costs more, but it may increase the probability of success by enough to offset the increased cost. We build our model in such a way that we can evaluate both options.

Figures 11.9 and 11.10 show a version of M1 adapted for modeling codevelopment deals. In cells G4:I9, we create the inputs for the two options: Develop with one team or two. We calculate the Probability of Success in cells H8:I8 and the expected Development Costs in cells H9:I9. In cell G5, we put a selection variable with which we choose one or the other option.

The calculations in this model are familiar. In row 34, we calculate Total Market Revenues. In row 36, we calculate Expected Contribution to the codevelopment partners by multiplying Market Revenues by the 75 percent margin

	A	B	C	D	E	F	G	H	I	J	K	L	M	N	O	P	Q	R	S	T	U	V	W
1	Pharma X-Y		M1 - Co-Development																				
2																							
3																							
4	Parameters						Case	Alternatives															
5		Discount Rate		10%			1	1	2														
6								Develop	Develop														
7		Costs						with One	with Both														
8			Phase II	10			Success Probability	0.40	0.46														
9			Phase III	40			Dev Costs	30.8	60.9														
10			Regulatory Approval	2																			
11																							
12		Outcome Probabilities																					
13			Pharma X Succeeds																				
14			Phase II	0.5																			
15			Phase III	0.8																			
16																							
17			Pharma Y																				
18			Succeeds if X Succeeds	0.9																			
19			Succeeds if X Fails	0.1																			
20																							
21		Market Size																					
22			Start	2012																			
23			Peak	2015																			
24			Patent Expires	2022																			
25			End	2030																			
26			Max	500																			
27			Upslope	166.7																			
28			Downslope	-62.5																			
29																							
30		Margins																					
31			US Margin	0.75																			
32	Calculations																						
33				2012	2013	2014	2015	2016	2017	2018	2019	2020	2021	2022	2023	2024	2025	2026	2027	2028	2029	2030	
34		Total Market Revenue		0	167	333	500	500	500	500	500	500	500	500	438	375	313	250	188	125	63	0	
35																							
36		EV Co-Dev Contribution		0	50	100	150	150	150	150	150	150	150	150	131	113	94	75	56	38	19	0	
37																							
38		PV Co-Dev Contribution	$858																				
39																							
40		Co-Dev EV Costs	30.8																				
41																							
42		Co-Dev NPI	27.8																				
43																							

Figure 11.9. M1 for codevelopment—numerical view.

	A	B	C	D	E	F	G	H	I	J
1	Pharma X-Y		M1 - Co-Development							
2										
3										
4	Parameters						Case	Alternatives		
5		Discount Rate		0.1			1	1	2	
6								Develop	Develop	
7		Costs						with One	with Both	
8			Phase II	10			Success Probability	=D14*D15	=D14*D15+D14*(1-D15)*D19+(1-D14)*D19	
9			Phase III	40			Dev Costs	=D8+D14*D9+D14*D15*D10	=2*D8+2*D9*D14+D10*I8	
10			Regulatory Approval	2						
11										
12		Outcome Probabilities								
13			Pharma X Succeeds							
14			Phase II	0.5						
15			Phase III	0.8						
16										
17			Pharma Y							
18			Succeeds if X Succeeds	0.9						
19			Succeeds if X Fails	0.1						
20										
21		Market Size								
22			Start	2012						
23			Peak	2015						
24			Patent Expires	2022						
25			End	2030						
26			Max	500						
27			Upslope	=(D26-0)/(D23-D22)						
28			Downslope	=(0-D26)/(D25-D24)						
29										
30		Margins								
31			US Margin	0.75						
32	Calculations									
33				=D22	=D33+1	=E33+1	=F33+1	=G33+1	=H33+1	
34		Total Market Revenue		=IF(D33<=$D$22,0,IF(D33<=$D$23,($D$27*(D33-$D$33)), IF(D33<=$D$24,$D$26,IF(D33<=$D$25,$D$26-(($D$24-D33)*$D$28),0))))	=IF(E33<=$D$22,0,IF(E33<=$D$23,($D$27*(E33-$D$33)),IF(E33<=$D$24,$D$26,IF(E33<=$D$25,$D$26-(($D$24-E33)*$D$28),0))))	=IF(F33<=$D$22,0,IF(F33<=$D$23,($D$27*(F33-$D$33)),IF(F33<=$D$24,$D$26,IF(F33<=$D$25,$D$26-(($D$24-F33)*$D$28),0))))	=IF(G33<=$D$22,0,IF(G33<=$D$23,($D$27*(G33-$D$33)),IF(G33<=$D$24,$D$26,IF(G33<=$D$25,$D$26-(($D$24-G33)*$D$28),0))))	=IF(H33<=$D$22,0,IF(H33<=$D$23,($D$27*(H33-$D$33)),IF(H33<=$D$24,$D$26,IF(H33<=$D$25,$D$26-(($D$24-H33)*$D$28),0))))	=IF(I33<=$D$22,0,IF(I33<=$D$23,($D$27*(I33-$D$33)),IF(I33<=$D$24,$D$26,IF(I33<=$D$25,$D$26-(($D$24-I33)*$D$28),0))))	
35										
36		EV Co-Dev Contribution		=$D$31*D34*INDEX($H$8:$I$8,$G$5)	=$D$31*E34*INDEX($H$8:$I$8,$G$5)	=$D$31*F34*INDEX($H$8:$I$8,$G$5)	=$D$31*G34*INDEX($H$8:$I$8,$G$5)	=$D$31*H34*INDEX($H$8:$I$8,$G$5)	=$D$31*I34*INDEX($H$8:$I$8,$G$5)	
37										
38		PV Co-Dev Contribution	=NPV(D5,D36:V36)							
39										
40		Co-Dev EV Costs	=INDEX(H9:I9,G5)							
41										
42		Co-Dev NPI	=C38/C40							
43										

Figure 11.10. M1 for codevelopment—relationship view (columns J through V omitted for clarity).

and the Probability of Success. Note the use of the INDEX function to choose the appropriate success probability from H8:I8, based on the value in G5. Finally, in cells C38:C42, we calculate Contribution, Costs, and NPI for the codevelopment partnership.

In the case shown in Figure 11.9, we assume a single development team is used and the joint venture receives $858M in contribution and expends $30.8M in costs, for an NPI of 27.8. By entering the value 2 in G5, we evaluate the alternative, which is to use two development teams. In this case, the joint venture has total contribution of $986M, costs of $60.9M, and an NPI of 16.2. It seems two teams are not better than one, at least as measured by NPI.

Since Pharma X faces an NPI of 9.7 without a deal, either of these codevelopment alternatives seems to be attractive. However, we have not yet considered the question of how Pharma X and Pharma Y would share the benefits of this arrangement.

Insight 7: A codevelopment deal in which both companies agree to develop the drug using a single team substantially increases the total contribution and the NPI.

Insight 8: A codevelopment deal using two development teams increases expected contribution more than a deal using one team but reduces the NPI.

Summary of M1

Our first model uses expected values for market share and for NPIs. It also focuses exclusively on Pharma X, leaving until later the analysis of Pharma Y. Nevertheless, it generates several valuable insights:

- Pharma X can achieve an NPI of 9.7 without any agreement with Pharma Y.
- Eliminating the possibility of a suit is worth $560M to Pharma X.
- No immediate buyout agreement is likely to be feasible.
- A license fee of up to 35 percent of revenues is advantageous to Pharma X.
- Under a codevelopment deal, the total contribution and NPI increase significantly.

The following table summarizes our findings to this point on the three options.

Option	Offer to Pharma Y	Pharma X NPI
No deal	N/A	9.7
Buyout(conditional contract)	$0–1.6B	27.8–9.7
License	0–35 percent	15.3–9.7
Codevelopment (1 team)	?	27.8

So far, a codevelopment deal is the best option for Pharma X. Codevelopment gives the company the potential for the highest NPI, and it requires no extra upfront expense. The next step in our analysis is to include Pharma Y in our model so we can evaluate possible agreements from its point of view.

To the Reader: Before reading on, expand M1 to include the outcomes for Pharma Y.

11.6 M2 MODEL AND ANALYSIS

We now have a good understanding of the situation facing Pharma X and which of the three deals looks most attractive to it. But to complete our analysis of the negotiation between the two companies, we must understand the economics of the situation facing Pharma Y. Fortunately, since Pharma Y faces a development task very similar to that of Pharma X, this is not difficult to do. We devote our efforts in M2 to incorporating Pharma Y's viewpoint while maintaining our focus on expected results. We thus continue to defer a discussion of risks and simulation to a later phase in the iterative modeling process.

11.6.1 Probability Tree

Our first task is to generalize the model in M1 to include the outcomes for Pharma Y. We begin by laying out the probability tree for Pharma Y, as we did for Pharma X. Figure 11.11 shows the same outcomes as in Figure 11.3 but

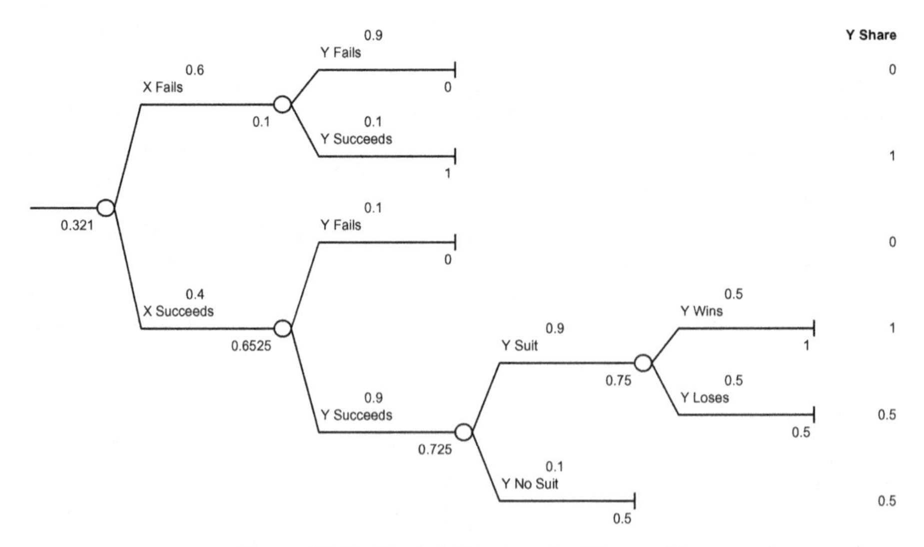

Figure 11.11. Probability tree for Pharma Y.

from the viewpoint of Pharma Y. The R&D outcomes for Pharma X are drawn first in the diagram to make the outcomes for Pharma Y *conditional* on those for Pharma X. The information we are given tells us that Pharma Y has a 90 percent chance of success overall if Pharma X succeeds in both phases, but it does not tell us the conditional probabilities if Pharma X fails in one or both stages. This requires us to make some assumptions. It seems consistent with what we know to assume that Pharma Y has only a 10 percent chance of success if Pharma X fails at *either* stage, although other assumptions would also be consistent with the given data.

This diagram generates six possible outcomes, leading to three possible values of market share for Pharma Y. Two paths lead to Pharma Y having 100 percent share; two more lead to 0 share; the remaining paths lead to 50 percent share. The overall expected market share for Pharma Y is 32.1 percent (versus 13.9 percent expected share for Pharma X).

Why does Pharma Y hold such an advantage in market share? It does not have any advantage in terms of the likelihood it actually will develop the drug. Its advantage lies in the possibility of suing Pharma X, since when it wins the suit, it gets 100 percent share and Pharma X gets nothing.

Insight 9: The expected market share for Pharma Y is more than twice as large as the expected market share for Pharma X.

11.6.2 Base Case

Figures 11.12 and 11.13 show our new model, M2, in which we have modified M1 to reflect the outcomes for Pharma Y. There are four new parameters: the Market Shares for the three main regions (cells D33:D35) and the Non-US Royalty for Pharma Y in Europe and Japan (cell D39).

In cell E52, we calculate the Expected Market Share for Pharma Y. In row 55, we calculate Total Market Revenues, as in M1; then, in rows 57–59, we calculate revenues by region for the United States, Europe, and Japan, under the assumption that total revenues are split 50/30/20 percent among these three regions. This level of detail is necessary because Pharma Y has a 75 percent margin in the United States but only receives a 10 percent royalty elsewhere. Next, we calculate Contribution for Pharma X and Pharma Y in rows 61–62. Finally, in cells D65, D68, and D71, we calculate Contribution, Costs, and NPI for Pharma Y, just as we did for Pharma X. Under our base-case assumptions, Pharma Y expects contribution of $390M, costs of $30.8M, and an NPI of 12.7. The advantage Pharma Y has in market share translates directly into a higher NPI than Pharma X. However, the advantage Pharma Y has in NPI is not proportional to its advantage in share.

Insight 10: The expected NPI for Pharma Y is higher than for Pharma X, even though it only receives a 10 percent royalty on sales outside the United States.

	A	B	C	D	E
1	Pharma X-Y		M2		
2					
3					
4	PARAMETERS				
5		Discount Rate		10%	
6					
7		Costs			
8			Phase II	10	
9			Phase III	40	
10			Regulatory Approval	2	
11					
12		Outcome Probabilities			
13			Pharma X Succeeds		
14			Phase II	0.5	
15			Phase III	0.8	
16					
17			Pharma Y		
18			Succeeds if X Succeeds	0.9	
19			Succeeds if X Fails	0.1	
20			Pharma Y Sues?	0.9	
21			Pharma Y Wins	0.5	
22					
23		Market Size			
24			Start	2012	
25			Peak	2015	
26			Patent Expires	2022	
27			End	2030	
28			Max	500	
29			Upslope	166.7	
30			Downslope	-62.5	
31					
32		Market Shares			
33			US	0.5	
34			Europe	0.3	
35			Japan	0.2	
36					
37		Margins			
38			US Margin	0.75	
39			Non-US Royalty (Pharma Y)	0.10	
40					
41	CALCULATIONS				
42		Pharma X Share			Share
43			Pharma X Fails	0.76	0.0
44			Pharma X Shares	0.20	0.5
45			Pharma X Alonve	0.04	1.0
46				1.000	0.139
47					
48		Pharma Y Share			Share
49			Pharma Y Fails	0.580	0.0
50			Pharma Y Shares	0.198	0.5
51			Pharma Y Alone	0.222	1.0
52				1.00	0.321
53					
54				2012	2013
55		Total Market Revenue		0	167
56					
57		Revenue By Region	US	0	83
58			Europe	0	50
59			Japan	0	33
60					
61		Pharma Y Contribution		0	17
62		Pharma Y Contribution		0	23
63					
64		PV Pharma X Contribution	$298		
65		PV Pharma Y Contribution	$390		
66					
67		Pharma X EV Costs	30.8		
68		Pharma Y EV Costs	30.8		
69					
70		PV Pharma X NPI	9.7		
71		PV Pharma Y NPI	12.7		
72					

	F	G	H	I	J	K	L	M	N	O	P	Q	R	S	T	U	V
54	2014	2015	2016	2017	2018	2019	2020	2021	2022	2023	2024	2025	2026	2027	2028	2029	
55	333	500	500	500	500	500	500	500	500	438	375	313	250	188	125	63	
57	167	250	250	250	250	250	250	250	250	219	188	156	125	94	63	31	
58	100	150	150	150	150	150	150	150	150	131	113	94	75	56	38	19	
59	67	100	100	100	100	100	100	100	100	88	75	63	50	38	25	13	
61	35	52	52	52	52	52	52	52	52	46	39	33	26	20	13	7	
62	45	68	68	68	68	68	68	68	68	60	51	43	34	26	17	9	

Figure 11.12. M2—numerical view.

	A	B	C	D	E	F
1	**Pharma X-Y**		**M2**			
2						
3						
4	**Parameters**					
5		**Discount Rate**		**0.1**		
6						
7		**Costs**				
8			Phase II	**10**		
9			Phase III	**40**		
10			Regulatory Approval	**2**		
11						
12		**Outcome Probabilities**				
13			**Pharma X Succeeds**			
14			Phase II	**0.5**		
15			Phase III	**0.8**		
16						
17			**Pharma Y**			
18			Succeeds if X Succeeds	**0.9**		
19			Succeeds if X Fails	**0.1**		
20			Pharma Y Sues?	**0.9**		
21			Pharma Y Wins	**0.5**		
22						
23		**Market Size**				
24			Start	**2012**		
25			Peak	**2015**		
26			Patent Expires	**2022**		
27			End	**2030**		
28			Max	**500**		
29			Upslope	=(D28-0)/(D25-D24)		
30			Downslope	=(0-D28)/(D27-D26)		
31						
32		**Market Shares**				
33			US	**0.5**		
34			Europe	**0.3**		
35			Japan	**0.2**		
36						
37		**Margins**				
38			US Margin	**0.75**		
39			Non-US Royalty (Pharma Y)	**0.1**		
40						
41	**Calculations**					
42		**Pharma X Share**			Share	
43			Pharma X Fails	=(1-D14)+D14*(1-D15)+D14*D15*D18*D20*D21	0	
44			Pharma X Shares	=D14*D15*D18*(D20*(1-D21)+(1-D20))	0.5	
45			Pharma X Alone	=D14*D15*(1-D18)	1	
46				=SUM(D43:D45)	=SUMPRODUCT(D43:D45,E43:E45)	
47						
48		**Pharma Y Share**			Share	
49			Pharma Y Fails	=(D14*(1-D15)+(1-D14)*D15+(1-D14)*(1-D15))*(1-D19)+(D14*D15*(1-D18))	0	
50			Pharma Y Shares	=D14*D15*D18*(D20*(1-D21)+(1-D20))	0.5	
51			Pharma Y Alone	=((1-D14)*(1-D15)+(1-D14)*D15+D14*(1-D15))*D19+D14*D15*D18*D20*D21	1	
52				=SUM(D49:D51)	=SUMPRODUCT(D49:D51,E49:E51)	
53						
54				**=D24**	**=1+D54**	
55		**Total Market Revenue**		=IF(D54<=$D$24,0,IF(D54<=$D$25,($D$29*(D54-$D$54)),IF(D54<=$D$26,$D$28,IF(D54<=$D$27,$D$28-(($D$26-D54)*$D$30),0))))	=IF(E54<=$D$24,0,IF(E54<=$D$25,($D$29*(E54-$D$54)),IF(E54<=$D$26,$D$28,IF(E54<=$D$27,$D$28-(($D$26-E54)*$D$30),0))))	
56						
57		**Revenue by Region**	US	=$D$33*D55	=$D$33*E55	
58			Europe	=$D$34*D55	=$D$34*E55	
59			Japan	=$D$35*D55	=$D$35*E55	
60						
61		**Pharma X Revenue**		=$D$38*D55*$E$46	=$D$38*E55*$E$46	
62		**Pharma Y Revenue**		=($D$38*D57+$D$39*(D58+D59))*$E$52	=($D$38*E57+$D$39*(E58+E59))*$E$52	
63						
64		**PV Pharma X Revenue**	=NPV(D5,D61:V61)			
65		**PV Pharma Y Revenue**	=NPV(D5,D62:V62)			
66						
67		**Pharma X EV Costs**	=D8+D14*D9+D14*D15*D10			
68		**Pharma Y EV Costs**	=D8+D14*D9+D14*D15*D10			
69						
70		**PV Pharma X NPI**	=C64/C67			
71		**PV Pharma Y NPI**	=C65/C68			
72						

Figure 11.13. M2—relationship view (columns F through V omitted for clarity).

We recall from the case that Pharma Y believes the market will be much bigger than $500M, perhaps as big as $900M. How sensitive are Pharma Y's results to this assumption? By substituting a market size of $900M in cell D28, we see that Pharma Y expects an NPI of 22.8 in this case, with expected contribution of $702M. This is important information, because if Pharma X is going to make an acceptable offer to Pharma Y, it must at least match the return that Pharma Y *believes* it can get without an agreement.

Insight 11: If Pharma Y believes the market size will be $900M, it expects contribution of $702M and an NPI of 22.8.

Figure 11.14 shows a tornado chart for Pharma Y's NPI that parallels Figure 11.7. The most important parameter is the overall size of the market (Max Market). However, for Pharma Y, U.S. Market Share and US Margin are critical. The probability that Pharma Y succeeds in developing the drug is also significant. All remaining parameters are less important.

11.6.3 Deal Analysis

Now that we have estimates of the return Pharma Y expects to receive, we deepen our analysis of the three possible types of deals. We begin with a possible buyout by Pharma X.

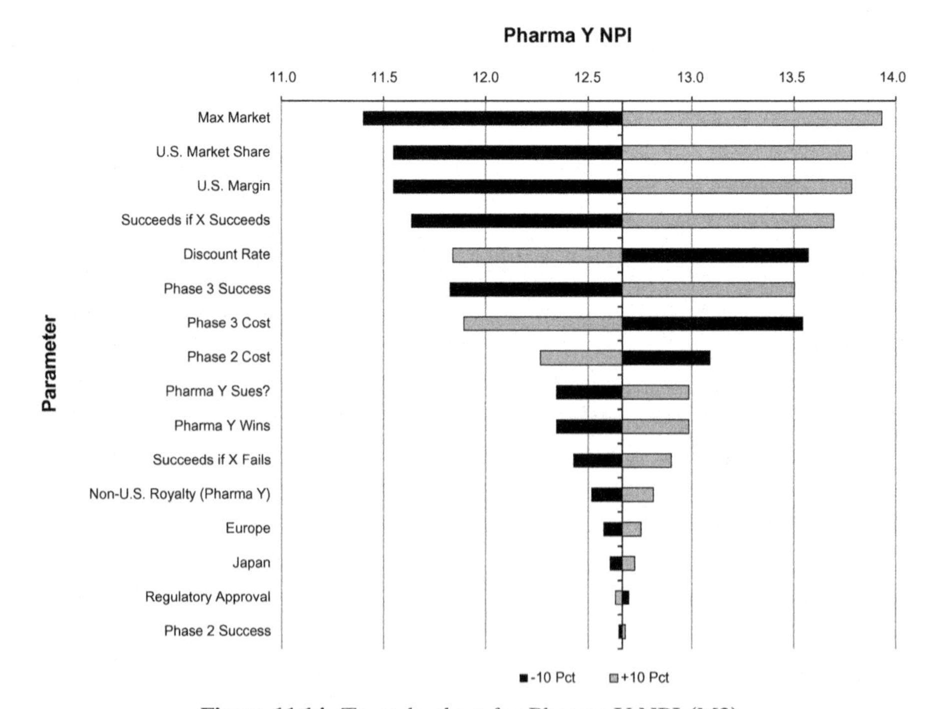

Figure 11.14. Tornado chart for Pharma Y NPI (M2).

11.6.3.1 Buyout Option In our analysis of M1, we determined that Pharma X can pay up to $560M immediately to Pharma Y and still have an expected NPI of 9.7. Would Pharma Y accept $560M? M2 shows that at a total market size of $500M, Pharma Y would expect contribution of $390M and an NPI of 12.7. In other words, it would expect to invest $30.8M and receive $390M. With a market size of $900M, Pharma Y would expect contribution of $702M and an NPI of 22.8. It seems unlikely, therefore, that Pharma Y would accept a buyout of $560M unless it is moderately risk averse or believes the market size to be significantly less than $900M.

Insight 12: Pharma Y will accept an immediate buyout that Pharma X would be willing to make only if Pharma Y is risk averse or believes the market size is about $700M.

The other option we considered was a conditional buyout triggered by both companies successfully developing the drug. In this case, Pharma X can offer as much as $1.6B and still reach the NPI of 9.7 it would get without an agreement. Would Pharma Y accept a lump sum of $1.6B to leave the market?

We look again at the probability tree for Pharma Y (Figure 11.11) and see that a conditional buyout agreement removes the last two event nodes (lawsuit and suit result) from the tree. In their place, the buyout contract gets executed when both Pharma X and Pharma Y succeed. When this happens, Pharma Y gives up the entire market to Pharma X but receives the contract amount (β) in return. Thus, there are three potential outcomes for Pharma Y:

- 36 percent chance of both companies having successful development leading to execution of the buyout contract for β
- 6 percent chance of only Pharma Y having successful development leading to 100 percent market share and contribution of $1.2B with a $500M market, or $2.2B with a $900M market
- 58 percent chance of Pharma Y having failed development leading to 0 share and $0 contribution

We express Pharma Y's expected profit in algebraic form as follows:

$$\text{E[Profit]} = 0.36 \times \beta + 0.06 \times \$1.2\text{B} + 0.58 \times \$0 \quad (\text{Market size} = \$500\text{M})$$

$$\text{E[Profit]} = 0.36 \times \beta + 0.06 \times \$2.2\text{B} + 0.58 \times \$0 \quad (\text{Market size} = \$900\text{M})$$

The minimum expected profit Pharma Y will accept depends on its market size assumption. If it assumes the market is $500M, the company will accept no less than $390M (its "no deal" expected contribution). If Pharma Y believes the market size is $900M, it will not accept an expected profit less than $702M. Solving for β under the two assumptions, we find that β equals $881M or $1.6B, respectively, for the small and large market assumptions.

If Pharma Y estimates the total market at $500M, it almost certainly would accept a $1.6B conditional buyout payment, since in that case its required β is only $881M. But if it really believes the market will grow to $900M, it might turn down an offer of $1.6B since its expected profit and NPI is the same either way.

Here is a case in which it is difficult to focus only on expected outcomes and to ignore risk. Consider the event node after both Pharma X and Pharma Y have succeeded with development. If Pharma Y believes the market is $900M and does not take the buyout contract, it stands to make an expected contribution of $1.6B at this point. But this expected value is an average over two possible outcomes: If it sues and wins (probability of 45 percent), it makes $2.2B; if it sues and loses or does not sue (combined probability of 55 percent), it makes $1.1B. The expected value of these two outcomes is $1.6B, which is the same as the maximum β Pharma X can afford to pay in a buyout. Would a rational company accept such an offer? If it were indifferent to risk, it would be indifferent to a choice between a guaranteed $1.6B and a gamble with an expected value is $1.6B. If it were even slightly risk averse, however, it would prefer the sure pay-off. We can say for certain that Pharma Y would not accept an offer below $1.1B, since it knows it can get at least that much on its own. (Pharma X would have an NPI of 15.1 and Pharma Y would have an NPI of 17.1 if β was set at $1.1B.) So it seems possible that a conditional buyout with a β of $1.1–1.6B would be mutually acceptable.

Insight 13: Pharma X could potentially buy out Pharma Y with a conditional payment of between $1.1B and $1.6B.

Although it is possible that Pharma X and Pharma Y could come to an agreement on a lump-sum buyout, $1.1B is an incredible amount of money for Pharma X to commit to pay upfront. Furthermore, Pharma X would be paying $1.1B for uncertain and risky future cash flows (we do not know exactly what future sales will be). Clearly, Pharma X would prefer a lower risk option.

11.6.3.2 License Option We now turn our attention to a license agreement. In our analysis of M1, we found that Pharma X could agree to pay Pharma Y up to about 35 percent of its contribution and still meet the NPI it would achieve without an agreement. But would Pharma Y accept a license deal of 35 percent? We answer this question by modifying M2 slightly to transfer contribution between Pharma X and Pharma Y according to the license agreement. Figure 11.15 shows the results of a Data Sensitivity analysis on this model variant in which we vary the License Fee from 0 to 100 percent. The top chart shows the NPI for Pharma X, which declines as the License Fee increases, reaching the no-agreement level of 9.7 at about 35 percent. The bottom chart shows the NPI for Pharma Y under the assumptions of total market being $500M and $900M. Remember that Pharma Y can achieve an NPI of 12.7 with no agreement if the market is $500M, and an NPI of 22.8 if

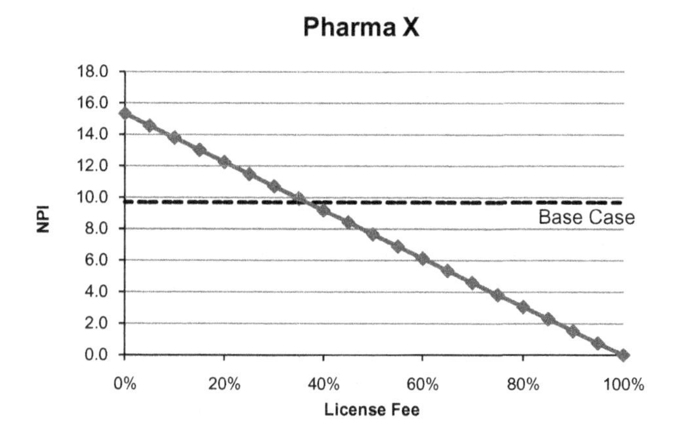

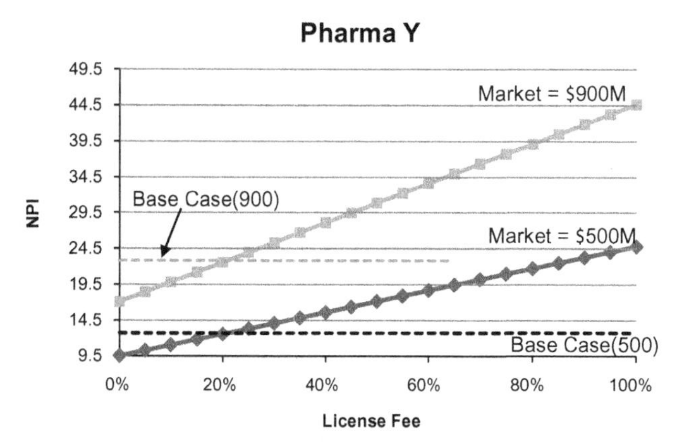

Figure 11.15. Sensitivity to License Fee (M2).

the market is $900M. We expect Pharma Y to demand at least similar returns. The sensitivity analysis shows that at a market size of $500M, Pharma Y can accept a license fee of about 25 percent or higher; if the market is actually $900M, it can accept license fees of 20 percent or higher.

These results suggest that there is a so-called *zone of agreement* between Pharma X and Pharma Y—that is to say, a range of possible license terms that both companies could accept. When the license fee is between about 20 percent and 35 percent, each party will see itself as better off (or, at least, not worse off) than it is without an agreement. Thus, unlike a buyout, a license agreement is definitely an attractive option for Pharma X to pursue with Pharma Y.

Insight 14: A license agreement with 20 to 35 percent of contribution going to Pharma Y creates higher NPIs for both companies than without an agreement.

11.6.3.3 Codevelopment Option We turn finally to an analysis of a possible codevelopment agreement. In M1, we found that a combined effort could return an NPI of 27.8 with an expected contribution of \$858M, but we did not analyze how this money might be shared between Pharma X and Pharma Y. To do this, we make minor modifications to the codevelopment version of M1 by adding parameters in cells H12 and H13 that represent the percentages the two companies use to share contribution and costs. It is not clear at this stage whether these two percentages should be equal. As a general rule, it is better to allow for flexibility in the early stages of modeling alternatives. If we discover that contribution and costs should always be shared equally, we can simplify the model later.

As a base case, we make two assumptions: The two companies share both contribution and costs equally, and one development team is used. (We ignore the option to use two teams since we learned in M1 that the NPI is lower in this case.) The model results are displayed in Figure 11.16. Under these assumptions, both companies expect a contribution of \$429M, costs of \$15.4M, and NPIs of 27.8. This sharing arrangement gives both firms an NPI in excess of what each would achieve without an agreement (9.7 and 12.7, respectively). Note that even if Pharma Y believed the ultimate market size was \$900M, in which case it could expect an NPI of 22.8 on its own, it would still do better under this codevelopment agreement. Furthermore, codevelopment appears more attractive than any of the acceptable licensing agreements. We immediately sense that codevelopment may be the most attractive type of agreement for both firms.

Insight 15: Under a codevelopment deal where Pharma X and Pharma Y share contribution and costs equally, both companies can achieve an NPI of 27.8 by using one development team.

The codevelopment option certainly appears to be worth exploring in depth. For instance, what happens when we change the two sharing percentages?

Figure 11.17 shows the results of a two-way Data Sensitivity analysis of the effects of changing the sharing percentages for costs and contribution. The upper panel shows the NPI for Pharma X, whereas the middle panel shows the NPI for Pharma Y. The unshaded regions are those combinations for which the NPI exceeds the base-case NPI (9.7 for Pharma X, 12.7 for Pharma Y). The upper panel shows that the higher the percentage of costs born by Pharma X, the higher the percentage of contribution it must take to ensure an NPI over 9.7. Likewise, the middle panel shows that the NPI for Pharma Y increases as Pharma X assumes more costs and decreases as Pharma X takes more contribution. For both firms, there are many cases in which the NPIs exceed the no-deal baseline.

The third panel in Figure 11.17 shows the zone of agreement: those combinations of sharing percentages in which *both* Pharma X and Pharma Y receive NPIs in excess of their no-deal values. Codevelopment obviously offers a wide

	A	B	C	D	E	F	G	H	I
1	Pharma X-Y		M2 - Co-Development						
2									
3									
4	Paramters						Case	Alternatives	
5		Discount Rate		10%			1	1	2
6								Develop	Develop
7		Costs						with One	with Both
8			Phase II	10			Success Probability	0.40	0.46
9			Phase III	40			Dev Costs	30.8	60.9
10			Regulatory Approval	2					
11							Pharma X		
12		Outcome Probabilities					Cost Share	50%	
13			Pharma X Succeeds				Revenue Share	50%	
14			Phase II	0.5					
15			Phase III	0.8					
16									
17			Pharma Y						
18			Succeeds if X Succeeds	0.9					
19			Succeeds if X Fails	0.1					
20									
21		Market Size							
22			Start	2012					
23			Peak	2015					
24			Patent Expires	2022					
25			End	2030					
26			Max	500					
27			Upslope	166.7					
28			Downslope	-62.5					
29									
30		Margins							
31			US Margin	0.75					
32	Calculations								
33				2012	2013	2014	2015	2016	2017
34		Total Market Revenue		0	167	333	500	500	500
35									
36		EV Co-Dev Contribution		0	50	100	150	150	150
37									
38		PV Co-Dev Contribution	$858						
39									
40		Co-Dev EV costs	30.8						
41									
42		Co-Dev NPI	27.8						
43									
44									
45		Pharma X							
46		PV Revenue	$429						
47		EV Costs	15.4						
48		NPI	27.8						
49									
50		Pharma Y							
51		PV Revenue	$772						
52		EV Costs	15.4						
53		NPI	50.1						
54									

	J	K	L	M	N	O	P	Q	R	S	T	U	V	W
33	2018	2019	2020	2021	2022	2023	2024	2025	2026	2027	2028	2029	2030	
34	500	500	500	500	500	438	375	313	250	188	125	63	0	
36	150	150	150	150	150	131	113	94	75	56	38	19	0	

Figure 11.16. M2 for Codevelopment—numerical view.

Pharma X NPI

Cost Share	Revenue Share 0%	10%	20%	30%	40%	50%	60%	70%	80%	90%	100%
0%	-	-	-	-	-	-	-	-	-	-	-
10%	0.0	27.8	55.7	83.5	111.4	139.2	167.1	194.9	222.8	250.6	278.5
20%	0.0	13.9	27.8	41.8	55.7	69.6	83.5	97.5	111.4	125.3	139.2
30%	0.0	9.3	18.6	27.8	37.1	46.4	55.7	65.0	74.3	83.5	92.8
40%	0.0	7.0	13.9	20.9	27.8	34.8	41.8	48.7	55.7	62.7	69.6
50%	0.0	5.6	11.1	16.7	22.3	27.8	33.4	39.0	44.6	50.1	55.7
60%	0.0	4.6	9.3	13.9	18.6	23.2	27.8	32.5	37.1	41.8	46.4
70%	0.0	4.0	8.0	11.9	15.9	19.9	23.9	27.8	31.8	35.8	39.8
80%	0.0	3.5	7.0	10.4	13.9	17.4	20.9	24.4	27.8	31.3	34.8
90%	0.0	3.1	6.2	9.3	12.4	15.5	18.6	21.7	24.8	27.8	30.9
100%	0.0	2.8	5.6	8.4	11.1	13.9	16.7	19.5	22.3	25.1	27.8

Pharma Y NPI

Cost Share	Revenue Share 0%	10%	20%	30%	40%	50%	60%	70%	80%	90%	100%
0%	27.8	25.1	22.3	19.5	16.7	13.9	11.1	8.4	5.6	2.8	0.0
10%	30.9	27.8	24.8	21.7	18.6	15.5	12.4	9.3	6.2	3.1	0.0
20%	34.8	31.3	27.8	24.4	20.9	17.4	13.9	10.4	7.0	3.5	0.0
30%	39.8	35.8	31.8	27.8	23.9	19.9	15.9	11.9	8.0	4.0	0.0
40%	46.4	41.8	37.1	32.5	27.8	23.2	18.6	13.9	9.3	4.6	0.0
50%	55.7	50.1	44.6	39.0	33.4	27.8	22.3	16.7	11.1	5.6	0.0
60%	69.6	62.7	55.7	48.7	41.8	34.8	27.8	20.9	13.9	7.0	0.0
70%	92.8	83.5	74.3	65.0	55.7	46.4	37.1	27.8	18.6	9.3	0.0
80%	139.2	125.3	111.4	97.5	83.5	69.6	55.7	41.8	27.8	13.9	0.0
90%	278.5	250.6	222.8	194.9	167.1	139.2	111.4	83.5	55.7	27.8	0.0
100%	-	-	-	-	-	-	-	-	-	-	-

Zone of Agreement

Cost Share	Revenue Share 0%	10%	20%	30%	40%	50%	60%	70%	80%	90%	100%
0%	N	Y	Y	Y	Y	Y	N	N	N	N	N
10%	N	Y	Y	Y	Y	Y	N	N	N	N	N
20%	N	N	Y	Y	Y	Y	Y	N	N	N	N
30%	N	N	Y	Y	Y	Y	Y	N	N	N	N
40%	N	N	Y	Y	Y	Y	Y	Y	N	N	N
50%	N	N	N	Y	Y	Y	Y	Y	N	N	N
60%	N	N	N	Y	Y	Y	Y	Y	Y	N	N
70%	N	N	N	Y	Y	Y	Y	Y	Y	N	N
80%	N	N	N	N	Y	Y	Y	Y	Y	Y	N
90%	N	N	N	N	Y	Y	Y	Y	Y	Y	N
100%	-	-	-	-	-	-	-	-	-	-	-

Figure 11.17. Sensitivity to sharing percentages (M2).

range of sharing agreements that could be acceptable to both parties. Which of these options might Pharma X propose to Pharma Y?

We start with the option in which both costs and contribution are shared equally, or 50–50, the same as in Insight 15 (50–50 indicates Pharma X bears 50 percent of the costs and receives 50 percent of the contribution). In this case, both firms receive NPIs of 27.8. If we move diagonally from this point, either to 40–40 or to 60–60, we see that the NPIs remain fixed at 27.8. This reminds us that NPI does not reflect the *magnitude* of costs and contribution, just the *ratio* between them. As long as each company is taking identical proportions for the costs and revenues, the NPI will be 27.8.

Since Pharma Y is much smaller than Pharma X, we reasonably infer that Pharma Y might want to own a smaller portion of the codevelopment deal. This might be due to higher risk aversion or merely a lack of funds to invest. Thus, Pharma Y would be more likely to prefer a project in which it contributed a small portion of the costs and received an equally small portion of the contribution. If, for example, Pharma X proposed a 90–90 split, Pharma X would pay costs of $28M and expect contribution of $772M. Pharma Y would take on costs of only $3M and expect contribution of $86M. This might well be a more attractive alternative than a 50–50 split, in which Pharma Y would also receive the same NPI of 27.8 but with much more substantial upfront costs of $15M.

This analysis suggests that one approach Pharma X might take in negotiating with Pharma Y would be to offer a codevelopment deal in which costs and contribution are both shared in the same proportion. Each firm would receive the same NPI, but Pharma X would take on a larger proportion of the upfront development costs, while receiving in compensation a larger proportion of the contribution. The actual proportions could be between 50 and 90 percent, depending on how much of the development costs Pharma Y wants to take on.

Insight 16: Codevelopment deals in which Pharma X takes between 50 and 90 percent of both contribution and costs are advantageous to Pharma X and are likely to be acceptable to Pharma Y.

Summary of M2

Our second model retains all the essential features of the first model and simply adds in the results for Pharma Y. When we look at the situation from the viewpoint of Pharma Y, we see that it has a higher expected market share due to its patent position and can therefore demand substantial compensation in any deal Pharma X might offer. This analysis generates several new and revised insights:

- Pharma X would have to pay Pharma Y at least $1.1B for a buyout conditional on both companies being successful in development.
- Pharma Y would demand a license fee of at least 20 percent, and Pharma X could pay up to 35 percent.
- In a codevelopment deal in which both costs and contribution were shared evenly, both companies could achieve a high NPI (27.8).

We update our findings from the previous section by including some of our results for Pharma Y in the following table.

Option	Offer to Pharma Y	Pharma X NPI	Pharma Y NPI
No Deal	N/A	9.7	22.8
Buyout	$1.1–1.6B	15.1–9.7	17.1–22.8 ($900M market)
License	20–35 percent	12.3–9.7	22.8–26.7 ($900M market)
Codevelop	50–50 sharing	27.8	27.8

In M1 we found that a codevelopment deal looked like the best alternative from Pharma X's perspective. In M2, we calculated Pharma Y's results and found that a codevelopment deal looked like the best option for Pharma Y as well. This gives us hope that Pharma X and Pharma Y can reach a mutually beneficial agreement. The one remaining model improvement we have to discuss before finalizing our results is the use of simulation to reveal more about the risks of the various deals. We now turn to this task.

To the Reader: Before reading on, consider the pros and cons of developing a simulation model at this point to analyze the deals open to Pharma X. What information would a simulation model provide that an expected-value model does not, and how would you expect it to alter the conclusions we already have drawn?

11.7 M3 MODEL AND ANALYSIS

We made a decision early in the modeling process first to analyze the situation facing Pharma X and Pharma Y using expected outcomes and later to consider the possibility of developing a simulation model to deepen our risk analysis. That time has come.

In a simulation model for the drug-development process, we use Bernoulli random variables to represent success and failure at each stage and for the outcomes of the lawsuit. A Bernoulli random variable takes on one of two values—call them Success and Failure, or 1 and 0. The *probability* of Success (or 1) is the only parameter in the distribution. So, for example, we represent the Phase 2 development outcome for Pharma X using a Bernoulli distribution with a probability of Success of 0.5. And we represent the probability Pharma Y sues by using a Bernoulli distribution with a probability of Yes of 0.9. In Crystal Ball, we represent a Bernoulli random variable with the Yes/No distribution:

CB.YESNO(Probability Yes)

Figures 11.18 and 11.19 show M3, which is a modification of M2 that supports simulation. In cells D14:E23, we enter the Success probabilities for the eight required Bernoulli, or Yes/No, distributions in a form in which Success (or Yes) is coded as a "1" and Failure as a "0." We have eight distributions here: two for the success of Pharma X in development; four to represent the conditional distributions of success for Pharma Y, given outcomes for Pharma X; and two to represent whether Pharma Y sues and whether it wins the suit.

In cells C45:C57, we simulate the outcomes of these random events. For example, the success or failure of Pharma X at Phase 2 is modeled in cell D45 using the function:

CB.YESNO(D14)

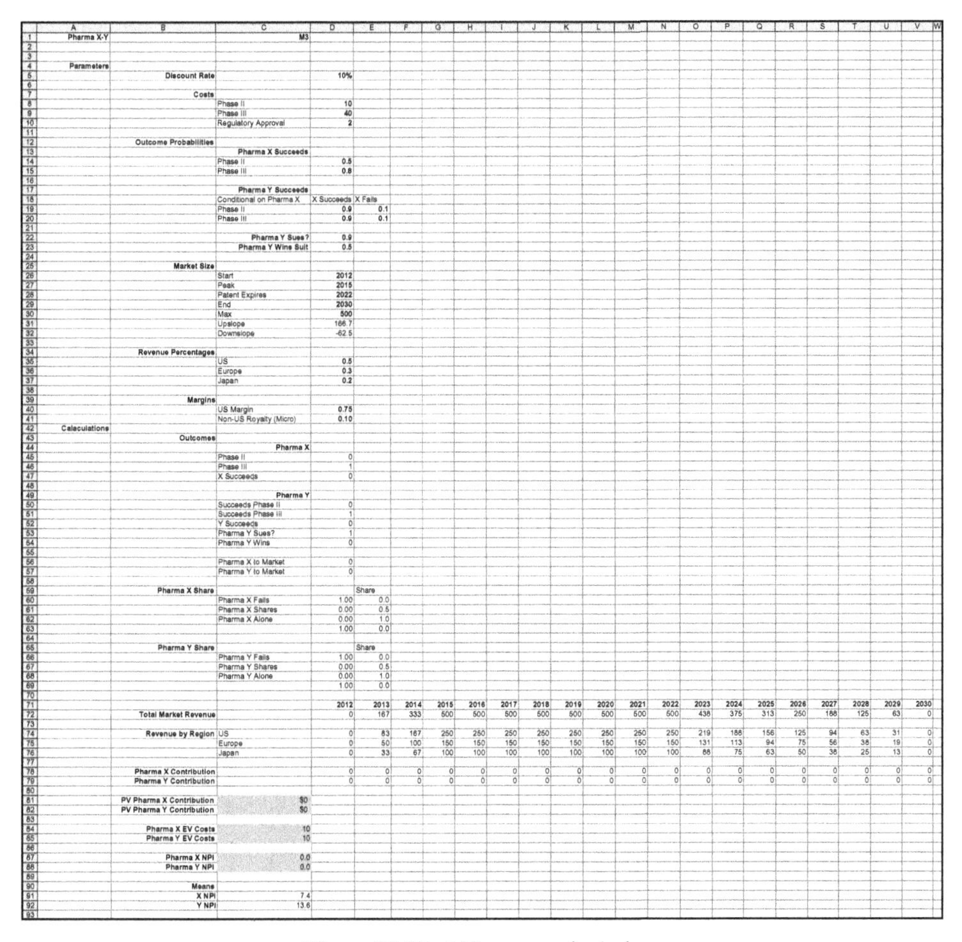

	A	B	C	D	E
1	Pharma X-Y		M3		
2					
3					
4	Parameters				
5		Discount Rate		10%	
6					
7		Costs			
8			Phase II	10	
9			Phase III	40	
10			Regulatory Approval	2	
11					
12		Outcome Probabilities			
13			Pharma X Succeeds		
14			Phase II	0.5	
15			Phase III	0.8	
16					
17			Pharma Y Succeeds		
18			Conditional on Pharma X	X Succeeds	X Fails
19			Phase II	0.9	0.1
20			Phase III	0.9	0.1
21					
22			Pharma Y Sues?	0.9	
23			Pharma Y Wins Suit	0.5	
24					
25		Market Size			
26			Start	2012	
27			Peak	2015	
28			Patent Expires	2022	
29			End	2030	
30			Max	500	
31			Upslope	166.7	
32			Downslope	-62.5	
33					
34		Revenue Percentages			
35			US	0.5	
36			Europe	0.3	
37			Japan	0.2	
38					
39		Margins			
40			US Margin	0.75	
41			Non-US Royalty (Micro)	0.10	
42	Calaculations				
43		Outcomes			
44			Pharma X		
45			Phase II	0	
46			Phase III	1	
47			X Succeeds	0	
48					
49			Pharma Y		
50			Succeeds Phase II	0	
51			Succeeds Phase III	1	
52			Y Succeeds	0	
53			Pharma Y Sues?	1	
54			Pharma Y Wins	0	
55					
56			Pharma X to Market	0	
57			Pharma Y to Market	0	
58					
59		Pharma X Share			Share
60			Pharma X Fails	1.00	0.0
61			Pharma X Shares	0.00	0.5
62			Pharma X Alone	0.00	1.0
63				1.00	0.0
64					
65		Pharma Y Share			Share
66			Pharma Y Fails	1.00	0.0
67			Pharma Y Shares	0.00	0.5
68			Pharma Y Alone	0.00	1.0
69				1.00	0.0
70					

	B	C	2012	2013	2014	2015	2016	2017	2018	2019	2020	2021	2022	2023	2024	2025	2026	2027	2028	2029	2030
72	Total Market Revenue		0	167	333	500	500	500	500	500	500	500	500	438	375	313	250	188	125	63	0
73																					
74	Revenue by Region	US	0	83	167	250	250	250	250	250	250	250	250	219	188	156	125	94	63	31	0
75		Europe	0	50	100	150	150	150	150	150	150	150	150	131	113	94	75	56	38	19	0
76		Japan	0	33	67	100	100	100	100	100	100	100	100	88	75	63	50	38	25	13	0
77																					
78	Pharma X Contribution		0	0	0	0	0	0	0	0	0	0	0	0	0	0	0	0	0	0	0
79	Pharma Y Contribution		0	0	0	0	0	0	0	0	0	0	0	0	0	0	0	0	0	0	0

	B	C
80		
81	PV Pharma X Contribution	$0
82	PV Pharma Y Contribution	$0
83		
84	Pharma X EV Costs	10
85	Pharma Y EV Costs	10
86		
87	Pharma X NPI	0.0
88	Pharma Y NPI	0.0
89		
90	Means	
91	X NPI	7.4
92	Y NPI	13.6
93		

Figure 11.18. M3—numerical view.

Note that the outcomes for Pharma Y are made conditional on those for Pharma X by using different distribution parameters depending on the outcome for Pharma X, as in the following formula (D50) for the success of Pharma Y in Phase 2:

```
IF(Pharma X succeeds in Phase 2)
        CB.YESNO(0.9),
ELSE
        CB.YESNO(0.1)
```

There are three possible market-share outcomes for Pharma X: 0, 50, and 100 percent. In cells D60:D62, we use logical functions (IF, AND, OR) to

	A	B	C	D	E	F
1	**Pharma X-Y**		**M3**			
2						
3						
4	**Parameters**					
5		**Discount Rate**		**0.1**		
6						
7		**Costs**				
8			Phase II	**10**		
9			Phase III	**40**		
10			Regulatory Approval	**2**		
11						
12		**Outcome Probabilities**				
13			**Pharma X Succeeds**			
14			Phase II	**0.5**		
15			Phase III	**0.8**		
16						
17			**Pharma Y Succeeds**			
18			conditional on Pharma X	X Succeeds	X Fails	
19			Phase II	**0.9**	**0.1**	
20			Phase III	**0.9**	**0.1**	
21						
22			**Pharma Y Sues?**	**0.9**		
23			**Pharma Y Wins Suit**	**0.5**		
24						
25		**Market Size**				
26			Start	**2012**		
27			Peak	**2015**		
28			Patent Expires	**2022**		
29			End	**2030**		
30			Max	**500**		
31			Upslope	=(D30-0)/(D27-D26)		
32			Downslope	=(0-500)/(D29-D28)		
33						
34		**Revenue Percentages**				
35			US	**0.5**		
36			Europe	**0.3**		
37			Japan	**0.2**		
38						
39		**Margins**				
40			US Margin	**0.75**		
41			Non-US Royalty (Micro)	**0.1**		
42	**Calaculations**					
43		**Outcomes**				
44			**Pharma X**			
45			Phase II	=CB.YesNo(D14)		
46			Phase III	=CB.YesNo(D15)		
47			X Succeeds	=D45*D46		
48						
49			**Pharma Y**			
50			Succeeds Phase II	=IF(D45=1,CB.YesNo(D19),CB.YesNo(E19))		
51			Succeeds Phase III	=IF(D46=1,CB.YesNo(D20),CB.YesNo(E20))		
52			Y Succeeds	=D50*D51		
53			Pharma Y Sues?	=CB.YesNo(D22)		
54			Pharma Y Wins	=CB.YesNo(D23)		
55						
56			Pharma X to Market	=D45*D46		
57			Pharma Y to Market	=D50		
58						
59		**Pharma X share**			Share	
60			Pharma X Fails	=IF(OR(D47=0, AND(D47=1,D52*D53*D54=1)),1,0)	0	
61			Pharma X Shares	=IF(OR(AND(D47=1,D52=1,D53=0),AND(D47=1,D52=1,D53=1,D54=0)),1,0)	0.5	
62			Pharma X Alone	=IF(AND(D47=1,D52=0),1,0)	1	
63				=SUM(D60:D62)	=SUMPRODUCT(D60:D62,E60:E62)	
64						
65		**Pharma Y share**			Share	
66			Pharma Y Fails	=IF(D52=0,1,0)	0	
67			Pharma Y Shares	=IF(OR(AND(D47=1,D52=1,D53=0),AND(D47=1,D52=1,D53=1,D54=0)),1,0)	0.5	
68			Pharma Y Alone	=IF(OR(AND(D52=1,D47=0),AND(D52=1,D47=1,D53=1,D54=1)),1,0)	1	
69				=SUM(D66:D68)	=SUMPRODUCT(D66:D68,E66:E68)	
70						
71				**=D26**	**=1+D71**	
72		**Total market revenue**		=IF(D71<=$D$26,0,IF(D71<=$D$27,($D$31*(D71-$D$71)),IF(D71<=$D$28,$D$30,IF(D71<=$D$29,$D$30-(($D$28-D71)*$D$32),0))))	=IF(E71<=$D$26,0,IF(E71<=$D$27,($D$31*(E71-$D$71)),IF(E71<=$D$28,$D$30,IF(E71<=$D$29,$D$30-(($D$28-E71)*$D$32),0))))	
73						
74		**Revenue by region**	US	=$D$35*D72	=$D$35*E72	
75			Europe	=$D$36*D72	=$D$36*E72	
76			Japan	=$D$37*D72	=$D$37*E72	
77						
78		**Pharma X Contribution**		=$D$40*D72*$E$63	=$D$40*E72*$E$63	
79		**Pharma Y Contribution**		=($D$40*D74+$D$41*(D75+D76))*$E$69	=($D$40*E74+$D$41*(E75+E76))*$E$69	
80						
81		**PV Pharma X Contribution**	=NPV(D5,D78:V78)			
82		**PV Pharma Y Contribution**	=NPV(D4,D79:V79)			
83						
84		**Pharma X EV Costs**	=D8+D9*D45+D10*D45*D46			
85		**Pharma Y EV Costs**	=D8+D9*D50+D10*D50*D51			
86						
87		**Pharma X NPI**	=C81/C84			
88		**Pharma Y NPI**	=C82/C85			
89						
90		**Means**				
91		**X NPI**	=CB.GetForeStatFN(C87,2)			
92		**Y NPI**	=CB.GetForeStatFN(C88,2)			
93						

determine which of these outcomes Pharma X gets, based on the outcomes of the Bernoulli trials. For example, Pharma X gets zero market share in two cases: A, when its own development fails, or B, when it succeeds but Pharma Y also succeeds, sues, and wins. This is captured in pseudo-code in the following formula:

```
IF(Case A OR Case B)
        1,
ELSE
        0.
```

A result of 1 indicates Pharma X has zero market share. A result of 0 indicates one of the other outcomes must be true. We compute the actual share in E63 using a SUMPRODUCT formula to multiply the outcome indicators (D60:D62, only one of which is 1 on a given trial) by the possible shares in E60:E62. Similar calculations determine the share for Pharma Y in E69.

The remaining calculations are essentially the same as those in M2, although the actual values change with every trial of the Bernoulli random variables. The simulated values for the two NPIs are in cells C87:C88. We use the Crystal Ball function CB.GETFORESTATFN in cells C91:C92 to capture the mean NPIs on the spreadsheet for each simulation run. The base-case NPIs are 6.8 for Pharma X and 14.3 for Pharma Y, which are similar but not identical to the deterministic results from M2.

Of course, the reason to build a simulation model in this case is to go beyond mean values to consider the entire distribution of outcomes. But at this point, we are in for a surprise. Figure 11.20 shows the distributions of Contribution, Costs, and NPIs for Pharma X in the base case. We see here the *same three outcomes* we saw in the probability tree in Figure 11.3. Since there are only three possible outcomes for market share, there are only three possible outcomes for Contribution, Costs, and NPI. Unfortunately, it looks as if our efforts in building a simulation model may have been wasted.

Mistakes and dead ends are unavoidable in modeling (however, in hindsight, we always wish we had seen them coming). We have included this section on a simulation model for two reasons: one is to demonstrate how such a model is built, and the other is to show how easy it is to build models that seem useful beforehand but actually offer few additional insights.

Before we leave this discussion of simulation models, it is worth observing that a simulation approach is not necessarily useless here. There are only a small number of random variables in the case, and each one refers to success or failure. This is what gives rise to the simple probability tree in Figure 11.3 and to the small number of discrete outcomes for NPI. But the problem contains many additional parameters that could be considered uncertain, most notably Total Market Size. If additional parameters were uncertain, there

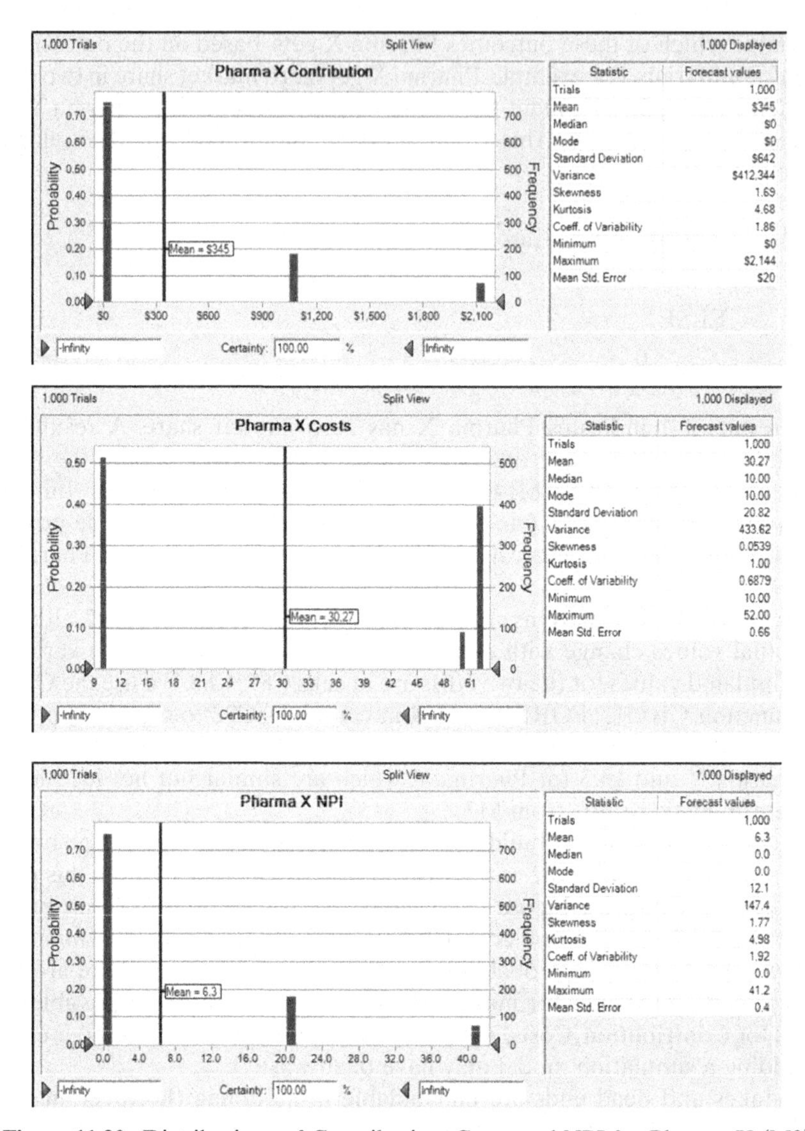

Figure 11.20. Distributions of Contribution, Costs, and NPI for Pharma X (M3).

would be many more possible outcomes, and a full risk analysis might be more revealing.

To pursue this idea a bit further, we show how additional uncertainties change the form of the results. We have no specific information on the distributions of the various parameters in the model (other than the fact that Pharma Y thinks total market share will be around $900M), but we vary each one around its base-case value to see what the impact of some degree of uncertainty is. To this end, we modify the model so that the three cost parameters, the three market share parameters, and the Pharma Y royalty rate all are governed by normal distributions with a mean equal to the base-case value and a standard deviation of 10 percent of the mean. Similarly, we allow the four dates in the market-share function to vary with equal probability between the base-case value and dates one year later and one year earlier.

The results of this exercise are shown in Figure 11.21, which contrasts with Figure 11.20. In Figure 11.20, we see three outcomes for NPI: one at 0, a second

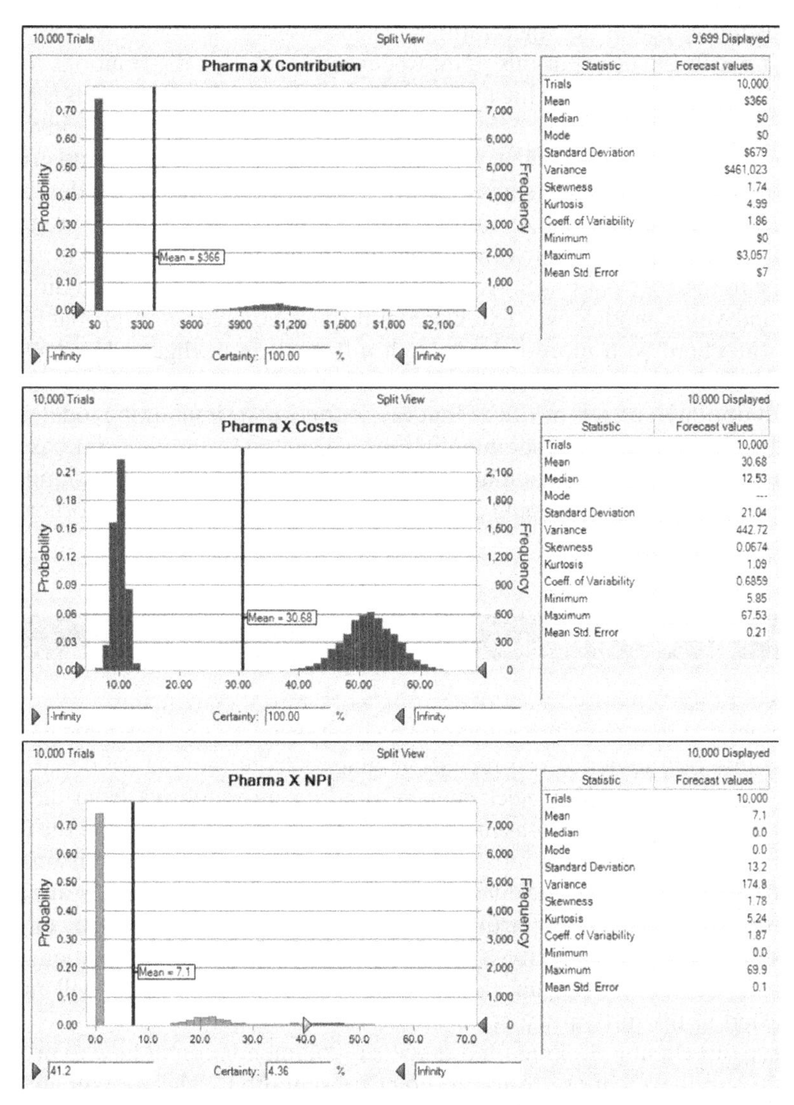

Figure 11.21. Distributions of Contribution, Costs, and NPI for Pharma X (M3 with additional uncertainties).

at about 20, and a third at about 40. The NPI of 20 corresponds to a market share of 50 percent; the NPI of 40 corresponds to a market share of 100 percent. When we introduce uncertainty in all the parameters (Figure 11.21), there are more outcomes but there are still three clusters: one at 0, one at about 20, and one at about 40. Not surprisingly, market share continues to dominate the results. The other uncertainties add some degree of variation, but the three distinct possibilities for share continue to tell the story. However, we do learn that the maximum NPI could be as high as 70, although there is only a 4 percent chance of it exceeding 41.2, the maximum value when only the yes/no outcomes are uncertain.

Is it worth the effort at this point to get more realistic distributions for the 11 parameters to which we have assigned arbitrary distributions? This is a hard question to answer without being able to talk to the decision makers involved. Our hunch is that it is not. First, we suspect the distributions they could provide would be equally arbitrary since they are not likely to know a great deal about such factors as the future size of the market. Generally, when clients are forced to give probabilities in which they have little confidence, they have little confidence in the results as well. A second factor is that the analysis we did using expected value models was extensive and quite complex and probably provides the client with more than enough information to digest. Third, we did some rudimentary risk analysis already (in particular, when we considered the buyout option), and it is not clear that the outputs of a simulation model would significantly change the recommendations we can make at this point based on a simpler model. This reasoning leads us to conclude that a full simulation model is not worth developing and that we should wrap up our efforts at this point.

11.8 PRESENTATION OF RESULTS

One vital task remains: to condense our work into a convincing presentation to our client. Since our work is focused on deal structure and risk management, we assume our audience will include managers from the finance department as well as the XCardia project manager and a representative from the legal department. A successful presentation must be short and focused on the issues that interest the audience most, not on technical details about our analysis. This transition from the technical to the managerial point of view is often difficult for analysts to make, particularly in a case like this with so many alternatives and so much complexity. It is wise to keep in mind that our audience will probably not have a technical background and most certainly will not be interested in the nitty-gritty of our analysis.

Since our client is looking for a specific recommendation of how to proceed with XCardia, we choose to use a direct presentation style. We will give our results and recommendations early on in the presentation after just a few introductory comments. The rest of the presentation will explain Pharma X's situation in more detail and explore each of the potential deal styles before returning again to our recommendations. Here is the outline for our presentation:

- Introduction
  - What's at risk?
  - What are the alternatives?
- Results in brief
- Base case
  - What outcomes does Pharma X face?
  - What are the alternatives?
  - What can Pharma X expect with no deal?
  - What can Pharma Y expect with no deal?
- Deal results
  - Buyout
  - License
  - Codevelopment
- Conclusions

Our presentation slides are in Figures 11.22– 11.32. The first slide introduces the two companies and compares their most important features. Since our audience consists of managers at Pharma X who know their own company

Can Opposites Attract?

- Pharma X and Pharma Y are developing competing cardiovascular disease medications

	Pharma X	Pharma Y
Size	Large	Small
Prob. of Development	40%	40%
International Marketing	Strong	Weak
Patent Position	Weak	Strong

Figure 11.22. Slide 1.

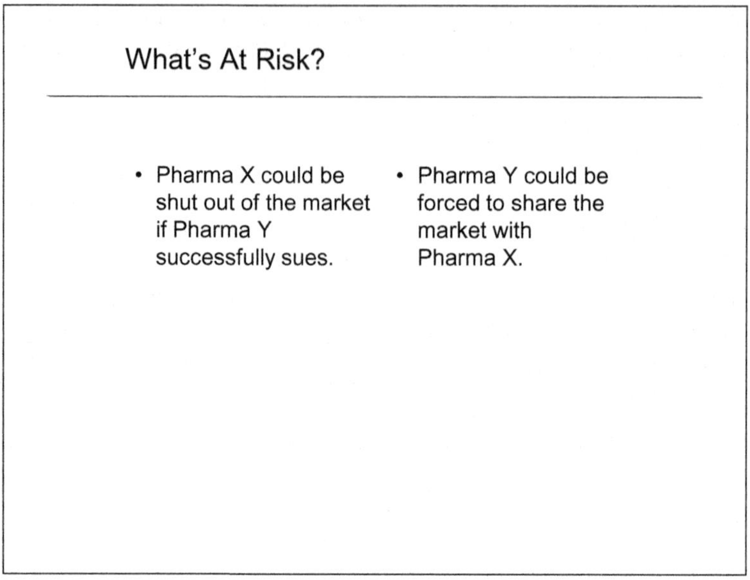

What's At Risk?
• Pharma X could be shut out of the market if Pharma Y successfully sues.
• Pharma Y could be forced to share the market with Pharma X.

Figure 11.23. Slide 2.

3 Deal Styles
Buyout
• Pharma X pays a lump sum to Pharma Y to withdraw from development and to not sue.
License
• Pharma X pays Pharma Y a set percentage of contribution; Pharma Y does not sue.
Codevelopment
• Both agree to develop the drug jointly and to share costs and contribution. Pharma Y will not sue and Pharma X will market outside the U.S.

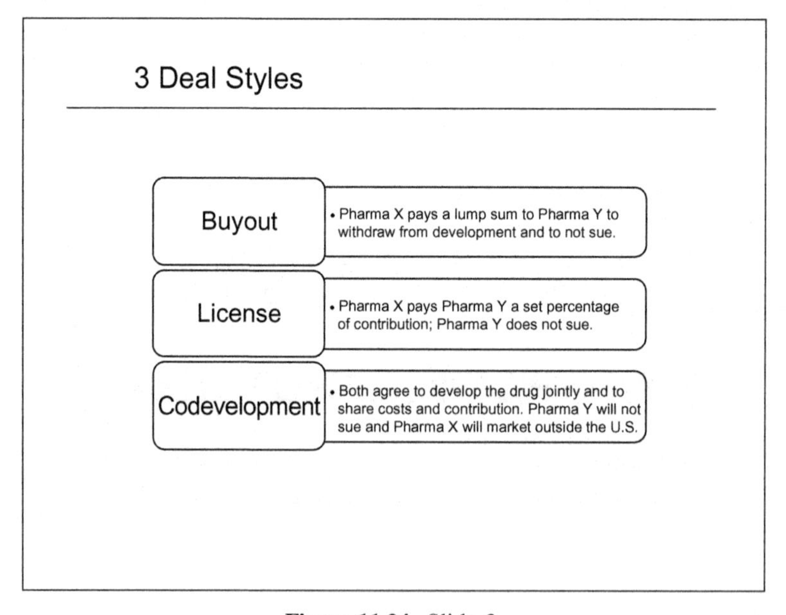

Figure 11.24. Slide 3.

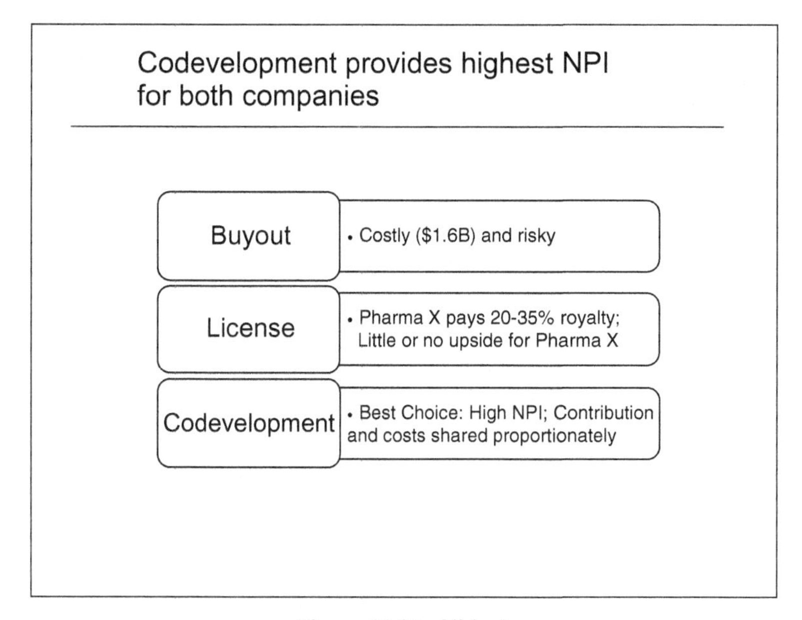

Figure 11.25. Slide 4.

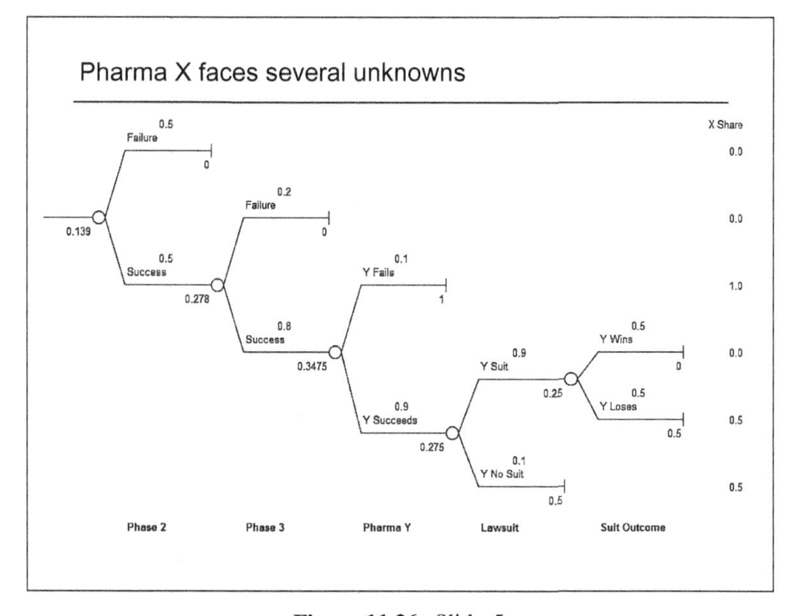

Figure 11.26. Slide 5.

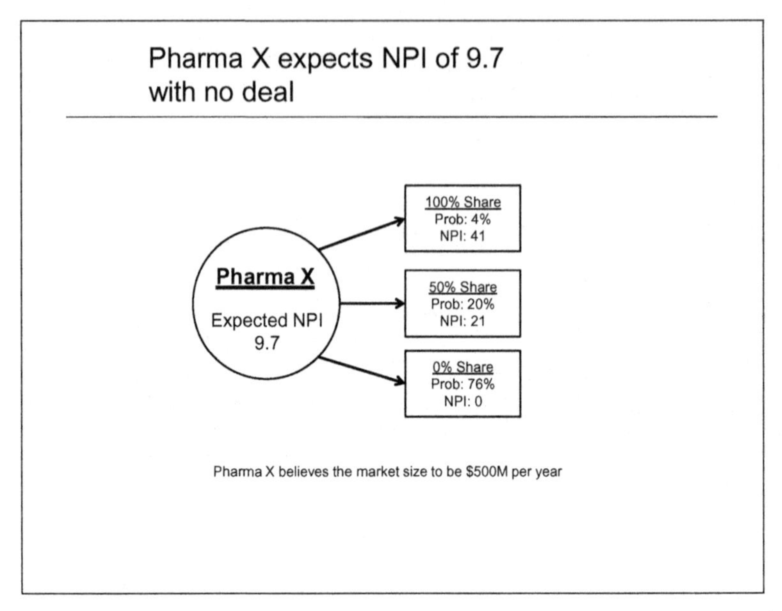

Figure 11.27. Slide 6.

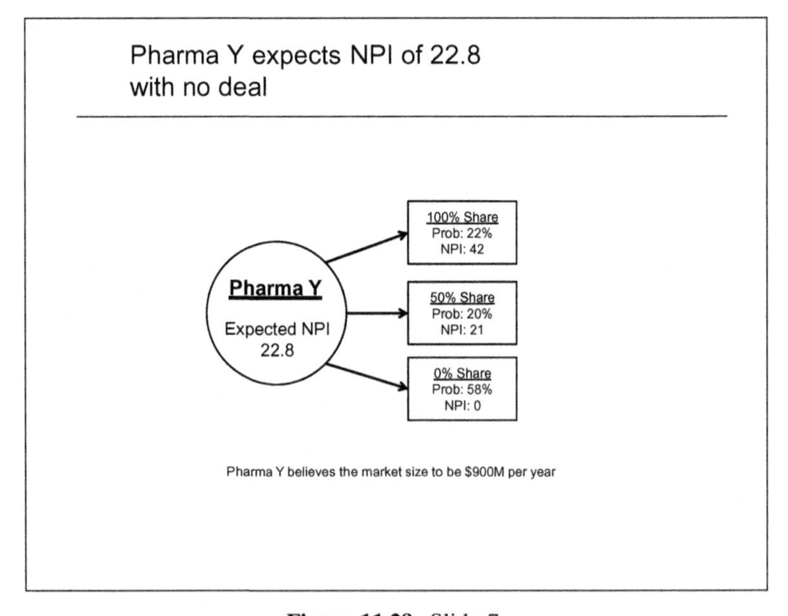

Figure 11.28. Slide 7.

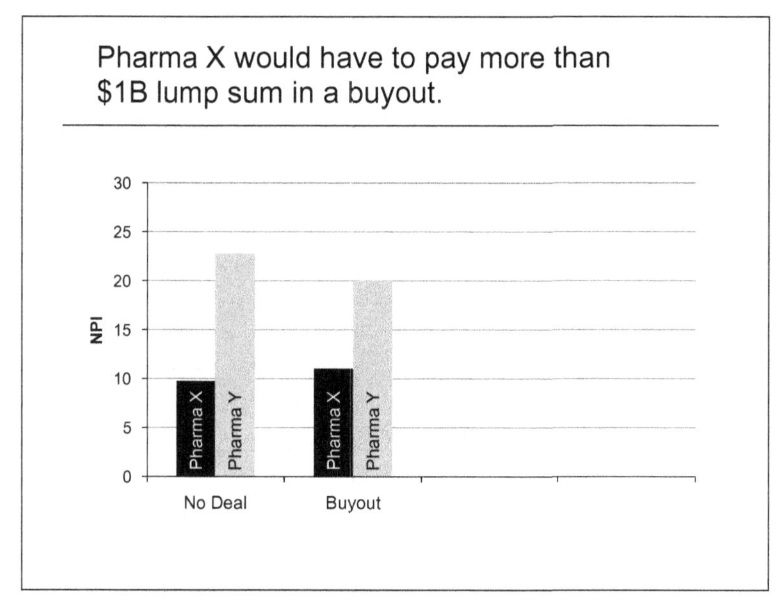

Figure 11.29. Slide 8.

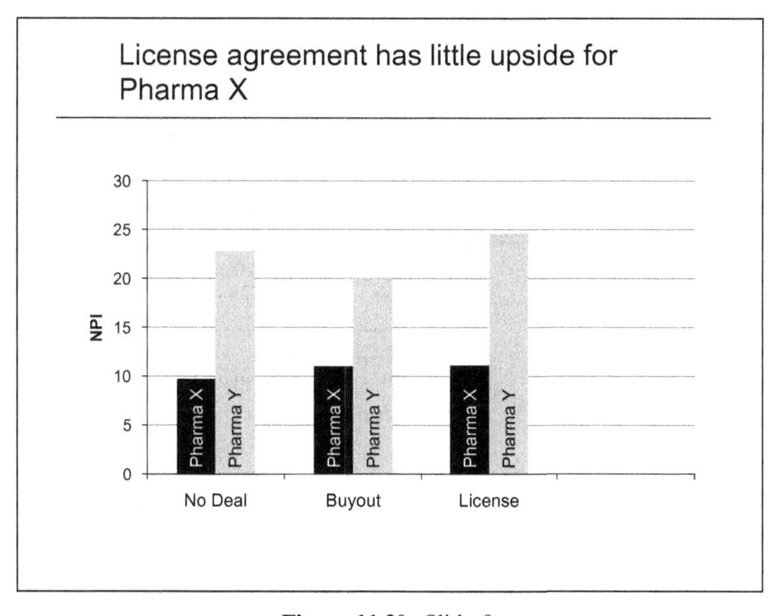

Figure 11.30. Slide 9.

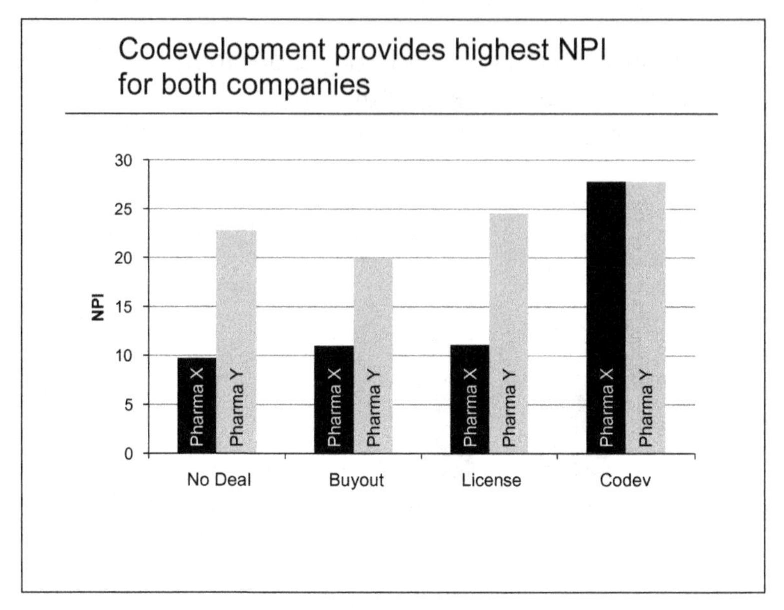

Figure 11.31. Slide 10.

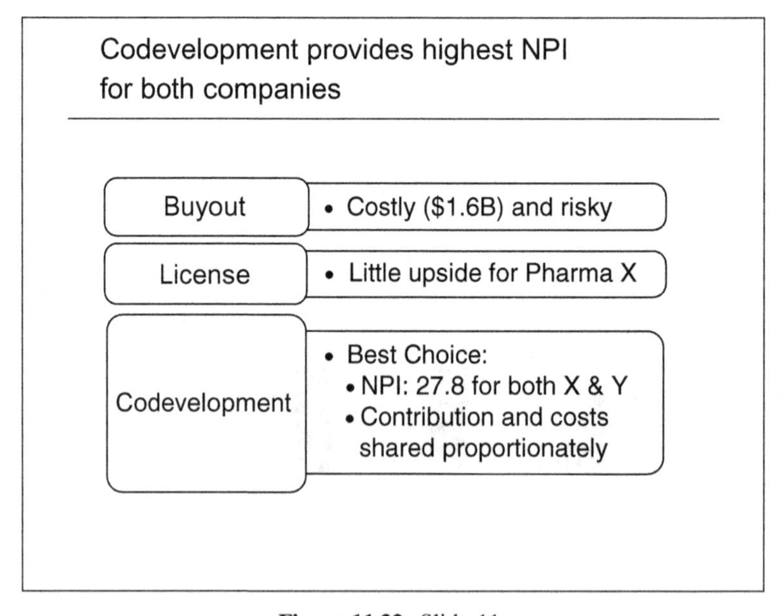

Figure 11.32. Slide 11.

well, describing the company might seem redundant. We have two purposes in doing this: one is to help the audience see the problem kernel as we have defined it, and the other is to help Pharma X management begin to think about Pharma Y as a potential partner, rather than just as a threat. Thus, we point out the strengths and weaknesses of both companies in this situation.

Our second slide (Figure 11.23) continues this theme by describing what each company could lose if it developed this drug on its own without a deal. We want our audience to see that both companies could either prosper or suffer. The third slide (Figure 11.24) describes the three types of deals we analyzed. The fourth slide (Figure 11.25) presents our conclusion that a codevelopment agreement is the best option for Pharma X.

Slide 5 (Figure 11.26) starts the next section of the presentation where we dive deeper into the analysis. This slide shows the probability tree for Pharma X, showing all the possible outcomes and probabilities. In many presentations, we would not show this level of detail, but it is necessary for managers in the audience to understand the possible outcomes before they can understand why certain deals are more attractive than others. Since the audience members are pharmaceutical industry experts, they are already familiar with the uncertainty of drug development and marketing. So, we carefully outline the alternatives now to lay the groundwork for our argument later.

The next two slides (Figures 11.27 and 11.28) establish the base case, which, in our analysis, consists of the results each company could achieve in the absence of an agreement. It is here that we put on the table the all-important numbers 9.7 and 22.8, which are the NPIs Pharma X and Pharma Y can achieve without an agreement. These are critical because we base all of our analysis on the assumption that neither company will accept a deal that gives it less than this. We show that Pharma Y has much higher probabilities of good outcomes due to its strong patent position. We also point out that Pharma Y believes the market will be much larger than Pharma X's estimate. This contributes to the large spread between base-case NPIs.

The next three slides (Figures 11.29– 11.31) communicate the results of our analysis of each of the deals: buyout, license, and codevelopment. Despite all the information we have assembled, we believe it is essential to boil down the material to a graph that compares all the options. These three slides are based on the same graph. We reveal the graph piece by piece to allow us to draw attention to each alternative in turn. With Slide 8 (Figure 11.29), we show that a buyout is possible but expensive. The next slide (Figure 11.30) shows that a license agreement is beneficial for both companies, but not dramatically so. Finally, Slide 10 (Figure 11.31) shows that codevelopment is by far the best alternative for Pharma X and that it is beneficial for Pharma Y as well.

We elaborate on the codevelopment alternative more than the others, since it is our recommended option. We explain that the potential revenues under codevelopment are very large. This is for two reasons: First, Pharma Y would not sue, and second, the joint venture would benefit from Pharma X's 75 percent margin outside the United States. So a codevelopment deal extracts all the potential economic value possible from the new drug. On that ground alone, it must be considered carefully. The real question about codevelopment is how the two companies should share the contribution and costs.

The compelling reason for Pharma X and Pharma Y to share contribution and costs in the same proportion is fairness. It seems fair for a partner who pays for a given percentage of the costs to claim the same percentage of the contribution. This proportional sharing also leads to both companies having identical NPIs (27.8). It is up to the companies to decide what percentage they want to claim. It is likely that Pharma Y will want a smaller percentage of costs and contribution since it is a small company and may not be able to afford large costs or large risks.

Our last slide (Figure 11.32) distills our results and reiterates our conclusion that a properly structured codevelopment deal is the best alternative.

11.9 SUMMARY

In this chapter we analyzed the financial attractiveness of potential sharing agreements between two pharmaceutical companies developing competing drugs. Our first modeling decision was to start with a deterministic model and to consider expected values of possible outcomes. We used probability trees to map out the possible outcomes of market share, along with their probabilities.

We also decided to focus our initial efforts exclusively on our client company, Pharma X. This kept the analysis simple as we began to explore and gain insight into the situation facing Pharma X. Furthermore, this choice allowed us to identify options that were acceptable to Pharma X, and therefore, it reduced the number of options to consider later on when looking at Pharma Y.

Once we added Pharma Y to our model, we were able to examine how various agreements affected both companies. We found that for all the agreement styles, there was a range of contract terms that would be acceptable to both companies. This range is called the zone of agreement, and it will be up to the contract negotiators to come to agreement on the exact terms within the zone.

Lastly, we modified our model to simulate uncertainty in event outcomes and various other parameters. We discovered that this added little value to our analysis since the market-share outcomes dominated the results, and the deterministic model accounted for this adequately.

CHAPTER 12

Invivo Diagnostics, Inc.

12.0 INTRODUCTION

Invivo Diagnostics, Inc. is a pharmaceutical company facing a major challenge to its viability. Most of the firm's revenues come from sales of a single drug on which the patent protection will soon expire. Although Invivo has six new drugs under development, revenues could decline significantly in the near future. Invivo needs credible forecasts of the revenues it will receive from the drugs it has under development as part of its efforts to prepare for the future. In the actual situation on which this case is based, models were developed to forecast revenues from each of the six drugs being developed. We concentrate in this chapter on just one drug, but the model we develop should be general enough to apply to other drugs as well.

In previous chapters, we have created models and generated insights with little or no data. We often find that we can learn a surprising amount by modeling without data. We encourage modelers to model first and collect data later to avoid wasting time and money collecting data that may have little or no bearing on insights that the model generates. Invivo presents a different modeling challenge because some data on how prices and market shares evolve over time are available. Of course, these data represent the histories of other drugs, not the drug we are considering, so we cannot expect the data to apply perfectly. Data may serve to guide our modeling and help to refine a model structure, but the common-sense and modeling skills we have developed to this point will be our main tools.

12.1 INVIVO DIAGNOSTICS CASE

Invivo Diagnostics

Invivo Diagnostics is a \$300M pharmaceutical company built on a single product that accounts for over 75 percent of revenues. A year from now, in June 2001, the patent for this product will expire, and the company wants to understand

Modeling for Insight: A Master Class for Business Analysts, by Stephen G. Powell & Robert J. Batt

how new products currently in development will help fill the expected $100–200M annual revenue gap that will begin to open as revenues from this product decline.

Invivo currently has under development six major products; each (if successful) will come on the market within the next six years. The most promising of these products, Product 1, represents a technological innovation in the diagnosis of heart disease through ultrasound imaging. The question Invivo faces is how much revenue will Product 1 bring to the company over the next 10 to 15 years?

A typical pharmaceutical product goes through a well-defined development process consisting of the following sequential phases (the times given are typical of most products):

- Preclinical phase: animal trials focusing on safety (typically 13 weeks)
- Phase 1: safety studies conducted on 50 to 100 normal healthy male volunteers (typically three to six months)
- Phase 2: efficacy and safety trials on a diseased target population under controlled medical conditions (typically six to nine months)
- Phase 3: efficacy and safety trials on the target population under actual conditions of use (typically six to nine months and may involve thousands of patients)
- FDA submission: preparation of a new drug application (NDA), involving extensive statistical analysis and report writing (typically six months)
- FDA review: The FDA evaluates the NDA based on the preclinical and clinical data (typically 17 to 24 months)

Although expectations are that Invivo will file with the FDA sometime in early 2002, the actual filing date could differ by one or two quarters, depending on the vagaries of the development process. As for the length of time it will take for FDA review, recent pressure by the PhRMA (an industry trade association) on the FDA has caused review times to fall. Although two years is still considered possible, Invivo now thinks the review time for Product 1 could be as little as 19 months.

A complicating factor is that many compounds fail to pass one of these stages and are dropped from development. Product 1 is about to begin Phase 2 studies. Company experts estimate that there is an 80 percent chance Product 1 will successfully pass Phase 2 and a 90 percent chance it will subsequently pass Phase 3. Essentially all drugs intended for diagnostic purposes pass FDA approval if they pass Phase 3 testing.

Any revenue forecast for this product will necessarily hinge on the demand for the product, as well as on the price and market share Invivo can command. The demand for Product 1 is closely related to the demand for ultrasound procedures in general. Ultrasound is a rapidly growing diagnostic tool because it is both cheaper and less invasive than the competing modalities, which include X-rays, MRI, and nuclear medicine. The demand for imaging agents for ultrasound procedures for the heart is projected to be 1.176M treatments in 2002 and is expected to grow 11 percent annually for three to four years, thereafter growing 3 percent annually for the next decade or so. Invivo expects the market price to be between $90 and $115 per treatment when the first product is introduced in 2002.

Competition in all drug categories is intense. The dates competitors introduce their products and those products' relative strengths both influence market share and, therefore, sales revenue. *Strength* depends on three factors: efficacy, safety, and ease of use. Invivo has competitive intelligence (Figure 12.1) on six potential competitors, including their expected launch dates and the relative strengths of their products (Invivo's product has a strength of 1.0).

Invivo itself has little expertise in forecasting prices and market shares. There is a widely held belief in the industry that prices of a given category of compounds generally decline over time, depending on the number of competing products and the number of years since the first product was introduced. Market shares seem to depend on the number of competitors in a category and on the relative strength of the products. Figures 12.2 and 12.3 give historical price and market-share data for six different drug families (labeled A–F) in a variety of medical categories.

We have been hired by the senior management of Invivo Diagnostics to develop a credible forecast of revenues over time for Product 1. Note that Product 1 is not expected to close the entire revenue gap by itself, but management hopes it will make a substantial contribution to revenues.

To the Reader: Before reading on, frame this problem by setting boundaries for the analysis; making assumptions; listing questions; defining outputs, inputs, and decisions; and so on.

Competitor	Date of Entry	Expected Strength	Range of Strength
1	1Q, 2002	1.0	NA
2	1Q, 2003	1.1	NA
3	2Q, 2003	1.0	NA
4	3Q, 2005	1.2	1.0–1.4
5	4Q, 2005	1.1	0.95–1.25
6	4Q, 2005	2.0	1.8–2.4

Notes:

1. Invivo's intelligence is solid regarding competitors 1 through 3, as they are already well into Phase 3 trials. Information about the dates of entry and strengths for Products 4 through 6 is more uncertain. The entry dates could be off by one to three quarters. Estimated ranges for the actual strengths of the products are provided.
2. Invivo believes competitors 1 through 3 are certain to pass FDA review successfully. Competitors 4 and 5 have about a 90 percent chance of making it to market. Competitor 6 is working on a product that involves a new technology; Invivo's experts estimate a 40 percent chance of failure for this product.

Figure 12.1. Date of entry and strength of competing products.

	Product Family A		*Product Family B*		*Product Family C*	
	Number of Competitors	Market Price	Number of Competitors	Market Price	Number of Competitors	Market Price
Year						
1	1	100	2	100	1	100
2	1	100	2	100	2	100
3	2	95	2	94	3	94
4	4	87	3	90	3	94
5	4	85	4	84	3	89
6	4	80	4	80	3	84
7	5	67	4	80	3	76
8	5	60	4	74	3	70
9	5	50	4	68	3	70
10	5	40	4	60	3	70

Notes:
1. Price data are normalized to a value of 100 when the first viable product comes to market.
2. Number of competitors represents the total number of competing products in the market.

Figure 12.2. Historical price data.

12.2 FRAME AND DIAGRAM THE PROBLEM

12.2.1 The Outcome Measure

Invivo is looking to fill a revenue gap created by its main product going off patent. This suggests that our outcome measure should involve revenue dollars, but exactly how to summarize revenues is unclear. We could consider revenues in a particular year, or we could take average revenues over some time period as our outcome measure. We could also use a net present value of revenues (which would require both a time horizon and a discount rate), or we could even use the entire time series of revenues as our outcome.

Annual revenue is a logical choice due to its simplicity. Furthermore, we are told that the expected annual revenue gap is between \$100M and \$200M. This serves as a good reference point for judging outcomes. Keeping in mind that we are modeling only one of six products that are in the pipeline, a more appropriate target is \$17M to \$33M in revenue per year, about one sixth of the total revenue gap.

But revenue in any single year is not sufficient. We are concerned about the next 10 to 15 years. This suggests that we should look at the net present value (NPV) of the revenues over that time period. Using NPV allows us to take into account both the time value of money and the riskiness of future revenues.

We consider both annual revenues and the NPV of revenues in our analysis. By looking at annual revenues, we can see whether we meet our goal in any

Product Family D			*Entrant*	
	Entrant		Share of Market (%)	
Year	Number	Strength	First	Second
1	1	1.0	100	–
2			100	–
3			100	–
4	2	1.8	80	20
5			70	30
6			62	38
7			50	50
8			44	56
9			39	61
10			40	60

Product Family E			*Entrant*		
	Entrant		Share of Market (%)		
Year	Number	Strength	First	Second	Third
1	1	1.0	100	–	–
2			100	–	–
3	2	0.95	80	20	–
4			67	33	–
5	3	0.5	50	35	15
6			40	35	25
7			40	38	22
8			40	39	21
9			40	39	21
10			40	40	20

Product Family F			*Entrant*				
	Entrant		Share of Market (%)				
Year	Number	Strength	First	Second	Third	Fourth	Fifth
1	1	1.0	100	–	–	–	–
2			100	–	–	–	–
3	2	1.0	65	35	–	–	–
4			55	45	–	–	–
5			50	50	–	–	–
6	3	0.6	45	45	10	–	–
7	4	1.0	35	40	10	15	–
8	5	0.7	30	30	12	15	13
9			27	25	14	20	14
10			24	23	14	23	16

Notes:

1. These data represent products in the same general category as Product 1.
2. Data are normalized so that the first viable product is introduced in Year 1; subsequent products are introduced by competitors in the years, and with the relative strengths, listed.

Figure 12.3. Historical market-share data.

given year. By looking at the year-to-year trend in revenues, we can make qualitative judgments about the forces at play in our model (e.g., is demand going up fast enough to offset a decline in unit price?). Measuring the NPV of revenues, on the other hand, allows us to compare easily the results of different models.

It is helpful to have a target in mind for our NPV measurement, just as we have the $17–33M target for annual revenue. Consider a constant stream of cash flows of $17M a year for 15 years. The NPV of this cash flow is $129M, using an arbitrary discount rate of 10 percent. Likewise, the NPV of $33M a year for 15 years is $251M. The target ranges, then, are annual revenues between $17M and $33M and an NPV of revenues between $130M and $250M. Of course, these are only targets, not requirements. Our job is to build a model that Invivo's management can use to plan for the future, regardless of the model's outcome.

To the Reader: Before reading on, develop the problem kernel and sketch an influence diagram.

12.2.2 Setting Boundaries and Diagramming the Model

As presented to us, this is a complex problem with lots of details about the likelihood of passing the various development stages, competitors' actions, pricing, market share, and so on. We start by placing most of these details outside the boundaries of our model. Later, we add them back in a controlled manner. At this point, we develop the problem kernel.

> Invivo has a 72 percent chance of successfully developing a new drug having a strength of 1.0. Six competitors are known to be developing drugs for the same market, with differing dates of entry and product strengths. Drug prices tend to decline as time passes and as the number of competitors increases. Market shares also evolve over time, depending on the number of competitors, the timing of entry, drug strength, and, perhaps, other factors. Invivo's management wants to know how much revenue the company can expect from its new drug.

Since price and market share clearly are important in this problem, we look first for a way to relate these factors to revenue. The simplest way to accomplish this is to decompose revenue into three elements: Total Demand, Market Share, and Price (Figure 12.4).

Revenue is simply the product of these three values:

$$\text{Revenue} = \text{Total Demand} \times \text{Market Share} \times \text{Price}$$

This formula provides the basic structure of our model from first to final iteration. Any complication or detail we add to the model influences one or more of the three variables that determine Revenues. Since we are also considering the NPV of Annual Revenues, we refine our influence diagram to show this and the addition of parameters for the Discount Rate and time horizon (Figure 12.5).

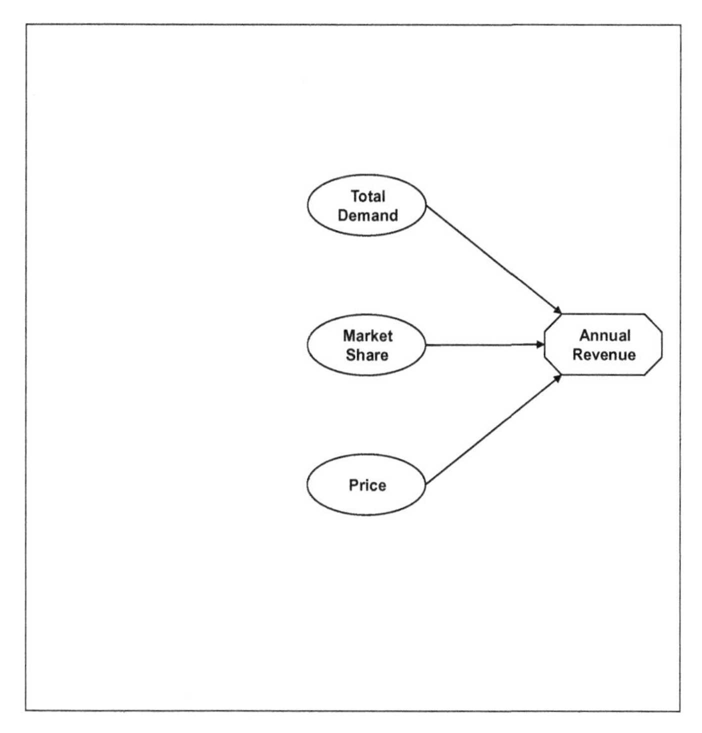

Figure 12.4. Basic influence diagram.

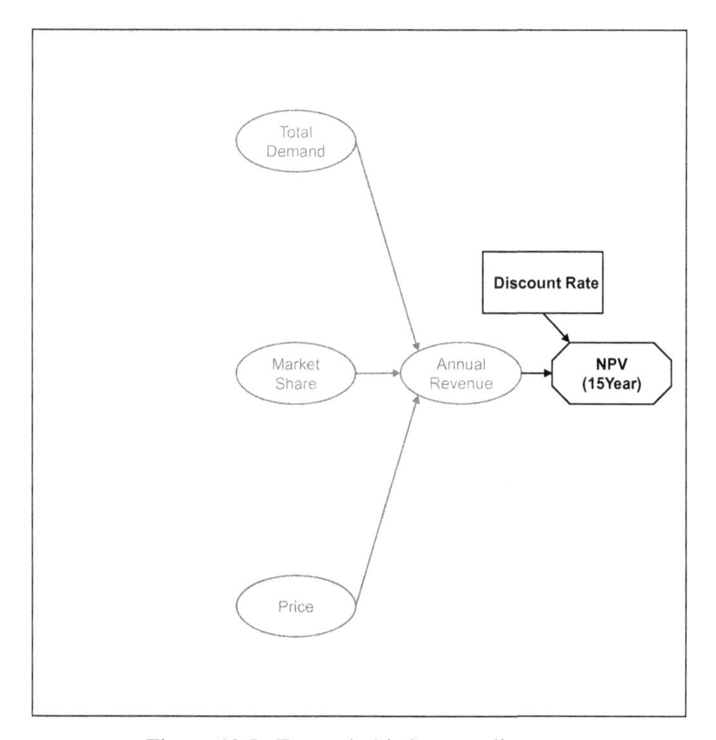

Figure 12.5. Expanded influence diagram.

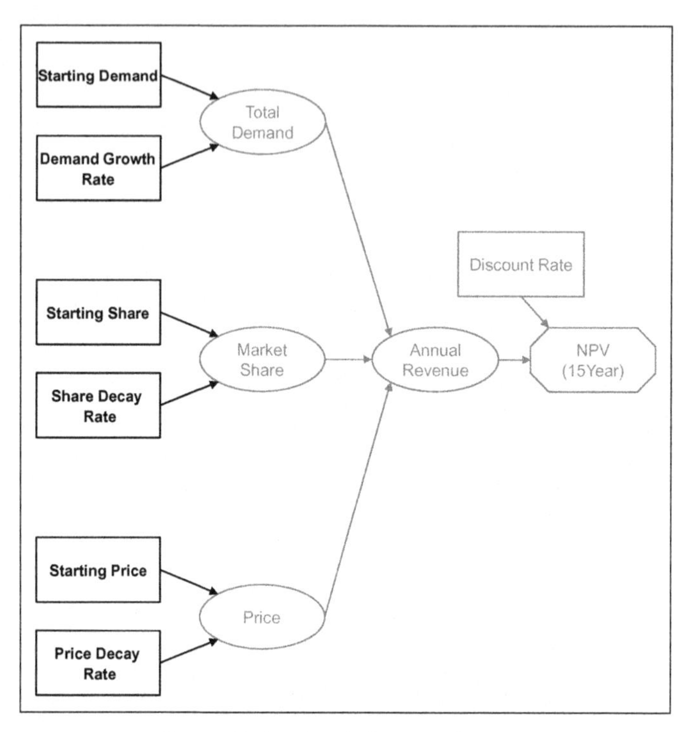

Figure 12.6. Final influence diagram.

As stated, we begin by eliminating most of the complicating factors and interactions from our model. One way to do this is to assume that all three of our key variables—Total Demand, Market Share, and Price—follow constant percentage growth or decay processes. In other words, each of these variables will have a starting value and then will grow or decline by a set percentage each year (Figure 12.6). We also assume there is one price for all the drugs competing in this market. Finally, we assume for the moment that Invivo is first to market.

To the Reader: Before reading on, design and build a first model based on the problem kernel and influence diagram.

12.3 M1 MODEL AND ANALYSIS

Figures 12.7 and 12.8 show M1 in numerical and relationship views. The Total Demand parameters are based on information in the case. We simplify the growth rate to a constant 6 percent rather than include the two different

	A	B	C	D	E	F	G	H
1	**Invivo Model 1**		**M1**					
2								
3								
4								
5								
6	**Parameters**							
7								
8		Total Demand						
9		Base	1.176	million				
10		Growth	6%					
11		Market Share						
12		Base	100%					
13		Growth	-5%					
14		Price						
15		Base	$100					
16		Growth	-5%					
17								
18		r	10%					
19								
20								
21								
22	**Calculations**		**Demand (M)**	**Share**	**Price**	**Revenue ($M)**	**Base Case**	
23								
24		2002	1.18	100%	$ 100	117.60	$ 117.60	
25		2003	1.25	95%	$ 95	112.50	$ 112.50	
26		2004	1.32	90%	$ 90	107.63	$ 107.63	
27		2005	1.40	86%	$ 86	102.96	$ 102.96	
28		2006	1.48	81%	$ 81	98.50	$ 98.50	
29		2007	1.57	77%	$ 77	94.23	$ 94.23	
30		2008	1.67	74%	$ 74	90.14	$ 90.14	
31		2009	1.77	70%	$ 70	86.23	$ 86.23	
32		2010	1.87	66%	$ 66	82.50	$ 82.50	
33		2011	1.99	63%	$ 63	78.92	$ 78.92	
34		2012	2.11	60%	$ 60	75.50	$ 75.50	
35		2013	2.23	57%	$ 57	72.23	$ 72.23	
36		2014	2.37	54%	$ 54	69.09	$ 69.09	
37		2015	2.51	51%	$ 51	66.10	$ 66.10	
38		2016	2.66	49%	$ 49	63.23	$ 63.23	
39								
40						**NPV**	NPV	
41						$719.35	$ 719.35	
42								

Figure 12.7. M1—numerical view.

	A	B	C	D	E	F	G	H
1	**Invivo Model 1**		**M1**					
2								
3								
4								
5								
6	**Parameters**							
7								
8		Total Demand						
9		Base	1.176	million				
10		Growth	0.06					
11		Market Share						
12		Base	1					
13		Growth	-0.05					
14		Price						
15		Base	100					
16		Growth	-0.05					
17								
18		r	0.1					
19								
20								
21								
22	**Calculations**		**Demand (M)**	**Share**	**Price**	**Revenue ($M)**	**Base Case**	
23								
24		2002	=C9	=C12	=C15	=C24*D24*E24	117.60	
25		2003	=C24*(1+$C$10)	=D24*(1+$C$13)	=E24*(1+$C$16)	=C25*D25*E25	112.50	
26		2004	=C25*(1+$C$10)	=D25*(1+$C$13)	=E25*(1+$C$16)	=C26*D26*E26	107.63	
27		2005	=C26*(1+$C$10)	=D26*(1+$C$13)	=E26*(1+$C$16)	=C27*D27*E27	102.96	
28		2006	=C27*(1+$C$10)	=D27*(1+$C$13)	=E27*(1+$C$16)	=C28*D28*E28	98.50	
29		2007	=C28*(1+$C$10)	=D28*(1+$C$13)	=E28*(1+$C$16)	=C29*D29*E29	94.23	
30		2008	=C29*(1+$C$10)	=D29*(1+$C$13)	=E29*(1+$C$16)	=C30*D30*E30	90.14	
31		2009	=C30*(1+$C$10)	=D30*(1+$C$13)	=E30*(1+$C$16)	=C31*D31*E31	86.23	
32		2010	=C31*(1+$C$10)	=D31*(1+$C$13)	=E31*(1+$C$16)	=C32*D32*E32	82.50	
33		2011	=C32*(1+$C$10)	=D32*(1+$C$13)	=E32*(1+$C$16)	=C33*D33*E33	78.92	
34		2012	=C33*(1+$C$10)	=D33*(1+$C$13)	=E33*(1+$C$16)	=C34*D34*E34	75.50	
35		2013	=C34*(1+$C$10)	=D34*(1+$C$13)	=E34*(1+$C$16)	=C35*D35*E35	72.23	
36		2014	=C35*(1+$C$10)	=D35*(1+$C$13)	=E35*(1+$C$16)	=C36*D36*E36	69.09	
37		2015	=C36*(1+$C$10)	=D36*(1+$C$13)	=E36*(1+$C$16)	=C37*D37*E37	66.10	
38		2016	=C37*(1+$C$10)	=D37*(1+$C$13)	=E37*(1+$C$16)	=C38*D38*E38	63.23	
39								
40						**NPV**	NPV	
41						=NPV(C18,F24:F38)	719.35	
42								

Figure 12.8. M1—relationship view.

growth rates mentioned in the problem; we add this detail in later. Starting Market Share and Price are set at 100 percent and $100, respectively, based on the assumption that Invivo is first to enter the market. We choose plausible decay rates of 5 percent per year for Market Share and Price. A quick glance at the data in Figures 12.2 and 12.3 supports these rough estimates.

In column C, we calculate the total market Demand for each year. Demand starts at the base level of 1.176M treatments that Invivo has forecast and grows at 6 percent each year. Similar calculations are performed in columns D and E to calculate Invivo's Share and the market Price, respectively. Lastly, in column F, the annual Revenue is calculated by multiplying Demand, Share, and Price from columns C, D, and E.

To the Reader: Before reading on, use M1 to generate as many interesting insights into the situation as you can.

This first model gives an NPV of $719M, which is well above the upper target of $250M. Also, annual revenues are above the upper target of $33M in every year. Thus, we have our first conclusion:

Insight 1: Invivo can far surpass the revenue targets if it is first to enter the market.

These results are so overwhelmingly positive that it hardly seems worth refining the model. However, the results are based on the highly optimistic assumption that Invivo will be first to enter the market. Figure 12.9 shows the remaining process stages Invivo must complete to get Product 1 to market and the likely duration range of each stage. If every stage is completed in the minimum time, the earliest Product 1 would get to market would be May 2003. At the other extreme, the latest Invivo expects Product 1 to reach the market is June 2004. Summing the average duration of each of the remaining development stages, we find Invivo is most likely to enter the market in late 2003 or early 2004. The competitor intelligence shows with virtual certainty that at

Current Date: June-00

	Duration (months)		
	Minimum	Average	Maximum
Phase 2	6	7.5	9
Phase 3	6	7.5	9
Submission	6	6	6
FDA Review	17	20.5	24
TOTAL	35	41.5	48
Completion	May-03	November-03	June-04

Figure 12.9. Estimation of Invivo's market entry date (M1).

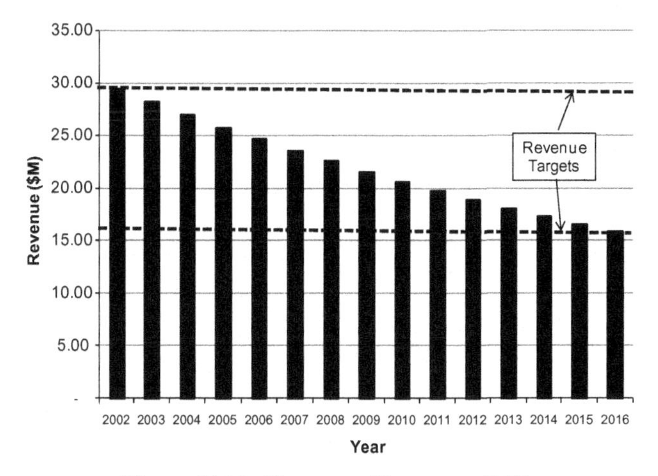

Figure 12.10. Base-case Revenues (M1).

least one competitor will enter before Invivo does. Thus, our current assumption of first entry for Invivo is overly optimistic.

The historical data in Figure 12.3 show that later entrants start with a small share of the market. Not wanting to spend much time analyzing data yet, we give the data a casual glance and assume a starting Market Share of 25 percent. With this change, M1 now reports an NPV of $180M, which is in the middle of the target zone. Annual revenue, which ranges from $29M to $16M, also falls between the targets (Figure 12.10). We consider these results our base case since the values based on 100 percent market share are too far-fetched to serve as a useful reference point.

Insight 2: Invivo can reach the revenue targets even if it is not first to market.

Despite M1's simplicity, there are several additional insights we can glean from it. Figure 12.10 shows that annual revenue is steadily decreasing. Why is this? In Figure 12.11, we display annual revenue and its three components in normalized terms. To normalize this data, we show each variable as a percentage of its first-year value. For example, in year 2014, Demand is at 200 percent, or twice its starting level, while Revenue is down to 60 percent of its starting level. The figure shows that despite significant growth in total demand, declining price and share lead to declining revenues. Management will need to be prepared with a stream of new products in order to counter this effect.

Insight 3: Market demand is not growing fast enough to counter the effects of falling share and price.

Since the decline rates of price and market share clearly play a central role in this model, we consider what combinations of these two parameters allow

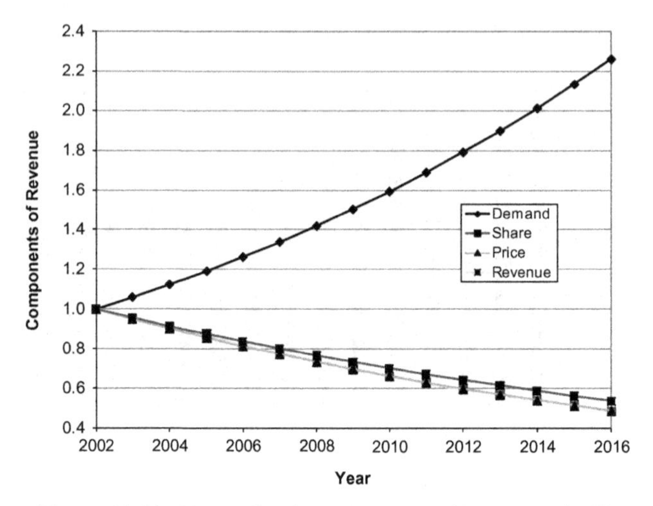

Figure 12.11. Normalized components of Revenue (M1).

	A	B	C	D	E	F	G	H	I	J	K
3	Share									Price	
4		**-16%**	**-14%**	**-12%**	**-10%**	**-8%**	**-6%**	**-4%**	**-2%**	**0%**	
5	**-16%**	$83	$88	$92	$98	$103	$110	$117	$125	$134	
6	**-14%**	$88	$92	$98	$104	$111	$118	$127	$136	$147	
7	**-12%**	$92	$98	$104	$111	$119	$127	$137	$148	$161	
8	**-10%**	$98	$104	$111	$119	$128	$138	$149	$163	$178	
9	**-8%**	$103	$111	$119	$128	$138	$150	$164	$179	$197	
10	**-6%**	$110	$118	$127	$138	$150	$164	$180	$198	$219	
11	**-4%**	$117	$127	$137	$149	$164	$180	$199	$220	$246	
12	**-2%**	$125	$136	$148	$163	$179	$198	$220	$246	$277	
13	**0%**	$134	$147	$161	$178	$197	$219	$246	$277	$313	
14											

Figure 12.12. Two-way sensitivity: Share Decay and Price Decay (M1) (Shaded region is below NPV target).

Invivo to achieve its revenue targets. This is a task for a two-way Data Sensitivity table (Figure 12.12). The table in Figure 12.12 shows that Invivo can reach the target revenues (the unshaded region) even if share and price both decay at about 8 percent per year. Furthermore, if Invivo can maintain its market share at 25 percent (0 percent decay), it can withstand a large annual price decay rate and still meet the lower NPV target. We convert these observations to more general insights as follows:

Insight 4: Invivo can reach the target revenues even when both price and market share decay at moderate rates.

Insight 5: Maintaining reasonable market share (around 25 percent) is key to withstanding large price drops and still meeting the NPV target.

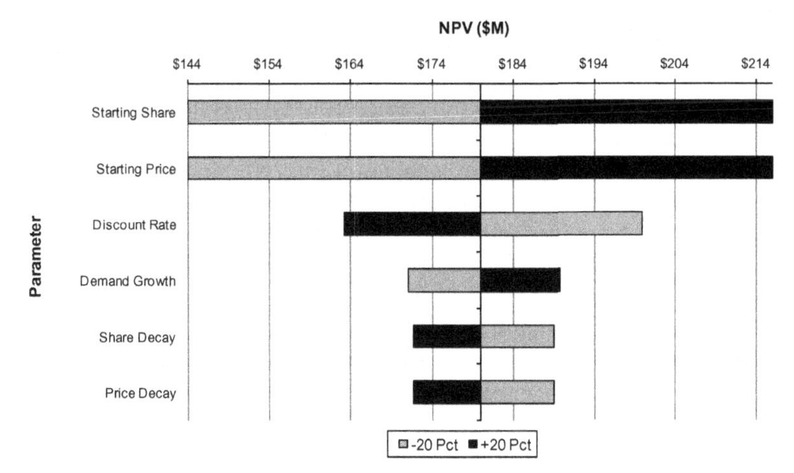

Figure 12.13. Tornado chart (M1).

Finally, the tornado chart in Figure 12.13 compares the relative effects of the parameters on the NPV of revenues.

Insight 6: The starting share and starting price have the biggest impacts on NPV, whereas the rates at which price, share, and demand change have more modest impacts.

In subsequent models, we explore just how Invivo can affect these parameters, but for now, it is enough to know Starting Share and Starting Price are key drivers of NPV.

We summarize our analysis to this point as follows:

- Annual Revenue and NPV of Revenue are our outcome measures.
- The model has three modules: Total Demand, Share, and Price.
- With six products in the pipeline, Invivo can meet the revenue goals even if it is not first to market.
- Starting Market Share and Price are major factors in determining Revenue.

We focus our subsequent efforts on improving our modeling of Demand, Share, and Price and on how Invivo can influence these factors in its favor.

To the Reader: Before reading on, decide how best to improve M1.

12.4 M2 MODEL AND ANALYSIS

We now consider where to focus our efforts for the next model iteration. Each of the three modules—Demand, Share, and Price—can be improved. In M1, we assumed that demand grows at a constant rate, whereas the case states that growth will be strong at first and slower later. Since this is a simple change to make to the model, we incorporate it into M2.

We learned from M1 that the starting market share and price are both key determinants of revenues, so the Share and Price modules are also good candidates for further refinement. In M1, we simplified the Market Share and Price modules by ignoring details related to the approval process, including timing of entry and the strength of products on the market. We now redraw our problem boundaries to include some of these details.

Invivo believes share and price are affected by market entry. A cursory look at the data confirms this impression. However, we do not know exactly what the interactions are. Figure 12.14 shows our current state of knowledge in an influence diagram. The Price module seems easier to deal with since there is only one market price per time period to calculate rather than shares for all competitors. Therefore, we focus on improving the Price module in M2 and postpone refining the Share module.

To the Reader: Before reading on, modify M1 to improve the Price module.

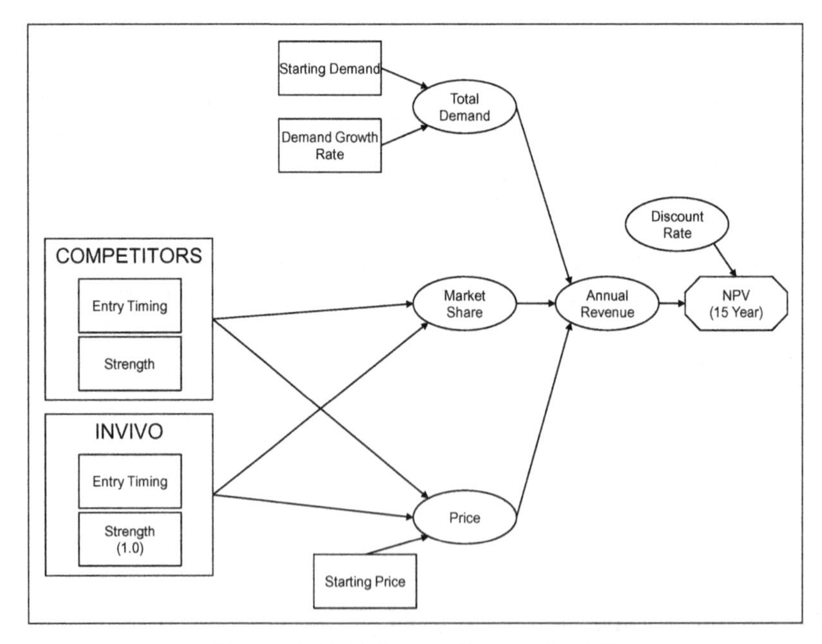

Figure 12.14. Influence diagram for M2.

12.4.1 Using the Data to Improve the Price Module

Our task is to develop a model (technically, a module of a larger model) for forecasting the market price. Before we go any further, it is important to note that even here we are simplifying. We are assuming there is one market price for all products in a therapeutic category. This suggests either a commodity market or a highly regulated market. It does not make sense to assume a commodity market since there is evidence in the data that products have different strengths and different market shares. It does, however, seem likely that this is a highly regulated market. Perhaps the per-unit revenue a company can receive is set by insurance companies and Medicare. These payers set a price for a given treatment and do not differentiate between the competing products.

We should also reiterate that our goal here is not just to model the evolution of price for a single drug but, ultimately, to develop a modeling *framework* that Invivo could apply to all six of the drugs it has under development. This leads us to develop a more general and flexible model structure than would be necessary if we were concerned with just one drug.

As we develop our model for price, we look to all information sources for suggestions about the nature of the relationships that determine price. Industry wisdom, common sense, and data are all valid sources. We prefer to state our ideas in the form of explicit *hypotheses* so they are clear and testable. Since some of our hypotheses may be wrong, we reserve the right to change or remove them from consideration. But clearly stated hypotheses help us formulate models that behave realistically.

Experts at Invivo believe price generally declines over time but is also affected by the number of competitors in the market. These seem like reasonable hypotheses based on common sense. A quick scan of the data in Figure 12.2 confirms these notions. Therefore, we state two hypotheses:

Hypothesis 1: Price falls as time passes.

Hypothesis 2: Price falls as the total number of competitors increases.

If we had no other information, we would be free to invent any relationship among price, time, and number of competitors that was consistent with these hypotheses. Perhaps price drops exponentially with time and an additional $10 for every competitor that enters. Or perhaps price goes down $1 a year, plus 30 percent for the first competitor, 20 percent for the second competitor, and so on. The possibilities are endless. Fortunately, we have some data in hand to guide our work.

Before jumping into a mathematical analysis of the data, it is best to take some time to just *look* at the data. In Figure 12.2, we see that price stays at 100 for two years for all the product families and then declines steadily. Of course, we have no data about what happens when a product leaves the market, so we need to be careful about assuming price always goes down.

Beyond that, it is hard to draw other conclusions from the data. Perhaps some graphs can reveal useful patterns.

In Figure 12.15a we have graphed price against time for Products A, B, and C. This figure shows that price does decline with time, and after the first two years, it does so in a roughly linear fashion. Furthermore, the linear decline seems to have approximately the same slope for all three products. The exception to this rule is that price stays at 100 for the first two years before beginning to decline. This linear pattern is extremely informative. Instead of considering all possible relationships between price and time, we now believe we can focus on a linear relationship. The fact that the pattern holds for all three product families suggests that a model that works well for these families may indeed create a good forecast for our new product.

In Figure 12.15b, we have graphed price against number of competitors for Products A, B, and C. This figure may seem a little confusing due to the vertical drops in the graph for each product family. These vertical drops occur because price is declining with time even when the number of competitors stays constant. For example, the price of Product C drops four times while the number

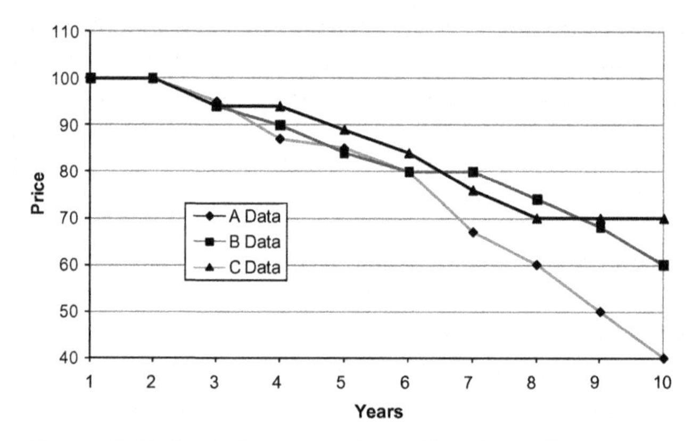

Figure 12.15. (a) Price versus Time (data from Figure 12.2).

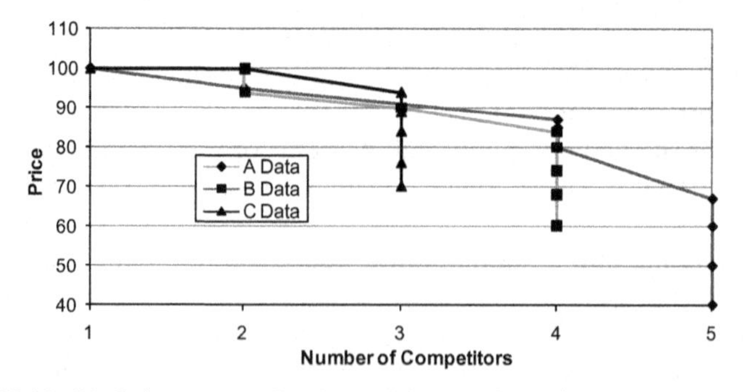

Figure 12.15. (b) Price versus Number of Competitors (data from Figure 12.2).

of competitors in this category stays at three. The important feature of the graph is that price also seems to decline as competitors enter. As with price versus time, the price versus competitor effect seems to be similar in magnitude for all families. We must admit, however, that the linearity and the similarity among the three families is not as strong as with price and time. Keep in mind that we are trying to simplify a very complex marketplace into just two variables. Therefore, it is not surprising that reality does not fall into a simple pattern.

We now propose a model that is consistent with our hypotheses and our observations of the data:

$$P_t = P_0 + \alpha ET + \beta COMPS$$

where

P_t: Price at time t

P_0: Price at time 0

α: Coefficient of elapsed time

ET: Elapsed time ($ET = 1$ for year of first entry)

β: Coefficient of competition

$COMPS$: Total number of competing products in the market

According to this model, price changes by α for every year that passes and changes by β for every competitor that enters the market. Furthermore, these two effects are additive. As long as both α and β are negative, this model agrees with both hypotheses stated above.

To test our model, we create a spreadsheet in which we plot the output of our linear model from above along with the historical data. We then experiment with values for P_0, α, and β to see whether we can match the historical data. Since our goal is for our model to approximate reality as best we can, we want to find parameter values that align the model results to the data as closely as possible. After some experimentation, we realize that it is difficult to match the price versus time data over the final few years. The price of Family A drops quickly in the later years, whereas the price of Family C levels off in the later years. We have to choose a middle ground when choosing parameters for our model. Figures 12.16a and 12.16b show the historical data as points and the price calculated by our model as solid lines for the parameters $P_0 = 115$, $\alpha = -5$, and $\beta = -3$. With these parameters, the fit between the model and the data is good but not perfect.

We also have to deal with the nonlinearity in price in the first two years. The data show that price is always 100 in the first two years. To incorporate this, we nest our linear equation inside a MIN function that caps the price at 100:

$$P_t = \text{MIN}(P_0 + \alpha ET + \beta COMPS, 100)$$

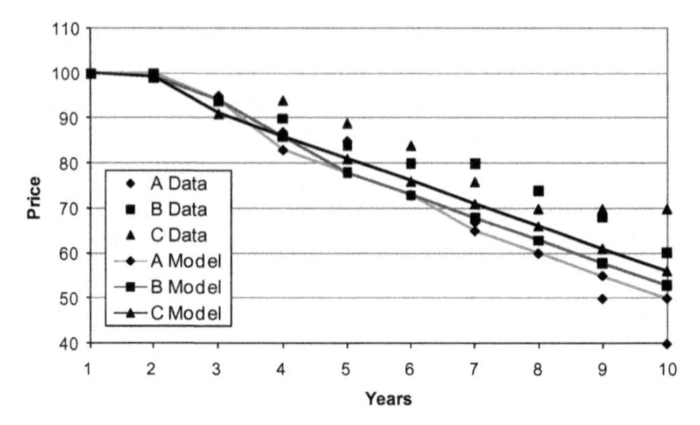

Figure 12.16. (a) Price versus Time with linear model approximation ($P_0 = 115$, $\alpha = -5$, $\beta = -3$).

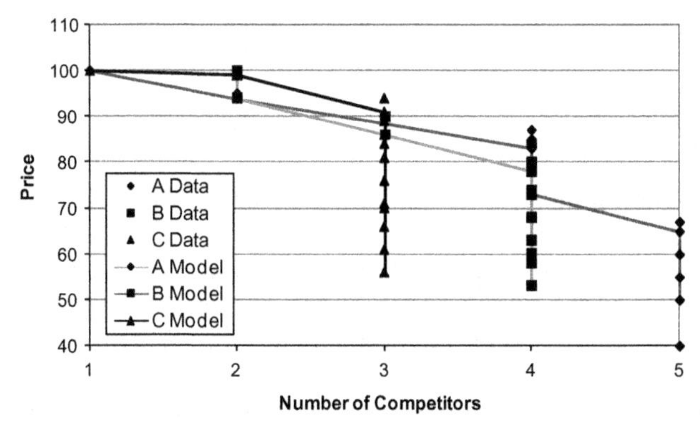

Figure 12.16. (b) Price versus Number of Competitors with linear model approximation ($P_0 = 115$, $\alpha = -5$, $\beta = -3$).

This allows us to select coefficients that fit the linear part of the data without being distracted by the first two years.

After experimenting with our model to get a feel for its behavior, we use linear regression to fine-tune the parameters for the coefficients α and β. Recall that we showed in Figures 12.16a and 12.16b that our model can fit the data fairly well with $P_0 = 115$, $\alpha = -5$, and $\beta = -3$. Once again, we must decide how to handle the kink in the data in years 1 and 2. If we use data from all years in the regression, the coefficients will be artificially low as the regression tries to accommodate the first two years of constant price. We could throw out data from both years 1 and 2 under the assumption that the price for those years is fixed and not part of the data's linear section. Or we could throw out only the data from year 1. This would allow the regression model to take into account the year 2 data point as the starting point of the linear data. We test all three alternatives. Figure 12.17 shows that all three have similar coefficients

	All Data	Drop Y1	Drop Y1&Y2
P_o: y-Intercept	114.44	119.76	125.86
α: Years	−4.00	−4.23	−4.41
β: Comps	−3.45	−4.42	−5.64
R^2	88.5%	89.2%	88.2%
Adjusted R^2	87.6%	88.3%	87.1%

Figure 12.17. Summary of regression models for Price.

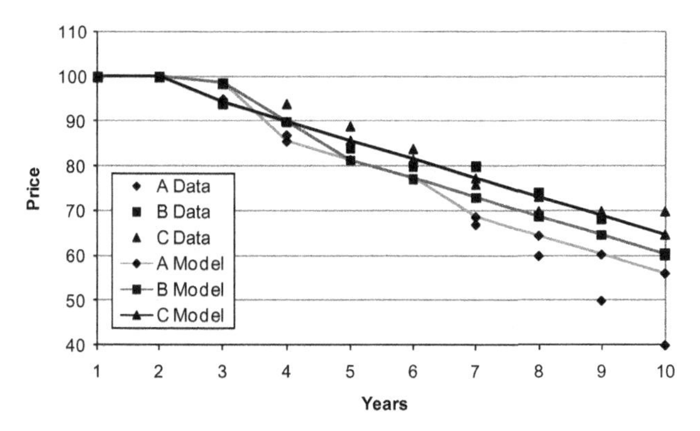

Figure 12.18. (a) Price versus Time with regression model (P_0 = 120, α = −4.2, β = −4.4).

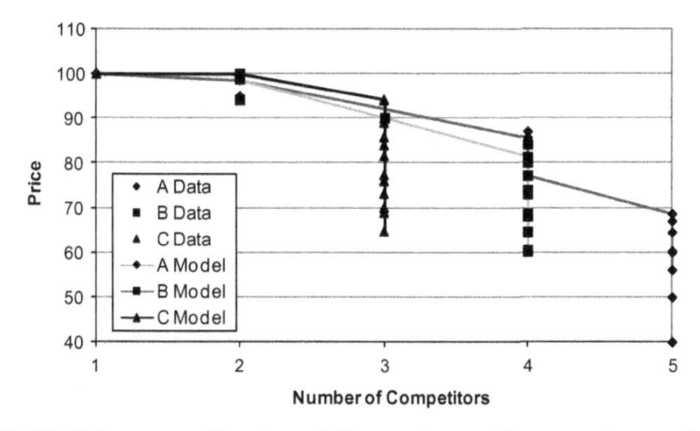

Figure 12.18. (b) Price versus Number of Competitors with regression model (P_0 = 120, α = −4.2, β = −4.4).

and virtually identical R^2. The parameter α is around −4, and β ranges from −3.5 to −5.5. We choose the Drop Y1 option, in which data from only the first year is dropped, because it makes the most intuitive sense.

Figures 12.18a and 12.18b show the historical data and our model forecasts using the coefficient values from the regression labeled Drop Y1.

The forecast certainly does not perfectly match the data, but it is a good approximation.

Our model is, after all, a simplification of a complex reality, so we need to keep its limitations in mind. The greatest limitation is that the model allows negative prices, which is nonsensical. If we allow the Time and/or Competitors variables to become large, the model will forecast a negative price. Common sense suggests that price would instead decline to some asymptote above zero. However, for the sake of simplicity, we use a MAX function to keep price from going below 0. With this addition, the pricing model is bounded by 0 and 100 and is a linear function of time and number of competitors between these two bounds:

$$P_t = \text{MAX}(\text{MIN}(P_0 + \alpha ET + \beta COMPS, 100)), 0)$$

As stated, many other relationships are possible between price and the two independent variables Time and Competitors. In this case, a linear relationship seems plausible based on the data we have, and our experiments show that it works reasonably well. It is up to us, as the modelers, to decide whether the fit is "good enough" or whether other options should be explored. Although this discussion has only covered a linear model for this module, we also explored (but do not report) relationships involving percentage changes and exponential relationships because the linear model is simplest and fit well enough.

We find that many novice modelers learn one type of model or tool and try to make every situation fit that mold. Linear relationships and linear regression are frequently abused in this way. Novice modelers frequently take any data they can get their hands on and run it through linear regression analysis, blindly looking for coefficients and constants. This approach will result in a model—maybe even a useful model—but it can just as easily violate basic common sense.

In contrast, we first formulated hypotheses our model should follow. Then, we created a simple model that could conform to these hypotheses. Next, we experimented with this model to see whether it could approximate the data at hand. Finally, we refined the model parameters using regression. In this case, we believe the linear model is a reasonable representation of the price/time/competitors relationship. The emphasis throughout this process is on hypothesis making and model exploration, not on statistical techniques *per se*.

12.4.2 Changes to the Model

There are just a few changes to make going from M1 to M2 (Figures 12.19 and 12.20). The most significant is to substitute our new pricing module for the constant decay-rate relationship used in M1. Because our new pricing module requires the number of competitors in the market as an input, we also create

	A	B	C	D	E	F	G	H	I
1	Invivo M2	Price Decay Function							
2									
3									
4									
5									
6	Parameters								
7		Total Demand							
8		Base	1.176						
9		Growth 1	11%						
10		Growth 2	3%						
11		First Year of G2	2006						
12									
13		Market Share							
14		Base	25%						
15		Growth	-5%						
16		Price							
17		Base	$100						
18		y-intercept	119.76						
19		Time Coefficient	-4.23						
20		Players Coefficient	-4.42						
21									
22		r	10%						
23									
24									
25	Calculations								
26		Competitors							
27		Number	Entry Year	Strength					
28		Us	2004	1					
29		1	2002	1.0					
30		2	2003	1.1					
31		3	2003	1.0					
32		4	2005	1.2					
33		5	2005	1.1					
34		6	2005	2.0					
35									
36									
37		Year Num		Players	Demand	Share	Price	Revenue ($M)	
38									
39		1	2002	1	1.18	25%	$ 100	29.40	
40		2	2003	3	1.31	24%	$ 98	30.39	
41		3	2004	4	1.45	23%	$ 89	29.22	
42		4	2005	7	1.61	21%	$ 72	24.78	
43		5	2006	7	1.66	20%	$ 68	22.82	
44		6	2007	7	1.71	19%	$ 63	20.93	
45		7	2008	7	1.76	18%	$ 59	19.11	
46		8	2009	7	1.81	17%	$ 55	17.36	
47		9	2010	7	1.86	17%	$ 51	15.68	
48		10	2011	7	1.92	16%	$ 46	14.06	
49		11	2012	7	1.98	15%	$ 42	12.51	
50		12	2013	7	2.04	14%	$ 38	11.01	
51		13	2014	7	2.10	14%	$ 34	9.58	
52		14	2015	7	2.16	13%	$ 30	8.20	
53		15	2016	7	2.23	12%	$ 25	6.87	
54									
55								NPV	
56								$161.15	
57									

Figure 12.19. M2—numerical view.

a module that generates this information. For simplicity, we ignore the quarter of entry and consider only the year of entry. Also, since we are working with a deterministic model, we ignore all uncertainty. The other change to incorporate is the assumption that demand grows rapidly at first and then slows down. We use Invivo's estimates that demand will grow at 11 percent for the first three years and at 3 percent per year after that.

As mentioned, we use the coefficients from the linear regression for our price-model parameters. The data suggested coefficients of $\alpha = -4.2$ and $\beta = -4.4$, which means that the price drops by \$4.20 each year and by \$4.40 with the entry of each new competitor.

To the Reader: Before reading on, use M2 to generate as many interesting insights into the situation as you can.

	A	B	C	D	E	F	G	H
1	Invivo M2	Add Price Decay Function						
2								
3								
4								
5								
6	Parameters							
7		Total Demand						
8		Base	1.176					
9		Growth 1	0.11					
10		Growth 2	0.03					
11		First Year of G2	2006					
12								
13		Market Share						
14		Base	0.25					
15		Growth	-0.05					
16		Price						
17		Base	100					
18		y-intercept	119.757461514295					
19		Time Coefficient	-4.23159493140643					
20		Players Coefficient	-4.42318567389255					
21								
22		r	0.1					
23								
24								
25	Calculations							
26		Competitors						
27		Number	Entry Year	Strength				
28		Us	2004	1				
29		1	2002	1				
30		2	2003	1.1				
31		3	2003	1				
32		4	2005	1.2				
33		5	2005	1.1				
34		6	2005	2				
35								
36								
37		Year Num		Players	Demand	Share	Price	Revenue ($M)
38								
39		1	2002	=COUNTIF($C$28:$C$34,"<="&C39)	=C8	=C14	=MAX(MIN($C$17,$C$18+$C$19*B39+$C$20*D39),0)	=E39*F39*G39
40		2	2003	=COUNTIF($C$28:$C$34,"<="&C40)	=E39*(1+IF(C40<$C$11,$C$9,$C$10))	=F39*(1+$C$15)	=MAX(MIN($C$17,$C$18+$C$19*B40+$C$20*D40),0)	=E40*F40*G40
41		3	2004	=COUNTIF($C$28:$C$34,"<="&C41)	=E40*(1+IF(C41<$C$11,$C$9,$C$10))	=F40*(1+$C$15)	=MAX(MIN($C$17,$C$18+$C$19*B41+$C$20*D41),0)	=E41*F41*G41
42		4	2005	=COUNTIF($C$28:$C$34,"<="&C42)	=E41*(1+IF(C42<$C$11,$C$9,$C$10))	=F41*(1+$C$15)	=MAX(MIN($C$17,$C$18+$C$19*B42+$C$20*D42),0)	=E42*F42*G42
43		5	2006	=COUNTIF($C$28:$C$34,"<="&C43)	=E42*(1+IF(C43<$C$11,$C$9,$C$10))	=F42*(1+$C$15)	=MAX(MIN($C$17,$C$18+$C$19*B43+$C$20*D43),0)	=E43*F43*G43
44		6	2007	=COUNTIF($C$28:$C$34,"<="&C44)	=E43*(1+IF(C44<$C$11,$C$9,$C$10))	=F43*(1+$C$15)	=MAX(MIN($C$17,$C$18+$C$19*B44+$C$20*D44),0)	=E44*F44*G44
45		7	2008	=COUNTIF($C$28:$C$34,"<="&C45)	=E44*(1+IF(C45<$C$11,$C$9,$C$10))	=F44*(1+$C$15)	=MAX(MIN($C$17,$C$18+$C$19*B45+$C$20*D45),0)	=E45*F45*G45
46		8	2009	=COUNTIF($C$28:$C$34,"<="&C46)	=E45*(1+IF(C46<$C$11,$C$9,$C$10))	=F45*(1+$C$15)	=MAX(MIN($C$17,$C$18+$C$19*B46+$C$20*D46),0)	=E46*F46*G46
47		9	2010	=COUNTIF($C$28:$C$34,"<="&C47)	=E46*(1+IF(C47<$C$11,$C$9,$C$10))	=F46*(1+$C$15)	=MAX(MIN($C$17,$C$18+$C$19*B47+$C$20*D47),0)	=E47*F47*G47
48		10	2011	=COUNTIF($C$28:$C$34,"<="&C48)	=E47*(1+IF(C48<$C$11,$C$9,$C$10))	=F47*(1+$C$15)	=MAX(MIN($C$17,$C$18+$C$19*B48+$C$20*D48),0)	=E48*F48*G48
49		11	2012	=COUNTIF($C$28:$C$34,"<="&C49)	=E48*(1+IF(C49<$C$11,$C$9,$C$10))	=F48*(1+$C$15)	=MAX(MIN($C$17,$C$18+$C$19*B49+$C$20*D49),0)	=E49*F49*G49
50		12	2013	=COUNTIF($C$28:$C$34,"<="&C50)	=E49*(1+IF(C50<$C$11,$C$9,$C$10))	=F49*(1+$C$15)	=MAX(MIN($C$17,$C$18+$C$19*B50+$C$20*D50),0)	=E50*F50*G50
51		13	2014	=COUNTIF($C$28:$C$34,"<="&C51)	=E50*(1+IF(C51<$C$11,$C$9,$C$10))	=F50*(1+$C$15)	=MAX(MIN($C$17,$C$18+$C$19*B51+$C$20*D51),0)	=E51*F51*G51
52		14	2015	=COUNTIF($C$28:$C$34,"<="&C52)	=E51*(1+IF(C52<$C$11,$C$9,$C$10))	=F51*(1+$C$15)	=MAX(MIN($C$17,$C$18+$C$19*B52+$C$20*D52),0)	=E52*F52*G52
53		15	2016	=COUNTIF($C$28:$C$34,"<="&C53)	=E52*(1+IF(C53<$C$11,$C$9,$C$10))	=F52*(1+$C$15)	=MAX(MIN($C$17,$C$18+$C$19*B53+$C$20*D53),0)	=E53*F53*G53
54								
55								NPV
56								=NPV(C22,H39:H53)
57								

Figure 12.20. M2—relationship view.

12.4.3 Insights from M2

Using base-case parameters, M2 generates an NPV of revenues of $161M, approximately $19M below the M1 result but still well above our lower NPV target of $129M. However, we see that annual revenues drop below our lower target of $17M by the ninth year. To understand this result, we examine the two changes in M2: Total Demand and Price. Figure 12.21 shows the total demand curves for M1 and M2. M2 demand quickly rises above M1 demand due to the faster growth rate through 2005, but by 2010, M1 demand has surpassed M2 demand.

Figure 12.22 compares the price curves for M1 and M2. M2 has a higher price for a couple of years and then drops below the M1 price curve. However,

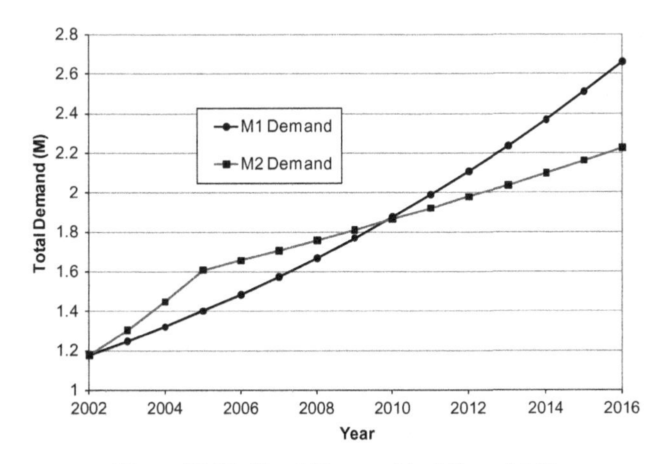

Figure 12.21. Total Demand in M1 and M2.

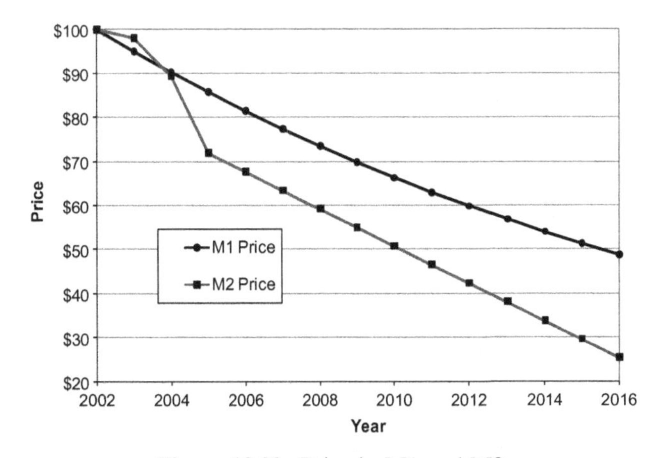

Figure 12.22. Price in M1 and M2.

the slopes of the lines are similar for most of the time period. M2 is significantly below M1 because of the rapid increase in competitors in 2004 and 2005.

The overall effect of these differences is captured in the annual revenues shown in Figure 12.23. M2 revenue stays above M1 for the first few years due to the higher demand and price but then drops below M1 for the remaining years. Fortunately, the downward effect of this difference on NPV is attenuated by these later years being discounted more heavily in the NPV calculation.

Insight 7: It is important to be in the market early when the price is high and the discounting effect is small.

This insight suggests Invivo should push to get to market as soon as possible to capture high prices. We revisit this issue when we look at market share and timing.

How sensitive is the NPV to the slope coefficients α and β? Figure 12.24 shows the results of a two-way Data Sensitivity analysis on these two param-

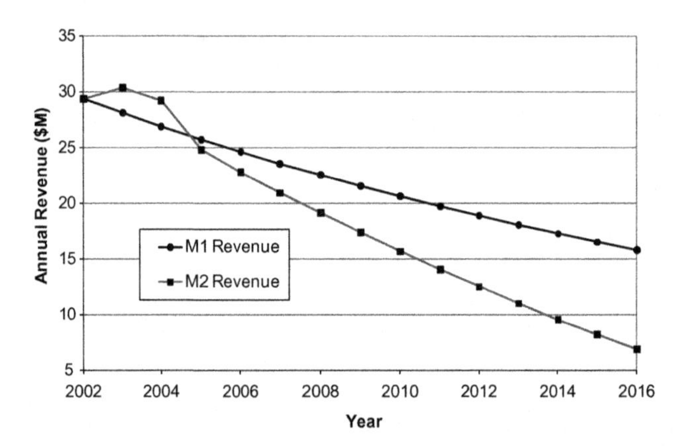

Figure 12.23. Annual Revenue in M1 and M2.

	A	B	C	D	E	F	G	H	I	J	K	L	M
1	NPV: Time Coefficient by Competitors Coefficient												
2													
3	α: Time Coefficient											β: Competitors Coefficient	
4		$ (10)	$ (9)	$ (8)	$ (7)	$ (6)	$ (5)	$ (4)	$ (3)	$ (2)	$ (1)	$ -	
5	$ (10)	$59	$64	$70	$76	$83	$91	$100	$109	$119	$129	$140	
6	$ (9)	$62	$68	$74	$81	$89	$98	$108	$118	$129	$140	$151	
7	$ (8)	$66	$72	$80	$88	$97	$107	$117	$128	$140	$152	$164	
8	$ (7)	$70	$78	$86	$96	$106	$117	$128	$140	$153	$165	$177	
9	$ (6)	$76	$85	$95	$105	$117	$129	$141	$154	$167	$179	$191	
10	$ (5)	$83	$94	$105	$117	$130	$143	$156	$168	$181	$193	$204	
11	$ (4)	$93	$105	$118	$131	$144	$157	$170	$182	$195	$206	$216	
12	$ (3)	$106	$119	$132	$145	$158	$171	$184	$196	$208	$219	$227	
13	$ (2)	$121	$134	$147	$160	$173	$185	$198	$210	$221	$230	$235	
14	$ (1)	$135	$148	$161	$174	$187	$199	$211	$223	$233	$238	$238	
15	$ -	$149	$162	$175	$188	$201	$213	$225	$236	$238	$238	$238	
16													

Figure 12.24. Two-way sensitivity on price equation coefficients (M2).

eters. Although the M2 base-case result of $161M is well above the lower target, the sensitivity table shows that if either coefficient approaches –$6, Invivo misses the target NPV.

Insight 8: A small reduction of the price coefficients below the base-case values results in an NPV below the target NPV.

The tornado chart shown in Figure 12.25 shows the importance of the parameters in the price equation. All three of the price equation parameters (P_0, α, β) are among the top five most influential parameters. Although it is clear that P_0 has the strongest effect on NPV, P_0 is not a truly independent variable. Its value is effectively determined by α and β. Given α and β, P_0 must be set so that the price function returns a price of $100 for the first two years. Thus, since the value of P_0 is dependent on the values of α and β, the best interpretation of the tornado chart is that NPV is highly sensitive to the estimates of α and β.

Perhaps the most important insight to gain from Figure 12.25 is that Starting Share is still an important factor. This reminds us that we have yet to tackle the question of forecasting market share.

Insight 9: The price coefficients have a strong impact on NPV, but starting share is still important.

We summarize our analysis to this point as follows:

- More realistic models for demand and price give results that are still above the NPV target.

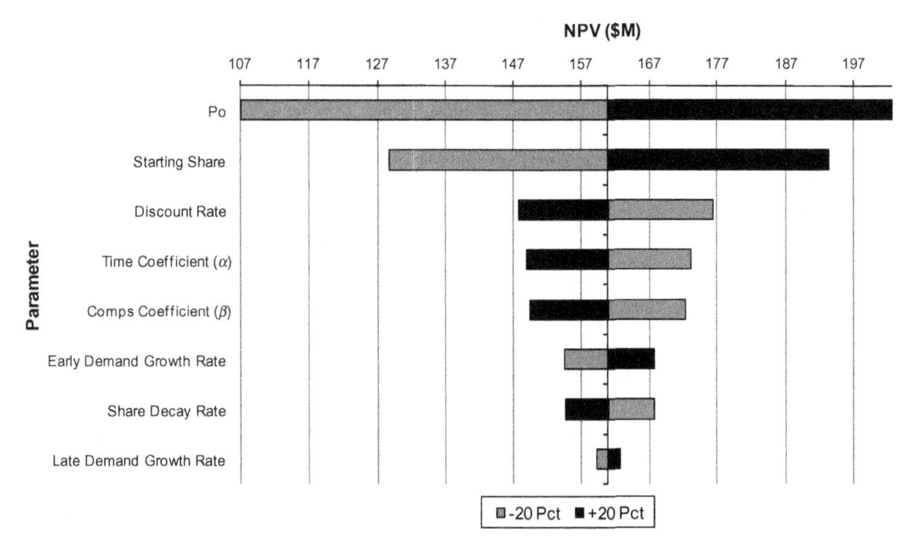

Figure 12.25. Tornado chart (M2).

- A small reduction in the price parameters causes Invivo to miss the target.
- The shapes of the demand, price, and revenue curves in M2 are not dramatically different from M1.

To the Reader: Before reading on, modify and improve M2 in whatever ways seem beneficial to you. Pay particular attention to improving the Market Share module.

12.5 M3 MODEL AND ANALYSIS

We now turn to the Market Share module. In M1 and M2, we calculated share based on a starting share value and a constant decay rate. This method ignores the effects of the number of competitors, their order of entry, and the products' strength. We now set out to refine this module to incorporate these effects.

12.5.1 Using the Data to Improve the Market Share Module

Invivo believes market share is linked to both the number of competitors in a category and the relative strength of the products. As with the Price module, there are any number of market-share models that satisfy these beliefs. Fortunately, we have again been given some data to guide our work.

We start by scanning the data on product families D–F (Figure 12.3) and by looking for patterns that hold true in all or most cases. We observe the following:

- The first product to market has 100 percent share until another product enters. There is no case in which the first two products entered in the same year.
- Products that enter after the category has been established get a very small starting share.
- It is hard to tell whether strength has any bearing on starting share.
- In general, the later a product enters, the lower its starting share.
- Products with high strength seem to have higher shares in year 10.
- When a new product enters the market, it must take share from other products. Thus, the entry of a product must affect the share of at least one incumbent.
- When a new product enters the market, it does not take an equal amount of share from each of the incumbents. For example, if Product Z's starting share is 20 percent, it might take 5 percent from Product X and 15 percent from Product Y.

We also graph the market share data to see whether a visual display suggests other hypotheses (Figures 12.26a–c). We observe more patterns from examining the data in this way:

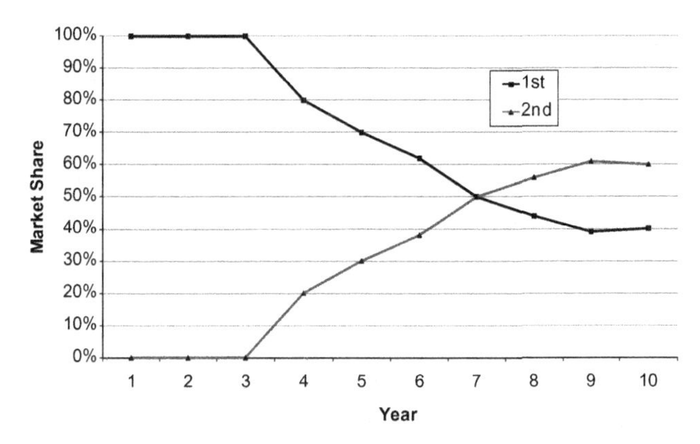

Figure 12.26. (a) Market share for product family D (data from Figure 12.3).

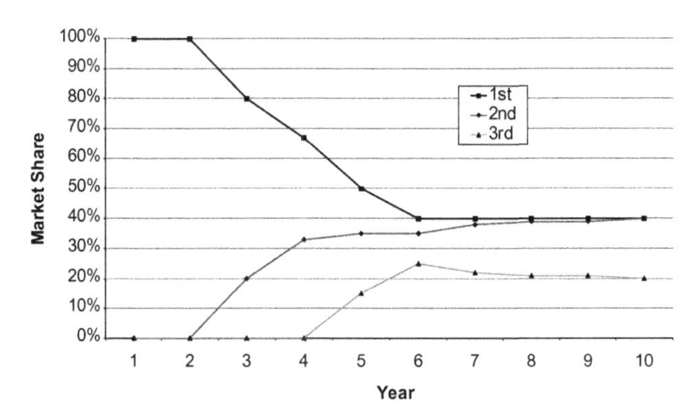

Figure 12.26. (b) Market share for product family E (data from Figure 12.3).

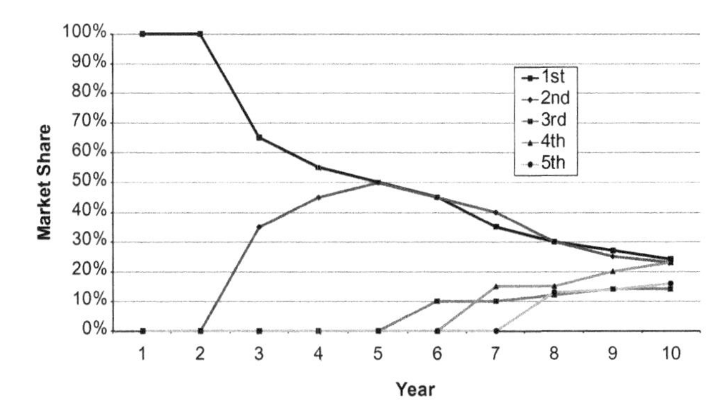

Figure 12.26. (c) Market share for product family F (data from Figure 12.3).

- Market share shifts gradually from year to year.
- The more competitors, the lower the average final share.
- The year-to-year changes are smaller as time passes. Each product's share seems to move toward a long-run-equilibrium value. This value is not necessarily the same for all products.
- The market share for a given product can go up or down from one year to the next.

The data in table and graph form tell us a lot about the behavior of market share and help us develop a sense of market-share changes over time. Note that we have not yet made a single calculation or run a regression!

We now develop a set of hypotheses based on our observations and common sense. Remember that a hypothesis does not have to be satisfied by all the data. Hypotheses are principles we believe to be generally true, but we recognize that randomness and other factors not considered here could cause a hypothesis to be violated in some circumstances.

Hypothesis 1: Starting share is higher for products with higher strength.
Hypothesis 2: Starting share is lower for each successive entrant.
Hypothesis 3: Shares shift gradually.
Hypothesis 4: All incumbent shares are affected by new entrants.
Hypothesis 5: Shares move toward a long-run equilibrium.
  a. Long-run equilibrium increases with strength.
  b. Long-run equilibrium decreases with the number of competitors.

Our hypotheses suggest we have two tasks in building a model. We need to determine the starting share for new products and then determine the year-to-year change for incumbent products. H1 and H2 deal with the first issue, whereas H3–H5 deal with the second issue.

12.5.1.1 Determining Starting Share We came up with H1 and H2 by using common sense and by examining the data on market share. In Figures 12.27a and b, we graph starting share as a function of strength and entry number. Graphing the data certainly does not add much support to either hypothesis. With so little data and results so widely scattered, it looks as if both hypotheses are questionable. We draw a linear best-fit line through the two graphs and see that the lines do slope in the direction we predicted (i.e., higher strength leads to higher share, and more entrants leads to lower share). However, the explanatory power of the regressions is very low, and the regression statistics suggest that this model is weak. A multiple linear regression of starting share on both entry number and strength again shows coefficients with the predicted signs, but the p-values (0.29 and 0.79, respectively) do not fall below the customary 5 percent cutoff (Figure 12.28).

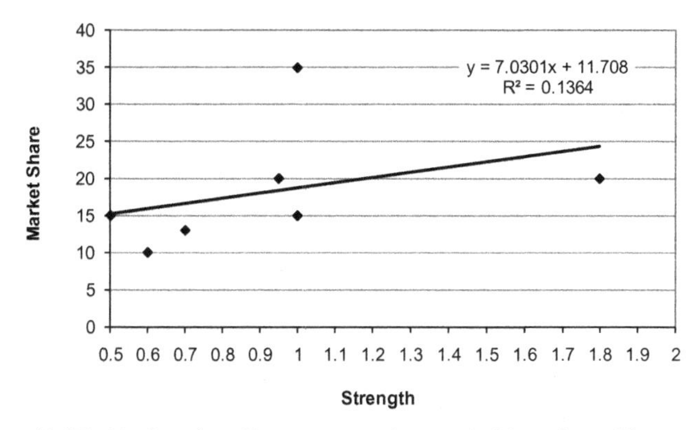

Figure 12.27. (a) Starting Share versus Strength (data from Figure 12.3).

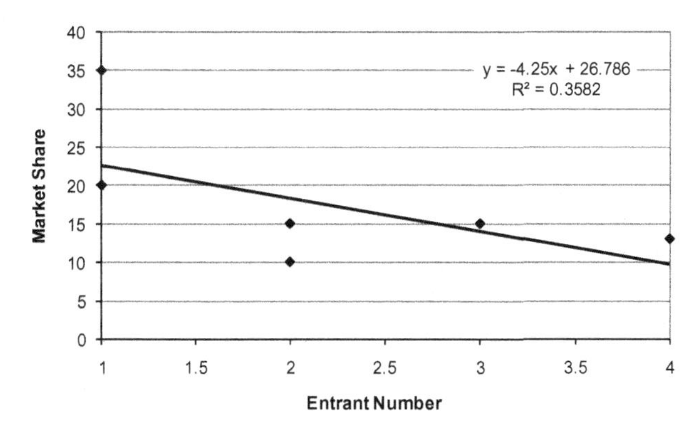

Figure 12.27. (b) Starting Share versus Entrant Number (data from Figure 12.3).

	A	B	C	D	E	F	G
1		Summary Output					
2							
3	*Regression Statistics*						
4	R Square	0.37					
5	Adjusted R Square	0.06					
6	Standard Error	7.97					
7	Observations	7					
8							
9	Anova						
10		*df*	*SS*	*MS*	*F*	*Significance F*	
11	Regression	2	149.44	74.72	1.18	0.40	
12	Residual	4	253.99	63.50			
13	Total	6	403.43				
14							
15		*Coefficients*	*Standard Error*	*t Stat*	*P-value*		
16	Intercept	23.78	12.53	1.90	0.13		
17	Entrant #	-3.85	3.16	-1.22	0.29		
18	Strength	2.36	8.47	0.28	0.79		
19							

Figure 12.28. Multiple regression of Starting Share on Strength and Entrant Number.

With little supporting evidence for H1 and H2, some modelers might be tempted to throw them out. But that seems a bit extreme since both H1 and H2 conform to common sense, and we do not have enough data to override their common-sense appeal. For now, we choose to set aside H1 and hang on to H2.

We jettison H1 because there is no hint of it in the data. Specifically, in Family D, the second entrant has a strength of 1.8 and gets 20 percent share on entry; however, in Family E, the second entrant has a much lower strength (0.95) and also gets 20 percent, whereas in Family F, the second entrant has a lower strength of 1.0 and gets a much higher share (35 percent). The main reason we include H2 in our model now is that the data show H2 is true in every case except one (Family F, Product 4).

We need to create a model that reflects our decision to set aside H1 but include H2. With so little data to work with, we are quite free in deciding how to model the starting share. Also, we recognize that whether a late entrant starts with 15 percent or 10 percent share will have very little effect on Invivo's overall NPV. Thus, we propose the following model for the starting share of each succeeding entrant:

```
IF(Entrant is first)
        100 percent,
ELSE IF (Entrant is second)
        Second Entrant Share,
ELSE
        (1-Share Fraction Decay Rate)×Previous Entrant Share.
```

There are two parameters in this model design: Second Entrant Share and Share Fraction Decay Rate.

In this model formulation, the first player to enter the market gets 100 percent. The second player to enter the market gets a starting share that we set with the Second Entrant Share parameter, say, 20 percent. Each successive entrant gets a smaller fraction of the Second Entrant Share parameter. If, for example, the Share Fraction Decay Rate is 25 percent, the third entrant starts with 15 percent ((1–25%) × 20%), the fourth starts with 11.25 percent ((1–25%) × 15%), and so on. If multiple players enter the market in the same year, the entrants evenly split the total share available to all new entrants that year. In summary, there are four required values to determine a new entrant's share: year of entry, order of entry, Second Entrant Share, and Share Fraction Decay Rate.

12.5.1.2 Determining Year-to-Year Changes The previous section describes our model for forecasting a new entrant's starting share. The next challenge is to figure out how to forecast the year-to-year change in share. We refer back to H4 and H5 and see that there seem to be two forces acting to change share from year to year. H4 states that entry of new products affects the incumbents' share. H5 states that share moves toward a long-run-equilibrium value. Add H3 to the mix, that share shifts gradually, and then we conclude that products do not instantly shift to their equilibrium values. Thus, the year-to-year shift should be the sum of two effects: an instantaneous share loss due to new entrants and a gradual shift toward a long-run-equilibrium value.

12.5.1.3 Share Lost to New Entrants In section 12.5.1.1, we created a model for determining a new entrant's share. We now need to know which incumbent products that share is taken from. If there is only one incumbent, the answer is easy: The incumbent's loss is equal to the new entrant's starting share. For example, assume the first competitor has 100 percent share. If the second product's starting share is 20 percent, the incumbent's share drops to 80 percent.

The situation is more complicated when there are multiple incumbents. It is hard to infer a pattern from the historical data because we cannot easily separate the effect of new entrants from the effect of moving toward long-run equilibrium. One possibility is that all incumbents share equally the loss to the new entrant. In this case, a new entrant starting with 20 percent would take 10 percent from each of two incumbents. We could also create a model that distributes the loss based on the strength of the incumbent products. The difficulty with both ideas is that products with small market shares potentially could be wiped out. Continuing the example from above, if a new entrant is taking 10 percent share from each incumbent, an incumbent that only had 10 percent would be taken down to 0 percent share and, thus, knocked out of the market. Since we do not observe this pattern in the data, we choose to build our model in such a manner that a product cannot be eliminated from the market.

A third alternative is to split the loss based on current shares. For example, assume two products—Product X and Product Y—are in the market with 80 percent and 20 percent shares, respectively. A third product—Product Z—enters with a starting share of 10 percent. Product X, with 80 percent, provides 80 percent of Product Z's share, or 8 percent (80 percent of 10 percent). Product Y, with 20 percent share, provides 20 percent, or 2 percent (20 percent of 10 percent). Therefore, after entry, Product X will be at 72 percent, Product Y at 18 percent, and Product Z at 10 percent. This model seems plausible and meets the need of not allowing any competitor to be taken out of the market, which is enough justification for including it in M3.

12.5.1.4 Shift Toward Long-Run Equilibrium We now tackle the question of determining long-run-equilibrium (LRE) shares and the shift toward the LRE. H5 states that LRE increases with strength and decreases with the number of competitors. Remember, the hypothesis was created based on examining the data. A second look now reveals additional patterns about strength and final share. In Family D, we see the second entrant's strength is almost double that of the first entrant. Therefore, we might expect to see the second product with a final share twice as big as the first product. However, we see that the second product's share is only 50 percent larger than the first product. In Family E, we see that the first two products have almost identical strengths and end with identical shares. We also see that the third product has a strength half that of the first two and ends with half the share. That seems different from what we saw with Family D. The story with Family F gets pretty complicated, but we do note that the three products with strengths of 1.0 end with almost identical shares, and the two products with strengths of less than 1.0 end with less share than the stronger products.

One plausible possibility is that a product's long-run-equilibrium share is determined by that product's strength relative to the strength of all the products in the market. Although this hypothesis is consistent with much of what we observe in the data, it is too vague to use to calculate long-run shares. We make the hypothesis more precise by specifying that the long-run-equilibrium share is determined by the *ratio* of a product's strength to the market's total strength. This formulation leaves open the question of how we might measure the "market's total strength." There are many possibilities. Perhaps the total strength is given by the strength of the best product. Alternatively, it could be given by the average strength of all products. A simple alternative, and the one we adopt, is to measure the market's total strength by the *sum* of the strengths of all competing products.

Let's stop at this point and examine the process we have used to formulate this hypothesis. We began with a weak initial idea, which we call Hypothesis 6:

> **Hypothesis 6:** The long-run-equilibrium share of a product is determined by its own strength relative to the strength of all products in the market.

Then, we refined the hypothesis in stages, trying to make it more concrete and operational. The first step was to specify what the words "relative to" mean.

> **Hypothesis 6′:** The long-run-equilibrium share of a product is determined by the *ratio* of its own strength to the market's total strength.

Then, we specified what we meant by "the market's total strength."

> **Hypothesis 6″:** The long-run-equilibrium share of a product is determined by the ratio of its own strength to the *sum* of the strengths of all products in the market.

Here is how this hypothesis works: If there are two products in the market, one with a strength of 1.0 and the other with a strength of 2.0, the total strength of the market is 3.0. The first product contributes one third of that strength

	Extrant	Strength	Actual LRE Share	Projected LRE Share	Projected-Actual LRE Share
Family D	1	1.00	40	36	–4
	2	1.80	60	64	4
Family E	1	1.00	40	41	1
	2	0.95	40	39	–1
	3	0.50	20	20	0
Family F	1	1.00	24	23	–1
	2	1.00	23	23	0
	3	0.60	14	24	0
	4	1.00	23	23	0
	5	0.70	16	16	0

Figure 12.29. Long-run equilibrium share data versus forecasts from model.

and should have an equilibrium share of approximately 33 percent, whereas the second product has the remaining 67 percent. More formally, with n products, long-run share is calculated as follows:

$$\text{LRE Share}_i = \frac{\text{Strength}_i}{\sum_{i=1}^{n} \text{Strength}_i}$$

In Figure 12.29, we use this relationship to project LRE shares at year 10 for all ten products represented in the data. In most cases, the model comes close to predicting the actual shares. The maximum error (entrant 1 in Family D) is four percentage points or about 10 percent of the actual share.

Now that we know how to calculate the LRE, we need to know how market shares evolve toward it. H3 suggests that shares change gradually. We see from the data in Figures 12.26a–c that changes in share tend to be large at first and then smaller as the LRE is approached. This is quite similar to the constant-growth/decay-rate models we used in M1. However, instead of moving toward infinity or zero, we now want share to move toward a specific goal: the LRE. A model in which shares move a set percentage toward the LRE each period accomplishes what we want. We refer to this type of relationship as a *gap-closing model*. Of course, the LRE also changes every time a new product enters the market, so we expect complex dynamics as a result.

12.5.1.5 Putting It Together Figure 12.30 shows an influence diagram for the Market Share module. Market Share in any given year is determined by Market Share in the previous year, Loss to New Entrants, and Gap Close. Loss to New Entrants depends on New Entrant Share and the Fraction of New

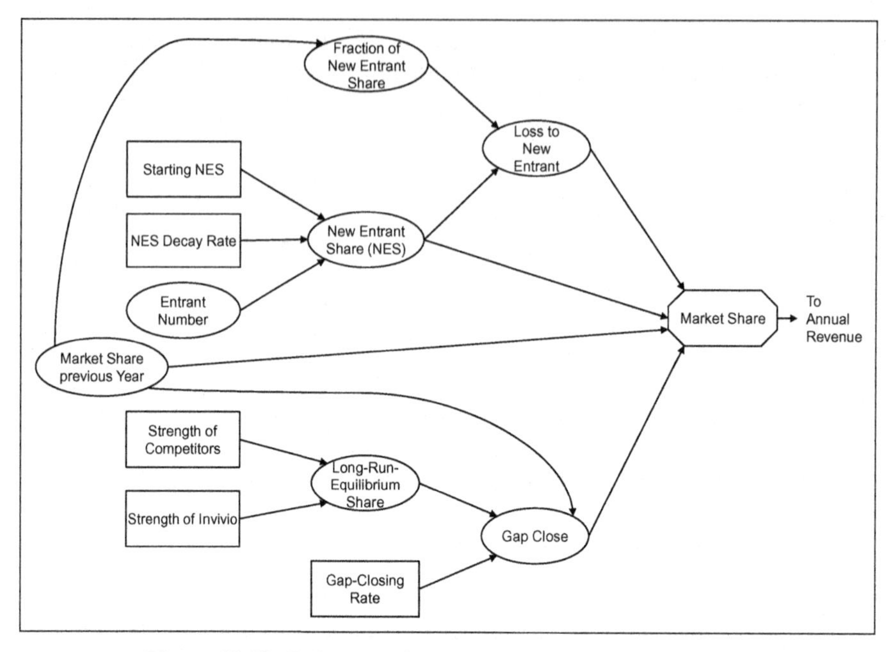

Figure 12.30. Influence diagram for Market Share module.

Entrant Share. Gap Close depends on LRE Share and the Gap-Closing Rate. The three new parameters in this module are Starting New Entrant Share (Starting NES), New Entrant Share Decay Rate (NES Decay Rate), and the Gap-Closing Rate. All the other inputs come from the competitive analysis of when others will enter or from intermediate calculations within the model itself.

As we consider how to build our Market-Share module in the spreadsheet, we must keep in mind that the model needs to be robust enough to apply to several different drugs, not just to this one particular product. By "robust," we mean that the model needs to be able to function correctly even if the parameters are changed or the product entry dates and strengths change from the deterministic base-case values. For example, the model needs to work regardless of whether Product B enters in 2003 or 2008. The other compelling reason to make the model robust is that, at some point, we will want to simulate the uncertainty around product entry dates and product strength, and this will cause many parameters to change. There are, however, practical limits to how robust we can make the model. For example, the case at hand has six potential competitors in addition to Invivo. We build the model to allow for no more than six competitors.

The actual implementation of this model is quite complex. We describe its components in steps as follows:

1. Assign a rank (R) to each product based on year of entry. Products that enter in the same year are ranked in the order they appear on the spreadsheet.
2. Calculate the starting share (SS) for each product as it enters.
   a. All products that enter in the first year share the entire market equally (i.e., SS = 100/Number of Entrants).
   b. For all products that enter after the first year, calculate a theoretical starting share (SS') and an actual starting share (SS). We need both variables because it is possible that multiple products enter in the same year. Rather than try to predict which product actually entered first, we assume all products that enter in a given year evenly split the share that is forecast to be taken by new entrants that year. Thus, we need to forecast the share each new entrant would receive if it were the only product entering that year (SS'). Then, we calculate an even split of that share among all new entrants that year (SS).

      We have already discussed the conceptual model for the share of the second and later entrants that is based on two parameters: Starting New Entrant Share and New Entrant Share Decay Rate. In essence, this model requires that the second entrant receive a fixed share (Starting New Entrant Share) and that later entrants receive a share equal to (1– New Entrant Share Decay Rate) times the previous entrant's share. We write this relationship as follows:

$$SS' = NES \times (1-d)^{(R-2)}$$

      where

      NES = New Entrant Share
      d = New Entrant Share Decay Rate
      R = Rank

      This function assigns the second entrant (R = 2) the NES value as its starting share. Each successive entrant has a starting share of (1 – d) times the previous entrant's share. For example, if d equals 20 percent, the third entrant will start with 80 percent of what the second entrant started with. Finally, the actual share (SS) is determined by adding all the theoretical starting shares (SS') for entrants and dividing by the number of entrants in that year.
3. Calculate the end of year share (S_t)
   a. If t = entry year for a given product, then $S_t = SS$.
   b. For years after the entry year,

$$S_t = S_{t-1} - \text{Loss to New Entrants} + \text{Gap Close to LRE}.$$

In other words, the year-to-year change for an incumbent product is affected by both new entrants and movement toward the long-run equilibrium.

c. Loss to New Entrant = $S_{t-1} \times$ Total SS for year t.

This shows that the loss to new entrants is proportional to the share in the previous year.

d. Gap Close to LRE = Gap Closing Rate $\times$ $(LRE_{t-1} - S_{t-1})$.

This formula calculates the gap between the product's share last year and the long-run equilibrium last year and then calculates how much to move toward the long-run equilibrium. We use values from year $t - 1$ to calculate the gap close to ensure market shares do not exceed 100 percent. Also, calculating the gap close on year $t - 1$ values means that the gap close in year t is not affected by the new entrants that year, which seems like a plausible assumption.

e. Long-Run Equilibrium is equal to the Strength of the product, divided by the sum of the strengths of all the products currently in the market.

12.5.2 Changes to the Model

Our market-share module contains three parameters for which we need to estimate values. However, before we do that, we must translate the module described above into a functioning spreadsheet.

12.5.2.1 Modifying the Model In M2, we added a Competitors module to calculate the number of products in the market and thereby to calculate Price. For M3, we expand on that module (rows 29–37) to rank order the competitors and calculate their theoretical starting shares (Figures 12.31 and 12.32). The RANK formula is used to rank the competitors based on year of entry. However, this function assigns the same value to products that enter in the same year. Because the Starting Share calculation relies on each product having a unique sequence number, we use the Order column (column F) to make the additional calculations to give each product its sequence number.

	A	B	C	D	E	F	G	H
28	Calculations							
29		Competitors						
30		Name	Entry Year	Strength	=RANK()	Order	Start Share	
31		Invivo	2004	1	4	4	11%	
32		A	2002	1.0	1	1	100%	
33		B	2003	1.1	2	2	20%	
34		C	2003	1.0	2	3	15%	
35		D	2005	1.2	5	5	8%	
36		E	2005	1.1	5	6	6%	
37		F	2005	2.0	5	7	5%	
38								

Figure 12.31. M3—Competitors module–numerical view.

	A	B	C	D	E	F	G	H
28	**Calculations**							
29		**Competitors**						
30		**Name**	**Entry Year**	**Strength**	**=rank()**	**Order**	**Start Share**	
31		Invivo	2004	1	=RANK(C31,$C$31:$C$37,1)	=E31	=IF(E31=1,1/COUNTIF($C$31:$C$37,C31),$C$21*(1-$C$22)^(F31-2))	
32		A	2002	1	=RANK(C32,$C$31:$C$37,1)	=IF(COUNTIF($E$31:E31,E32),E32+COUNTIF($E$31:E31,E32),E32)	=IF(E32=1,1/COUNTIF($C$31:$C$37,C32),$C$21*(1-$C$22)^(F32-2))	
33		B	2003	1.1	=RANK(C33,$C$31:$C$37,1)	=IF(COUNTIF($E$31:E32,E33),E33+COUNTIF($E$31:E32,E33),E33)	=IF(E33=1,1/COUNTIF($C$31:$C$37,C33),$C$21*(1-$C$22)^(F33-2))	
34		C	2003	1	=RANK(C34,$C$31:$C$37,1)	=IF(COUNTIF($E$31:E33,E34),E34+COUNTIF($E$31:E33,E34),E34)	=IF(E34=1,1/COUNTIF($C$31:$C$37,C34),$C$21*(1-$C$22)^(F34-2))	
35		D	2005	1.2	=RANK(C35,$C$31:$C$37,1)	=IF(COUNTIF($E$31:E34,E35),E35+COUNTIF($E$31:E34,E35),E35)	=IF(E35=1,1/COUNTIF($C$31:$C$37,C35),$C$21*(1-$C$22)^(F35-2))	
36		E	2005	1.1	=RANK(C36,$C$31:$C$37,1)	=IF(COUNTIF($E$31:E35,E36),E36+COUNTIF($E$31:E35,E36),E36)	=IF(E36=1,1/COUNTIF($C$31:$C$37,C36),$C$21*(1-$C$22)^(F36-2))	
37		F	2005	2	=RANK(C37,$C$31:$C$37,1)	=IF(COUNTIF($E$31:E36,E37),E37+COUNTIF($E$31:E36,E37),E37)	=IF(E37=1,1/COUNTIF($C$31:$C$37,C37),$C$21*(1-$C$22)^(F37-2))	
38								

Figure 12.32. M3—Competitors module–relationship view.

This calculation uses the COUNTIF function to check whether a product's rank number appears in any of the products listed above the current product. If so, the rank number is increased by the number of times the rank has appeared previously. Thus, if the rank is three and just one prior product had a rank of three, the product under consideration would be given an Order number of four (3 + 1). If a Rank number has not been used previously, then the Rank and Order numbers are the same.

In column G, we calculate the Starting Shares by first checking whether a product has a rank of one. This indicates that the product enters in the first year of the market. If so, the starting share is equal to 100 percent divided by the total number of products entering in that first year (see 2a from Section 12.5.1.5 above). If the product is not a first-year product, the starting share is calculated based on the Order number in the prior column and on the equation shown in 2b above (Section 12.5.1.5).

The Share Calculations module (rows 39–58) has two sections (Figures 12.33 and 12.34). The first section (columns C–E) calculates the total market share new entrants will claim each year and then divides that evenly among the new entrants. Column C uses the COUNTIF function to count how many new entrants there are each year. Column D uses SUMIF to add the theoretical starting share of all the entrants for the year. The MIN function is also used to prevent the total share taken by new entrants from being greater than 100 percent (this is unlikely to happen, but the use of the MIN function guarantees it will not occur). Column E calculates the share taken by each new entrant by dividing column D by column C.

The second section of the Share module (columns F–AP) calculates each product's actual end-of-year market share. Since we are interested in projecting only Invivo's NPV, we need to project only Invivo's market share. However, projecting the share of each product is useful for both error checking and understanding the interactions of the various competitors. By projecting all market shares, we can verify that total market share across all products equals 100 percent. Another reason to project market share for all products is that it helps us better understand market dynamics.

We use five columns to calculate the share over time for each entrant. For example, Invivo's share is calculated in columns F through J. The first column (start) for each product is the year's starting market share and is normally equal to the prior year's ending share. The only time this changes is for a product's first year. In that case, the starting column is equal to the starting share calculated in column E. The New Entrant column calculates the market share that new entrants take away from an incumbent. This equals the ending share of the prior year, multiplied by the total share taken by new entrants (see 3c above). The Gap Close column calculates the movement toward the prior year's long-run equilibrium. The End column adds the three prior columns to calculate the product's market share at year's end. The LRE column divides the product's strength by the total strength in the market at the time. The SUMIF function is used here to add the strength of all products in the market.

	B	C	D	E	F	G	H	I	J	K	L	M	N	O	P
39	**SHARE CALCS**														
40					**Invivo**					**A**					
41					Entry Year	2004				Entry Year	2002				
42					Strength	1.00				Strength	1.00				
43		Entrants	TotalShareGrab	GrabPer	Start	NewEntrant	GapClose	End	LRE	Start	NewEntrant	GapClose	End	LRE	
44	2002	1	100%	100%	0%	0%	0%	0%		100%	0%	0%	100%	100%	
45	2003	2	35%	18%	0%	0%	0%	0%		100%	-35%	0%	65%	32%	
46	2004	1	11%	11%	11%	0%	0%	11%	24%	65%	-7%	-16%	41%	24%	
47	2005	3	20%	7%	11%	-2%	7%	16%	12%	41%	-8%	-8%	25%	12%	
48	2006	0	0%		16%	0%	-2%	14%	12%	25%	0%	-6%	18%	12%	
49	2007	0	0%		14%	0%	-1%	13%	12%	18%	0%	-3%	15%	12%	
50	2008	0	0%		13%	0%	0%	12%	12%	15%	0%	-2%	14%	12%	
51	2009	0	0%		12%	0%	0%	12%	12%	14%	0%	-1%	13%	12%	
52	2010	0	0%		12%	0%	0%	12%	12%	13%	0%	0%	12%	12%	
53	2011	0	0%		12%	0%	0%	12%	12%	12%	0%	0%	12%	12%	
54	2012	0	0%		12%	0%	0%	12%	12%	12%	0%	0%	12%	12%	
55	2013	0	0%		12%	0%	0%	12%	12%	12%	0%	0%	12%	12%	
56	2014	0	0%		12%	0%	0%	12%	12%	12%	0%	0%	12%	12%	
57	2015	0	0%		12%	0%	0%	12%	12%	12%	0%	0%	12%	12%	
58	2016	0	0%		12%	0%	0%	12%	12%	12%	0%	0%	12%	12%	
59															
60															
61	**Year Num**	**Year**	**Players**	**Demand**	**Share**	**Price**	**Revenue ($MM)**			**M1 Base 25%**	**M2 Base Case**	**M3 Base Case**			
62															
63	1	2002	1	1.18	0%	$ 100	-			$ 29	$ 29	$ -			
64	2	2003	3	1.31	0%	$ 98	-			$ 28	$ 30	$ -			
65	3	2004	4	1.45	11%	$ 89	14.57			$ 27	$ 29	$ 15			
66	4	2005	7	1.61	16%	$ 72	18.06			$ 26	$ 25	$ 18			
67	5	2006	7	1.66	14%	$ 68	15.42			$ 25	$ 23	$ 15			
68	6	2007	7	1.71	13%	$ 63	13.89			$ 24	$ 21	$ 14			
69	7	2008	7	1.76	12%	$ 59	12.86			$ 23	$ 19	$ 13			
70	8	2009	7	1.81	12%	$ 55	12.07			$ 22	$ 17	$ 12			
71	9	2010	7	1.86	12%	$ 51	11.37			$ 21	$ 16	$ 11			
72	10	2011	7	1.92	12%	$ 46	10.68			$ 20	$ 14	$ 11			
73	11	2012	7	1.98	12%	$ 42	9.97			$ 19	$ 13	$ 10			
74	12	2013	7	2.04	12%	$ 38	9.23			$ 18	$ 11	$ 9			
75	13	2014	7	2.10	12%	$ 34	8.45			$ 17	$ 10	$ 8			
76	14	2015	7	2.16	12%	$ 30	7.61			$ 17	$ 8	$ 8			
77	15	2016	7	2.23	12%	$ 25	6.71			$ 16	$ 7	$ 7			
78															
79							**NPV**			NPV	NPV	NPV			
80							$74.36			$ 180	$ 161	$ 74			
81															

Figure 12.33. M3—Share module–numerical view (columns P through AN omitted for clarity).

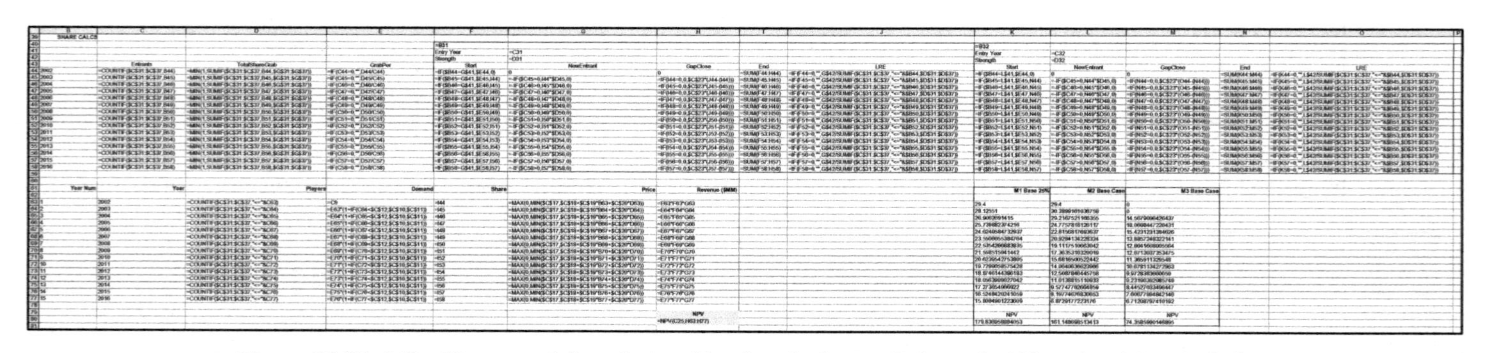

Figure 12.34. M3—Share module–relationship view (columns P through AN omitted for clarity).

This module is complex, and the extensive use of IF functions makes errors likely. It is essential to take time to debug the module thoroughly before moving on. The use of simple input values, such as 1, 10, and 100, makes this easier. Also, it is important to change the input values to test both the true and the false conditions in the IF statements.

12.5.2.2 Selecting Parameter Values We now need to select values for the three parameters in our share module: Starting New Entrant Share, New Entrant Share Decay Rate, and Gap-Closing Rate. As we did with the pricing module, we start by graphing model results against actual data and by varying the parameters in the model to try to match its output to the data. Figure 12.35 shows typical results.

We also use Solver to help find reasonable parameter values. To do this, we calculate the sum of squared deviations between the market share data and the model and use Solver to minimize the sum of deviations by changing the three parameters. Figure 12.36 shows the resulting parameter values when we use Solver to optimize parameters for each of the three product families independently. The results across product families are not particularly similar. Keep in mind that these three product families all have different histories. It

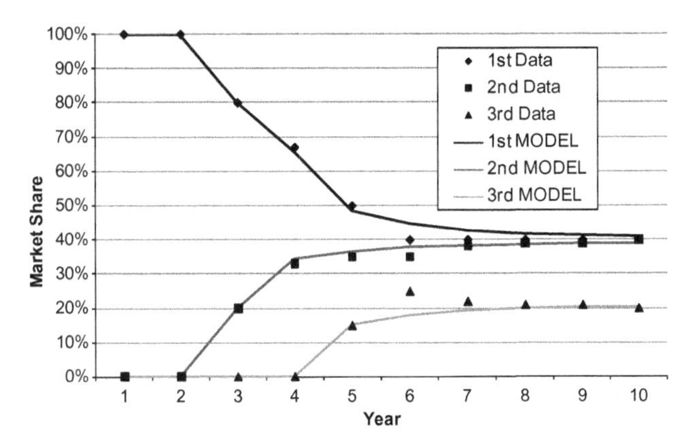

Figure 12.35. Historical share data plotted against market-share module results. (Starting NES = 30%, NES Decay Rate = 10%, Gap-Close Rate = 20%).

	Family D	Family E	Family F
Starting NES	18%	20%	34%
NES Decay	N/A	5%	51%
Gap Close Rate	31%	49%	60%

Figure 12.36. Market Share parameter values as determined by Solver.

is not worth our time at this point to try to pinpoint a best-fit set of values for all three products. We could just average across the three products, but this does not seem to be a better approach than simply choosing plausible parameter values. For now, we just pick mid-range values and make a note to come back to refine them later if we find they are important. Thus, we choose 20 percent Starting NES, 25 percent NES Decay Rate, and 50 percent Gap-Closing Rate. Figure 12.37a–c shows the match of the data to our model forecasts using our chosen values. Although our model does not perfectly predict the results of any of the three historical product families, it certainly provides a reasonable approximation.

Having decided on a set of Market Share parameters, we now put M3 to work for the current Invivo product.

To the Reader Before reading on, use M3 to generate as many interesting insights into the situation as you can.

12.5.3 Insights from M3

The base-case result of M3 is an NPV of $74 M. For the first time in our study, the result is significantly below the target NPV of $129 M. We also see that only in 2005 do we have revenues above the $17 M/year target. These results differ dramatically from what we saw in M1 and M2. Figure 12.38 clearly shows why M3 has such different results: Invivo's market share in M3 is lower than in M2 in every year. Furthermore, Invivo does not even enter the market until 2004. With M2, we generated Insight 7, which said that it was important to be in the market early while prices are high and the discounting effect is small. In M3, Invivo misses those first two years and the NPV dramatically suffers because of it. Insight 10 strengthens Insight 7.

Insight 10: Late entry for Invivo leads to low market share and dramatically lower revenues.

Figure 12.39 shows the sensitivity of NPV to Invivo's entry year. If Invivo enters in 2002, its NPV is above $180M. If its entry is delayed until 2008, its NPV is only $20M. This result strengthens the insight that getting to market early has huge benefits. Figure 12.40 shows a tornado chart for the parameters in M3. (Note that Entry Year is varied up and down by one year, not by 20 percent.) These results suggest that the impact of a one-year change in entry year dwarfs the effects of varying the other parameters by 20 percent. Entry Year and Strength are the two parameters with the strongest effect on NPV. This is good news because Invivo can influence both of these values through its R&D efforts.

Insight 11: Entry year and product strength, both of which can be influenced by R&D, have a strong effect on NPV.

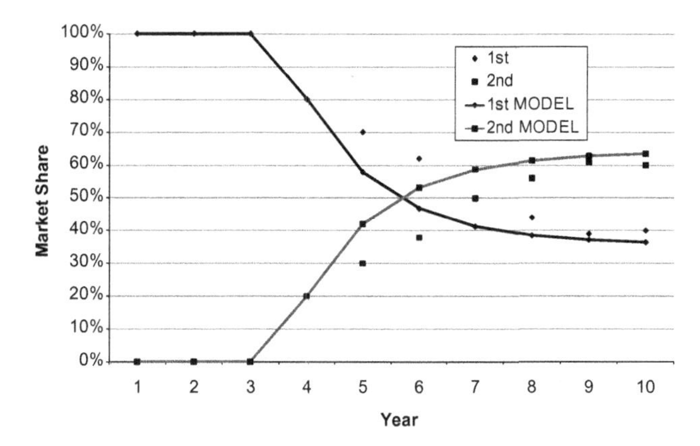

Figure 12.37. (a) Market-share data plotted against model results for product family D (Starting NES = 20%, NES Decay Rate = 25%, Gap-Close Rate = 50%).

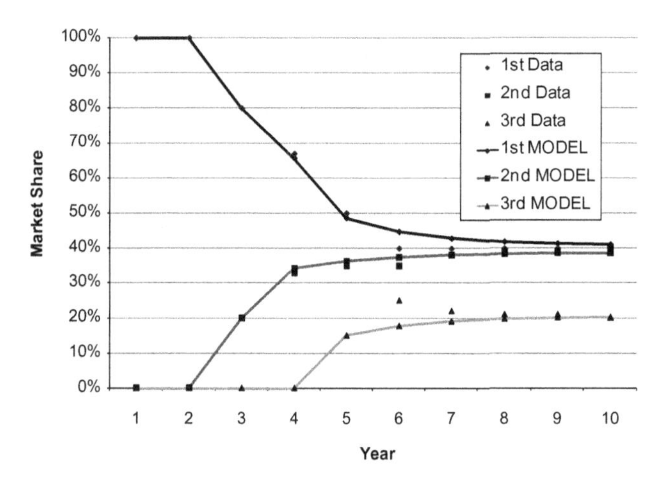

Figure 12.37. (b) Market-share data plotted against model results for product family E (Starting NES = 20%, NES Decay Rate = 25%, Gap-Close Rate = 50%).

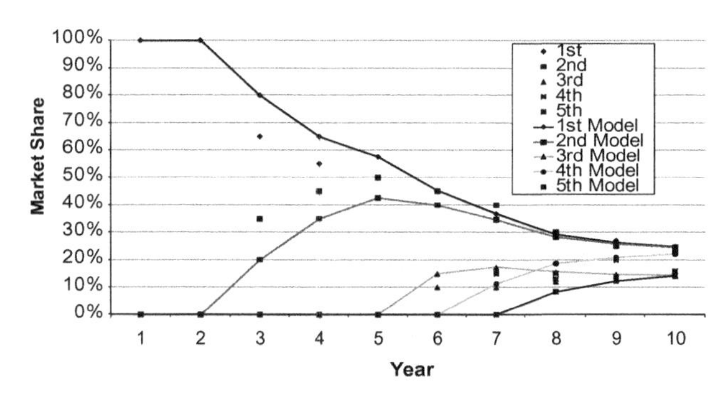

Figure 12.37. (c) Market-share data plotted against model results for product family F (Starting NES = 20%, NES Decay Rate = 25%, Gap-Close Rate = 50%).

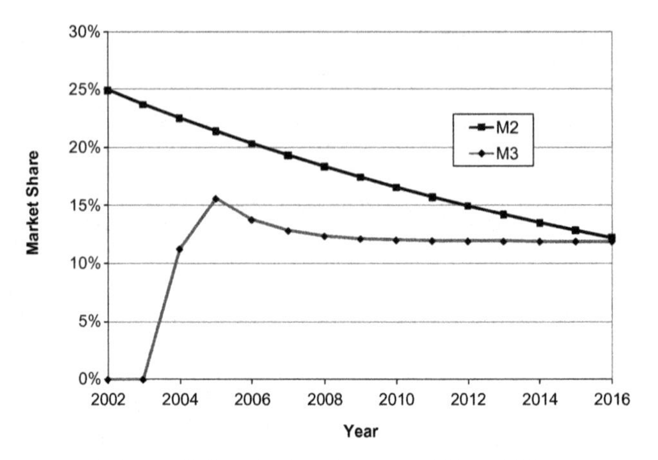

Figure 12.38. Market Share for M2 and M3.

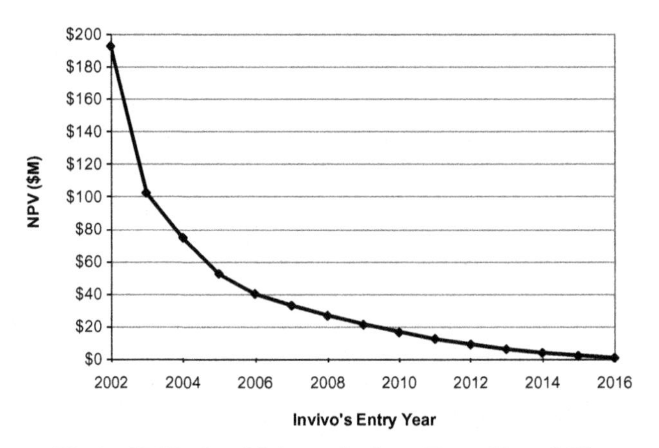

Figure 12.39. Sensitivity to Invivo's Entry Year (M3).

We also see in the tornado chart that the three market-share-module parameters (Starting Share, Share Decay Rate, and Gap-Close Rate) have relatively *little* effect on NPV. This finding suggests we do not need to revisit the values we are using for these parameters. Spending additional time refining the values would not change the outcome appreciably.

A two-way Data Sensitivity table in which we vary Invivo's entry year and product strength allows us to see what conditions are necessary to reach the lower NPV target of $129M (Figure 12.41). Invivo's expected entry year is 2004, and the sensitivity table shows that even with a product strength of 2.0 and entry in 2004, Invivo would fail to meet the NPV target. On the other hand, getting to market in 2002, the year the first competitor is expected to enter, leads to an above-target NPV, even with a strength of only 0.5.

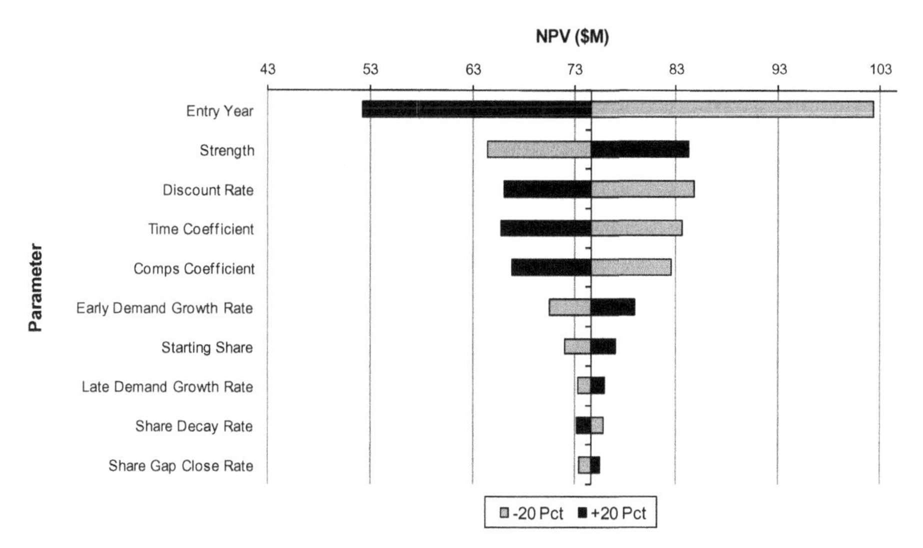

Figure 12.40. Tornado chart (M3). Note: "Entry Year" was varied ± one year. All other variables were varied by ± 20% of the base-case value.

	A	B	C	D	E	F	G	H	I	J	K	L	M	N	O	P	Q	R
1	**M3 NPV: Entry Year by Strength**																	
2																		
3	**Entry Year**									**Strength**								
4		**0.5**	**0.6**	**0.7**	**0.8**	**0.9**	**1**	**1.1**	**1.2**	**1.3**	**1.4**	**1.5**	**1.6**	**1.7**	**1.8**	**1.9**	**2**	
5	**2002**	$144	$155	$165	$175	$184	$192	$200	$208	$216	$223	$230	$236	$242	$248	$254	$260	
6	**2003**	$68	$75	$82	$89	$96	$102	$108	$114	$120	$125	$131	$136	$141	$146	$151	$155	
7	**2004**	$48	$53	$59	$64	$69	$74	$79	$84	$89	$93	$97	$102	$106	$110	$114	$117	
8	**2005**	$33	$37	$41	$45	$49	$52	$56	$59	$63	$66	$69	$73	$76	$79	$82	$85	
9	**2006**	$24	$27	$31	$34	$37	$40	$43	$46	$49	$51	$54	$57	$59	$62	$64	$67	
10	**2007**	$20	$23	$25	$28	$30	$33	$35	$38	$40	$42	$44	$46	$48	$51	$53	$55	
11	**2008**	$16	$18	$20	$23	$25	$27	$28	$30	$32	$34	$36	$37	$39	$41	$42	$44	
12	**2009**	$13	$15	$16	$18	$20	$21	$23	$24	$25	$27	$28	$30	$31	$32	$33	$35	
13	**2010**	$10	$12	$13	$14	$15	$16	$17	$19	$20	$21	$22	$23	$24	$25	$26	$27	
14																		

Figure 12.41. Two-way sensitivity: Entry Year and Strength (M3) (shaded region is below NPV target).

Insight 12: Entering the market in the first year, even with low strength, allows Invivo to meet its NPV goal.

This insight presents Invivo's management with an interesting question: Can the product be ready sooner if it has a lower strength? For the situation at hand, it seems that even the most optimistic timeline would only get the product to market in 2003, and nothing can be done to improve that. However, management should keep this strength versus entry year trade-off in mind for future products.

By now, we see a common theme emerging in the insights. Several of the insights point to the idea that getting to market early is important. As far back as Insight 1 in M1, we saw that getting to market first would guarantee Invivo above-target revenues. With Insight 7 from M2, we saw that getting to market early was important because the price is higher and the effects of discounting are reduced. Insight 10 from M3 again points to entry year as a key to improving NPV. This convergence of insights gives us confidence in the results of our analysis.

We summarize our analysis to this point as follows:

- Invivo is likely to enter the market two years after the first product enters the market.
- Entering late results in a below-target NPV because Invivo has no revenue for the first two years and low market share once it does enter.
- If Invivo can improve its entry year and product strength, it can greatly increase its NPV.

Until now, we have dealt with uncertainty by changing parameters one at a time. For example, the observations from the tornado chart and Insight 11 are based on all the other values in the model holding constant. However, we know that there is uncertainty regarding not only Invivo's entry date but also the entry dates and strengths of all the competitors' products, as well as the market price for the product category. Uncertainty potentially could change the results dramatically. For example, if the competitors enter late or fail to enter, Invivo might be the first product to market, in which case it would reap huge benefits. We need to simulate the uncertainty in the problem to account fully for all the combinations of parameter values.

To the Reader: Before reading on, modify and improve M3 to reflect the uncertainty in the approval process and competitor entry.

12.6 M4 MODEL AND ANALYSIS

By adding uncertainty to the model, we learn not only what is most likely to happen but also what extreme outcomes are possible. In this case, we expect the most dramatic change is the possibility that Invivo's product will not pass FDA approval and NPV will be zero. But other possibilities exist, such as the late entry of competitors, which could make Invivo's NPV very high.

12.6.1 Modeling Uncertainty

There are three areas of uncertainty in the Invivo case: price, Invivo's entry date, and the strength and entry dates of the competitors' products.

So far, all our models have been based on a starting price of \$100. However, Invivo expects the market price to be between \$90 and \$115 when the first product is introduced in 2002. We originally picked the \$100 price because it was close to the middle of the expected range and because it made comparison with the historical data easy. We now need to incorporate the uncertainty in the price. Since we have no other information about price besides the expected range, we use a uniform distribution between \$90 and \$115. The mean of this distribution is \$102.50, \$2.50 above the starting price we have been using. We expect this change alone to increase our estimate of Invivo's NPV.

There are two sources of uncertainty in the FDA approval process: how long each stage takes to complete and whether a product successfully passes each stage. Invivo's product is just entering Phase 2 in June of 2000. Based on past experience, the shortest possible time to get through the approval process is three years. The longest time to approval is four years. We could model the time to approval as a uniform distribution between these two extremes. However, this would probably overstate the variability in the outcome because the total time to approval is the *sum* of the duration of each step. We see in the data in Figure 12.1 that there is a range of uncertainty for each of the steps. It seems best, therefore, to assume a uniform distribution for the duration of each step. If we assume the duration of each step is independent, then it is highly unlikely the worst-case or best-case total duration will actually occur. That is, it is unlikely that all four steps would be the best case or worst case for one product. This explains why modeling the total duration as a uniform distribution overstates the variance of the process. We explore the actual distribution of total duration later.

The uncertainty of passing a phase of the process leads to the possibility of having no product and, thus, to an NPV of zero. We use a Bernoulli distribution (the Yes–No distribution in Crystal Ball) to simulate whether the product passes a given phase.

There are three potential sources of uncertainty regarding Invivo's competitors: whether they enter, their date of entry, and their product strength. Also, the degree of uncertainty differs for different competitors. Invivo's competitive intelligence is better for competitors 1 through 3 than for the later competitors. For both the entry date and strength, we use a uniform distribution to model the uncertainty for the same reasons we do with Invivo. We use the Bernoulli distribution to model whether each competitor enters, just as we did for determining whether Invivo passes each phase.

12.6.2 Changes to the Model

Remember that the pricing module is built on a linear relationship with time and competitors, bounded by 0 at the low end and 100 at the high end. Two parameters of this module need to be changed as we allow the maximum price to change: Max Price and P_0 (see cells C86:C89 in M4; Figures 12.42 and 12.43). Price was capped at \$100 in M2 and M3. We now simulate maximum price

	B	C
85	PRICE CALCS	
86	Actual Max Price	$ 112.53
87	Difference from Base	$ 12.53
88	Shifted Base	$ 112.53
89	Shifted y-incpt	$ 132.29
90		

Figure 12.42. Maximum Price module—numerical view (M4).

	B	C
85	PRICE CALCS	
86	Actual Max Price	=CB.Uniform(C17,C18)
87	Difference from Base	=C86-C19
88	Shifted Base	=C19+C87
89	Shifted y-incpt	=C20+C87
90		

Figure 12.43. Maximum Price module—relationship view (M4).

based on a uniform distribution between $90 and $115. We also need to shift P_0, the y-intercept of the linear equation, based on the new maximum price. To accomplish this, we calculate the deviation between the new, random maximum price, and the deterministic maximum price of $100. P_0 is then shifted by this amount. For example, if the new maximum price is $110, the deviation from the deterministic maximum price is $10. P_0 is thus increased by $10. This has the effect of allowing the entire price curve to shift up or down in lockstep with the change in maximum price.

In the earlier models, Invivo's entry year was a parameter (2004 in the base case). We now need a separate module and additional parameters to calculate the entry year based on uncertainty (see cells G7:M25 and rows 92–115 in M4; Figures 12.44 and 12.45). The new parameters include the current date, the high and low duration expectations for each phase, and the outcome of each phase. As mentioned, we use the Bernoulli distribution to simulate the event of passing a phase. We use a uniform distribution to simulate the duration of each process and add the resulting simulated durations to get the total time until approval. The CEILING function is used to round up the sum to the next month. The data regarding the FDA approval process duration is given in months, but we decided back in M1 to build our model based on years. It would take a lot of work at this point to convert the model to months, and it is not clear that that would really be an improvement. Rather, we simulate the durations in months and then convert the result to years.

Working with calendar dates in Excel is tricky because Excel converts dates to serial numbers, where January 1, 1900, is one. (This is the case on Windows computers; on Macintosh computers, the numbering starts at 1904.) Note that the current year and month are shown as separate parameters, which allows

	B	C	D	E	F	G	H
92	**Invivo Entry**						
93		Pass?	Time (Months)				
94	Phase 2	1	7.2				
95	Ohase 3	1	6.4				
96	Subission	1	6.0				
97	Review	1	19.6				
98							
99	Result	1	40.0				
100							
101	Entry Date		10/31/2003				
102	Entry Year		2003				
103							
104	**Competitor Entry**						
105		Pass?	Entry Offset	Entry Date	Entry Year	Strength	
106	A	1	0.00	1/31/2002	2002	1.00	
107	B	1	0.00	1/31/2003	2003	1.10	
108	C	1	0.00	4/30/2003	2003	1.00	
109	D	1	-0.95	7/31/2005	2005	1.06	
110	E	1	1.50	11/30/2005	2005	1.00	
111	F	1	8.83	6/30/2006	2006	1.82	
112							

Figure 12.44. Entry and Strength module—numerical view (M4).

	B	C	D	E	F	G	H
92	**Invivo Entry**						
93		Pass?	Time (Months)				
94	Phase 2	=CB.YesNo(H12)	=IF(I12=J12,I12,CB.Uniform(I12,J12))				
95	Ohase 3	=CB.YesNo(H13)	=IF(I13=J13,I13,CB.Uniform(I13,J13))				
96	Subission	=CB.YesNo(H14)	=IF(I14=J14,I14,CB.Uniform(I14,J14))				
97	Review	=CB.YesNo(H15)	=IF(I15=J15,I15,CB.Uniform(I15,J15))				
98							
99	Result	=PRODUCT(C94:C97)	=CEILING(SUM(D94:D97),1)				
100							
101	Entry Date		=EOMONTH(DATE(H8,H9,1),D99)				
102	Entry Year		=IF(C99,YEAR(D101),9999)				
103							
104	**Competitor Entry**						
105		Pass?	Entry Offset	Entry Date	Entry Year	Strength	
106	A	=CB.YesNo(H20)	=IF(K20=0,K20,CB.Uniform(-K20,K20))	=EOMONTH(DATE(J20,I20,1),D106)	=IF(C106,YEAR(E106),9999)	=IF(L20=M20,L20,CB.Uniform(L20,M20))	
107	B	=CB.YesNo(H21)	=IF(K21=0,K21,CB.Uniform(-K21,K21))	=EOMONTH(DATE(J21,I21,1),D107)	=IF(C107,YEAR(E107),9999)	=IF(L21=M21,L21,CB.Uniform(L21,M21))	
108	C	=CB.YesNo(H22)	=IF(K22=0,K22,CB.Uniform(-K22,K22))	=EOMONTH(DATE(J22,I22,1),D108)	=IF(C108,YEAR(E108),9999)	=IF(L22=M22,L22,CB.Uniform(L22,M22))	
109	D	=CB.YesNo(H23)	=IF(K23=0,K23,CB.Uniform(-K23,K23))	=EOMONTH(DATE(J23,I23,1),D109)	=IF(C109,YEAR(E109),9999)	=IF(L23=M23,L23,CB.Uniform(L23,M23))	
110	E	=CB.YesNo(H24)	=IF(K24=0,K24,CB.Uniform(-K24,K24))	=EOMONTH(DATE(J24,I24,1),D110)	=IF(C110,YEAR(E110),9999)	=IF(L24=M24,L24,CB.Uniform(L24,M24))	
111	F	=CB.YesNo(H25)	=IF(K25=0,K25,CB.Uniform(-K25,K25))	=EOMONTH(DATE(J25,I25,1),D111)	=IF(C111,YEAR(E111),9999)	=IF(L25=M25,L25,CB.Uniform(L25,M25))	
112							

Figure 12.45. Entry and Strength module—relationship view (M4).

for easier sensitivity analysis later on if needed. The EOMONTH function is used to add the total process duration (in months) to the current date parameters. Note that the DATE function is nested in the EOMONTH function to convert the two current date parameters to the Excel serial value. Lastly, the YEAR function is used to identify the year when Invivo completes the approval process. An IF function is used to set the Year of Entry to 9999 when Invivo does not pass all of the process steps. The 9999 value is beyond the scope of the main module calculations, and thus, Invivo is treated as never entering the market.

Including the uncertainty in competitor entry in the model requires the same steps as modeling Invivo's own uncertainty. Instead of entry and strength being deterministic parameters, we now need several parameters to determine whether a product enters the market and, assuming it does, its date of entry and strength. As with Invivo, we use a Bernoulli distribution to simulate the entry outcome, and we use a uniform distribution to simulate the deviation from expected entry date. Again, we work in months first because that is how the data are reported, and we then convert the resulting entry date to a year of entry. We use the same IF statement and 9999 year value to signal a product that did not make it to market. Strength is calculated based on a uniform distribution. The results of these calculations are copied into the Competitors module we first created in M2, which subsequently drives the Price, Share, and Revenue calculations.

If we take a step back and look at the M4 spreadsheet, we see that it is stunningly complex. Imagine how overwhelming it would have been to build this model from scratch, without any of the work we did in M1, M2, and M3. Furthermore, note that even in this complex model, the basic structure and calculations we built for M1 are still the heart of the model (Revenue = Demand × Share × Price). By starting with a simple model, it was easy to add complexity in steps. With each iteration, we added the necessary modules and connected them to existing modules. By building and testing each model carefully as we went along, we eventually built a complex model, one that we have high confidence is functioning as we intend. Furthermore, we have arrived at M4 with a strong understanding of the behavior of the model already in hand. We are much better off for having worked iteratively than we would be if we had jumped straight to M4.

To the Reader: Before reading on, use M4 to generate as many interesting insights into the situation as you can.

12.6.3 Insights from M4

The first question we explore is how NPV changes with the addition of uncertainty to the model. Figure 12.46 shows that the mean NPV is just under $78M. (With 10,000 trials, the mean standard error is less than 1 percent of the mean NPV, which is more than sufficient for our purposes.) The M4 result is slightly

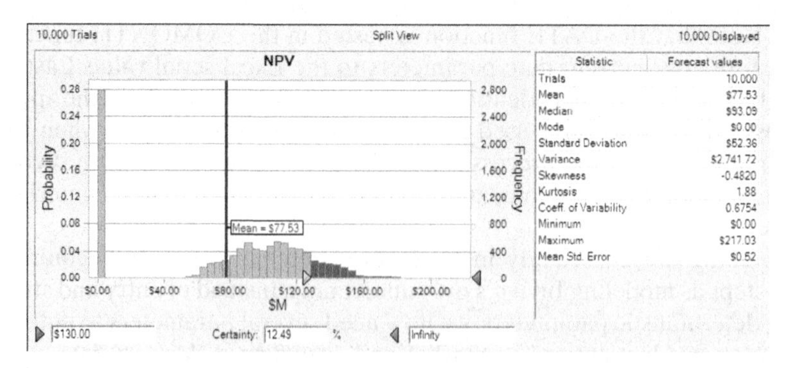

Figure 12.46. NPV distribution (M4).

better than the M3 result of $74M. This is quite surprising given that there is a 28 percent probability that Invivo will not even make it to market. With the zero values filtered out of the NPV distribution (Figure 12.47), we see that the conditional mean is about $108M, which is significantly higher than the M3 result!

We noted above that some of the upward movement in NPV is due to the price distribution having a mean above the $100 we used for Price in earlier models. But that only explains about $4M of the increase in the conditional mean. Part of the increase probably comes from the uncertainty in the entry year. We know from M3 that Invivo's entry year has a large effect on NPV. Figure 12.48 shows the distribution of Invivo entry dates in M4. The span of possible entry dates is 13 months, from June 2003 to June 2004. However, the

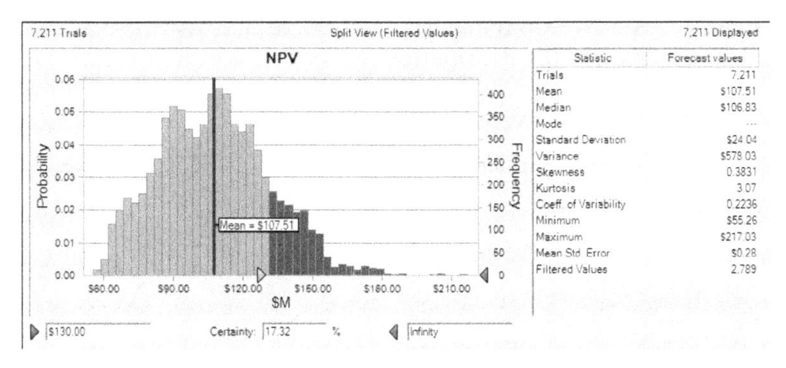

Figure 12.47. NPV distribution with zero values filtered out (M4).

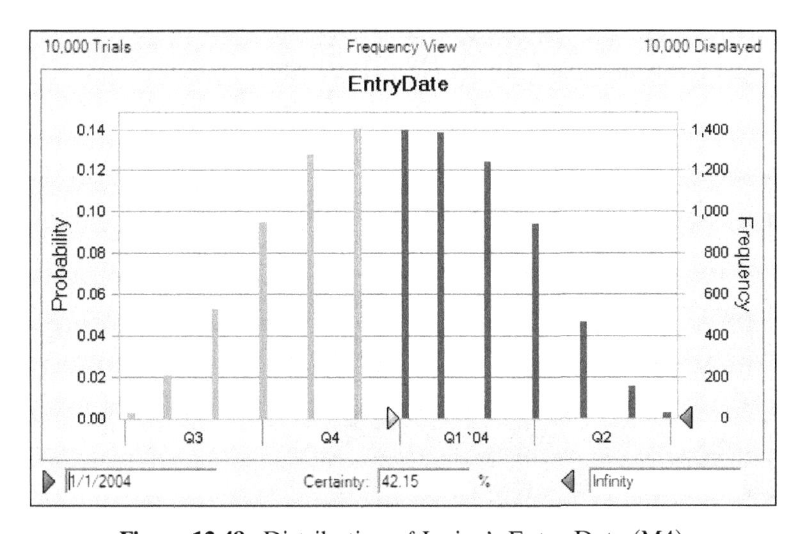

Figure 12.48. Distribution of Invivo's Entry Date (M4).

model only uses the *year* of entry in the calculations; therefore, there are only two potential entry dates in the model: 2003 or 2004, with a 58 percent probability of 2003 entry. Looking back at Figure 12.41, we see that the M3 result improves to $102M if the entry year is 2003. Thus, much of the NPV increase in M4 comes from the possibility of entering early.

Insight 13: The uncertainty in the model allows for much higher NPVs than seen in previous models.

The rest of the increase likely comes from the potential for a larger long-run-equilibrium share. In M3, Invivo's LRE was 12 percent. Figure 12.49 shows the distribution of Invivo's LRE. We see there is a greater than 60 percent probability that Invivo's LRE will be above 12 percent. A higher LRE suggests that Invivo's market share is larger in many years than in the corresponding years in M3. This drives up the mean NPV. (Note that LRE is based on total strength in the market, the distribution of which is shown in Figure 12.50.)

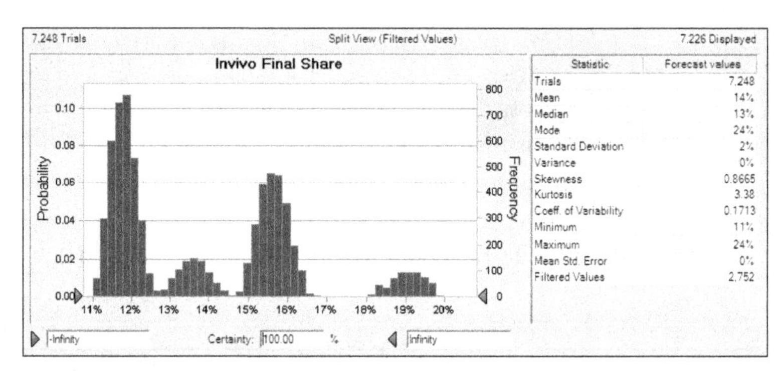

Figure 12.49. Distribution of Invivo's Final Market Share ignoring zero values (M4).

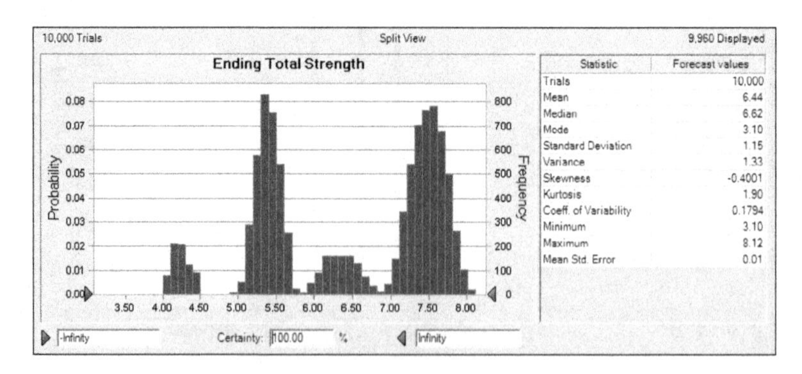

Figure 12.50. Distribution of Ending Total Strength in the market (M4).

Insight 14: Uncertainty allows for earlier entry and higher LRE, both of which drive up the mean NPV.

Since M4's mean NPV is higher than M3's mean NPV, and there is a very strong upside potential, we conclude that the opportunity for early entry and larger LRE is strong enough to outweigh the risk of failing to enter. Simply stated, the future looks considerably brighter for Invivo under M4 than under M3.

Insight 15: The potential for a large NPV more than outweighs the risk of not making it through the approval process.

Looking again at Figure 12.46, we see that Invivo only has a 13 percent chance of earning an NPV greater than $130M, its lower target. That is not as rosy an outlook as we would hope. To improve this probability, Invivo must increase its probability of entering the market, its probability of entering in 2003, its product strength, or some combination of the above.

To get a sense of what Invivo must do to improve its situation, we consider what would happen if Invivo was guaranteed market entry (in other words, passing all testing phases and FDA approval). Figure 12.47 shows that in this guaranteed entry scenario, the company has a 17 percent chance of meeting or exceeding the NPV target. This is only four percentage points better than the risky-entry scenario.

What must Invivo do to achieve a mean NPV of $130M? Figure 12.51 shows the results of a Crystal Ball Sensitivity run that helps answer this question. For both sensitivities, we assume Invivo is guaranteed market entry. The lower line shows the mean NPV with guaranteed entry but random entry year. We see that Invivo would need a product strength of just over 1.3 in order to have a mean NPV of $130M. The upper line shows the mean NPV assuming guaran-

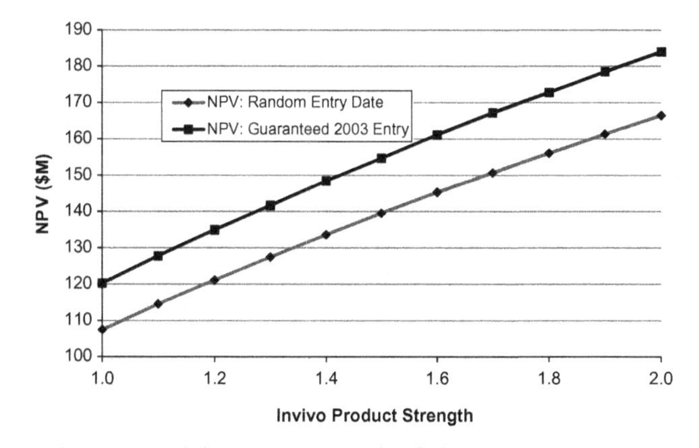

Figure 12.51. Sensitivity for Strength (M4) (100% probability of entry).

teed entry in 2003. Even in this highly optimistic scenario, Invivo needs a product strength of more than 1.1 to achieve a mean NPV of $130M.

Insight 16: Invivo would need significant improvements in its probability of entry, time until entry, and product strength in order to have a mean NPV of $130M.

12.7 PRESENTATION OF RESULTS

As we prepare to present our results, we recognize that we have two main objectives. The first is to convince Invivo that we have built a credible forecasting model as requested. The second is to convey the key insights about Product 1 that we discovered by exploring our forecasting model. These two objectives are tightly linked: The more credible our general approach, the more credible our results on Product 1; the more insightful our results on Product 1, the more credible our overall approach.

Who can we expect to be in our audience? We know that Invivo is run by scientists, many of them presumably doctors, who will be comfortable with an analytic approach to forecasting. We assume they understand the overall financial situation facing their company and the importance of the six products in the pipeline for replacing the revenues soon to be lost. However, we also assume they are less familiar with business forecasting and therefore will be particularly interested in how we think through the issues surrounding future demand, price, and market share. Of course, they will expect us to convince them that our modeling approach is credible before they will accept our forecasts.

Since our study was as much about developing a logical approach to forecasting as it was about developing specific forecasts for Product 1, we choose an indirect approach to this presentation. Rather than focusing on specific results, an indirect approach allows us to build up the logic of our approach first. Once the audience understands the overall logical structure, they will be receptive to considering our specific analyses of Product 1.

The overall outline of our presentation is simple. After an introduction we will develop the logic of our models through a series of influence diagrams. Then we will present our forecasts for Product 1 in enough detail so that our audience understands the future of that product as we see it, but also appreciates the power of our model.

We have 4 models and 16 insights to choose from for our presentation. This is far too much material to include in any one presentation. How can we select the most important results from our analysis? First, we focus on the results of M3 and M4, since M1 and M2 are partially developed versions of M3. M4 adds randomness to M3, so we will use it only when we wish to discuss uncertainty.

Since our audience is interested in our forecasts for Product 1, we plan to include a series of graphs showing how our forecasts of demand, price, and share combine into a forecast of revenue. We begin with forecasts for total demand and market price. Then we show how Product 1's share will evolve over time, to make the point that Invivo's share of this market will rise to a modest level and then taper off. Then we show the shares of all six competitors, to make the point that Product 1 will be third or fourth into a crowded market. Finally, we show a plot of Product 1 annual revenue over time, which rises to about $18M and then declines to about $6M.

To round out our presentation, we offer two additional types of results. These additional results will be of interest for their own sake but will also show how our modeling approach can provide critical insights not available from any other source. First we display a distribution of the NPV of revenues (from M4). This allows us to discuss how we modeled uncertainty using simulation. It also allows us to report the mean NPV and the probability the NPV exceeds the $130M benchmark, both of which we expect to be critical pieces of information for management. Second, we discuss the sensitivity of NPV to the entry date and strength, both of which can be influenced by R&D.

Our first slide (Figure 12.52) introduces the problem and explains our task. The next four slides develop the influence diagram. Figure 12.53 shows how annual revenues are calculated from demand, price, and share. Each of the next three slides (Figures 12.54–56) shows how one of those components is calculated. During the presentation we plan to take quite a bit of time to

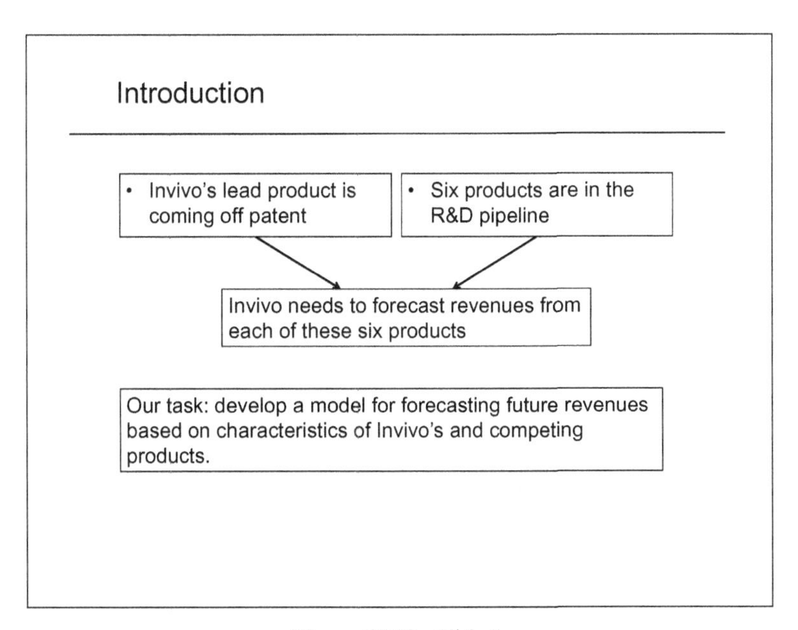

Figure 12.52. Slide 1.

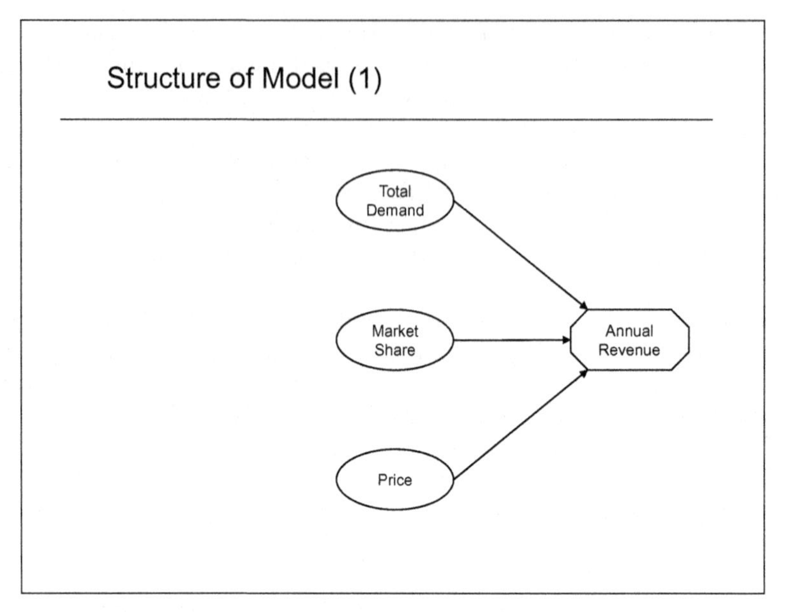

Figure 12.53. Slide 2.

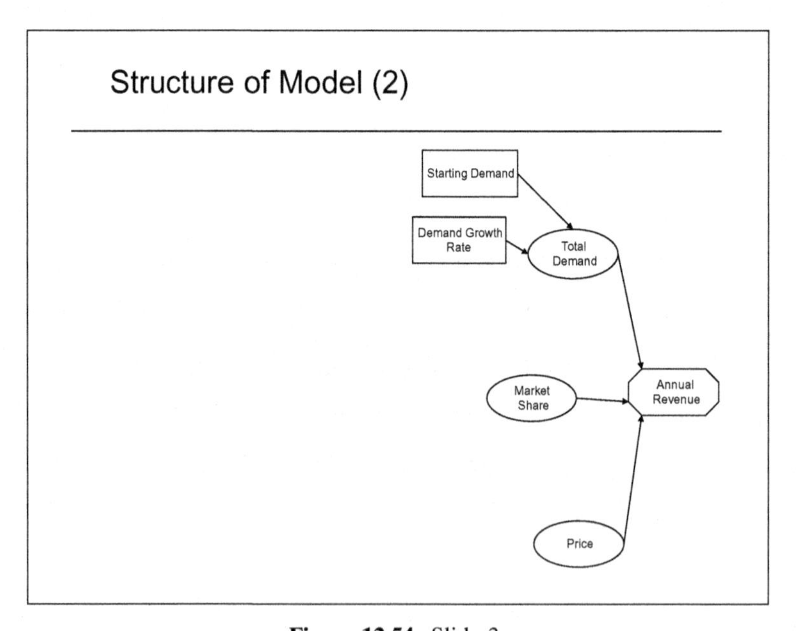

Figure 12.54. Slide 3.

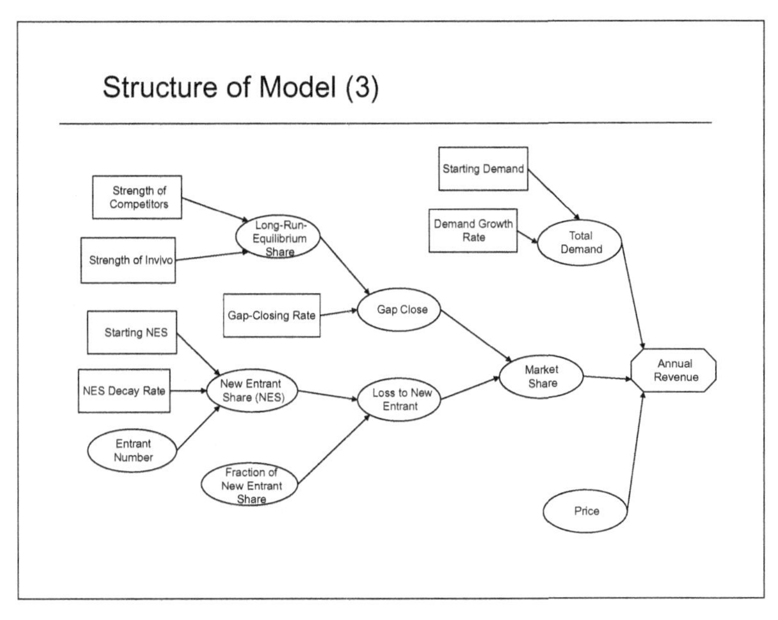

Figure 12.55. Slide 4.

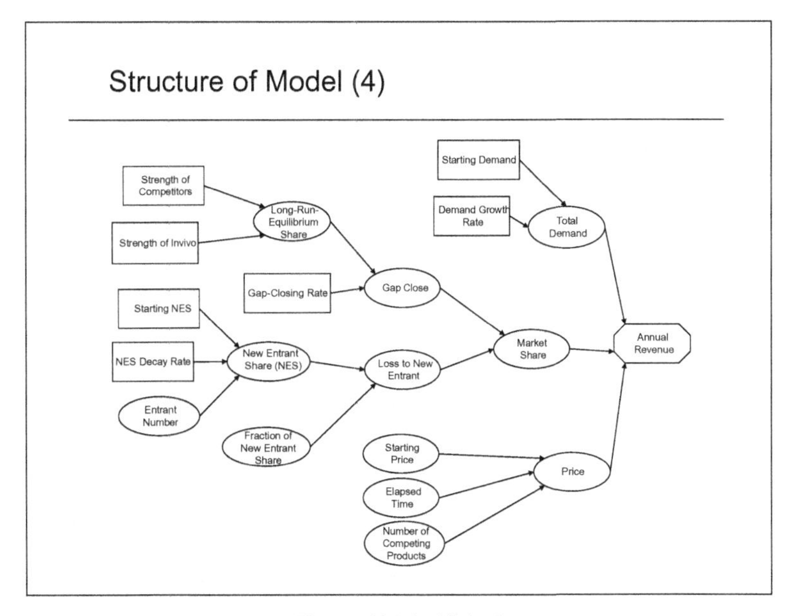

Figure 12.56. Slide 5.

explain each of these slides, since it is critical that the audience understands the underlying logic of our models before we show our forecasts.

We now present our results. We begin with forecasts of demand and price (Figures 12.57 and 12.58). Then we show how Product 1's market share will evolve, rising to about 16 percent and then declining toward the long-run equilibrium (Figure 12.59). To show this evolution in the context of the market, we show the shares for all six products expected in the market (Figure 12.60). This allows us to point out that Product 1 is likely to enter third or fourth into a market that will eventually become quite crowded.

This sequence of forecasts culminates in our forecasts for annual revenues, Figure 12.61. Annual revenues will rise initially upon entry into the market to a peak around \$18M, but they will quickly decline to around \$6M after 15 years.

All of the results we have presented up to this point have been based on a deterministic model. We now turn our attention to the uncertainties Invivo faces, specifically in the initial market price, and the date of entry and strength of competing products. Figure 12.62 shows the distribution of NPVs from Product 1. About 25 percent of the time revenues are zero as the product fails to come to market. The overall mean NPV is \$78M, but the range goes as high as about \$160M. There is, however, only a 12 percent chance of exceeding the \$130M benchmark.

The final two substantive slides, Figures 12.63 and 12.64, show the sensitivity of the NPV of revenues to changes in entry date and strength, respectively.

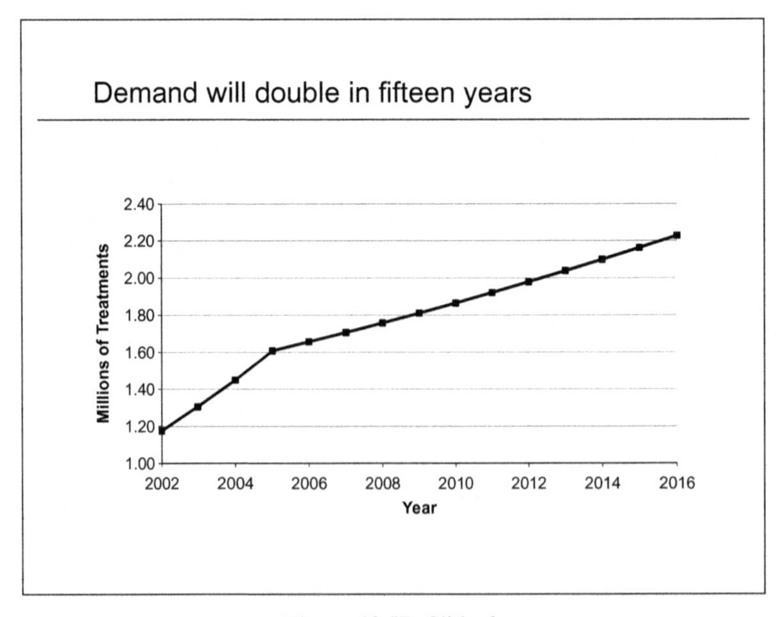

Figure 12.57. Slide 6.

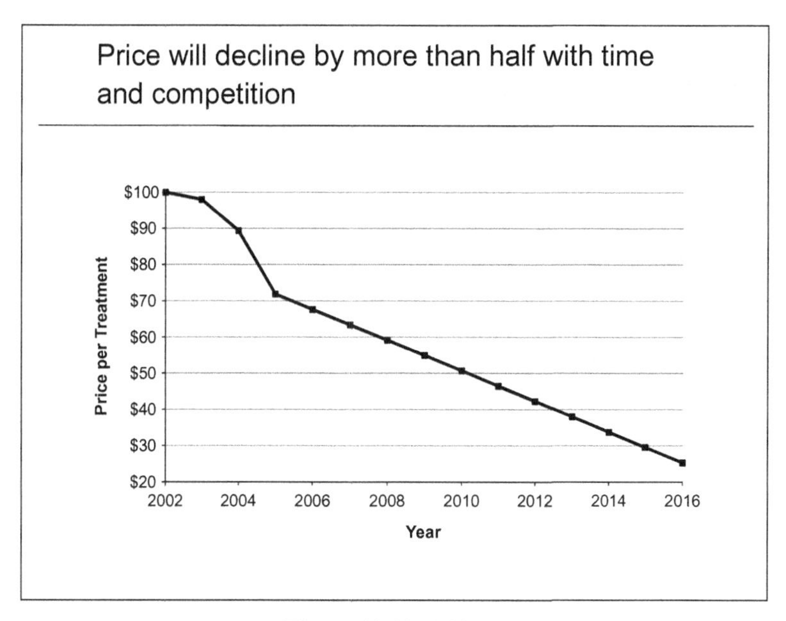

Figure 12.58. Slide 7.

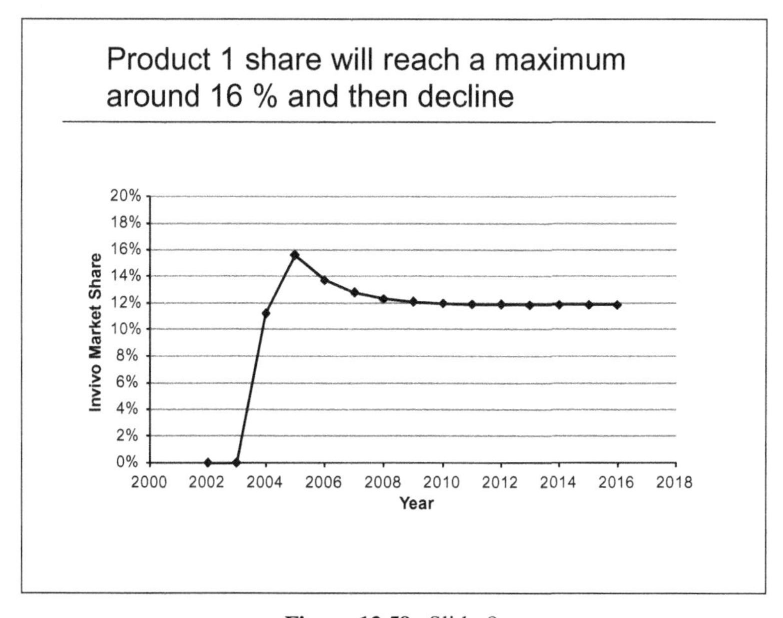

Figure 12.59. Slide 8.

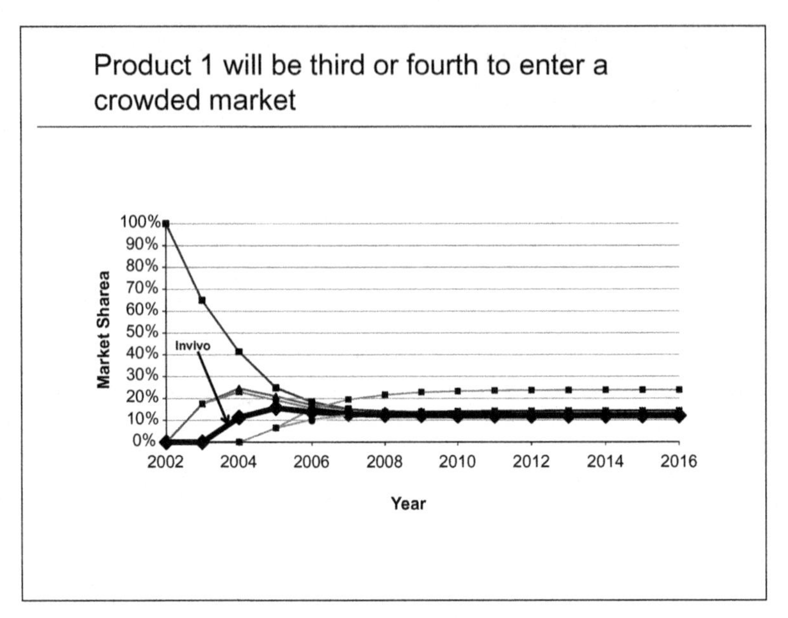

Figure 12.60. Slide 9.

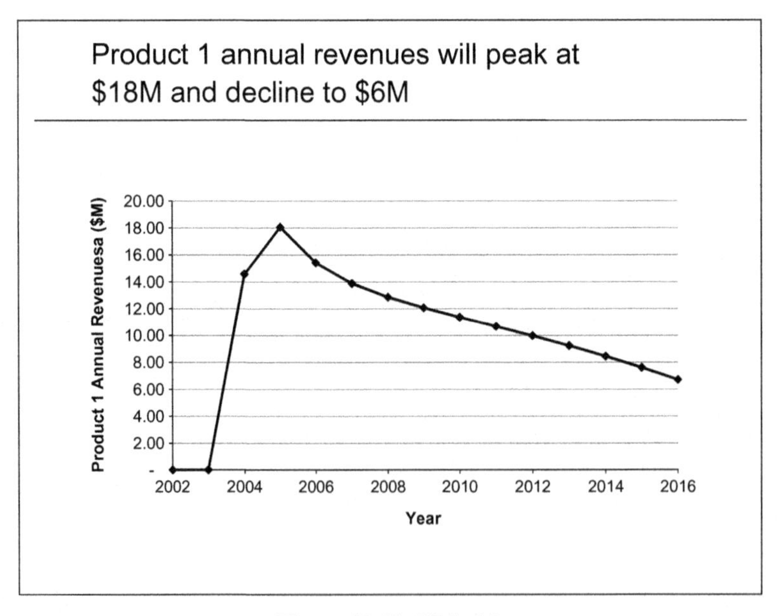

Figure 12.61. Slide 10.

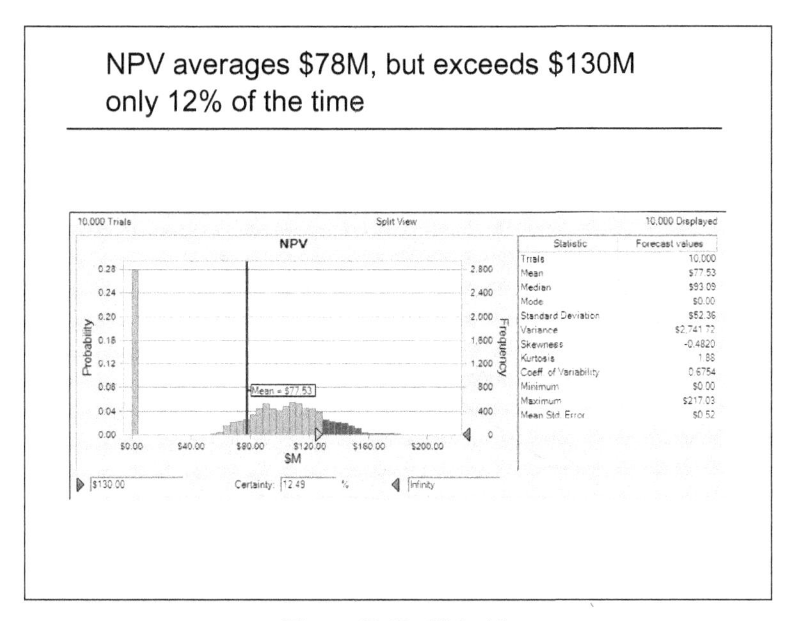

Figure 12.62. Slide 11.

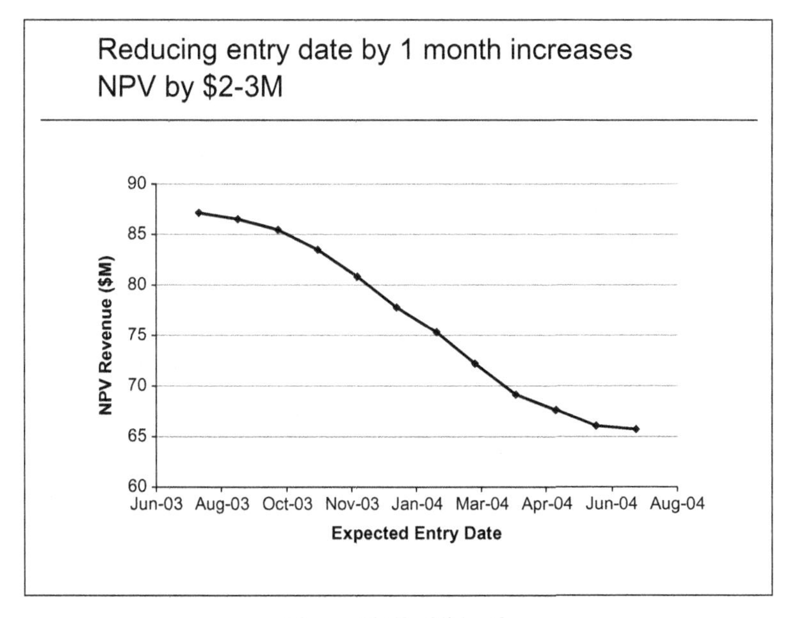

Figure 12.63. Slide 12.

Increasing strength by 0.1 increases NPV by $6M

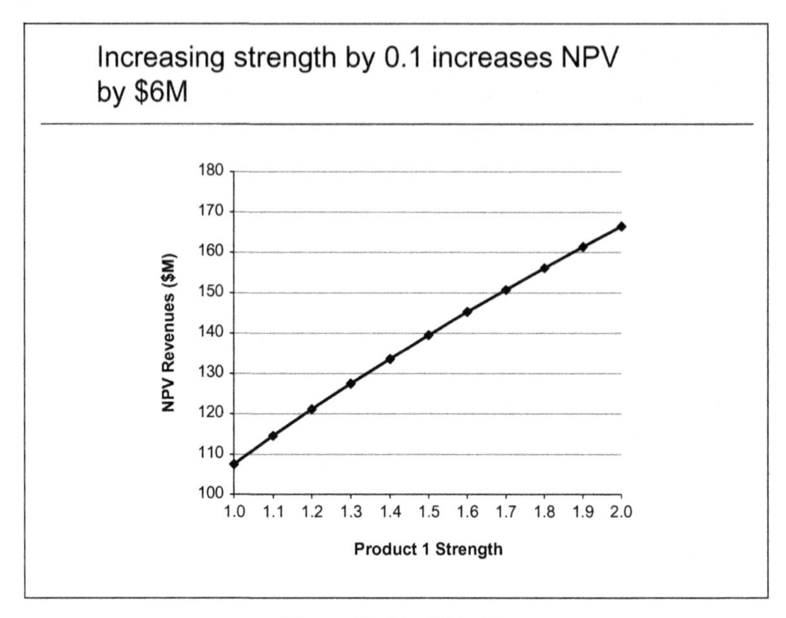

Figure 12.64. Slide 13.

Summary

- Purpose of study: develop method for forecasting revenues on six products in pipeline
- Results for Product 1
  - Share averages 15%
  - Annual revenues peak at $18M
  - NPV averages $78M
  - 12% chance NPV exceeds $130M
  - Earlier entry increases NPV
  - Increased strength increases NPV

Figure 12.65. Slide 14.

Note how we have expressed the overall meaning of each graph in terms of a rule of thumb for how much the NPV changes for a unit change in the independent variable, either date of entry or strength. These simple rules are easier for managers to remember than complex graphs. The final slide, Figure 12.65, summarizes our results.

We admit we hope for a bit of a "gee-whiz" effect with these results. We hope the managers will be impressed with our ability to forecast not only Invivo's market share but the shares of all of the competitors. We have given them more than they asked for. Invivo asked for an estimate of future revenues. Through good modeling, we provided forecasts of demand, price, market share, and revenue. We related both annual revenues and the NPV of revenues to upper and lower benchmarks to make the results more easily digested. And we showed how changes in the timing or quality of their product, both of which they can influence, would affect the results.

12.8 SUMMARY

Without a doubt, this is the most complex case we have explored in this book. While M1 was quick and simple, we spent a lot of time with M2 and M3, crafting modules that could both approximate the historical data we were given and forecast values for the new product. We developed hypotheses and then built and tuned models to match the data as best we could. In M4, we added uncertainty to the model to understand the risks in the process.

Through this progression, the numeric results changed quite a bit. In M1, it looked as if Invivo could easily hit the targets for both NPV and annual revenue. This confidence softened some with M2 as we added a more realistic pricing module. With M3, we added a market share module that considered the entry year and strength of not only Invivo but also of all the competitors. Suddenly the NPV dropped, and achieving the target NPV seemed out of reach. The results of M4 were slightly better as the inclusion of uncertainty allowed for earlier entry and a higher long-run equilibrium for Invivo.

As we have shown in previous cases, insights from the various model iterations tend to build on one another and zero in on a few key conclusions. It is unusual for insights to change dramatically from one iteration to the next even if the numeric results change substantially (as they did in this case). Here are the first three insights of this case:

- Invivo can far surpass the targets by being first to enter.
- Invivo can surpass the target even if it is not first to enter.
- Total demand is not increasing fast enough to counter the falling price and share.

These initial insights held true as we elaborated the model. We refined these insights as we discovered through our work how important early entry was in order to have a larger starting share and to take advantage of higher prices with little discounting. By the time we reached M4, we were able to quantify the probability of reaching the goal and to estimate the magnitude of the increase in NPV from improvements in the R&D outcomes.

It is important to keep our modeling efforts proportional to the benefit attainable. In this case, Invivo requested a forecasting model that could be used for several products for several years. This required more careful, robust programming but also made the effort justifiable. Had Invivo only been concerned with one product, we might have stopped with M1. With just that model alone, we would have known enough to tell Invivo that attaining the revenue target was possible and that the company should do whatever it can to get to market as soon as possible.

APPENDIX A

Guide to Solver

A.1 INTRODUCTION

Solver is an Excel add-in for *optimization*, which involves finding the values of a set of decision variables that maximize or minimize an objective function. A basic version of Solver comes with every copy of Excel, but more advanced and powerful versions are available from Frontline Systems (www.solver.com).

A.2 CONCEPTS

Four concepts are essential in optimization:

- Decision variable
- Objective function
- Constraint
- Linearity

A *decision variable* is an input value chosen by the decision maker in a problem represented by a model. Typical examples include the number of employees to hire, the amount of raw materials to buy, or the price to charge for a service.

An *objective function* is a mathematical formula that calculates the result the decision maker uses to determine how well their choices work out. Examples include profit, NPV, or cost. The objective function is either maximized or minimized. Thus, one would want to maximize profit but minimize cost.

A *constraint* is a restriction or requirement that decision variables must meet. Typical constraints include a budget that cannot be exceeded, or a requirement that at least 25 percent of a portfolio is invested in stocks.

Linearity refers to the mathematical form of the objective function and constraints. Linear functions in this context involve nothing more complex than multiplying decision variables by constants and adding up. So, for example, the following budget constraint is linear:

$$\text{Budget} \geq \text{Travel expenses} + \text{Software expenses} + \text{Phone expenses}$$

Likewise, the following objective function is linear:

$$\text{Profit} = \$32 \times \text{Product A Sales} + \$45 \times \text{Product B Sales} + \$55 \times \text{Product C Sales}$$

A.3 ALGORITHMS

Solver uses a variety of mathematical approaches to find the optimal values of the decision variables. These methods are known as algorithms. Three main algorithms are available in most versions of Solver:

- LP Simplex
- GRG Nonlinear
- Evolutionary

The LP Simplex method is used strictly for problems in which the objective function and all constraints are linear. Essentially, it solves different subsets of constraints to find the combination of decision variables that optimizes the objective function.

The GRG Nonlinear method will work on linear problems, but it is designed for problems with nonlinear objectives, nonlinear constraints, or both. (A problem is considered nonlinear for Solver if the objective function or even one constraint is not linear.) It takes the existing values of the decision variables residing in the spreadsheet as its initial solution and considers small changes in those variables that improve the objective. In this way, it gradually marches "uphill" if the goal is to maximize, or "downhill" if the objective is to minimize, until it reaches an optimal solution. Because this so-called "hill-climbing" search procedure cannot see the entire mountain it is trying to climb, it is possible for it to get stuck on a small peak while a higher peak sits in the distance. This is the problem of local optima, which can be dealt with by trying different starting values for the decision variables and testing whether different final solutions are reached.

The Evolutionary method uses the genetic algorithm approach to find optimal or near-optimal solutions. In this approach, a group (or population) of solutions is generated and this population is then subjected to random mutation and natural selection. Good solutions—those with high values of the objective if the goal is to maximize or low values if the goal is to minimize—are preserved from generation to generation, whereas bad solutions are killed off. In this way, the fitness of the population improves over time and eventually the best remaining solutions will be optimal or near optimal.

A.4 EXAMPLE

Figure A.1 shows a simple advertising planning model in which the goal is to maximize profits by choosing how much to advertise in each of the four quarters of the year. The decision variables are Advertising Expenditures in cells D18:G18. The objective is to maximize Total Profit in cell C21. There are two sets of constraints in the problem. One constraint is that advertising expenditures cannot be negative. The other is that the total amount spent on advertising (cell H18) must be less than the advertising budget (cell C15).

To invoke the Solver add-in, go to Add-ins—Premium Solver. The Solver Parameters window shown in Figure A.2 will open. The "Set Target Cell" is the objective function in cell C21. Choose "Equal To: Max" to maximize this cell. Enter the cell addresses of the decision variables (D18:H18) in the window

	A	B	C	D	E	F	G	H	I	J
1	Advertising Budget Model									
2	PARAMETERS									
3				Q1	Q2	Q3	Q4		Notes	
4		Price	$40.00						Current price	
5		Cost	$25.00						Accounting	
6		Seasonal		0.9	1.1	0.8	1.2		Data analysis	
7		OHD rate	0.15						Accounting	
8		Sales Parameters								
9			35						Consultants	
10			3000							
11		Sales Expense		8000	8000	9000	9000		Consultants	
12		Ad Budget	$40,000						Current budget	
13										
14	DECISIONS							Total		
15		Ad Expenditures		$10,000	$10,000	$10,000	$10,000	$40,000	sum	
16	OUTPUTS									
17		Profit	$69,662							
18	CALCULATIONS									
19		Quarter		Q1	Q2	Q3	Q4	Total		
20		Seasonal		0.9	1.1	0.8	1.2			
21										
22		Units Sold		3592	4390	3192	4789	15962	given formula	
23		Revenue		143662	175587	127700	191549	638498	price*units	
24		Cost of Goods		89789	109742	79812	119718	399061	cost*units	
25		Gross Margin		53873	65845	47887	71831	239437	subtraction	
26										
27		Sales Expense		8000	8000	9000	9000	34000	given	
28		Advertising		10000	10000	10000	10000	40000	decisions	
29		Overhead		21549	26338	19155	28732	95775	rate*revenue	
30		Total Fixed Cost		39549	44338	38155	47732	169775	sum	
31										
32		Profit		14324	21507	9732	24099	69662	GM -TFC	
33		Profit Margin		9.97%	12.25%	7.62%	12.58%	10.91%	pct of revenue	
34										

Figure A.1. Advertising Budget model.

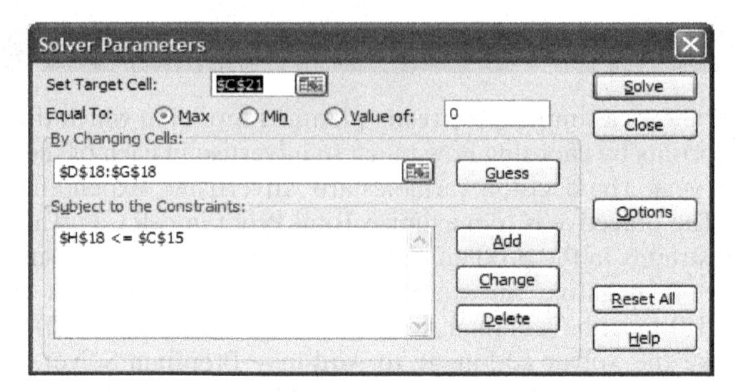

Figure A.2. Solver parameters window.

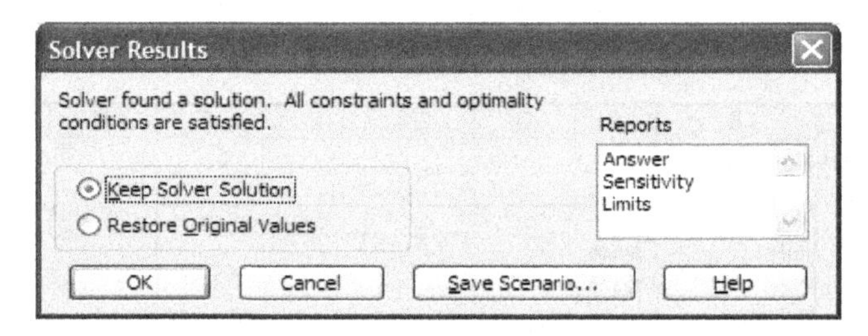

Figure A.3. Solver results window.

under "By Changing Cells." Finally, the budget constraint is entered by choosing Add and then by entering H18 ≤ C15. We can ensure that the decision variables are not negative by choosing Options and checking the box for "Assume Non-Negative." Make sure "Assume Linear Model" is not checked. Click on OK to return to the Solver Parameters window.

Choose Solve to run Solver with these specifications. A few seconds later the window in Figure A.3 will appear, signifying the successful termination of the algorithm and offering the choice to keep the original values of the decision variables in the spreadsheet or to keep the Solver solution.

Figure A.4 shows the model after optimization. The optimal advertising plan involves high expenditures in quarters 2 and 4 and low expenditures in quarters 1 and 3. Total profit has increased from \$69,600 to \$71,447.

The objective function in this model is nonlinear. We can demonstrate this by creating a graph of Profit as a function of, say, Q1 Advertising. Or we can examine the formula in row 27, where we see that the model assumes decreasing returns to advertising. Had we chosen the option "Assume Linear Model," Solver would have detected the nonlinearity and returned the warning message shown in Figure A.5. (Note: different versions of Solver have somewhat different inputs and windows.)

	A	B	C	D	E	F	G	H	I	J
1	**Advertising Budget Model**									
2	PARAMETERS									
3				Q1	Q2	Q3	Q4		**Notes**	
4		Price	$40.00						Current price	
5		Cost	$25.00						Accounting	
6		Seasonal		0.9	1.1	0.8	1.2		Data analysis	
7		OHD rate	0.15						Accounting	
8		Sales Parameters								
9			35						Consultants	
10			3000							
11		Sales Expense		8000	8000	9000	9000		Consultants	
12		Ad Budget	$40,000						Current budget	
13	DECISIONS							Total		
14		Ad Expenditures		$7,273	$12,346	$5,117	$15,263	$40,000	sum	
15	OUTPUTS									
16		Profit	$71,447							
17										
18	CALCULATIONS									
19		Quarter		Q1	Q2	Q3	Q4	Total		
20		Seasonal		0.9	1.1	0.8	1.2			
21										
22		Units Sold		3193	4769	2523	5676	16161	given formula	
23		Revenue		127709	190776	100906	227039	646430	price*units	
24		Cost of Goods		79818	119235	63066	141899	404019	cost*units	
25		Gross Margin		47891	71541	37840	85140	242411	subtraction	
26										
27		Sales Expense		8000	8000	9000	9000	34000	given	
28		Advertising		7273	12346	5117	15263	40000	decisions	
29		Overhead		19156	28616	15136	34056	96965	rate*revenue	
30		Total Fixed Cost		34430	48963	29253	58319	170965	sum	
31										
32		Profit		13461	22578	8587	26820	71447	GM -TFC	
33		Profit Margin		10.54%	11.83%	8.51%	11.81%	11.05%	pct of revenue	
34										

Figure A.4. Optimized version of Advertising Budget model.

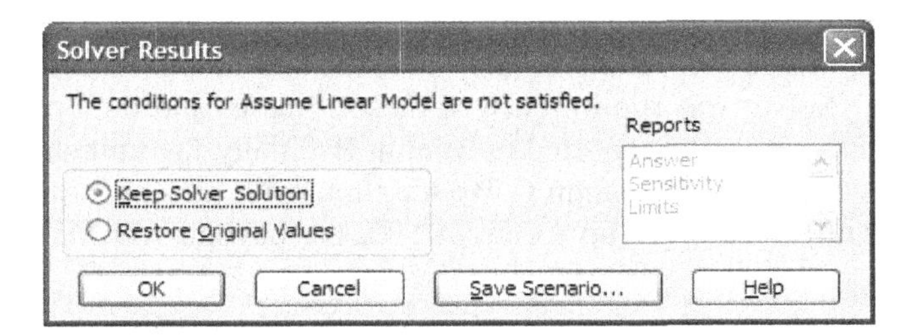

Figure A.5. Solver warning when model is not linear.

	A	B	C	D
1	Ad Budget	Objective: Profit	Change	
2	40,000	$71,447		
3	45,000	$73,279	0.36643	
4	50,000	$74,817	0.30771	
5	55,000	$76,097	0.25596	
6	60,000	$77,147	0.20990	
7	65,000	$77,990	0.16857	
8	70,000	$78,646	0.13121	
9	75,000	$79,132	0.09722	
10	80,000	$79,462	0.06611	
11	85,000	$79,650	0.03752	
12	90,000	$79,706	0.01115	
13	95,000	$79,706	0.00000	
14	100,000	$79,706	0.00000	
15				

Figure A.6. Solver Sensitivity results.

A.5 SOLVER SENSITIVITY

We often want to know how the optimal solution to a problem changes as we vary an input parameter. For example, in the advertising budget example discussed above, we might ask how the profit would change if we could increase the budget. This requires running Solver for a variety of different budget values. This can be done by changing the budget input, running Solver, and recording the results—a tedious procedure if we want to test a large number of inputs. The Solver Sensitivity tool in the Sensitivity Toolkit can be used to automate this process. For more information on using this tool, see Appendix C.

The results of a Solver Sensitivity run are shown in Figure A.6. We varied the Budget in cell C15 from $40,000 to $100,000 in increments of $5,000 (column A). Solver was run for each of these 13 cases and the optimal value of the objective function reported in column B. Finally, the unit change in the objective is calculated in column C. We see that Profit increases (at a decreasing rate) with the Budget up to about $90,000, beyond which it no longer increases.

REFERENCES

1. Baker K. *Optimization Modeling with Spreadsheets*. Duxbury, 2006.
2. Powell S, Baker K. *Management Science: The Art of Modeling with Spreadsheets*. Hoboken, NJ: Wiley, 2007.

APPENDIX B

Guide to Crystal Ball

B.1 INTRODUCTION

Crystal Ball is an Excel add-in for *simulation*, which involves creating a probability distribution for one or more outputs given probability distributions for one or more inputs. Purchasers of this book can download a trial version of Crystal Ball from the web (www.crystalball.com).

B.2 CONCEPTS

There are two essential concepts in Crystal Ball: Assumptions and Forecasts.

An *Assumption cell* is an input cell into which a probability distribution has been entered. A *Forecast cell* is an output cell for which Crystal Ball will develop a probability distribution.

B.3 ALGORITHM

The algorithm Crystal Ball uses to generate output distributions is very simple. Random samples are first taken from the probability distribution in each Assumption cell, and these values are entered into the spreadsheet. Then the spreadsheet is recalculated and Crystal Ball stores the values of each of the Forecast cells. These two steps are repeated as often as the user wants, and histograms are created for each of the Forecast cells. These histograms are estimates of the true probability distributions of those cells.

B.4 EXAMPLE

We use the Advertising Budget model (Figure A.1) to illustrate the workings of Crystal Ball. The question we ask is: How does Profit vary if both Price and Cost are uncertain? In particular, if Price is governed by a normal distribution

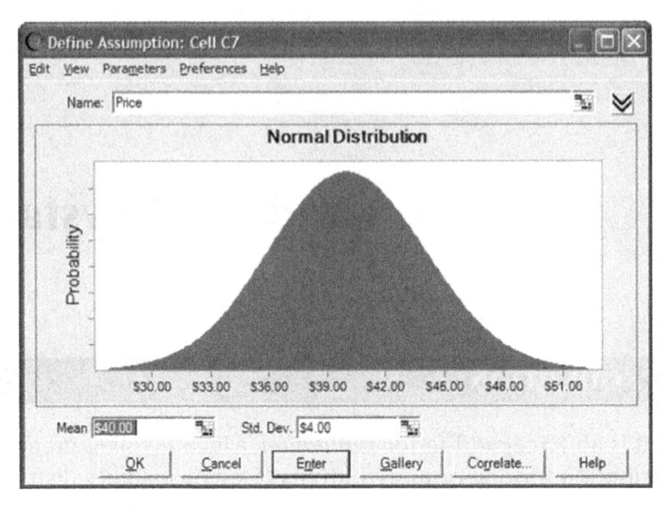

Figure B.1. Normal distribution for Price.

and Cost by a uniform distribution, what is the distribution of Profit? We are particularly interested in the mean Profit and the probability that Profit will be negative.

The first task is to enter a probability distribution for Price in cell C7. Place the cursor on cell C7. Then choose Crystal Ball and select Define Assumption. A menu of distributions appears from which we choose the normal distribution. We input the mean (40) and the standard deviation (4), as shown in Figure B.1. Note that Assumption cells are colored green by Crystal Ball.

Next we enter a distribution for Cost. Place the cursor on cell C8. Choose Define Assumption, and select the Uniform distribution. Input a minimum value of 20 and a maximum value of 30, as shown in Figure B.2.

The next step is to define our Forecast cell. Crystal Ball will save the value in this cell each time we generate a new value for the random inputs. The single output in this example is Profit, in cell C21. We specify that this is a Forecast cell by selecting it and choosing Define Forecast. The cell is colored blue by Crystal Ball. Figure B.3 shows the spreadsheet along with cell comments for the Assumption and Forecast cells.

The final step before running a simulation is to specify the number of times we want to repeat the sampling process. In Crystal Ball language, we must specify the number of trials. We do this by selecting Run Preferences, going to the Trials tab, and entering a number like 10,000 under "Number of trials to run."

We click on Start to run the simulation, and when it is done, we click on View Charts—Forecast Charts to select a Forecast cell to view. Figure B.4 shows the results. This is a histogram of 10,000 trials for Profit. We have used the various chart-editing capabilities in Crystal Ball to display on this chart both the mean value ($69,670) and the probability of an outcome above zero (82.94 percent).

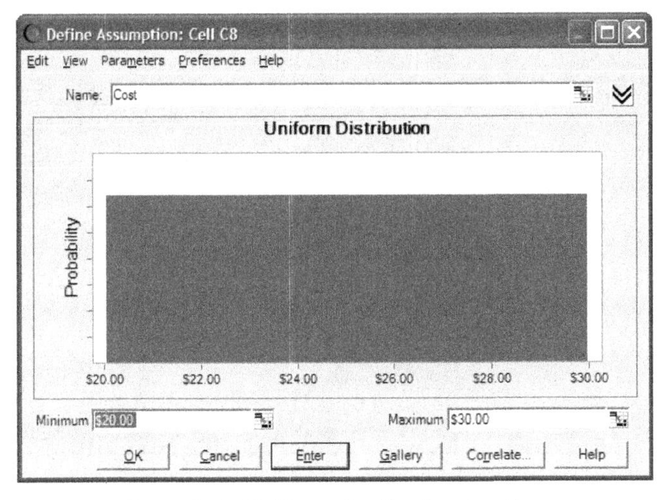

Figure B.2. Uniform distribution for Cost.

	A	B	C	D	E	F	G	H	I	J
1	**Advertising Budget Model**									
2										
3										
4	**PARAMETERS**									
5				**Q1**	**Q2**	**Q3**	**Q4**		**Notes**	
6		**Price**	**$40.00**						Current price	
7		**Cost**	**$25.00**						Accounting	
8		**Seasonal**		**0.9**		8	**1.2**		Data analysis	
9		**OHD rate**	**0.15**						Accounting	
10		**Sales Parameters**								
11			**35**						Consultants	
12			**3000**							
13		**Sales Expense**		**8000**	**8000**	**9000**	**9000**		Consultants	
14		**Ad Budget**	**$40,000**						Current budget	
15	**DECISIONS**							**Total**		
16		**Ad Expenditures**		**$10,000**	**$10,000**	**$10,000**	**$10,000**	**$40,000**	sum	
17	**OUTPUTS**									
18		**Profit**	**$69,662**							
19	**CALCULATIONS**									
20		**Quarter**		**Q1**	**Q2**	**Q3**	**Q4**	**Total**		
21		**Seasonal**		0.9	1.1	0.8	1.2			
22										
23		**Units Sold**		3592	4390	3192	4789	15962	given formula	
24		**Revenue**		143662	175587	127700	191549	638498	price*units	
25		**Cost of Goods**		89789	109742	79812	119718	399061	cost*units	
26		**Gross Margin**		53873	65845	47887	71831	239437	subtraction	
27										
28		**Sales Expense**		8000	8000	9000	9000	34000	given	
29		**Advertising**		10000	10000	10000	10000	40000	decisions	
30		**Overhead**		21549	26338	19155	28732	95775	rate*revenue	
31		**Total Fixed Cost**		39549	44338	38155	47732	169775	sum	
32										
33		**Profit**		14324	21507	9732	24099	69662	GM -TFC	
34		**Profit Margin**		9.97%	12.25%	7.62%	12.58%	10.91%	pct of revenue	
35										

Figure B.3. Advertising Budget model set up for Crystal Ball.

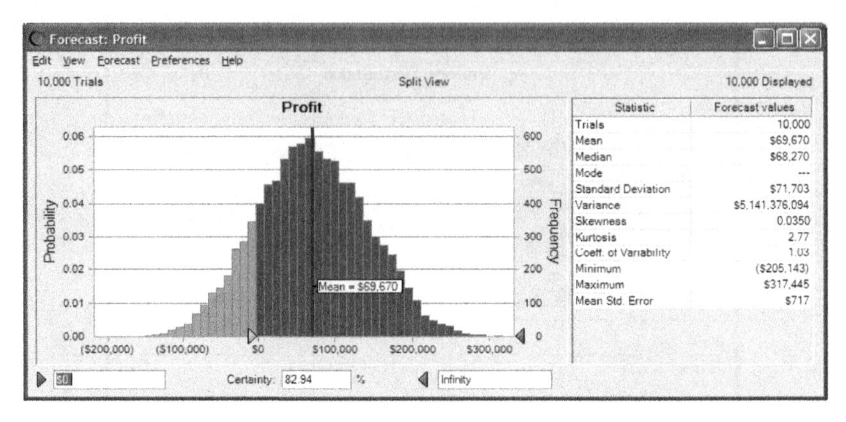

Figure B.4. Distribution of Profit with 10,000 trials.

	A	B	C	D
1	OHD rate	Profit: Mean	Profit: StDev	
2	0.15	69,664	70,814	
3	0.16	63,272	70,286	
4	0.17	56,903	70,328	
5	0.18	50,513	69,856	
6	0.19	44,134	68,983	
7	0.20	37,738	68,667	
8	0.21	31,356	68,566	
9	0.22	24,966	67,444	
10	0.23	18,602	67,521	
11	0.24	12,200	67,349	
12	0.25	5,819	66,250	
13				

Figure B.5. Crystal Ball Sensitivity to Overhead Rate.

B.5 CRYSTAL BALL SENSITIVITY

We often want to know how simulation results vary as we change an input parameter. For example, we might want to know how the mean and standard deviation of Profit change as the Overhead Rate changes (cell C10). The Crystal Ball Sensitivity tool in the Sensitivity Toolkit is designed to make it easy to answer this kind of question. For more information on using this tool, see Appendix C.

The results of a Crystal Ball Sensitivity run are shown in Figure B.5. Here we vary the Overhead Rate from 15 percent to 25 percent in steps of 1 percent (column A). For each of these 11 cases, the Sensitivity tool has run Crystal Ball with the appropriate value for the Overhead Rate. The results show that the mean value declines steeply as the Overhead Rate increases (column B) but that the rate of decline is constant. The standard deviation declines much less (column C), also at a constant rate.

REFERENCE

1. Evans J, Olsen D. *Introduction to Simulation and Risk Analysis*. Prentice Hall, 2002.

APPENDIX C

Guide to the Sensitivity Toolkit

C.1 INTRODUCTION

The Sensitivity Toolkit is an Excel add-in for *sensitivity analysis*, which involves varying one or more inputs and determining the effect on the outputs. The Toolkit includes four sensitivity tools:

- Data Sensitivity
- Tornado Chart
- Solver Sensitivity
- Crystal Ball Sensitivity

The Sensitivity Toolkit was created by Bob Burnham at the Tuck School of Business and is provided free on the school's website (http://mba.tuck.dartmouth.edu/toolkit/).

C.2 DATA SENSITIVITY

The most basic sensitivity question we can ask is how the output of a model varies as one or more inputs vary. For example, we might ask in the Advertising Budget model how Price (in cell C7) or Q1 Advertising (in cell D18) affect Profit (in cell C21).

The Data Sensitivity tool is designed to answer this type of question quickly. Select Add-ins—Sensitivity Toolkit—Data Sensitivity. The window shown in Figure C.1 will appear. Select One-way Table and enter the cell address of the desired output (Toolkit!C21) in the field labeled "Results Cell(s)." Select Next and the window in Figure C.2 will appear. Enter the cell address of the parameter you want to change (Toolkit!D18) under "Cell to Vary," select "Begin, End, Increment" in the field labeled "Input Type," and enter "First Value" (10,000), "Last Value" (20,000), and "Increment/N" (500). Select Finish and the results will appear as in Figure C.3. Because of diminishing returns, Profit increases to a peak and then declines with Q1 Advertising.

Modeling for Insight: A Master Class for Business Analysts, by Stephen G. Powell & Robert J. Batt

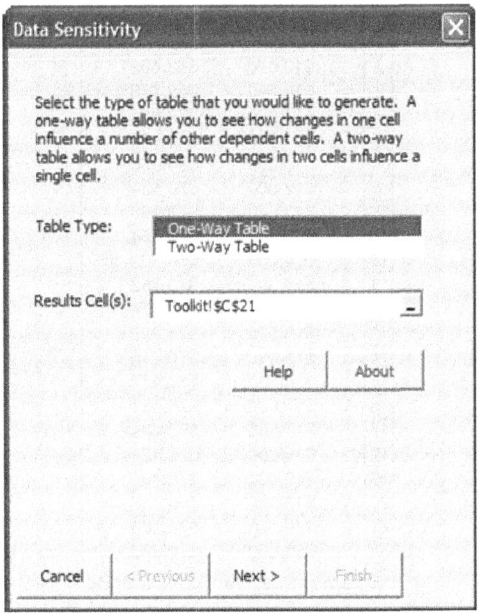

Data Sensitivity

Select the type of table that you would like to generate. A one-way table allows you to see how changes in one cell influence a number of other dependent cells. A two-way table allows you to see how changes in two cells influence a single cell.

Table Type: One-Way Table / Two-Way Table

Results Cell(s): Toolkit!$C$21

Help | About

Cancel | < Previous | Next > | Finish

Figure C.1. First input window for Data Sensitivity.

One-Way Inputs

Enter the address of the cell to be modified and the method for generating the input.

Cell to Vary: Toolkit!$D$18

Input Type: Begin, End, Increment / Begin, End, Num Obs

First Value: 10000

Last Value: 20000

Increment/N: 500

Cancel | < Previous | Next > | Finish

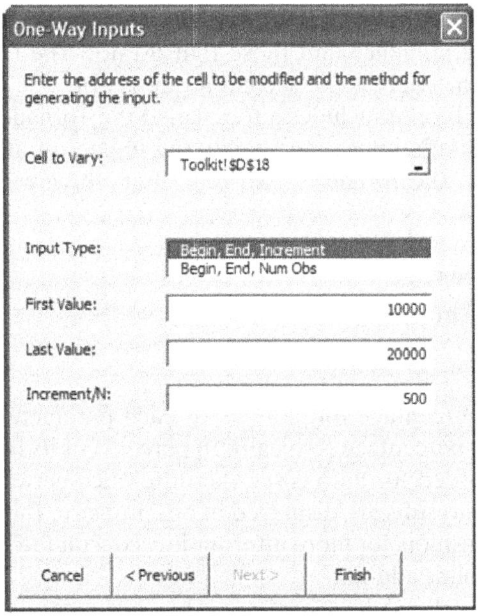

Figure C.2. Second input window for Data Sensitivity.

	A	B	C
1	**$D$18**	**Profit**	
2	$10,000	$69,662	
3	$10,500	$69,778	
4	$11,000	$69,882	
5	$11,500	$69,976	
6	$12,000	$70,060	
7	$12,500	$70,134	
8	$13,000	$70,198	
9	$13,500	$70,254	
10	$14,000	$70,302	
11	$14,500	$70,342	
12	$15,000	$70,374	
13	$15,500	$70,398	
14	$16,000	$70,416	
15	$16,500	$70,427	
16	$17,000	$70,431	
17	$17,500	$70,429	
18	$18,000	$70,421	
19	$18,500	$70,407	
20	$19,000	$70,388	
21	$19,500	$70,363	
22	$20,000	$70,333	
23			

Figure C.3. Results of Data Sensitivity.

C.3 TORNADO CHART

Most models have dozens or even hundreds of inputs, many of which do not have a substantial effect on the output. It is often helpful to identify inputs that have a strong influence and those that do not. The first group is worth additional research, whereas the second group is not.

The Sensitivity Toolkit includes a tool called the Tornado Chart, which displays each parameter's impact and ranks the impact of all parameters from biggest to smallest. The Tornado Chart tool offers three ways to generate this type of graph:

- Constant percentage
- Variable percentage
- Percentiles

In a constant percentage tornado chart, each parameter is varied up and down by the same percentage (typically 10 percent) of its base case value. The output cell is recorded for both cases, the difference between these results is calculated, and the results are displayed in order of size. The other two options work in a similar fashion; for more information, consult the Help option within the Sensitivity Toolkit add-in.

We invoke the Tornado Chart tool by selecting Add-ins—Sensitivity Toolkit—Tornado Chart. The first window that appears (Figure C.4) requires

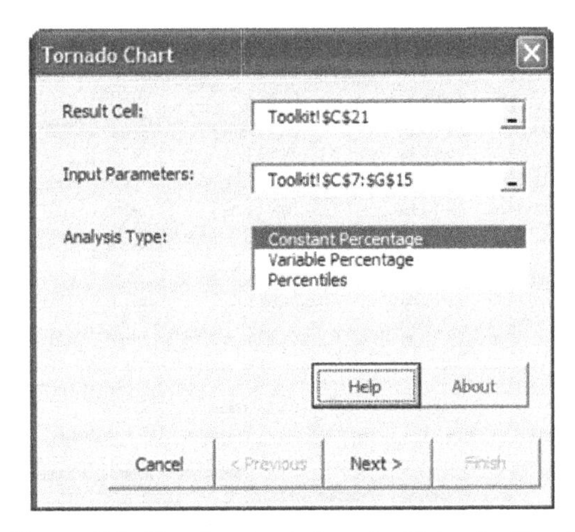

Figure C.4. First input window for Tornado Chart.

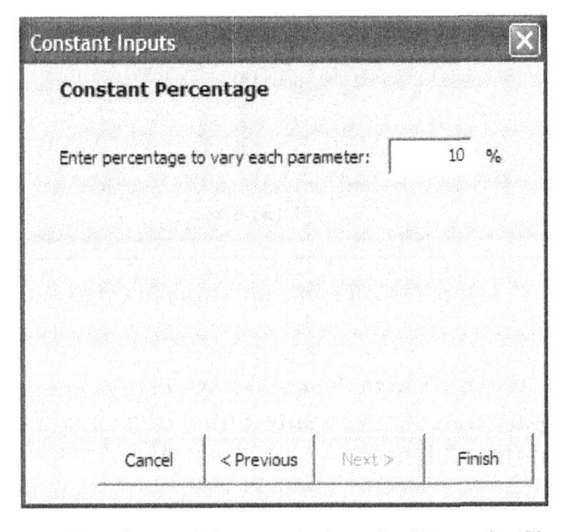

Figure C.5. Second input window for Tornado Chart.

us to enter the "Result Cell" (Profit: Toolkit!C21), the "Input Parameters" (Toolkit!C7:G15), and the "Analysis Type" (Constant Percentage). Select Next and the second input window appears (Figure C.5). Choose the percentage change for each parameter (10 percent in this case). Select Finish and the results will be created in a separate sheet as shown in Figure C.6.

The table is created first, by varying each parameter by +10 percent and recording the output, and then by –10 percent and recording the output. Then the range between the first and second values is calculated, and finally the parameters are sorted in decreasing order by range. The chart is then created showing the ranges around the base case output value. Since the parameters are sorted by range, the chart tends to look like a tornado. In the example

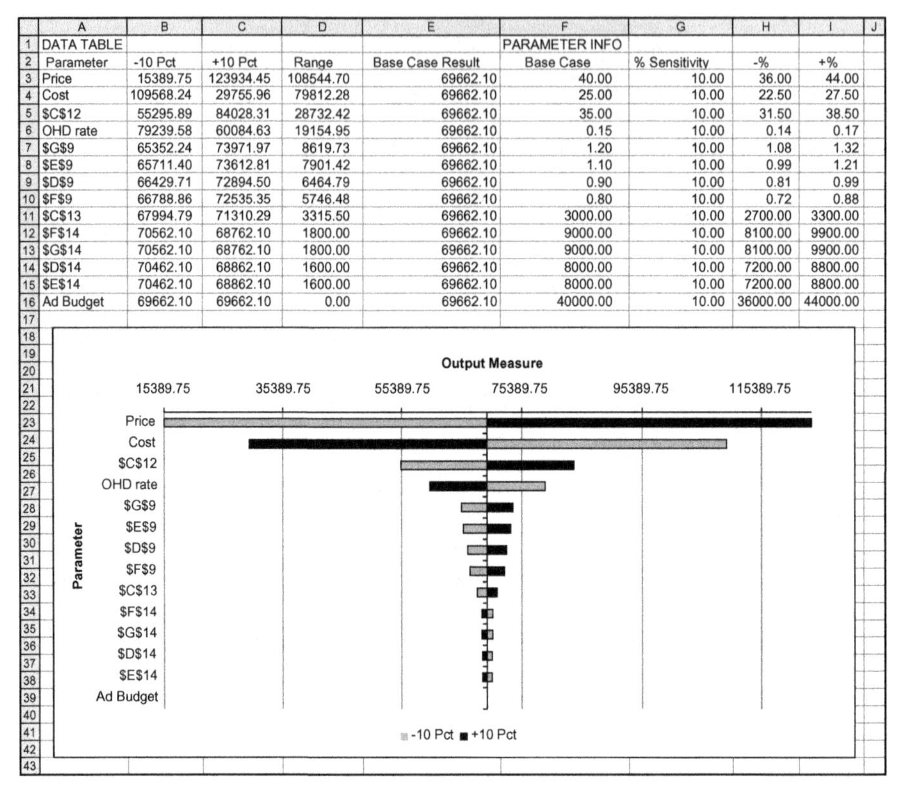

DATA TABLE					PARAMETER INFO			
Parameter	-10 Pct	+10 Pct	Range	Base Case Result	Base Case	% Sensitivity	-%	+%
Price	15389.75	123934.45	108544.70	69662.10	40.00	10.00	36.00	44.00
Cost	109568.24	29755.96	79812.28	69662.10	25.00	10.00	22.50	27.50
$C$12	55295.89	84028.31	28732.42	69662.10	35.00	10.00	31.50	38.50
OHD rate	79239.58	60084.63	19154.95	69662.10	0.15	10.00	0.14	0.17
$G$9	65352.24	73971.97	8619.73	69662.10	1.20	10.00	1.08	1.32
$E$9	65711.40	73612.81	7901.42	69662.10	1.10	10.00	0.99	1.21
$D$9	66429.71	72894.50	6464.79	69662.10	0.90	10.00	0.81	0.99
$F$9	66788.86	72535.35	5746.48	69662.10	0.80	10.00	0.72	0.88
$C$13	67994.79	71310.29	3315.50	69662.10	3000.00	10.00	2700.00	3300.00
$F$14	70562.10	68762.10	1800.00	69662.10	9000.00	10.00	8100.00	9900.00
$G$14	70562.10	68762.10	1800.00	69662.10	9000.00	10.00	8100.00	9900.00
$D$14	70462.10	68862.10	1600.00	69662.10	8000.00	10.00	7200.00	8800.00
$E$14	70462.10	68862.10	1600.00	69662.10	8000.00	10.00	7200.00	8800.00
Ad Budget	69662.10	69662.10	0.00	69662.10	40000.00	10.00	36000.00	44000.00

Figure C.6. Results for Tornado Chart.

here, the top four parameters have a substantial impact, whereas the remaining parameters have little influence.

C.4 SOLVER SENSITIVITY

Solver identifies the values of decision variables in a model that optimize a single cell called the objective function. If we wish to determine the sensitivity of the optimal solution to changes in other parameters, we must run Solver several times. The Solver Sensitivity tool in the Sensitivity Toolkit automates this process.

In Appendix A, we illustrated how Solver could be used to identify the optimal pattern of advertising expenditures when total spending was constrained by a budget. Then, we asked how changes in the budget would affect the optimal results. This is an application for Solver Sensitivity.

We select Add-ins—Sensitivity Toolkit—Solver Sensitivity and the window shown in Figure C.7 appears. Solver Sensitivity automatically identifies the

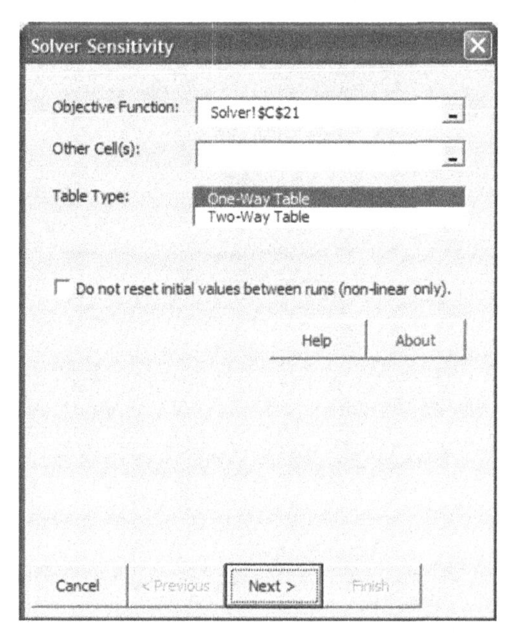

Figure C.7. First input window for Solver Sensitivity.

objective function if Solver has been run previously on the model. If the results of additional cells are needed, they can be input under "Other Cell(s)." Select Next and the window shown in Figure C.8 will appear. Input the cell address of the budget (Solver!C15) under "Cell to Vary." Then select "Begin, End, Increment," and input the "First Value" (40,000), "Last Value" (100,000), and "Increment/N" (5,000). Select Finish and the results will appear as in Figure A.6.

C.5 CRYSTAL BALL SENSITIVITY

Crystal Ball estimates the probability distribution of one or more Forecast cells given probability distributions for one or more Assumption cells. When we want to determine how some aspect of a Forecast cell distribution, such as the mean or maximum value, varies with an input parameter, we must run Crystal Ball many times. The Crystal Ball Sensitivity tool in the Sensitivity Toolkit automates this process.

In Appendix B, we showed how to estimate the distribution of Profit using Crystal Ball. We then asked how the mean and standard deviation of Profit change as the Overhead Rate changes (cell C10). This is an application for Crystal Ball Sensitivity.

We select Add-ins—Sensitivity Toolkit—Crystal Ball Sensitivity and the window shown in Figure C.9 appears. Crystal Ball Sensitivity automatically

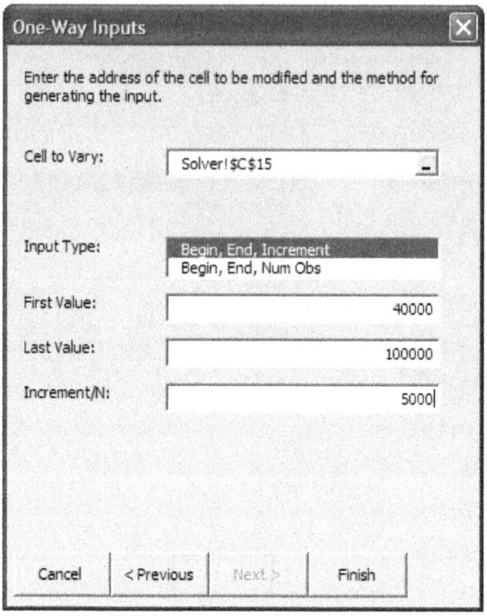

One-Way Inputs

Enter the address of the cell to be modified and the method for generating the input.

Cell to Vary: Solver!$C$15

Input Type: Begin, End, Increment / Begin, End, Num Obs

First Value: 40000

Last Value: 100000

Increment/N: 5000

Cancel | < Previous | Next > | Finish

Figure C.8. Second input window for Solver Sensitivity.

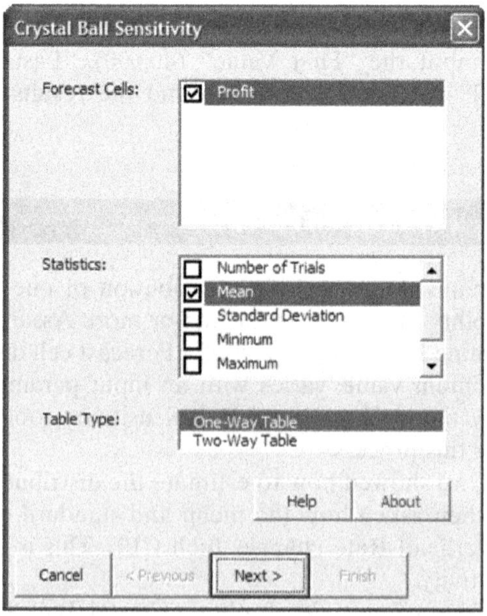

Crystal Ball Sensitivity

Forecast Cells: Profit

Statistics: Number of Trials / Mean / Standard Deviation / Minimum / Maximum

Table Type: One-Way Table / Two-Way Table

Help | About

Cancel | < Previous | Next > | Finish

Figure C.9. First input window for Crystal Ball Sensitivity.

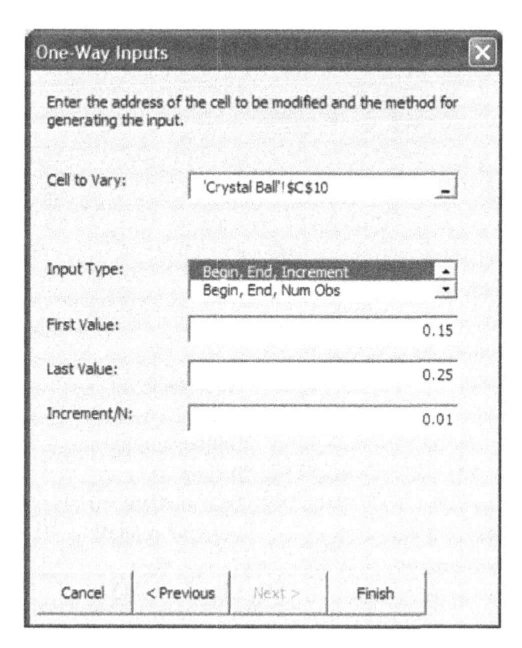

Figure C.10. Second input window for Crystal Ball Sensitivity.

identifies the Forecast cells in the model. It also offers to capture any of the following six statistics for each Forecast cell:

- Number of trials
- Mean
- Standard deviation
- Minimum
- Maximum
- Mean standard error

Finally, an option is presented for a one-way or a two-way table.

Select Next and the window shown in Figure C.10 appears. In this window, enter the "Cell to Vary" (the budget in C10), and then select "Begin, End, Increment/N," "First Value" (0.15), "Last Value" (0.25), and "Increment/N" (0.01). The results appear as in Figure B5.

INDEX